AAPS Introductions in the Pharmaceutical Sciences

Volume 16

The *AAPS Introductions in the Pharmaceutical Sciences* book series is designed to support pharmaceutical scientists at the point of knowledge transition. Springer and the American Association of Pharmaceutical Scientists (AAPS) have partnered again to produce a second series that juxtaposes the *AAPS Advances in the Pharmaceutical Sciences* series. Whether shifting between positions, business models, research project objectives, or at a crossroad in professional development, scientists need to retool to meet the needs of the new scientific challenges ahead of them. These educational pivot points require the learner to develop new vocabulary in order to effectively communicate across disciplines, appreciate historical evolution within the knowledge area with the aim of appreciating the current limitations and potential for growth, learn new skills and evaluation metrics so that project planning and subsequent evolution are evidence-based, as well as to simply "dust the rust off" content learned in previous educational or employment settings, or utilized during former scientific explorations. The *Introductions* book series will meet these needs and serve as a quick and easy-to-digest resource for contemporary science.

Sougata Jana
Editor

Chitosan-Based Drug Delivery Systems

Springer

aaps®

Editor
Sougata Jana
Department of Health and Family Welfare
Directorate of Health Services
Kolkata, West Bengal, India

ISSN 2522-834X ISSN 2522-8358 (electronic)
AAPS Introductions in the Pharmaceutical Sciences
ISBN 978-3-032-08346-3 ISBN 978-3-032-08347-0 (eBook)
https://doi.org/10.1007/978-3-032-08347-0

This Springer imprint is published by the registered company Springer Nature Switzerland AG
The registered company address is: Gewerbestrasse 11, 6330 Cham, Switzerland

Preface

This book provides an overview of current advances in chitosan-based drug delivery systems, focusing on fabrication technology, methods, and applications. Chitosan, a renewable polysaccharide obtained by chitin deacetylation, is suitable as a biodegradable matrix for the production of different drug delivery systems owing to its biocompatibility, bioadhesive, antibacterial activity, hemostatic action, adsorption, properties, and functional capabilities, particularly the possibility of being covalently cross-linked through its amino groups, allowing its stabilization as a matrix for controlled release systems. The mucoadhesive properties of chitosan improve the mucous membranes absorption of drugs on the gastrointestinal tract, respiratory tract, eyes, etc.

The current research trends are to continue innovative modifications of chitosan derivatives, nanotechnologies, stimuli-responsive delivery, and theragnostic techniques (combine therapeutic and diagnostic capabilities) to improve promising chitosan-based drug delivery systems. This book also covers various important chitosan-based approaches such as interpenetrating polymer network matrix, polyelectrolyte complexes, nanogels, nanohybrids, micelles, liposomes, pH-responsive, light-activated drug delivery, electrospun, 3D/4D (bio) printing for drugs, genes, and RNA, wound dressing, tissue engineering for drugs, and targeted therapeutic delivery.

Each chapter develops a particular aspect of recent advances of chitosan-based delivery systems to cover the importance, fabrication technology, characterization, evaluation, delivery of therapeutic and biomedical applications, and future perspectives.

Chapter "Introduction of Chitosan: Chemistry, Properties, Synthesis, Character ization, and Evaluation" provides a comprehensive overview of the chemistry, properties, synthesis, characterization, and analytical techniques and also summarizes the most frequently reported applications of chitosan and its derivatives in biomedicine.

Chapter "Interpenetrating Polymer Networks of Chitosan-Based Matrix in Drug Delivery" focuses on Interpenetrating Polymer Networks (IPNs) of Chitosan-based matrix in advanced drug delivery systems, highlighting their design, synthesis, and functional advantages. IPNs have unique properties, such as structural mimicry of the extracellular matrix and inherent antimicrobial activity.

Chapter "Polyelectrolyte Complexes of Chitosan as Potential Drug Delivery Systems" highlights on chitosan-based polyelectrolyte complexes (CH PECs) as multifunctional carriers for advanced drug delivery. This chapter provides an in-depth overview of CH PECs, focusing on their formation mechanisms, physico-chemical properties, preparation techniques, and characterization methods.

Chapter "Chitosan-Based Nanogels in Drug Delivery Applications" delves into the synthesis methods of chitosan-based nanogels and also highlights their potential as drug delivery carriers, such as in antitumor drugs, protein, bone, dental tissue engineering, and bioimaging.

Chapter "Chitosan-Based Micelles in Drug Delivery and Bioimaging" is composed of recent advancements in chitosan-based stimuli-responsive micelles and hybrid systems, along with challenges related to scalability, clinical translation, and promising tool for the development of next-generation drug delivery and imaging technologies.

Chapter "Advances of Chitosan-Based Transdermal Drug Delivery Systems" focuses on the chitosan-based transdermal drug delivery systems. This chapter composed of chitosan-based different transdermal dosage forms, including gels, patches, microneedles, etc. for better understanding of its chemical modifications and physicochemical properties can help to enhance the efficacy of chitosan-based transdermal formulations.

Chapter "pH-Responsive Chitosan-Based System for Targeted Drug Delivery" focuses on chitosan-based pH-responsive system targeted drug delivery in cancer therapy. This chapter explores the structural and functional attributes of chitosan, focusing on the protonation-deprotonation dynamics of its amino groups, which enable precise pH-triggered drug release in acidic tumor microenvironments.

Chapter "Chitosan-Based Hydrogels for Stimuli-Responsive Drug Delivery" focuses on chitosan-based hydrogel for stimuli-responsive drug delivery. Chitosan-based hydrogels have drawn interest due to their superior antibacterial, biodegradable, pH-responsive, temperature, light, and electric and magnetic responsive and biocompatible qualities, which make them ideal for targeted and oral delivery of drug.

Chapter "Chitosan-Based Light-Activated Drug Delivery System" also addresses the challenges facing clinical translation, including stability, targeting specificity, and scalability, and outlines recent innovations in chemical modification of chitosan to enhance its performance in chitosan-based light-activated systems as a compelling strategy for biomedicine applications.

Chapter "Electrospun Chitosan-Based Systems in Drug Delivery" aims to comprehensively analyze chitosan-based electrospun DDS, focusing on their fabrication techniques, critical process parameters, properties, and biomedical applications in targeted and sustained drug delivery, wound dressings, tissue engineering, bone regeneration, etc.

Chapter "Chitosan-Based Nanoparticles for Brain Drug Delivery" highlights on chitosan-based nanoparticles for brain drug delivery, focusing on their ability to cross the blood-brain barrier (BBB) and improving treatment for neurological disorders.

Chapter "Chitosan-Based Gene and RNA Delivery in Therapeutic Applications" provides the mechanisms of chitosan-based gene and RNA delivery, highlighting the interactions between chitosan and nucleic acids, as well as the role of chitosan in cellular uptake and intracellular trafficking.

Chapter "Chitosan-Based Liposomes in Cancer Therapy" explores the design, formulation, therapeutic applications, and current progress of chitosan-based liposomes in cancer therapy, while highlighting the challenges and future directions for clinical translation.

Chapter "Chitosan-Based System in 3D/4D (Bio) Printing for Drug Delivery and Tissue Engineering Applications" fabricates for chitosan-based system for three-dimensional (3D)/four-dimensional (4D) (bio) printing in drug delivery and tissue-engineered scaffolds. These bio-fabrication techniques facilitate the development of intricate, cellular structures with exceptional precision and reproducibility, meeting the requirements of personalized treatment.

Chapter "Chitosan-Based Systems in Wound Dressing Applications" represents recent advances in wound dressing preparations with chitosan in composite with natural or synthetic polymers and bioactive agents. These composite biomaterials have improved antimicrobial, structural, and regenerative properties, making them potentially useful for the treatment of acute and chronic wounds.

This reference edited book is developed by different groups of renowned scientists, chemists, biologists, and engineers from around the world. This book will be an important reference resource for the scientists, researchers, and industrialists who are working on chitosan and chitosan-based drug delivery systems in the fields of pharmaceuticals, biomedical science, biomedical engineering, nanotechnology, and material science and engineering for the most updated findings and future research trends.

We acknowledge that this book would not have been possible without the support and contribution of all the authors and their respective teams. Finally, the editor would also like to acknowledge the very efficient and friendly staff of Springer, Dr. Charlotte Nunes and others staffs, for providing support from starting to course of the entire production process.

Kolkata, India Sougata Jana

Contents

About the Editor

Sougata Jana received BPharm (Gold Medalist) from West Bengal University of Technology (WBUT), Kolkata, MPharm (Pharmaceutics) from Biju Patnaik University of Technology (BPUT), and PhD from Maulana Abul Kalam Azad University of Technology (MAKAUT), West Bengal (Formerly Known as WBUT), India. He received, "M.N Dev Memorial Award" for securing the highest marks in the state of West Bengal in the year 2005 from IPA Bengal branch, Kolkata, India. He has been engaged for 18 years in the field of pharmacy including teaching, research & health services, etc. He received "Best Poster Presentation Award" at 21st West Bengal State Science and Technology Congress, 2014, Burdwan, and "Outstanding Paper Award" at 1st Regional Science and Technology Congress, 2016, Bardhaman Division, organized by DST, Govt. of West Bengal and Bankura Christan College, West Bengal, India. He has published above 45 research and review articles in different national and international peer-reviewed journals. He has edited 18 books in Springer, Elsevier, and Wiley and published more than 50 book chapters in different edited books from Elsevier, Springer, Wiley VCH, CRC Press, and Taylor & Francis group. He is the reviewer for various peer-reviewed international journals of Elsevier, Wiley, Springer, Taylor and Francis, Dove Press, etc. He is a life member of Association of Pharmaceutical Teachers of India (APTI) and Associateship from Institution of Chemists (AIC), India. He successfully guided 17 postgraduate students for their research projects. He is working in the field of drug delivery science and technology including modification of synthetic and natural biopolymers, microparticles, nanoparticles, semisolids, and interpenetrating network (IPN) systems for controlled drug delivery.

About the Editor

Introduction of Chitosan: Chemistry, Properties, Synthesis, Characterization, and Evaluation

Gema Díaz Bukvic, María Marta Fidalgo, Ezequiel Rossi, and María Inés Errea

Abstract Chitosan is composed of β-(1 → 4)-*N*-acetyl-D-glucosamine and β-(1 → 4)-D-glucosamine units. It is a partially deacetylated form of chitin, the second most abundant polysaccharide on Earth. This chapter includes a deep analysis of the most commonly used processes for its obtention from the shells of crustaceans, which are the most abundant natural source of chitin. Chitosan's renewable origin, natural abundance, biocompatibility, and chemical versatility, together with the distinctive properties conferred by its polycationic structure, make it an attractive material for a wide range of applications. Considering that key properties of this biopolymer, such as its feasibility for chemical modification and processability, are strongly influenced by its molecular weight (MW) and degree of deacetylation (DD), their accurate determination is essential. In this context, this chapter provides a comprehensive overview of the analytical techniques most frequently employed for the determination of these structural parameters. Furthermore, the influence of MW and DD on key physical properties of this polysaccharide, such as solubility and viscosity, is discussed in detail. The chapter also

G. D. Bukvic
Instituto Tecnológico de Buenos Aires (ITBA), Buenos Aires, Argentina

CONICET (Consejo Nacional de Investigaciones Científicas y Técnicas), Buenos Aires, Argentina

Department of Civil and Environmental Engineering, University of Missouri, Columbia, MO, USA
e-mail: gediaz@itba.edu.ar

M. M. Fidalgo
Department of Civil and Environmental Engineering, University of Missouri, Columbia, MO, USA

E. Rossi · M. I. Errea (✉)
Instituto Tecnológico de Buenos Aires (ITBA), Buenos Aires, Argentina

CONICET (Consejo Nacional de Investigaciones Científicas y Técnicas), Buenos Aires, Argentina
e-mail: ezrossi@itba.edu.ar; merrea@itba.edu.ar

S. Jana (ed.), *Chitosan-Based Drug Delivery Systems*, AAPS Introductions in the Pharmaceutical Sciences 16, https://doi.org/10.1007/978-3-032-08347-0_1

summarizes the most frequently reported applications of chitosan and its derivatives in fields including food packaging, environmental remediation, and biomedicine.

Keywords Chitosan's structural properties · Analytical characterization techniques · Chemical structure · function relationship · Chitosan-based materials' applications · Extraction and deacetylation methods

1 Chitosan Sources

Polysaccharides are highly abundant and widely distributed natural polymers found in every living organism. They fulfill diverse essential roles in sustaining life, including functions related to energy storage (e.g., starch in plants and glycogen in animals) and structure (e.g., cellulose and chitin). Chitin (β-(1 → 4)-*N*-acetyl-D-glucosamine) (Fig. 1) is the second most abundant polysaccharide after cellulose, being a core constituent of the fungal cell walls, arthropods' (e.g., insects and crustaceans) exoskeletons, cephalopod beaks, mollusks' cartilages and radula, fish scales and lissamphibians' skin, among other organisms' tissues, muscles or structures, including many in corals and marine sponges [1]. It is estimated that chitin's natural production is 10^{10} to 10^{12} annual tons worldwide, and it can either be present at trace levels in the marine source or constitute up to 85% of its structure, depending on the phylum [2, 3].

As mentioned above, chitin primarily comes from marine species, linking its production to the fishing industry, which is key in the global economy. In fact, according to the Food and Agriculture Organization of the United Nations of 2024 [4], the main source of chitin, crustaceans, were the second most traded aquatic animal products in 2022, with about ten million tons produced that year. Unfortunately, this is associated with the generation of millions of tons of biowaste, including exoskeletons of lobsters, crabs, and shrimps. Besides, their slow biodegradation leads to accumulation and negative environmental impacts, including visual pollution and unpleasant odors.

Therefore, the scientific community is encouraged to search for ways to valorize such waste. One approach is to use them to extract chitin and further transform it into chitosan, a more versatile polymer, allowing chemical modifications to tailor it for

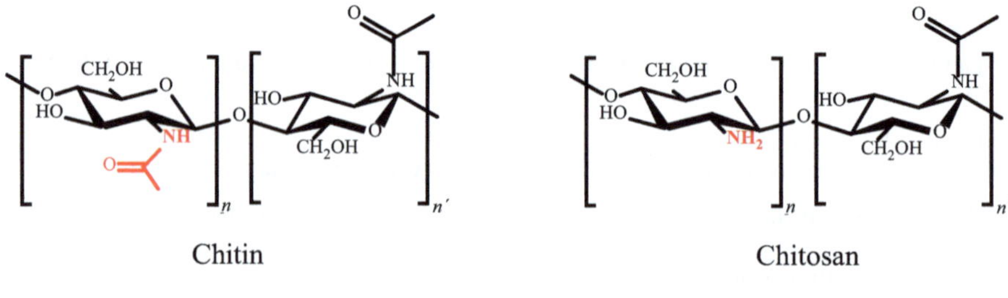

Fig. 1 Chitin and chitosan chemical structures

multiple applications [5]. Chitosan is the partially deacetylated form of chitin, constituted by units of β-(1 → 4)-*N*-acetyl-D-glucosamine andβ-(1 → 4)-D-glucosamine units (Fig. 1), and it is the only polysaccharide with a polycationic nature, because of its free amino groups in the structure, which confer distinctive properties. Moreover, its well-established biocompatibility and biodegradability make it appealing for a broad range of applications across diverse fields.

Considering all the above, this chapter intends to explore the processes involved in the obtention of chitosan, and how they impact the structure and physicochemical properties of this biopolymer. Furthermore, commonly used characterization techniques will also be provided. Finally, an overview of chitosan-based materials' applications is also presented.

2 Chitosan Synthesis from Chitin

Naturally occurring chitosan is negligible relative to how prevalent chitin is in nature [6]. As a result, chitosan is predominantly commercially synthesized through the deacetylation of chitin, as previously discussed. The following section describes the most commonly used techniques for chitin extraction, followed by a summary of the principal deacetylation techniques utilized in the production of chitosan from chitin.

Briefly, the first step for chitosan synthesis is chitin's obtention, which involves a sequence of steps that vary depending on whether the biomass source is crustaceans, fungi, or insects. Given that the primary source is the shells of shrimps and crabs, the extraction process will be detailed for that case. Shells are mainly composed of chitin-protein fibrils, typically arranged in a tightly packed, helicoidal structure that constitutes the exoskeleton. This configuration is further reinforced by mineral layers, which act as a barrier for proteins. Therefore, to obtain chitosan from crustaceans, after crushing, grinding, and drying the waste, demineralization, deproteination, decolorization and deacetylation processes are carried out, as summarized next (Fig. 2).

2.1 *Chitin Extraction*

The quality of chitin, particularly in terms of purity, is influenced by the sequence in which demineralization and deproteinization are carried out, as the order of these steps affects the efficiency of each process. When demineralization is conducted prior to deproteinization, the dissolution of $CaCO_3$ disrupts the packed matrix arrangement, increases the surface area of exposed proteins, and thereby facilitates their removal in the subsequent step [7]. This sequence also enhances the exposure of pigmented compounds, allowing for their simultaneous removal during

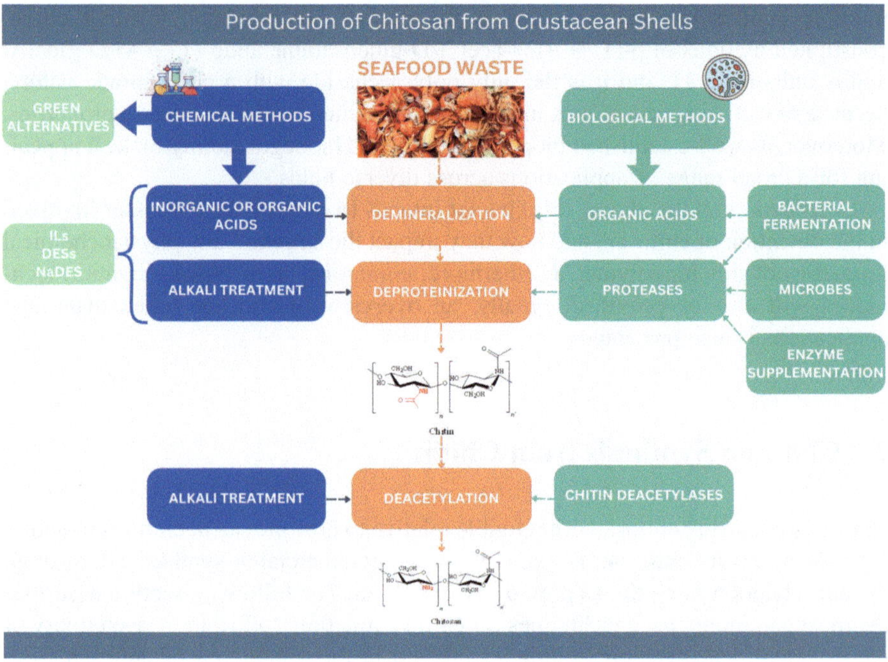

Fig. 2 General scheme of the chitosan production process from crustacean shells

deproteinization. Conversely, if deproteinization is performed first, an additional step is often required to eliminate residual pigments (i.e., decolorization) [8].

Next, the state-of-the-art of each process in chitin extraction (i.e., demineralization, deproteinization, decolorization) is presented.

2.1.1 Chemical Methods

Although chemical methods are often questioned for their potential environmental impact, they remain the most used ones for chitin extraction. Moreover, green alternatives, such as bacterial fermentation, enzymatic treatments, ionic liquids (ILs), or deep eutectic solvents (DESs), generally do not yet match the efficiency of conventional chemical approaches or are less economically viable, as discussed next.

Chemical demineralization is carried out by the use of an acidic solution (e.g., hydrochloric, sulfuric, nitric, acetic, oxalic, or formic acids) to react with the calcium carbonate, yielding calcium salts and carbon dioxide, and recovering the remaining solids by filtration. According to the literature, the most conventional chemical demineralization treatments use mild conditions: HCl 2 M, stirring at room temperature for an hour, and a waste:acid ratio of (1:4 w/v) [9]. Although there are reports of other conditions being used for this treatment, both depolymerization and deacetylation can take place if the demineralization time is prolonged, or if higher temperatures or concentrations are used [8, 10].

On the other hand, for deproteinization, various alkalis can be employed (e.g., NaOH, KOH, Na_2CO_3, $NaHCO_3$, K_2CO_3), with NaOH and KOH being the most commonly used, typically at concentrations of up to 5 M, with reaction times varying from some minutes to a couple of days and temperatures of up to 160 °C. However, these conditions may also induce partial deacetylation and depolymerization of chitin, and a consequent decrease in molecular weight [11, 12].

Considering that coloration is mainly due to pigments (e.g., astaxanthin, canthaxanthin, lutein, astacene, and β-carotene), when the decolorization step is required, it is usually carried out by extraction at room temperature using solvents like acetone, ethanol, chloroform, ethyl acetate, or mixtures of solvents. Conventional oxidizing agents, such as hydrogen peroxide (0.5–3%) and sodium hypochlorite (0.32%), have also been used; however, it should be noted that these agents tend to attack free amino groups and may introduce modifications into the polymer [13]. In heavily pigmented shells, such as those of lobsters, treatments at room temperature using mixtures of acetone and sodium hypochlorite have been reported to be successful [14].

2.1.2 Biological Methods

Biological methods for demineralization and deproteinization have been reported to be more effective in obtaining highly pure chitin, while minimizing molecular weight loss compared to conventional chemical processes [15]. However, they may face limitations in efficiency and require longer processing times compared to chemical methods.

The two main strategies commonly explored are microbial fermentation and enzymatic treatment. On the one hand, fermentation of shells of crustaceans with lactic acid-producing bacteria offers a dual-function strategy: the produced lactic acid reacts with the exoskeleton's calcium carbonate, promoting efficient demineralization, while bacterial proteases contribute to protein removal. Although not all lactic acid bacteria exhibit high proteolytic activity, certain strains can achieve partial deproteinization. To enhance efficiency, co-cultivation with protease-producing microbes or enzyme supplementation may be required. In this context, regarding the use of non-lactic acid bacteria for chitin recovery, *Bacillus subtilis* has been investigated for shrimp fermentation, showing that the organic acids produced, together with its strong proteolytic activity, enable simultaneous demineralization and deproteinization, reaching up to 72% removal of proteins and minerals. This highlights the potential of alternative microbial strains in optimizing biological chitin extraction processes.

On the other hand, enzymatic treatments focus solely on deproteinization and do not affect the mineral fraction, requiring a complementary chemical treatment for demineralization. These methods rely on proteases that hydrolyze peptide bonds, with commonly used enzymes including alcalase, pepsin, papain, pancreatin, devolvase, and trypsin. However, enzymatic extraction is particularly challenging due to the complexity of the multilayered structure of crustacean shells, which limits enzyme accessibility to embedded proteins [8]. Consequently, even under optimized conditions, enzymatic methods generally leave 5–10% of residual proteins [16].

2.1.3 Green Chemical Methods

Among green chemistry alternatives, there are two main approaches, employing either ionic liquids or deep eutectic solvents. Both ILs and DESs have emerged as sustainable substitutes for volatile organic solvents because of their non-volatility and high extraction efficiency. However, the widespread application of ILs is limited by high production costs, complex synthesis processes, and potential toxicity in certain formulations. On the contrary, DESs are generally biodegradable, exhibit lower toxicity, and are composed of readily available, low-cost components [9], making them an environmentally and economically more favorable option overall [18].

Structurally, ILs are organic salts that remain liquid at room temperature, whereas DESs are formed through eutectic mixtures of hydrogen bond acceptors (HBAs) and donors (HBDs), resulting in a melting point lower than that of the individual components. A subclass of DESs, known as natural deep eutectic solvents (NaDES), is derived from plant-based metabolites and offers additional environmental compatibility.

For chitin extraction, some examples of ILs include 1-ethyl-3-methylimidazolium acetate 1-allyl-3-methylimidazolium bromide, diisopropylethyl ammonium acetate diisopropylethylammoniumopanoate, and dimethylbutylammonium acetate. Herein, a one-pot process simultaneously enables demineralization and deproteinization. In this approach, the basic functionality of the ionic liquid facilitates protein removal, while its acidic component reacts with calcium carbonate, promoting efficient demineralization [20, 21].

In the case of DESs, some formulations include combinations of choline chloride (ChCl) or betaine (Bet) as HBAs with hydrogen bond donors such as lactic acid, n-butyric acid, glycerol, or urea. Representative examples include ChCl or Bet with lactic acid, glycerol, butyric acid or urea. Demineralization is primarily driven by proton release from the solvent system, which reacts with calcium carbonate to promote its dissolution [22]. Once the mineral content is removed, the chitin–protein fibril network becomes more accessible, as discussed before, and swells in the presence of DESs. This swelling facilitates the establishment of new hydrogen bonds between the HBA components of the DES and the proteins' hydroxyl, carboxylic, and amino groups. Thusly, both intra- and intermolecular hydrogen bonds within the chitin–protein matrix are disrupted, which results in an effective separation of the two components [6].

2.2 *Chitin Deacetylation*

To obtain chitosan from chitin, the acetamide groups of the *N*-acetylglucosamine must by hydrolyzed [23]. The deacetylation procedure can be conducted by either conventional chemical methods or green alternative ones (biological or chemical), being the first option generally preferred for industrial-scale production, due to its lower cost and more straightforward implementation. The specific details of the deacetylation approaches are described in the next sections.

2.2.1 Conventional Chemical Treatments

Although the amido group hydrolysis is more favorable in acidic than alkaline media, considering the sensitivity of glycosidic bonds to cleavage at acidic pH, chemical deacetylation should be performed in alkaline media to minimize the depolymerization of the polysaccharide [16].

The hydrolysis of the *N*-acetyl groups is most commonly carried out under heterogeneous conditions using hot, concentrated sodium hydroxide (NaOH) solutions, but other deacetylation strategies involving different solvents, catalysts, or reaction environments have also been reported in the literature [17, 18]. The extent of deacetylation and the overall quality of the resulting product are strongly influenced by the conditions under which the deacetylation reaction is performed [19]. Several experimental studies have systematically evaluated the influence of NaOH concentration (up to 60% w/v) and temperature (requiring up to 150 °C) and time (from minutes to days) on heterogeneous alkaline deacetylation. The results indicate that all parameters significantly impact the deacetylation process [20–22].

Lastly, it is worth noting that multistage alkaline treatments can lead to a higher extent of deacetylation. The increased efficiency of this approach is attributed to the progressively enhanced swelling of the polymer that results from the gradual increase in deacetylation percentage after each treatment. This improves the accessibility of acetylated sites along the polymer chains and thereby facilitates more effective deacetylation in subsequent steps [23, 24].

2.2.2 Biological Treatments

In nature, chitin deacetylases catalyze the hydrolytic removal of *N*-acetyl groups from chitin. These enzymes have been identified in marine bacteria, various fungi, and certain insect species [25, 26]. Enzymatic deacetylation offers a way to overcome the environmental drawbacks of chemical deacetylation and prevent depolymerization [27]. The main drawback of this option is the lack of industrial methods for generating chitin deacetylases, making it hard to scale up.

2.2.3 Green Chemical Methods

A green chemical option is using deep eutectic solvents, which unfortunately fail to achieve a high extent of deacetylation. To address both the limited efficiency of DESs and the environmental concerns related to high NaOH usage, a hybrid system combining DESs (specifically betaine/glycerol) with a reduced NaOH concentration has been developed [28]. DES scan efficiently interferes with chitin's hydrogen bonding, making the *N*-acetyl groups more susceptible to NaOH attack. In this way, milder conditions (NaOH, 25% w/v, 100 °C) are enough to yield chitosan, making it a more environmentally friendly alternative to alkali treatment. Furthermore, the presence of DESs in the reaction medium contributes to mitigating depolymerization, a common drawback when using strong bases, high temperatures, or high-pressure conditions [28].

3 Chitosan Characterization

As mentioned before, chitosan is a copolymer constituted by acetylated and deacetylated units of D-glucosamine, with percentages of deacetylation above 50% [29, 30]. Chitosan's molecular weight and the proportion of deacetylated units that it contains (deacetylation degree, DD) can be significantly different depending on i) the source from which chitin was obtained and the season variability affecting it and ii) the methodologies employed for chitin's extraction and deacetylation processes.

Considering that chitosan structure highly impacts its properties, its characterization is crucial to guarantee consistent performance and reproducibility in any application [31, 32]. The next section outlines the techniques commonly employed for its structural analysis.

3.1 Chemical Characterization

3.1.1 *FT*-IR Spectroscopy

The main signals observed in a typical chitosan *FT*-IR spectrum are at 1657 cm^{-1} the amide type I carbonyl stretching; at ~1600 cm^{-1}, the scissors bending of the primary amine of the deacetylated units; and at~1550 cm^{-1} the amide band type II, corresponding to the N-H bending [33–35]. The region between ~1550 and 1600 cm^{-1} presents significant differences according to the deacetylation degree of chitosan (Fig. 3). As can be observed in the figure, as the deacetylation degree increases, so does the intensity of the signal at ~1600 cm^{-1}, with respect to the one at ~1550 cm^{-1}, due to the improved ratio of free amino groups relative to the amide groups.

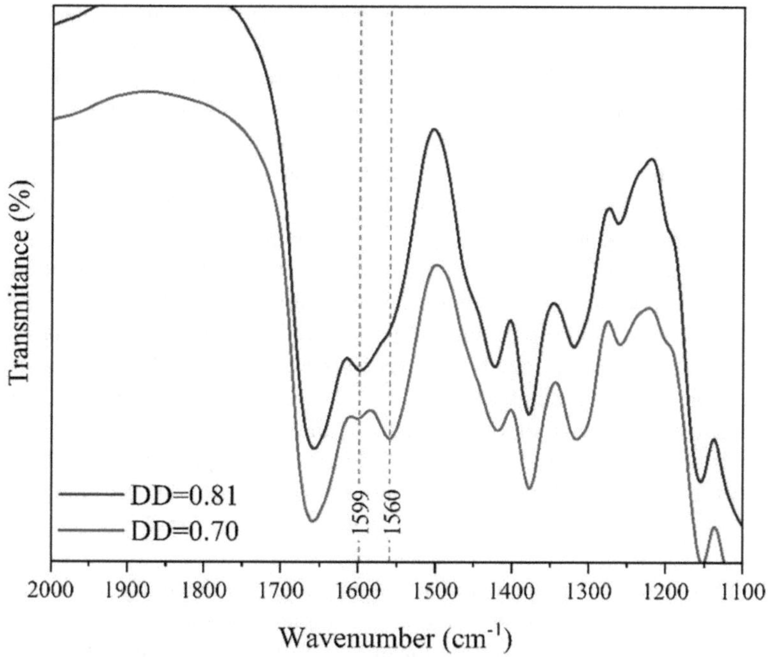

Fig. 3 *FT*-IR spectra of chitosan samples that differ in their degree of deacetylation. Spectra were recorded by the authors [33, 35], and only a magnified part is presented to highlight the relevant features

Additionally, the O-H and N-H bond stretching are observed at 3420 cm^{-1}. In the range between 2850 and 3000 cm^{-1}, the signals corresponding to C_{sp3}-H stretching are observed. Below 1500 cm^{-1}, the stretching of C_{sp3}-O (between 1000 and 1180 cm^{-1}) and C_{sp2}-N at 1262 cm^{-1} attributed to the amide band type III were observed, together with those corresponding to the C_{sp3}-H bending of methyl and methylene groups (between 1300 and 1450 cm^{-1}).

3.1.2 NMR-Spectroscopy

Chitosan's NMR spectra can be acquired for samples in solution, or for solid or semi-solid samples. To record solution-state spectra, chitosan must be in its water-soluble form, with its amino groups protonated. An approach to protonate chitosan is with aqueous HCl, then freeze-drying the resulting salt, dissolving it in deuterated water, and freeze-drying it again for proton exchange. Alternatively, protonation can be carried out directly using deuterium chloride or deuterated acetic acid, yielding the corresponding deuterated acidic salt.

The advantage of acquiring NMR spectra in solution is the higher resolution, allowing not only ^{1}H and ^{13}C spectra to be recorded, but also bidimensional spectra.

A typical chitosan ^{1}H-NMR spectrum shows the glucosamine H-1 shift at δ ~ 4.9 ppm, the H-1 of *N*-acetyl-D-glucosamine at δ ~4.7 ppm, H-2 at δ ~3.2 ppm, and the methyl group of the acetylated unit at δ ~2.1 ppm. The signals of H-3, H-4, H-5, and H-6 typically overlap in the δ ~3.6–4.1 ppm region [36]. In the case of the ^{13}C spectrum, the anomeric carbon signal appears at approximately 100 ppm, C-2 at ~56 ppm, C-3 at ~71 ppm, C-4 at ~77 ppm, C-5 at ~75 ppm, and C-6 at ~61 ppm. The signals corresponding to the acetyl groups are observed at ~23 ppm (methyl group) and ~ 176 ppm (carbonyl group) [37].

The main drawback of solution-state NMR is that many chemical modifications can make the products insoluble over the entire pH range, making solid-state analysis necessary to compare the native material with the modified product. In this regard, Magic-Angle Spinning (MAS) solid-state NMR spectroscopy is a versatile tool, but it is generally limited to ^{13}C spectra, due to resolution constraints and the lower sensitivity of other nuclei in the solid state. The chemical shifts commonly observed in solid-state ^{13}C-NMR are δ ~25 ppm for the methyl groups in the acetyl units, δ ~100 ppm for the anomeric carbon, δ ~70–83 ppm for C-3 to C-6, δ ~50 ppm for C-2 in glucosamine, and δ ~55 ppm for C-2 in the acetylated units [38].

An alternative that has gained increasing relevance in recent years is the possibility of acquiring spectra from semi-solid samples, such as hydrogels or swollen forms of chitosan or chitosan-based materials. In this case, high-resolution magic-angle spinning (HR-MAS) NMR spectroscopy is used, offering the advantage of enabling ^{1}H and bidimensional spectra. Although its resolution is lower than that of solution-state NMR, it allows the analysis of samples that cannot be dissolved but can be hydrated in D_2O or other deuterated solvents [34].

3.1.3 Deacetylation Degree (DD)

The three most extensively used techniques to determine the DD are: ^{1}H-NMR spectroscopy, acid–base titration, and elemental analyses. The main pros of these techniques are their high versatility and the fact that none of them require the use of chitosan standards.

- *^{1}H-NMR Spectroscopy*

The deacetylation degree can be calculated using ^{1}H-NMR spectroscopy with the following equation:

$$DD = \left[1 - \frac{1/3 \cdot I_{CH_3}}{1/6 \cdot I_{H2-H6}}\right] \tag{1}$$

where I_{CH_3} is the integral of the signal corresponding to the protons of the methyl groups of the acetylated units (~2 ppm); I_{H2-H6} is the integration of the region between ~3 and ~ 4.2 ppm, in which the resonance of the protons H-2 to H-6 is typically observed [39].

- *Acid-Base Titration*

Acid–base titration can be conducted by either potentiometric or conductometric methods [33]. In both cases, a specific mass of previously dried chitosan (m_{chit}) is dissolved in an aqueous solution of HCl of concentration C_1 and titrated with a NaOH aqueous solution of concentration C_2.

In the case of potentiometric titration, by plotting the pH against the consumed volume of NaOH, two inflection points can be determined (V_1 and V_2). Analogously, in the case of conductometric titration, the solution's conductivity is plotted against the NaOH volume consumed, and the resulting linear sections obtained will intersect in V_1 and V_2. These correspond to the volumes of NaOH required for neutralizing the HCl in excess (V_1), and deprotonating the amino groups present in chitosan (V_2). Therefore, the %DD can be calculated as follows, considering the molecular weight of the acetylated (molar fraction, X_{Ac}) and deacetylated units (molar fraction, X_{NH_2}) are 203 and 161, respectively, and the repetitive units' average molecular weight and moles (MW_{chit} and n_{chit}, respectively), the moles ofdeacetylated units (n_{NH_2}) result from the subtraction of the moles of NaOH used for titrating the HCl in excess ($C_1 \cdot V_1$), to the total moles of NaOH ($C_2 \cdot V_2$).

$$X_{NH_2} = \frac{n_{NH_2}}{n_{chit}} = \frac{(C_2 \cdot V_2 - C_1 \cdot V_1)}{m_{chit} \cdot MW_{chit}} \tag{2}$$

$$MW_{chit} = X_{NH_2} \cdot 161 + X_{Ac} \cdot 203 = X_{NH_2} \cdot 161 + (1 - X_{NH_2}) \cdot 203 \tag{3}$$

$$\%DD = \frac{(C_2 \cdot V_2 - C_1 \cdot V_1) \cdot 203}{m_{chit} + (203 - 161) \cdot (C_2 \cdot V_2 - C_1 \cdot V_1)} \cdot 100\% \tag{4}$$

- *Elemental Analysis*

It is also possible to calculate chitosan's deacetylation degree by elemental analysis, from which the percentages of carbon and nitrogen are obtained. Then, the carbon/nitrogen ratio is calculated, and the DD is obtained by the following equation:

$$DD = 1 - \left[\frac{C/N - 5.145}{6.861 - 5.145}\right] \tag{5}$$

The corresponding values of the C/N of fully *N*-deacetylated and fully *N*-acetylated polysaccharides are 5.145 and 6.861, respectively.

3.2 Crystallinity

Chitosan's crystallinity is determined by the crystalline structure of the chitin from which it is derived [40]. Chitin occurs naturally in the polymorphic forms α, β, and γ. These types differ in the orientation of their molecular chains, which gives rise to distinct physical properties [41].

In α-Chitin, the most abundant one [11], chains have an antiparallel arrangement, stabilized by strong hydrogen bonding that results in a compact and rigid structure [42]. β-Chitin, with a parallel chain orientation, has a more flexible and loosely packed configuration [38]. γ-Chitin has both parallel and antiparallel chains, typically with a 2:1 ratio of chains in one direction versus the opposite direction [43]. This arrangement confers intermediate structural and physical properties of α- and β-chitin, combining features of both. The structural differences mentioned influence the flexibility, toughness, and hardness of chitin-derived materials. For example, α-chitin is the predominant crystalline form in crustacean shells, providing mechanical stiffness.

Chitosan generally exhibits lower crystallinity than chitin. This is mainly due to the partial deacetylation process, which introduces structural irregularities in the chains and disrupts the ordered packing of the crystalline domains. Both chitosan's degree of deacetylation (DD) and the molecular weight (MW) significantly affect its crystallinity. As far as the DD is concerned, in the case of low to moderate DD values, the disruption of the original chitin structure tends to reduce crystallinity; contrarily, at higher DD levels, the increased uniformity and reduced steric hindrance promote chain regularity, which could enhance crystallinity to some extent [44]. Regarding molecular weight, crystallinity increases with increasing MW due to improved chain alignment and organization as the polymer chains become longer [45].

To assess the crystalline structure and degree of crystallinity, X-ray diffraction (XRD) is the most common technique. Typically, the main diffraction peaks appear at $2\theta \approx 11°$ and $2\theta \approx 20°$, associated with hydrated crystalline domains. A broad halo centered at $2\theta \approx 16°$ is attributed to amorphous regions. These reflections are characteristic of chitosan's semi-crystalline nature, which results from the structural irregularities introduced by partial deacetylation of chitin.

The crystallinity index (CrI) can be estimated using different approaches. A common method is the peak height method, which compares the intensity of the crystalline peak at $2\theta \approx 20°$ with that of the amorphous halo at $2\theta \approx 16°$ [46]. Alternatively, the amorphous subtraction method involves integrating the total area under the diffractogram and subtracting the amorphous part for the crystalline fraction estimation [47].

Although α-, β-, and γ-chitin polymorphs are well characterized, identifying these forms in chitosan is more complex due to its altered chain structure and lower crystallinity [48]. Nonetheless, recent studies have attempted to infer the presence of α- or β-like crystalline domains in chitosan based on peak positions, interplanar spacing, and hydrogen bonding patterns obtained from XRD and complementary techniques such as *FT*-IR and solid-state NMR spectroscopies [49].

3.3 Molecular Weight

Like all polymeric materials, chitosan exhibits a range of molecular weights rather than a single defined value. Due to the polydisperse nature of macromolecular systems, polymers consist of distinct molecular weight fractions, where each fraction '*i*' contains *Ni* molecules with a molar mass *Mi*. To describe this distribution, different molecular weight averages are used, each providing different insights into the polymer's overall molecular characteristics [50]:

- *Number-average molecular weight (M_n)*: Defined by Eq. 6, it represents the total mass of all molecules divided by the total number of molecules, meaning it is based on molar fractions (n_i).

$$M_n = \sum n_i \cdot M_i = \frac{\sum N_i \cdot M_i}{\sum N_i} \tag{6}$$

$$n_i = \frac{N_i}{\sum N_i} \tag{7}$$

- *Weight-average molecular weight (M_w)*: defined by Eq. 8, it gives more weight to larger molecules since it is calculated using mass fractions instead of molar fractions (w_i).

$$M_w = \sum w_i \cdot M_i = \frac{\sum N_i \cdot M_i^2}{\sum N_i \cdot M_i} \tag{8}$$

$$w_i = \frac{N_i \cdot M_i}{\sum N_i \cdot M_i} \tag{9}$$

- *Diffusion average molecular weight (M_z):* defined by Eq. 10, it further emphasizes the contribution of the highest molar mass fractions.

$$M_z = \frac{\sum w_i \cdot M_i^2}{\sum w_i \cdot M_i} = \frac{\sum N_i \cdot M_i^3}{\sum N_i \cdot M_i^2} \tag{10}$$

- *Viscosity-average molecular weight (M_v):* obtained from the intrinsic viscosity (η) of a polymer solution, which measures its ability to increase a solvent's viscosity when dissolved. η reflects he hydrodynamic volume of polymer chains in solutions and is related to molecular weight, as described by the

Mark-Houwink–Sakurada equation (Eq. 11), where a and K are empirical constants specific to the system.

$$[\eta] = K \cdot M_v^a \tag{11}$$

Since polymer chains do not have a single molecular weight but rather a distribution, a key parameter to describe their heterogeneity is the polydispersity index (PDI), defined as the ratio M_w/M_n. A PDI close to 1 indicates a narrow distribution, characteristic of uniform polymers, like the ones obtained via controlled polymerization techniques. Conversely, higher PDI values reflect broader distributions, typically found in natural polymers such as chitosan [51].

Several techniques are used to determine chitosan's molecular weight, each offering distinct advantages and limitations. The most used method is viscometry based on measuring the flow times of chitosan solutions at different concentrations (t_i), as well as for the pure solvent (t_0), at a controlled temperature, generally 25 °C, using a capillary viscometer (e.g., Ubbelohde type). The relative ($\eta_r = t_i/t_0$), specific ($\eta_{sp} = \eta_r - 1$), reduced ($\eta_{red} = \eta_{sp}/c$), and inherent viscosities ($\eta_{inh} = ln\ (\eta_r)/c$) are calculated, and the intrinsic viscosity (η) is obtained by extrapolation of the reduced and inherent viscosities to a concentration equal to zero. Finally, the viscosity-average molecular weight (M_v) is determined with the Mark–Houwink–Sakurada equation (Eq. 11) with appropriate constants (K, a) for the chitosan-solvent system [52].

Size exclusion chromatography (SEC) is another extensively utilized technique for chitosan's MW determination. This method separates polymer molecules based on their hydrodynamic volume, under the assumption that polymers with similar hydrodynamic volumes will elute at the same retention time. Molecular weight calibration requires the use of standards, with pullulan being the most commonly used due to its chemical similarity to chitosan, as well as its availability as a standard with a known molecular weight and narrow distribution [53]. Unlike viscometry, SEC provides a complete molecular weight distribution, enabling the calculation of number-average molecular weight (M_n), weight-average molecular weight (M_w), diffusion-average molecular weight (M_z), and polydispersity index.

Membrane osmometry is another technique used for the determination of the number-average molecular weight (M_n) of chitosan. It is based on measuring the osmotic pressure (π) of a polymer solution across a semipermeable membrane through which solvent molecules are allowed to pass, but polymer chains are not. A common choice for the membrane is cellulose acetate, given its compatibility with aqueous polymer solutions [50]. By plotting the reduced osmotic pressure ($\frac{\pi}{C}$) against concentration (C), M_n can be determined from the extrapolation to zero concentration using the equation:

$$\frac{\pi}{C} = RT\left(\frac{1}{M_n} + A_2C\right) \tag{12}$$

where R is the universal gas constant, T is the absolute temperature, and A_2 is the second virial coefficient, which accounts for polymer-solvent interactions [50]. Unlike SEC, which provides a full molecular weight distribution, membrane osmometry directly measures M_n and is particularly suitable for high-molecular-weight polymers. Nevertheless, it has limitations for low-molecular-weight fractions due to their lower osmotic pressure response [54].

Another technique that can be employed for chitosan's MW determination is light scattering, particularly for measuring the weight-average molecular weight (M_w). In this case, static light scattering (SLS) or multi-angle light scattering (MALS) are the most reliable approaches, as they measure the scattered light's intensity as a function of angle, which depends on molecular weight, concentration, and polymer-solvent interactions [54]. The Debye equation is used for M_w determination, incorporating the second virial coefficient (A_2) to account for polymer-solvent interactions:

$$\frac{K \cdot C}{R_\theta} = \frac{1}{M_w} + 2 \cdot A_2 \cdot C \tag{13}$$

where K is an optical constant dependent on the refractive index increment (*dn/dC*) and laser wavelength, C is the polymer concentration, R_θ is the Rayleigh ratio at scattering angle θ, and A_2 provides insight into solubility and intermolecular interactions [55].

To obtain a more accurate molecular weight determination, an interesting and powerful strategy is size-exclusion chromatography coupled with multi-angle light scattering (SEC-MALS). This technique separates chitosan molecules by size before light scattering analysis, providing direct M_w values independent of column calibration standards. SEC-MALS is particularly useful for polydisperse samples, as it enables the determination of the entire molecular weight distribution rather than a single average value [56].

On the other hand, dynamic light scattering (DLS) is mainly employed to measure chitosan's molecules' hydrodynamic radius (R_h) in the solution, by analyzing fluctuations in scattered light intensity due to Brownian motion [56]. The relationship between R_h and molecular weight is derived from the Stokes-Einstein equation, but estimating M_w from DLS requires empirical models that assume a correlation between hydrodynamic size and molecular weight, thus introducing variability and potential inaccuracies in molecular weight estimations [54].

It is important to note that all these techniques require chitosan to be in solution; consequently, molecular weight determination typically involves working in dilute acidic solutions (e.g., acetic or hydrochloric acids) to ensure complete solubilization of chitosan and reliable analytical measurements. The choice of solvent and its ionic strength can also influence measurements, particularly in viscometry and light scattering, making the selection of experimental conditions essential for an accurate MW determination.

4 Physicochemical Impact of Chitosan's Structural Characteristics

Chitosan's structural characteristics (i.e., MW and DD) have a substantial effect on its physicochemical properties and, consequently, on its suitability for specific applications.

Particularly, an increase in MW leads to stronger intermolecular and intramolecular interactions, as a higher number of available functional groups enhances hydrogen bonding and molecular cohesion. As a result, greater entanglement of the polymer chains occurs, contributing to a more complex and rigid network structure. This increase in molecular weight and chain interaction leads to several key changes in the material's properties:

- *Higher crystallinity,* as longer polymer chains tend to pack more efficiently, promoting ordered regions within the material.
- *Higher viscosity,* as a result of enhanced interchain interactions (see next section).
- *Lower permeability*, as the denser molecular arrangement reduces the diffusion of gases or liquids through the material [57].
- *Enhanced tensile and elongation strength*, as the molecular weight increases, the potential for molecular entanglement during film formation also rises, which in turn leads to improved tensile elongation and tensile strength [45].
- *Decreased water solubility*: Due to the increased interaction between chitosan chains, solubility decreases as molecular weight (MW) increases. For example, chitosan with an MW lower than 9 kDa exhibits high solubility in water without the need for acids. In contrast, chitosan with an MW of 30 kDa or higher requires an acidic medium for dissolution, as the free amino groups become protonated, forming ammonium salts that are soluble in this acidic aqueous environment [35]. The pKa of chitosan's amino groups has a value between 6.1 and 6.8, depending on its MW and DD [32]. Therefore, below those pH, the amino groups become protonated, making chitosan soluble.

Regarding the DD, considering constant MW, an increase in DD leads to stronger intermolecular and intramolecular interactions, which result in the following impacts on the properties:

- *Changes in crystallinity,* for low to moderate DD, the disruption of the original chitin structure tends to reduce crystallinity; while at higher DD, the increased uniformity and reduced steric hindrance promote chain regularity, leading to an enhanced crystallinity.
- *Changes in viscosity,* discussed in the next section.

- *Enhanced tensile and elongation strength:* attributed to an increase in crystallinity at higher DD [44].
- *Enhanced water solubility:* This happens because a higher DD reduces the number of acetyl groups, thus increasing the availability of free amino groups and their interaction with water molecules by hydrogen bonding.

Regarding the reactivity of chitosan, both the DD and MW significantly affect how its chains interact and how easily modifying agents can reach the hydroxyl and amino groups, thus affecting the feasibility of chemical functionalization. Additionally, since the viscosity of chitosan solutions impacts the diffusion of reagents, and this is affected by chitosan's MW and DD (see next section), this is also a factor to be considered.

- *Rheological Properties*

Chitosan solutions have complex rheological properties, generally exhibiting non-Newtonian shear-thinning behavior [58]. Its pseudoplastic response is attributed to the progressive disentanglement and alignment of polymer chains under shear stress, which reduces internal friction and facilitates flow. The extent of shear-thinning depends on chitosan concentration: at low concentrations (~0.05 g L^{-1}), solutions exhibit nearly Newtonian behavior, whereas at higher concentrations (~0.4 g L^{-1}), polymer–polymer interactions intensify, leading to a more pronounced shear-thinning behavior [59].

Regarding the viscosity, MW has a direct impact on this property: at constant concentration, higher MW chitosan results in higher viscosity due to increased chain entanglement and hydrodynamic volume [59]. Conversely, lower MW chitosan forms less entangled solutions with reduced viscosity, which may be advantageous in applications requiring lower resistance to flow.

On the other hand, DD has a more complex effect on viscosity. Higher DD generally enhances solubility but may lead to lower viscosity because of the amino groups' electrostatic repulsion, which reduces polymer-polymer interactions [60]. Conversely, very high DD is associated with a more entangled structure, leading to increased storage modulus (G') and loss modulus (G"), making the solution more elastic and resistant to flow, and thus increasing viscosity [61].

Temperature also influences viscosity, with increasing temperatures reducing shear stress due to weakened intermolecular interactions, and exponentially decreasing viscosity [59].

In summary, the interplay between MW, DD, concentration, and temperature defines the rheological behavior of chitosan, which determines the overall processability of its formulations. For instance, high viscosity can hinder processes such as filtration, mixing, and coating, while shear-thinning behavior (in which viscosity is lower for higher shear rates) is beneficial in spraying and extrusion [62].

5 Applications of Chitosan-Based Materials

The biodegradability, biocompatibility, and cost-effectiveness of chitosan make this polysaccharide highly attractive as a raw material for a wide broad of industries. Chitosan exhibits antibacterial activity against both Gram-positive and Gram-negative bacteria, including *Staphylococcus aureus*, *Escherichia coli*, and *Bacillus subtilis*.

Moreover, chitosan's versatility for chemical modification makes it possible to impart the required properties for specific applications, enhancing its performance.

In fact, there are a huge number of reports regarding chitosan-based products proposed for varied applications (Fig. 4), which are discussed next.

5.1 *Beverages and Food Industries*

Chitosan is recognized by the US FDA (Food and Drugs Administration) as a safe compound and, therefore, has been incorporated in the beverages and food industries. However, some reports are claiming that chitosan coming from crustaceans' shells may pose allergenic risks, likely due to residual impurities from their source [63, 64].

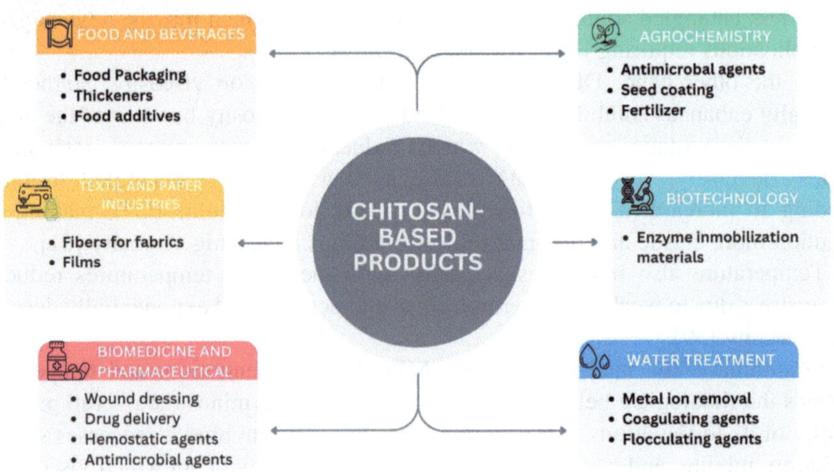

Fig. 4 Schematic representation of proposed applications for chitosan-based materials, including multiple commercial uses

5.1.1 Beverages

In the beverage industry, chitosan is incorporated for three primary purposes: i) clarification (achieved by electrostatic interaction with anionic compounds, like polyphenols), ii) preservation (due to its antimicrobial and antioxidant activities), and iii) encapsulation (of colored compounds, vitamins, or probiotics, which would otherwise degrade in time when beverages are stored). Besides the sanity aspects, chitosan can contribute to improving the organoleptic properties of beverages like beer, wine, green tea, or fruit juice. For instance, for green tea clarification, chitosan contributes to reducing turbidity and haze formation caused by compounds like the proline amino acid by adsorption [65]. Additionally, for green tea bitterness and odd taste removal, tannase enzymes, which are used to hydrolyze tannins—the main responsible of green tea astringency-can be coated with chitosan. Another case is its use in wine for toxin and odd taste compounds removal, and as an alternative for SO_2 due to its antimicrobial and antioxidant characteristics [64]. Moreover, compared to other traditional clarifying agents, such as bentonite, its flocculation capacity was more efficient.

5.1.2 Food

In this industry, chitosan is incorporated to prolong the shelf life and flavor of fresh goods in both edible and non-edible biofilms, coatings, and packaging. However, to use chitosan in such applications, some of its drawbacks, such as its hydrophilic nature, poor mechanical properties, and permeability to volatiles, must be overcome by chemically modifying it [66].

Beyond chitosan's inherent film-forming ability and structural protective role, providing a physical barrier against moisture, gases, and external contaminants, it exhibits remarkable bioactive characteristics (e.g., antimicrobial and antioxidant activities) that contribute significantly to the extension of food shelf life and the prevention of spoilage.

Additionally, chitosan-based materials significantly contribute to food safety and nutritional quality through their intrinsic antioxidant and antimicrobial activities. These properties help to delay microbial growth, inhibit oxidation processes, and prolong the perishable goods' shelf life. Furthermore, chitosan may offer dietary benefits by stimulating the gastrointestinal tract through prebiotic effects. Chitosan-based coatings are effective in reducing moisture transfer, limiting oxygen permeability, and delaying ethylene production, which are critical for upholding the freshness and visual features of fruits and vegetables during storage. Some examples of these chitosan-based coatings include: tomatoes, carrots, mangoes, strawberries, blueberries, bananas, apples, and broccoli [5]. Moreover, the incorporation of active agents such as gallic acid and sodium carbonate into chitosan films has successfully achieved oxygen-scavenging capabilities, enhancing the preservation performance of these biodegradable packaging systems, which highlights the chitosan's versatility as a key factor in active and sustainable food packaging technologies development [67].

5.2 Agrochemicals

Chitosan has greatly improved the efficiency in agrochemical applications and enhanced seed protection and germination when used as a coating agent. In this context, conventional pesticide application is often inefficient due to their formulations as emulsions or powders, which are susceptible to photolysis, hydrolysis, and microbial degradation. As a result, only 10–30% of their applied dose is effective. This low efficiency requires repeated applications to achieve effective pest control, thus increasing production costs and exacerbating environmental impact. To address this problem, significant efforts are focused on systems that enhance the efficacy of pesticide delivery. In this context, the scientific community has shown great interest in controlled-release technologies that enable the active ingredients of agrochemicals to reach their target at specific concentrations and specific times. A way to do so is by using chitosan in nanoformulations for encapsulation, entrapping the active ingredients via ionic or covalent inter- or intramolecular bonds with the polymeric chitosan matrix. Besides, chitosan nanoparticles can function as antimicrobial agents against pathogenic fungi [68, 69].

Regarding its use in seed coating, chitosan can be used to enhance the water absorption capacity, as well as the nutrient uptake, and consequently, positively affect seed germination and promote early growth. Besides, chitosan's hydrophilic nature allows the formation of a thin film around the seed, which provides a microenvironment that favors water absorption [70]. By being capable of maintaining a certain water balance, seed resilience is improved, which makes it more likely for the germination to be successful even in unfavorable environmental conditions. Besides, its amino and hydroxyl groups allow chitosan to form complexes with phytotoxic heavy metals, thereby averting the dysfunction of shoot and root cells [71].

5.3 Biotechnology

The industrial application of enzymes is often limited by their high cost and instability. To make them cost-effective, their reuse is essential. One approach involves immobilizing the agents by attaching them to a solid support, enabling their easy removal from the system and subsequent recovery for reuse [72]. Given its non-toxic and biocompatible nature, chitosan is widely recognized as an ideal enzyme immobilization support material, particularly in industrial-scale biocatalytic processes. Immobilization can be achieved through various strategies, including binding, entrapment, or cross-linking of enzymes [73]. Chitosan's amino groups have a key role in enabling these interactions, e.g., by enabling the attachment of cross-linking agents (e.g., glutaraldehyde) to immobilize enzymes, or by chelating metal ions that subsequently bind to enzymes [74].

5.4 *Textile Industry*

Chitosan's outstanding features (i.e., biocompatibility, biodegradability, film-forming ability, and antimicrobial activity) make it particularly attractive for textile applications. Among its various uses in the textile industry, one of the most extensively studied is its application in the antibacterial finishing of fabrics [75]. Because textiles can provide a favorable environment for microbial growth, potentially leading to unpleasant odors or even disease transmission, antimicrobial treatments have become increasingly important. In the textile industry, chitosan can be employed either as a finishing agent or as a part of antibacterial fibers [76]. In the first approach, chitosan is typically grafted or coated onto textile surfaces, imparting antimicrobial properties. In the second approach, chitosan is integrated directly into the fiber structure. For example, Japan Fuji Textile Co., Ltd. developed antibacterial fibers called *Chitopoly*, which are then woven into fabrics [77–79].

Beyond antimicrobial applications, another notable use of chitosan is in anti-wrinkle finishing, particularly for natural fabrics such as cotton, silk, and hemp, which are prone to deformation and loss of shape. Treatment with chitosan solutions through a common dipping-rolling-baking process creates a protective layer on the fabric surface while filling micropores within the fiber structure [79]. The strong hydrogen bonding interactions between chitosan and fiber molecules enhance structural stability, reduce molecular mobility, and increase wrinkle resistance [76].

Chitosan has also been employed to improve dye uptake, since its cationic nature effectively fixes anionic dyes, reducing the need for salt and improving dye uptake [76]. Finally, it has been used in antistatic finishing, especially for synthetic fabrics such as polyester, which are prone to static charge accumulation [79]. The positively charged ammonium groups and highly polar functional groups on chitosan molecules promote ionic conductivity and moisture retention on the fabric surface. This leads to the accumulation of water molecules on synthetic fabrics, which facilitates charge dissipation and increases surface conductivity, helping to prevent electrostatic discharge [79]. Such properties are particularly valuable in the production of antistatic workwear for electronic or industrial environments.

5.5 *Pulp and Paper Industry*

In the pulp and paper industry, chitosan is mainly employed in two key stages: surface coating and wet-end processing [80]. In surface coating applications, chitosan serves as a functional biopolymer to enhance the performance of paper products. Due to its structural similarity to cellulose and strong affinity for fiber surfaces, it adheres easily to cellulosic substrates. Chitosan can form a continuous film over the fiber network that provides a variety of functional benefits, including improved mechanical strength, enhanced surface properties, and acts as a barrier against water vapor, oxygen, and grease [81, 82]. Additionally, chitosan-coated papers exhibit antimicrobial activity, antistatic behavior, and increased dyeability, making them suitable for food packaging and specialty paper, such as printing substrates, medical-grade paper, smart packaging materials, barrier coatings for moisture-sensitive documents, and antibacterial paper for hygienic use [80].

At the wet-end of the papermaking process—where cellulose fibers are suspended in water and various additives are introduced—chitosan plays a different role. It is typically added to the pulp slurry in the form of a dilute acid solution prior to sheet formation [80]. In this case, chitosan acts as a multifunctional additive, improving fiber retention, promoting drainage, enhancing dry strength, and facilitating internal sizing [75]. Its cationic nature promotes interaction with anionic additives commonly used in papermaking, contributing to a more efficient production process and improved final product performance [83]. The use of chitosan in wet-end applications also supports the shift toward more sustainable and biodegradable additive systems in the industry.

5.6 Water Treatment

In the water treatment field, chitosan can be applied via two main mechanisms: coagulation/flocculation and adsorption. While coagulation and flocculation aim to aggregate and remove suspended particles or colloids, adsorption is used to capture dissolved pollutants through surface interactions.

Chitosan's effectiveness as a coagulant and/or flocculant for anionic pollutants is mainly due to its polycationic nature and its long polymer chains, which enable both charge neutralization and bridging mechanisms, which lead to particle aggregation and sedimentation [84]. These features, along with chitosan's nontoxicity, biodegradability, and strong metal-chelating ability, make it a promising alternative to conventional inorganic coagulants and synthetic polyelectrolytes often used in wastewater treatment.

To use chitosan as an adsorbent, efforts focus on achieving insoluble chitosan-based materials in the whole pH range [5]. In this context, various modification strategies have been developed, including functionalizing the C-6 hydroxyl or C-2 amino positions (or both), crosslinking through amide or ester bonds, and grafting of polymeric chains. Moreover, introducing functional groups with fixed charges, positive or negative, can enhance the electrostatic interaction of adsorbent and pollutants with the opposite charge, thereby improving adsorption efficiency [85].

However, industrial feasibility depends heavily on the reusability of the adsorbent. Although several regeneration cycles have been studied, the acidic conditions commonly employed tend to degrade the polymer. As a result, significant deterioration has been reported after just a few adsorption–desorption cycles, posing a challenge for long-term application [5, 35].

5.7 Cosmetics

Although chitosan is not as widely used in cosmetics as other biopolymers like collagen or hyaluronic acid, it is gaining increasing attention. Its use in the cosmetic industry has been approved by both the FDA and the European Union (Regulation EC No 1223/2009, 87]. According to the COSING database, it is unrestricted and listed for film-forming and hair-fixing functions. Its affinity for keratin and ability to form hydrophilic films make it a worthy ingredient in formulations targeting the skin, hair, nails, and oral cavities [87].

In cosmetic products, chitosan serves multiple functional roles in topical formulations. As a rheology modifier and thickener, it enhances the texture and consistency of creams, gels, foams, and emulsions. Its film-forming capacity helps to retain skin moisture by reducing transepidermal water loss, thereby improving hydration. Due to its antimicrobial activity, chitosan also acts as an active component in deodorants, anti-acne treatments, and oral hygiene formulations [34, 87]. This not only enhances efficacy but can also reduce the need for synthetic preservatives. In hair care, its cationic nature enables it to neutralize oppositely charged damaged hair surfaces, improving manageability and appearance by forming a protective film that enhances softness and volume [86]. Additionally, chitosan has shown promise as a carrier for active ingredients via encapsulation strategies [88].

In fact, several commercial products already incorporate chitosan in their formulation [75, 86, 89, 90].

5.8 Biomedicine

Chitosan has found widespread use in various medical applications due to its excellent biocompatibility, as well as its antimicrobial, antioxidant, anti-inflammatory, wound healing, antihypertensive, and anticancer properties. These biological features contribute to its versatility in therapeutic and biomedical contexts, and in this field, ongoing research continues to expand its potential applications, some of which are briefly outlined below [91].

- Anticancer agent: due to its ability to reduce tumor size by electrostatic interactions of its amino groups (positively charged) and the membranes of tumor cells (negatively charged) [92].
- Antihypertensive agent: due to its ability to inhibit angiotensin-converting enzyme activity [93].
- Wound healing promotion: by stimulation of cell proliferation, migration, and tissue regeneration [94, 95].
- Anti-inflammatory properties: by modulating inflammatory mediators and cytokine expression [96].
- Hemostatic activity: by accelerating blood clotting through platelet aggregation and interaction with red blood cells [97].

- Antioxidant agent: by free radicals' scavenging and oxidative stress reduction [98].
- Carrier for controlled drug delivery sensitive to temperature or pH changes: enabling targeted release and reduced systemic toxicity, by ionic interaction and complex formation to link anionic drugs to chitosan [99].
- Biomarkers: due to its ability to absorb organic molecules and metal ions, chitosan can be employed in the fabrication of biosensors [100].
- Immunomodulation: contributing to the activation or regulation of immune responses [101].
- Tissue engineering: as a scaffold material for cell adhesion, growth, and differentiation in regenerative medicine applications [102].
- Mucoadhesivity: enhancing the residence time of formulations in mucosal tissues such as oral, nasal, or ocular routes [103].

Some of the biomedical applications listed above are discussed more thoroughly in the subsequent chapters, and readers are therefore encouraged to explore them for deeper insight and a more comprehensive understanding of the topic.

5.9 Conclusions

Chitosan is a polysaccharide with distinctive structural features, derived from highly abundant and renewable natural sources. Its remarkable chemical versatility enables the design of derivatives tailored for a wide range of applications, representing a strategic advantage in developing functional materials. This tunability allows for the modulation of key physicochemical properties to meet specific functional requirements. However, it is important to emphasize that these properties are strongly influenced by the polymer's molecular weight and degree of deacetylation. Therefore, precise and reproducible characterization of these structural parameters is crucial to ensure consistent performance, effectively acknowledge, and leverage the most out of the material's potential for targeted applications.

In this context, the comprehensive overview of the most commonly employed analytical techniques for determining these parameters presented in this chapter offers valuable guidance for an accurate evaluation of the material. Among the wide range of applications for chitosan and its derivatives reported, those in the biomedical field appear to be the most advanced, with several products already on the market.

References

1. Lv, J., Lv, X., Ma, M., Oh, D.-H., Jiang, Z., & Fu, X. (2023). Chitin and chitin-based biomaterials: A review of advances in processing and food applications. *Carbohydrate Polymers, 299*, 120142. https://doi.org/10.1016/j.carbpol.2022.120142

2. Bakshi, P. S., Selvakumar, D., Kadirvelu, K., & Kumar, N. S. (2020). Chitosan as an environment friendly biomaterial—A review on recent modifications and applications. *International Journal of Biological Macromolecules, 150*, 1072–1083. https://doi.org/10.1016/j.ijbiomac.2019.10.113
3. Chakravarty, J., & Edwards, T. A. (2022). Innovation from waste with biomass-derived chitin and chitosan as green and sustainable polymer: A review. *Energy Nexus, 8*, 100149. https://doi.org/10.1016/j.nexus.2022.100149
4. The state of world fisheries and aquaculture 2024. (2024). https://doi.org/10.4060/cd0683en.
5. Errea, M. I., Rossi, E., Goyanes, S. N., & D'Accorso, N. B. (2017). Chitosan: From organic pollutants to high-value polymeric materials. In S. N. Goyanes & N. B. D'Accorso (Eds.), *Industrial applications of renewable biomass products: Past, present and future* (pp. 251–264). Springer.
6. Aranaz, I., Alcántara, A. R., Civera, M. C., Arias, C., Elorza, B., Heras Caballero, A., & Acosta, N. (2021). Chitosan: An overview of its properties and applications. *Polymers, 13*, 3256.
7. Li, Z., Liu, C., Hong, S., Lian, H., Mei, C., Lee, J., Wu, Q., Hubbe, M. A., & Li, M.-C. (2022). Recent advances in extraction and processing of chitin using deep eutectic solvents. *Chemical Engineering Journal, 446*, 136953. https://doi.org/10.1016/j.cej.2022.136953
8. Rasweefali, M. K., Sabu, S., Muhammed Azad, K. S., Raseel Rahman, M. K., Sunooj, K. V., Sasidharan, A., & Anoop, K. K. (2022). Influence of deproteinization and demineralization process sequences on the physicochemical and structural characteristics of chitin isolated from Deep-sea mud shrimp (Solenocera hextii). *Advances in Biomarker Sciences and Technology, 4*, 12–27. https://doi.org/10.1016/j.abst.2022.03.001
9. Al Shaqsi, N. H. K., Al Hoqani, H. A. S., Hossain, M. A., & Al Sibani, M. A. (2020). Optimization of the demineralization process for the extraction of chitin from Omani Portunidae segnis. *Biochemistry and Biophysics Reports, 23*, 100779. https://doi.org/10.1016/j.bbrep.2020.100779
10. Mahmoud, N. S., Ghaly, A. E., & Arab, F. (2007). Unconventional approach for demineralization of deproteinized crustacean shells for chitin production. *American Journal of Biochemistry and Biotechnology, 3*, 1–9.
11. Younes, I., & Rinaudo, M. (2015). Chitin and chitosan preparation from marine sources. Structure, properties and applications. *Marine Drugs, 13*, 1133–1174. https://doi.org/10.3390/md13031133
12. Percot, A., Viton, C., & Domard, A. (2003). Characterization of Shrimp Shell Deproteinization. *Biomacromolecules, 4*, 1380–1385. https://doi.org/10.1021/bm034115h
13. Pohling, J., Hawboldt, K., & Dave, D. (2025). Thermo-mechanical decolourization process for shrimp chitin (Pandalus borealis). *Sustainable Chemistry for the Environment, 9*, 100192. https://doi.org/10.1016/j.scenv.2024.100192
14. Azelee, N. I. W., Dahiya, D., Ayothiraman, S., Noor, N. M., Rasid, Z. I. A., Ramli, A. N. M., Ravindran, B., Iwuchukwu, F. U., & Selvasembian, R. (2023). Sustainable valorization approaches on crustacean wastes for the extraction of chitin, bioactive compounds and their applications – A review. *International Journal of Biological Macromolecules, 253*, 126492. https://doi.org/10.1016/j.ijbiomac.2023.126492
15. Pacheco, N., Garnica-Gonzalez, M., Gimeno, M., Bárzana, E., Trombotto, S., David, L., & Shirai, K. (2011). Structural characterization of chitin and chitosan obtained by biological and chemical methods. *Biomacromolecules, 12*, 3285–3290. https://doi.org/10.1021/bm200750t
16. No, H. K., & Meyers, S. P. (1995). Preparation and characterization of chitin and chitosan—A review. *Journal of Aquatic Food Product Technology, 4*, 27–52. https://doi.org/10.1300/J030v04n02_03
17. Kou, S., Peters, L. M., & Mucalo, M. R. (2021). Chitosan: A review of sources and preparation methods. *International Journal of Biological Macromolecules, 169*, 85–94. https://doi.org/10.1016/j.ijbiomac.2020.12.005

18. Huang, Z., Lv, X., Sun, G., Mao, X., Lu, W., Liu, Y., Li, J., Du, G., & Liu, L. (2022). Chitin deacetylase: From molecular structure to practical applications. *Systems Microbiology and Biomanufacturing, 2*, 271–284. https://doi.org/10.1007/s43393-022-00077-9
19. Narudin, N. A. H., Rosman, N. A., Shahrin, E. W. E., Sofyan, N., Hanif Mahadi, A., Kusrini, E., Hobley, J., & Usman, A. (2022). Extraction, characterization, and kinetics of N-deacetylation of chitin obtained from mud crab shells. *Polymers and Polymer Composites, 30*, 09673911221109611. https://doi.org/10.1177/09673911221109611
20. Chang, K. L. B., Tsai, G., Lee, J., & Fu, W.-R. (1997). Heterogeneous N-deacetylation of chitin in alkaline solution. *Carbohydrate Research, 303*, 327–332. https://doi.org/10.1016/S0008-6215(97)00179-1
21. Hasan, S., Boddu, V. M., Viswanath, D. S., & Ghosh, T. K. (2022). Chitin and Chitosan. In *Science and engineering*. Springer Nature Switzerland AG.
22. No, H. K., Cho, Y. I., Kim, H. R., & Meyers, S. P. (2000). Effective deacetylation of chitin under conditions of 15 psi/121 degrees C. *Journal of Agricultural and Food Chemistry, 48*, 2625–2627. https://doi.org/10.1021/jf9908421
23. Yaghobi, N., & Hormozi, F. (2010). Multistage deacetylation of chitin: Kinetics study. *Carbohydrate Polymers, 81*, 892–896. https://doi.org/10.1016/j.carbpol.2010.03.063
24. Novikov, V. Y., Derkach, S. R., Konovalova, I. N., Dolgopyatova, N. V., & Kuchina, Y. A. (2023). Mechanism of heterogeneous alkaline Deacetylation of chitin: A review. *Polymers (Basel), 15*. https://doi.org/10.3390/polym15071729
25. Harmsen, R. A. G., Tuveng, T. R., Antonsen, S. G., Eijsink, V. G. H., & Sørlie, M. (2019). Can we make chitosan by enzymatic Deacetylation of chitin? *Molecules, 24*. https://doi.org/10.3390/molecules24213862
26. Kaczmarek, M. B., Struszczyk-Swita, K., Li, X., Szczęsna-Antczak, M., & Daroch, M. (2019). Enzymatic modifications of Chitin, Chitosan, and Chitooligosaccharides. *Frontiers in Bioengineering and Biotechnology, 7*. https://doi.org/10.3389/fbioe.2019.00243
27. Liang, B., Song, W., Xing, R., Liu, S., Yu, H., & Li, P. (2023). The source, activity influencing factors and biological activities for future development of chitin deacetylase. *Carbohydrate Polymers, 321*, 121335. https://doi.org/10.1016/j.carbpol.2023.121335
28. Sun, X., Wang, Y., Yu, J., Wei, Q., & Ren, X. (2024). Study on the deacetylation and mechanism of chitin in natural deep eutectic solvent. *International Journal of Biological Macromolecules, 255*, 127698. https://doi.org/10.1016/j.ijbiomac.2023.127698
29. Sivashankari, P. R., & Prabaharan, M. (2017). 5 – Deacetylation modification techniques of chitin and chitosan. In J. A. Jennings & J. D. Bumgardner (Eds.), *Chitosan based biomaterials volume 1* (pp. 117–133). Woodhead Publishing.
30. Vicente, F. A., Huš, M., Likozar, B., & Novak, U. (2021). Chitin Deacetylation using deep eutectic solvents: Ab initio-supported process optimization. *ACS Sustainable Chemistry & Engineering, 9*, 3874–3886. https://doi.org/10.1021/acssuschemeng.0c08976
31. Nwabike Amitaye, A., Elemike, E. E., Akpeji, H. B., Amitaye, E., Hossain, I., Mbonu, J. I., & Aziza, A. E. (2024). Chitosan: A sustainable biobased material for diverse applications. *Journal of Environmental Chemical Engineering, 12*, 113208. https://doi.org/10.1016/j.jece.2024.113208
32. Xing, R., Liu, S., Guo, Z., Yu, H., Wang, P., Li, C., Li, Z., & Li, P. (2005). Relevance of molecular weight of chitosan and its derivatives and their antioxidant activities in vitro. *Bioorganic & Medicinal Chemistry, 13*, 1573–1577. https://doi.org/10.1016/j.bmc.2004.12.022
33. Rossi, E., Ramírez, J. A. Á., & Errea, M. I. (2020). Preparation of an environmentally friendly lead adsorbent. A contribution to the rational design of heavy metal adsorbents. *Journal Of Environmental Chemical Engineering, 8*, 104210. https://doi.org/10.1016/j.jece.2020.104210
34. Lombardo, G., Dorm, B. C., Salvay, A. G., Franzi, L., Gaffney, M. L., Peredo Camio, J. B., Trovatti, E., Rossi, E., & Errea, M. I. (2024). Novel chitosan-based hydrogels as promising wound dressing materials with advanced properties. *International Journal of Biological Macromolecules, 279*, 135423. https://doi.org/10.1016/j.ijbiomac.2024.135423

35. Díaz Bukvic, G., Ojeda Henriquez, M., Rodríguez Vannini, A. B., Fidalgo, M. M., Salvay, A. G., Rossi, E., & Errea, M. I. (2024). Impact of the three-dimensional arrangements of Polyhydroxylated crosslinkers on the resulting properties of chitosan-based hydrogels. *Polysaccharides, 5*, 358–379.
36. Pereira, A. G. B., Muniz, E. C., & Hsieh, Y.-L. (2015). 1H NMR and 1H–13C HSQC surface characterization of chitosan–chitin sheath-core nanowhiskers. *Carbohydrate Polymers, 123*, 46–52. https://doi.org/10.1016/j.carbpol.2015.01.017
37. Rinaudo, M., Dung, P. L., Gey, C., & Milas, M. (1992). Substituent distribution on O, N-carboxymethylchitosans by 1H and 13C n. m.r. *International Journal of Biological Macromolecules, 14*, 122–128. https://doi.org/10.1016/S0141-8130(05)80001-7
38. Facchinatto, W. M., dos Santos, D. M., Fiamingo, A., Bernardes-Filho, R., Campana-Filho, S. P., de Azevedo, E. R., & Colnago, L. A. (2020). Evaluation of chitosan crystallinity: A high-resolution solid-state NMR spectroscopy approach. *Carbohydrate Polymers, 250*, 116891. https://doi.org/10.1016/j.carbpol.2020.116891
39. Hirai, A., Odani, H., & Nakajima, A. (1991). Determination of degree of deacetylation of chitosan by 1H NMR spectroscopy. *Polymer Bulletin, 26*, 87–94. https://doi.org/10.1007/BF00299352
40. Thomas, S., Pius, A., & Gopi, S. (2020). *Handbook of Chitin and Chitosan: Volume 1: Preparation and properties*. Elsevier.
41. Iber, B. T., Kasan, N. A., Torsabo, D., & Omuwa, J. W. (2021). A review of various sources of chitin and chitosan in nature. *Journal of Renewable Materials, 10*, 1097–1123. https://doi.org/10.32604/jrm.2022.018142
42. Hou, J., Aydemir, B. E., & Dumanli, A. G. (2021). Understanding the structural diversity of chitins as a versatile biomaterial. *Philosophical Transactions of the Royal Society A: Mathematical, Physical and Engineering Sciences, 379*, 20200331. https://doi.org/10.1098/rsta.2020.0331
43. Ahmed, S., & Ikram, S. (2017). *Chitosan: Derivatives, composites and applications*. Wiley.
44. Trung, T. S., Thein-Han, W. W., Qui, N. T., Ng, C.-H., & Stevens, W. F. (2006). Functional characteristics of shrimp chitosan and its membranes as affected by the degree of deacetylation. *Bioresource Technology, 97*, 659–663. https://doi.org/10.1016/j.biortech.2005.03.023
45. Rong Huei, C., & Hwa, H.-D. (1996). Effect of molecular weight of chitosan with the same degree of deacetylation on the thermal, mechanical, and permeability properties of the prepared membrane. *Carbohydrate Polymers, 29*, 353–358. https://doi.org/10.1016/S0144-8617(96)00007-0
46. Jaworska, M., Sakurai, K., Gaudon, P., & Guibal, E. (2003). Influence of chitosan characteristics on polymer properties. I: Crystallographic properties. *Polymer International, 52*, 198–205. https://doi.org/10.1002/pi.1159
47. Osorio-Madrazo, A., David, L., Trombotto, S., Lucas, J.-M., Peniche-Covas, C., & Domard, A. (2011). Highly crystalline chitosan produced by multi-steps acid hydrolysis in the solid-state. *Carbohydrate Polymers, 83*, 1730–1739. https://doi.org/10.1016/j.carbpol.2010.10.035
48. Ogawa, K., Yui, T., & Okuyama, K. (2004). Three D structures of chitosan. *International Journal of Biological Macromolecules, 34*, 1–8. https://doi.org/10.1016/j.ijbiomac.2003.11.002
49. Kumirska, J., Czerwicka, M., Kaczyński, Z., Bychowska, A., Brzozowski, K., Thöming, J., & Stepnowski, P. (2010). Application of spectroscopic methods for structural analysis of chitin and chitosan. *Marine Drugs, 8*, 1567–1636.
50. Kasaai, M. R., Arul, J., & Charlet, G. (2000). Intrinsic viscosity–molecular weight relationship for chitosan. *Journal of Polymer Science Part B: Polymer Physics. The Physiologist, 38*, 2591–2598. https://doi.org/10.1002/1099-0488(20001001)38:19<2591::AID-POLB110>3.0.CO;2-6
51. Carraher Jr, C.E. Introduction to polymer chemistry; CRC press: 2017.

52. Czechowska-Biskup, R., Wach, R. A., Rosiak, J. M., & Ulański, P. (2018). Procedure for determination of the molecular weight of chitosan by viscometry. *Progress on Chemistry and Application of Chitin and Its Derivatives, XXIII*, 45–54.
53. Lyalina, T., Shagdarova, B., Zhuikova, Y., Il'ina, A., Lunkov, A., & Varlamov, V. (2023). Effect of seed priming with chitosan hydrolysate on lettuce (Lactuca sativa) growth parameters. *Molecules, 28*, 1915.
54. Young, R. J., & Lovell, P. A. (2011). *Introduction to polymers*. CRC press.
55. Shaheen, M. E., Ghazy, A. R., Kenawy, E.-R., & El-Mekawy, F. (2018). Application of laser light scattering to the determination of molecular weight, second virial coefficient, and radius of gyration of chitosan. *Polymer, 158*, 18–24. https://doi.org/10.1016/j.polymer.2018.10.044
56. Kang, Y., Wu, X., Ji, X., Bo, S., & Liu, Y. (2018). Strategy to improve the characterization of chitosan by size exclusion chromatography coupled with multi angle laser light scattering. *Carbohydrate Polymers, 202*, 99–105. https://doi.org/10.1016/j.carbpol.2018.08.125
57. Chen, X.-G., Zheng, L., Wang, Z., Lee, C.-Y., & Park, H.-J. (2002). Molecular affinity and permeability of different molecular weight chitosan membranes. *Journal of Agricultural and Food Chemistry, 50*, 5915–5918. https://doi.org/10.1021/jf020151g
58. Luo, Q., Wang, Y., Han, Q., Ji, L., Zhang, H., Fei, Z., & Wang, Y. (2019). Comparison of the physicochemical, rheological, and morphologic properties of chitosan from four insects. *Carbohydrate Polymers, 209*, 266–275. https://doi.org/10.1016/j.carbpol.2019.01.030
59. Mucha, M. (1997). Rheological characteristics of semi-dilute chitosan solutions. *Macromolecular Chemistry and Physics, 198*, 471–484. https://doi.org/10.1002/macp.1997.021980220
60. Chawla, S. P., Kanatt, S. R., & Sharma, A. K. (2015). Chitosan. In K. G. Ramawat & J.-M. Mérillon (Eds.), *Polysaccharides: Bioactivity and biotechnology* (pp. 219–246). Springer.
61. Wang, W., & Xu, D. (1994). Viscosity and flow properties of concentrated solutions of chitosan with different degrees of deacetylation. *International Journal of Biological Macromolecules, 16*, 149–152. https://doi.org/10.1016/0141-8130(94)90042-6
62. Ahmed, H. O. (2014). *Formulation and rheological characterization of chitosan films*. University of Georgia.
63. Cosme, F., & Vilela, A. (2021). Chitin and chitosan in the alcoholic and non-alcoholic beverage industry: An overview. *Applied Sciences, 11*. https://doi.org/10.3390/app112311427
64. Bertrand, M., Simonin, S., & Bach, B. (2024). Applications of chitosan in the agri-food sector: A review. *Carbohydrate Research, 543*, 109219. https://doi.org/10.1016/j.carres.2024.109219
65. Rao, L., Hayat, K., Lv, Y., Karangwa, E., Xia, S., Jia, C., Zhong, F., & Zhang, X. (2011). Effect of ultrafiltration and fining adsorbents on the clarification of green tea. *Journal of Food Engineering, 102*, 321–326. https://doi.org/10.1016/j.jfoodeng.2010.09.007
66. Flórez, M., Guerra-Rodríguez, E., Cazón, P., & Vázquez, M. (2022). Chitosan for food packaging: Recent advances in active and intelligent films. *Food Hydrocolloids, 124*, 107328. https://doi.org/10.1016/j.foodhyd.2021.107328
67. Liu, T., Li, J., Tang, Q., Qiu, P., Gou, D., & Zhao, J. (2022). Chitosan-based materials: An overview of potential applications in food packaging. *Food, 11*. https://doi.org/10.3390/foods11101490
68. Sharma, A., Sood, K., Kaur, J., & Khatri, M. (2019). Agrochemical loaded biocompatible chitosan nanoparticles for insect pest management. *Biocatalysis and Agricultural Biotechnology, 18*, 101079. https://doi.org/10.1016/j.bcab.2019.101079
69. Maluin, F. N., & Hussein, M. Z. (2020). Chitosan-based agronanochemicals as a sustainable alternative in crop protection. *Molecules, 25*. https://doi.org/10.3390/molecules25071611
70. Xu, D., Li, H., Lin, L., Liao, M. A., Deng, Q., Wang, J., Lv, X., Deng, H., Liang, D., & Xia, H. (2020). Effects of carboxymethyl chitosan on the growth and nutrient uptake in Prunus davidiana seedlings. *Physiology and Molecular Biology of Plants, 26*, 661–668. https://doi.org/10.1007/s12298-020-00791-5
71. Yang, J., Yang, J., & Huang, J. (2017). Role of co-planting and chitosan in phytoextraction of as and heavy metals by Pteris vittata and castor bean–A field case. *Ecological Engineering, 109*, 35–40. https://doi.org/10.1016/j.ecoleng.2017.09.001

72. Bai, Y., Jing, Z., Ma, R., Wan, X., Liu, J., & Huang, W. (2023). A critical review of enzymes immobilized on chitosan composites: Characterization and applications. *Bioprocess and Biosystems Engineering, 46*, 1539–1567. https://doi.org/10.1007/s00449-023-02914-0
73. Verma, M. L., Kumar, S., Das, A., Randhawa, J. S., & Chamundeeswari, M. (2019). Enzyme immobilization on chitin and chitosan-based supports for biotechnological applications. In G. Crini & E. Lichtfouse (Eds.), *Sustainable agriculture reviews 35: Chitin and Chitosan: History, fundamentals and innovations* (pp. 147–173). Springer International Publishing.
74. Verma, M. L., Kumar, S., Das, A., Randhawa, J. S., & Chamundeeswari, M. (2020). Chitin and chitosan-based support materials for enzyme immobilization and biotechnological applications. *Environmental Chemistry Letters, 18*, 315–323. https://doi.org/10.1007/s10311-019-00942-5
75. Morin-Crini, N., Lichtfouse, E., Torri, G., & Crini, G. (2019). Applications of chitosan in food, pharmaceuticals, medicine, cosmetics, agriculture, textiles, pulp and paper, biotechnology, and environmental chemistry. *Environmental Chemistry Letters, 17*, 1667–1692. https://doi.org/10.1007/s10311-019-00904-x
76. Zhou, B. C.-E., Kan, C.-w., Sun, C., Du, J., & Xu, C. (2019). A review of chitosan textile applications. *AATCC Journal Of Research, 6*, 8–14. https://doi.org/10.14504/ajr.6.S1.2
77. Hirano, S. (1999). Chitin and chitosan as novel biotechnological materials. *Polymer International, 48*, 732–734. https://doi.org/10.1002/(SICI)1097-0126(199908)48:8<732::AID-PI211>3.0.CO;2-U
78. Rinaudo, M. (2006). Chitin and chitosan: Properties and applications. *Progress in Polymer Science, 31*, 603–632. https://doi.org/10.1016/j.progpolymsci.2006.06.001
79. Huang, L., Xiao, L., & Yang, G. (2018). Chitosan application in textile processing. *Current Trends in Fashion Technology & Textile Engineering, 4*, 32–34.
80. Song, Z., Li, G., Guan, F., & Liu, W. (2018). Application of chitin/chitosan and their derivatives in the papermaking industry. *Polymers, 10*, 389.
81. Kittur, F. S., Kumar, K. R., & Tharanathan, R. N. (1998). Functional packaging properties of chitosan films. *Zeitschrift für Lebensmitteluntersuchung und -Forschung A, 206*, 44–47. https://doi.org/10.1007/s002170050211
82. Despond, S., Espuche, E., Cartier, N., & Domard, A. (2005). Barrier properties of paper–chitosan and paper–chitosan–carnauba wax films. *Journal of Applied Polymer Science, 98*, 704–710. https://doi.org/10.1002/app.21754
83. Fredheim, G. E., & Christensen, B. E. (2003). Polyelectrolyte complexes: Interactions between Lignosulfonate and Chitosan. *Biomacromolecules, 4*, 232–239. https://doi.org/10.1021/bm020091n
84. Yang, R., Li, H., Huang, M., Yang, H., & Li, A. (2016). A review on chitosan-based flocculants and their applications in water treatment. *Water Research, 95*, 59–89. https://doi.org/10.1016/j.watres.2016.02.068
85. Díaz Bukvic, G., Rossi, E., & Errea, M. I. (2023). Polysaccharides as economic and sustainable raw materials for the preparation of adsorbents for water treatment. *Polysaccharides, 4*, 219–255. https://doi.org/10.3390/polysaccharides4030016
86. Kulka, K., & Sionkowska, A. (2023). Chitosan based materials in cosmetic applications: A review. *Molecules, 28*, 1817.
87. Aranaz, I., Acosta, N., Civera, C., Elorza, B., Mingo, J., Castro, C., De los Llanos Gandía, M., & Heras Caballero, A. (2018). Cosmetics and cosmeceutical applications of Chitin, Chitosan and their derivatives. *Polymers, 10*, 213.
88. Siqueira, N. M., Contri, R. V., Paese, K., Beck, R. C. R., Pohlmann, A. R., & Guterres, S. S. (2011). Innovative sunscreen formulation based on Benzophenone-3-loaded chitosan-coated polymeric Nanocapsules. *Skin Pharmacology and Physiology, 24*, 166–174. https://doi.org/10.1159/000323273
89. https://www.hospwork.pt/product/clorohexidina-quitosano-bexident-post-gel. Available online: Accessed on 28 Apr 2025.

90. https://ninelife.sg/products/bianca-rosa-chitosan-50-cream-2-oz-zin-519714-2-pack?srsltid=AfmBOopxmFt5zA6tk7kQ0zMJp0HcV5jQCwQp8BqFZcwrp0dnro0A8Sua. Available online: Accessed on 28 Apr 2025.
91. Harugade, A., Sherje, A. P., & Pethe, A. (2023). Chitosan: A review on properties, biological activities and recent progress in biomedical applications. *Reactive and Functional Polymers, 191*, 105634. https://doi.org/10.1016/j.reactfunctpolym.2023.105634
92. Adhikari, H. S., & Yadav, P. N. (2018). Anticancer activity of chitosan, chitosan derivatives, and their mechanism of action. *International Journal of Biomaterials, 2018*, 2952085. https://doi.org/10.1155/2018/2952085
93. Huang, X., Jiao, Y., & Zhou, C. (2021). Impacts of chitosan oligosaccharide (COS) on angiogenic activities. *Microvascular Research, 134*, 104114. https://doi.org/10.1016/j.mvr.2020.104114
94. Zhang, X., Liang, Y., Huang, S., & Guo, B. (2024). Chitosan-based self-healing hydrogel dressing for wound healing. *Advances in Colloid and Interface Science, 332*, 103267. https://doi.org/10.1016/j.cis.2024.103267
95. Lopes, R.-C., Nossa, T., Lustri, W.-R., Lombardo, G., Errea, M.-I., & Trovatti, E. (2024). Renewable polymers in biomedical applications: From the bench to the market. *Journal of Renewable Materials, 12*, 643–666.
96. Xu, J., Chang, L., Xiong, Y., & Peng, Q. (2024). Chitosan-based hydrogels as antibacterial/antioxidant/anti-inflammation multifunctional dressings for chronic wound healing. *Advanced Healthcare Materials, 13*, 2401490. https://doi.org/10.1002/adhm.202401490
97. Cassano, R., Perri, P., Scarcello, E., Piro, P., Sole, R., Curcio, F., & Trombino, S. (2024). Chitosan hemostatic dressings: Properties and surgical applications. *Polymers, 16*. https://doi.org/10.3390/polym16131770
98. Canbolat, F., Demir, N., Yayıntas, O. T., Pehlivan, M., Eldem, A., Ayna, T. K., & Senel, M. (2024). Chitosan nanoparticles loaded with quercetin and Valproic acid: A novel approach for enhancing antioxidant activity against oxidative stress in the SH-SY5Y human Neuroblastoma cell line. *Biomedicine, 12*. https://doi.org/10.3390/biomedicines12020287
99. Khorasani, M. A., & Naghib, S. M. (2025). A review of chitosan-based multifunctional nanocomposites for drug/gene/protein delivery and gene therapy in cancer treatments: Promises, challenges and outlooks. *International Journal of Biological Macromolecules, 306*, 141394. https://doi.org/10.1016/j.ijbiomac.2025.141394
100. Ali, A., & Ahmed, S. (2018). A review on chitosan and its nanocomposites in drug delivery. *International Journal of Biological Macromolecules, 109*, 273–286. https://doi.org/10.1016/j.ijbiomac.2017.12.078
101. Isa, M. T., Abdulkarim, A. Y., Bello, A., Bello, T. K., & Adamu, Y. (2024). Synthesis and characterization of chitosan for medical applications: A review. *Journal of Biomaterials Applications, 38*, 1036–1057. https://doi.org/10.1177/08853282241243010
102. Haider, A., Khan, S., Iqbal, D. N., Khan, S. U., Haider, S., Mohammad, K., Mustfa, G., Rizwan, M., & Haider, A. (2024). Chitosan as a tool for tissue engineering and rehabilitation: Recent developments and future perspectives–A review. *International Journal of Biological Macromolecules, 278*, 134172. https://doi.org/10.1016/j.ijbiomac.2024.134172
103. Edo, G. I., Yousif, E., & Al-Mashhadani, M. H. (2024). Chitosan: An overview of biological activities, derivatives, properties, and current advancements in biomedical applications. *Carbohydrate Research, 542*, 109199. https://doi.org/10.1016/j.carres.2024.109199

Interpenetrating Polymer Networks of Chitosan-Based Matrix in Drug Delivery

Disha Ghosh and Kishor Sarkar

Abstract Interpenetrating polymer networks (IPNs) have emerged as a promising strategy in the design of advanced drug delivery systems, offering enhanced control over drug release, improved biocompatibility, and the ability to incorporate multiple therapeutic agents. Chitosan, a natural, biodegradable polymer, is gaining increasing attention as a matrix component in IPNs due to its unique properties, such as structural mimicry of the extracellular matrix, ease of chemical modification, and inherent antimicrobial activity. This chapter delves into the development and application of chitosan-based IPNs for drug delivery, highlighting their design, synthesis, and functional advantages.

The chapter begins with an overview of chitosan's chemical structure and its potential for integration with other natural and synthetic polymers to form IPNs. It explores various crosslinking strategies, fabrication techniques, and the mechanisms governing drug release, such as diffusion, swelling, and degradation. A detailed analysis of the factors influencing drug encapsulation efficiency, release kinetics, and therapeutic effectiveness is presented. Applications in oral, transdermal, and targeted drug delivery are discussed, along with recent advancements in stimuli-responsive and nanotechnology-based IPN systems. Finally, the chapter addresses current challenges, such as scalability and stability, and provides insights into future trends in the development of smart drug delivery platforms. This chapter serves as a comprehensive guide for researchers and practitioners seeking to harness the potential of chitosan-based IPNs for innovative drug delivery solutions.

Keywords Chitosan · Interpenetrating polymer networks (IPNs) · Drug delivery systems · Controlled drug release · Biocompatibility and biodegradability · Crosslinking strategies · Stimuli-responsive IPNs

D. Ghosh · K. Sarkar (✉)
Gene Therapy and Tissue Engineering Lab, Department of Polymer Science and Technology, University of Calcutta, Kolkata, West Bengal, India
e-mail: kspoly@caluniv.ac.in; http://www.kishorgttl.com/

S. Jana (ed.), *Chitosan-Based Drug Delivery Systems*, AAPS Introductions in the Pharmaceutical Sciences 16, https://doi.org/10.1007/978-3-032-08347-0_2

1 Introduction

Drug delivery systems (DDS), which concentrate on the controlled release of medicinal substances to increase treatment efficacy and decrease side effects, have become an important field of study throughout the last several decades. In order to maximize bioavailability and reduce side effects, the optimal DDS makes sure that medications are delivered precisely and at a controlled rate to particular body locations [1]. Problems with traditional medication administration techniques include systemic side effects, poor solubility, and quick drug degradation. The application of biopolymers has emerged as a crucial field of study to get beyond these restrictions, with special focus on chitosan, a biopolymer made from chitin that is plentiful, biodegradable, non-toxic, and biocompatible [2, 3].

Chitosan's potential in drug delivery applications has long been acknowledged because of its adaptable chemical structure, which makes functionalization and modification simple. Proteins, peptides, and tiny molecules can all be encapsulated and released under controlled conditions thanks to its cationic properties and capacity to produce gels, nanoparticles, and membranes. Notwithstanding its numerous benefits, chitosan's application in drug delivery may be constrained by its mechanical strength, drug release characteristics, and stability under specific physiological circumstances [4]. The creation of chitosan-based interpenetrating polymer networks (IPNs) has drawn a lot of interest in order to overcome these problems. An IPN is a kind of polymer system where two or more networks of polymers are molecularly entangled but do not share a covalent link. These systems create a material with improved performance in drug delivery applications by combining the characteristics of each particular polymer. Incorporating other polymers into an IPN can improve the matrix's mechanical strength, stability, and controlled release qualities while also enhancing chitosan's natural qualities of biodegradability, non-toxicity, and muco adhesiveness [5, 6]. IPNs' adaptability makes them perfect for uses requiring precise medication release, especially in targeted drug delivery or the treatment of chronic illnesses. It is also possible to modify the way chitosan-based IPNs form in response to environmental cues like variations in temperature, pH, or ionic strength. Because of its responsiveness, therapeutic compounds can be released at specific locations in the body, including the gastrointestinal system, or in response to conditions unique to a disease, such as the acidic environment of tumours. This offers up new possibilities for targeted drug delivery [7]. Moreover, chitosan-IPNs' flexibility makes it possible to combine several active ingredients into a single delivery system, which makes it easier to provide combination treatments at the same time.

The development of chitosan-based IPNs for drug administration is still fraught with difficulties, despite its encouraging promise. These include problems with drug-loading capacity optimization, production technique scalability, and reproducibility of drug release patterns. Ongoing developments in crosslinking methods, polymer chemistry, and our knowledge of how chitosan interacts with other polymers, however, offer a solid basis for overcoming these obstacles [8, 9]. This chapter explores the design principles, manufacturing processes, and performance-influencing aspects of interpenetrating polymer networks of chitosan-based matrices in drug delivery systems. We also go over the state of the research, applications, and possibilities for the future, providing information on how these cutting-edge systems can transform drug delivery technology [10].

2 Fundamentals of Chitosan-Based Matrices

2.1 Chemical Structure and Properties of Chitosan

Chitosan is a biopolymer that is found in the shells of crustaceans and the cuticles of insects. It is a copolymer of N-acetyl-D-glucosamine and D-glucosamine, joined by 1,4-glycosidic bonds. Chitosan is a linear polysaccharide that is obtained by N-deacetylation of chitin compound. The molecule has three rings and three functional groups: amino groups, primary hydroxyl groups, and secondary hydroxyl groups. The natural polymer chitosan is polycationic and contains reactive hydroxyl and amino groups. The structure can be represented as $[C_6H_{11}NO_4]_n$, [11] where n represents as degree of polymerization. The degree of deacetylation (typically 60–95%) influences the properties of chitosan, such as solubility, viscosity, and reactivity. A higher DD means more free amino groups, enhancing its reactivity and cationic nature. Chitosan's solubility, viscosity, and reactivity are all influenced by the degree of deacetylation, which is normally between 60% and 95%. Its reactivity and cationic character are enhanced by a larger DD, which indicates more free amino groups. Chitosan appears as a white to off-white powder. It is more soluble than chitin in acidic water settings due to the protonation of the $-NH_2$ at the D-glucosamine repeat unit's C-1 position. Due to the protonation of amino groups, it is soluble in acidic solutions (such as lactic acid and acetic acid), but insoluble in water, organic solvents, and neutral or alkaline environments. It can have an impact on viscosity and functionality and molecular weight ranges from 10 kDa to more than 1000 kDa [12]. A primary amino group (C2) and primary and secondary hydroxyl groups (C6, C3) are the reactive groups present in chitosan. The acetamide group and glycosidic linkages are considered as well as functional groups. Numerous alterations are possible according to these functional groups, resulting in polymers with novel characteristics and behaviours. Derivatives of chitosan have been introduced with the intention of enhancing its solubility or biodegradability or adding new features or functionalities. For example, among other procedures, deacetylation, de-polymerization, or quaternization have been used to increase solubility in water-aqueous conditions [13]. Following its alteration, new chitosan actions have been reported; for example, phosphorylated chitosan reduces corrosion, while 6-O-sulphated chitosan stimulates neural differentiation [14].

2.2 Biocompatibility and Biodegradability

Numerous biological activities, such as antitumoral, antibacterial, antioxidant, and anti-inflammatory properties, are exhibited by chitin, chitosan, oligosaccharides, and derivatives; these compounds may find application as therapeutic polymers. The capacity of a material to fulfil its intended purpose without causing any adverse biological reaction is known as biocompatibility. One of chitosan's biggest benefits, particularly for pharmaceutical and biomedical applications, is its biocompatibility. The compatibility of the chitosan-biomaterials is greatly influenced by their physical and chemical characteristics, including their hydrophilicity, hydrophobicity, ionic groups, multi-component frame of structurally amorphous and crystalline morphology, and topography, or surface roughness [15].

The ability of a material to decompose into simpler components through biological processes is known as biodegradability. Biodegradable chitosan breaks down into innocuous byproducts like glucosamine, which are naturally produced by biological systems. Polymeric biomaterials undergo biodegradation when hydrolytically or enzymatically sensitive links in the polymer are broken, resulting in polymer breakdown. According to their manner of breakdown, polymeric biomaterials can be divided into two categories: hydrolytically and enzymatically degradable polymers.

2.3 Functional Groups and Modifications

Chitosan is extremely flexible because it contains reactive amino (-NH_2) and hydroxyl (-OH) groups. By altering their solubility according to the intended uses, these changes seek to improve their chemical and biological characteristics. The primary CS modifications discussed in the literature—i.e., N-acyl modifications, cross-linked chitosan, quaternization, and C-6 oxidation, etc.—are highlighted in the section that follows:

Quaternized chitosan derivatives—Quaternized gives a hydrophilic nature to chitosan by modifying the positive charge (NH_4^+) on it. While quaternized Chitosan remains permanently positively charged at pH values above 6.5, Chitosan is positively charged at pH values below 6.5, which makes it soluble in a large range of pH values. TMC is the best example of quaternized chitosan. After two successive reactions, TMC modification is produced. The first occurs between methyl iodide CH_3I and chitosan in the presence of N-methyl-2-pyrrolidinone (NMP), which is utilized as a solvent in alkaline circumstances (NaOH). The second occurs when an anionic exchange resin converts iodide ions into chloride ions. Alkyl halides' carbon length can be readily changed to produce a variety of quaternized chitosan [16, 17].

N-acyl modifications-N-acylation grafts various fatty acids onto Chitosan to provide it with hydrophobic qualities. The reaction is a particular amidation between the -NH_2 groups from chitosan and the -COOH groups from fatty acids. Acid anhydride or acyl halide is the chemical reagent used for N-acylation. Many acylation reaction was observed by the use of different types of acid anhydrides [18].

Cross-linked chitosan derivatives—in order to create a three-dimensional macromolecular network, cross-linked CS must be made using particular chemical agents to join the chains. To create CS-based hydrogel, CS is often crosslinked by covalent bonds in an acidic or basic medium with aldehyde derivatives as glyoxal, formalin, or glutaraldehyde. Here, the condensation reaction is formed between the aldehyde function and the amine group of chitosan [19]. The most widely used chitosan cross-linker, glutaraldehyde, stabilizes CS nanoparticles (NPs) by cross-linking technique [20]. Glutaraldehyde can also be used for the conjugation of other polymer-containing amino acids to chitosan [21]. The resulting crosslinked chitosan shows better mechanical properties than Chitosan.

Oxidative cleavage—it is mainly used for downstream functionalization of chitosan by cleaving oxo bonds. Mostly used reagents for oxo cleaving are $NaIO_4$ and KIO_4 in the presence of an acetate buffer, which results in the smaller polymer and ammonia gas [22, 23] (Fig. 1).

Non-covalent modifications—chitosan can be modified by non-covalent bonds with polymer protein and smaller composites. Types of non-covalent bonds which are used in chitosan modification are mainly hydrogen bonding, ionic interaction, and chelate formation. For non-covalent modification, it is a much easier process with simpler purification techniques, while the information about the internal arrangements of chemical structure is much difficult to produce. Solid CS-based materials (nanofibers, NPs, and dry films) or hydrogels are the main topic of this section. It is interesting to note that CS derivatives' non-covalent bond with particular additives might greatly increase their mechanical resistance and stability. Applications of bone tissue engineering and wound dressings brought attention to this tactic [24, 25]. It was recently investigated by Nie et al. [26] that CS has a

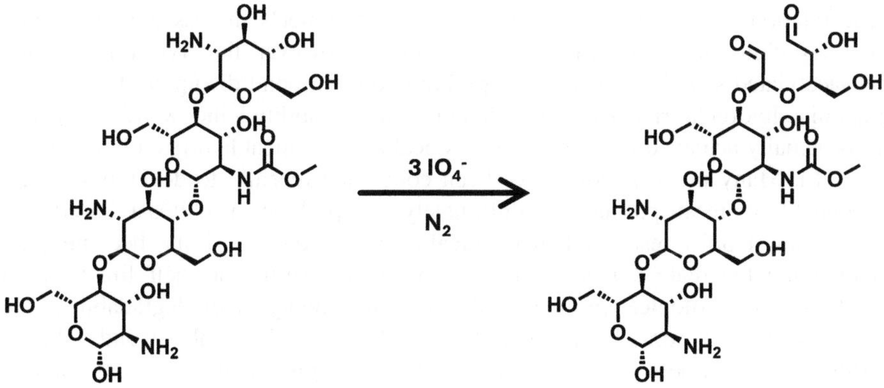

Fig. 1 Illustration of oxidative cleavage of chitosan using IO_4^-

considerable affinity for Cu^{2+}. Following the diffusion of the -OH ions within the hydrogel, they studied the structure of the hydrogel after immersing a mould containing CS–Cu solution in an alkaline gelation bath. Strong interactions between CS and metal ions produced a dense, rigid, multi-layered gel at high Cu^{2+} concentrations, whereas a composite gel with oriented fibres was produced at lower Cu^{2+} concentrations. Conversely, it is well known that calcium anions have a decreased affinity for CS. They found that, in contrast to the CS–Cu^{2+} system, raising the Ca^{2+} concentration did not result in a stronger gel or a change in structure. In fact, the CS–Ca^{2+} gel continued to be a fibrous substance.

2.4 *Challenges of Using Chitosan in Drug Deliver*

Chitin and chitosan have been shown to trigger an immunological response in certain individuals, which can result in allergic responses. As a result, this characteristic may limit its use in several biological applications. Unfortunately, occupational allergies are frequently caused by chitin-chitinase-stimulated hypersensitivity. Furthermore, recent studies on the immunologic effects of chitin in vitro and in vivo have revealed new aspects of how chitin controls both innate and adaptive immune responses. In addition to stimulating macrophages and other innate immune cells, exogenous chitin has been shown to regulate adaptive type 2 allergic inflammation. Additionally, these findings demonstrate that chitin activates macrophages via interacting with a variety of cell surface receptors, including the macrophage mannose receptor [27]. The incapacity of chitosan to dissolve in water with a neutral pH may limit its use in a variety of applications. However, solubility can be improved by changing their chemical composition or by employing the appropriate solvents. Temperature, pH, degree of acetylation, and polymer crystallinity are some of the factors that influence chitosan's solubility. Because of the polymer's enhanced crystallinity after deacetylation, which offsets the effect of the polymer's increased glucosamine moieties, chitosan was shown to be less soluble. However, the crystallinity of the half-acetylated sample decreased. Applying hydrogen bond disruptors like urea or guanidine hydrochloride also alters the chitosan's solubility window. Wide solubility is actually achieved by combining physical and chemical hydrogen bond breakage [28]. Many investigations have been conducted recently to develop safe and dependable chitosan products. Unfortunately, the problem of chitosan-based systems' poor stability restricts their practical use; consequently, it has become quite challenging to create chitosan formulations with a sufficient shelf life [29]. In addition to its biological properties like immunogenicity or biodegradability, the degree of chitosan purity has a major effect on the material's solubility and stability. Furthermore, a variety of factors, such as the molecular weight, moisture content, and degree of deacetylation, influence how stable chitosan is. Similar to this, chitin has a limited amount of stability; however, it can be improved by cross-linking it with enzymes or other chemicals [30]. Despite chitosan's FDA approval for oral medication delivery and wound healing, therapeutic formulations that contain chitosan may be toxic in vivo.

3 Interpenetrating Polymer Networks (IPNs)

3.1 Definition and Characteristics of IPNs

The term "Interpenetrating polymer network" denotes a polymeric blend that consists of two or more polymeric networks. In a broader picture, a polymer made up of two or more networks that are at least partially interlaced on a polymer scale but not covalently bound to one another is known as an interpenetrating polymer network (IPN) [31]. In IPN, two or more polymeric networks are entangled in such a way that they cannot be pulled apart from each other, and also, they are not bonded through any chemical bonding. It is well known that IPN systems improve the final product's phase stability. The network segments' permanent interlocking allows for the resolution of thermodynamic incompatibility. It shows high stability in terms of heat and favourable dielectric characteristics as well as transparency to radar. The refractive index was determined to be 1.35 and the percentage of light transmittance to be 90% which refers to clarity of vision [32]. The majority of IPNs are heterogeneous systems. Their mechanical and electrochemical qualities are superbly combined. A conductive polymer to transport anions and a hydrophobic polymer with good mechanical, chemical, and thermal properties make up their composition. IPN materials have attracted a lot of attention due to the unique characteristics that result from the interlocking of polymer chains [33]. IPN can be bent and cemented into a temporary shape and only return to its original permanent shape when exposed to external stimuli like heat, light, etc.,; it is said that IPN exhibit the shape memory effect.

3.2 Types of IPNs (Fig. 2)

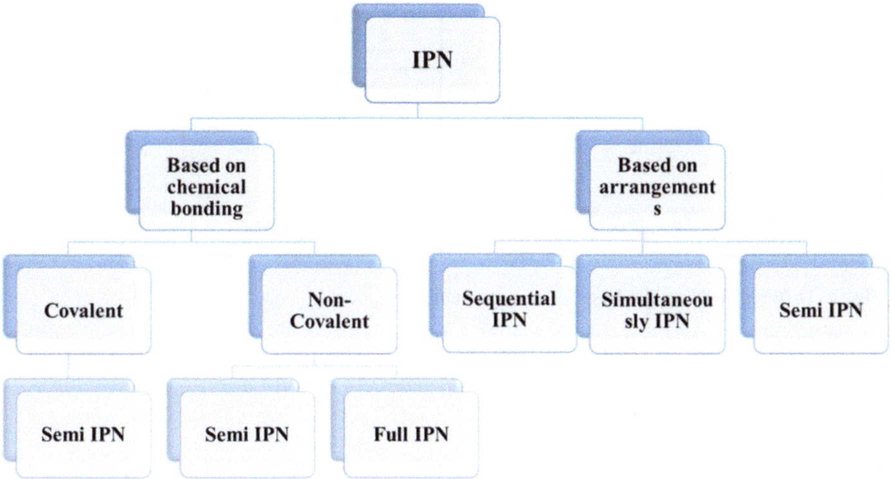

Fig. 2 Types of IPN

3.2.1 Based on Chemical Bonding

Because covalent cross-linking creates irreversible chemical bonds, hydrogels with a persistent network structure are produced. This kind of connection enables drug release through diffusion and the absorption of water and/or bioactive substances without disintegration.

Covalent semi-IPN—Two distinct polymer systems are cross-linked to create a single polymer network in a covalent semi-IPN.

Non-covalent semi-IPN—When only one polymer system is cross-linked, the IPN is said to be non-covalent semi-IPN.

Non-covalent full IPN—The two distinct polymers are individually cross-linked in a non-covalent full IPN.

3.2.2 Based on Arrangements Pattern (Fig. 3)

Sequential IPN—In sequential IPN, after the first component network's polymerization is complete, the second polymeric component network is polymerized in a successive IPN.

Semi IPN—In semi-IPN, the system is transmitted as semi-IPN if just one assembly component is cross-linked, leaving the other in a linear form.

Simultaneously IPN—When both component networks are polymerized simultaneously, the resulting IPN is created and can be called a simultaneous IPN [32].

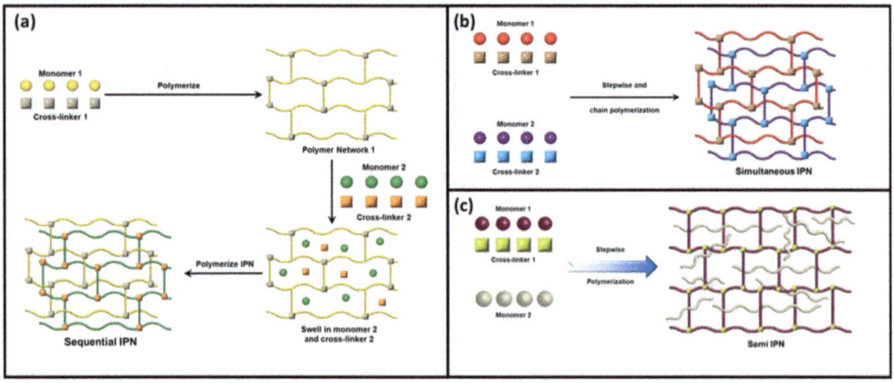

Fig. 3 Different IPN on the basis of arrangement

3.3 Advantages of IPNs for Drug Delivery Systems

Interpenetrating polymer networks (IPNs) offer numerous advantages in drug delivery systems due to their unique structural and functional properties, making them highly effective in controlled and sustained drug release. Their capacity to deliver accurate and extended drug release, which lowers dosage frequency and improves patient compliance, is one of their main advantages. IPNs improve mechanical strength, stability, and resistance to degradation by creating a cross-linked network of two or more polymers. This is especially advantageous for long-term biomedical applications. Because of their adjustable swelling and disintegration properties, scientists may create drug carriers that release medications at a precise rate, guaranteeing the best possible therapeutic outcome. Additionally, IPNs can accommodate both hydrophilic and hydrophobic molecules, making them ideal for combination therapies, and they can encapsulate and deliver multiple drugs at once.

3.4 Mechanisms of Drug Encapsulation and Release in IPNs

IPNs are perfect for drug delivery applications because of their special qualities, which include increased mechanical strength, stability, and controlled release characteristics. Multiple processes that can be adjusted to produce controlled and sustained drug delivery are involved in the mechanisms of drug encapsulation and release in IPNs.

Physical entrapment—during the development of the IPN, drugs may become physically trapped within the polymer network. The usual method for doing this is to dissolve or disperse the medication in the polymer solution prior to the crosslinking procedure. In a work of Banerjee, S, IPN hydrogel microspheres for the controlled release of DS (diclofenac sodium) were developed using two hydrophilic polymers, namely NaCMC and PVA. The release mechanism exhibited non-Fickian type behaviour, and the extent of crosslinking density, percentage drug loading, and NaCMC content into the IPN matrix influenced the release of DS. It was shown that the microspheres of this study were useful as controlled release devices to control the release rates of DS through the polymeric matrices developed [34]. In this instance, the medication is trapped inside the pores or interstitial spaces of the polymer network rather than chemically bonded to it. The diffusion of the drug molecules from the polymer network determines the drug's release.

Hydrophobic/hydrophilic interactions—Both hydrophilic and hydrophobic components can be used in IPN designs. While hydrophilic medications interact with the polymer's hydrophilic domains, hydrophobic drugs often interact with its hydrophobic domains. In another study, a new class of interpenetrating networks (IPNs) was created to study the delivery of hydrophobic therapeutic agents. It is made up of two networks: poly(n-butyl acrylate) (PBA) and methacrylic acid grafted with poly (ethylene glycol) tethers, P(MAA-g-EG). When the hydrophilic P(MAA-g-EG) networks move from the stomach (low pH) to the upper small intestine (neutral pH), they can trigger the release of an encapsulated therapeutic substance, such as a protein or a low molecular weight medication [35]. These interactions assist with controlling the release rate and support the drug's stability within the polymer network.

Ionic and electrostatic interactions—Ionic or electrostatic interactions can occasionally be used to introduce medicinal compounds having a charged or polar nature into the IPN. Ionic medicines, for instance, have the ability to create ionic connections with charged groups within the network of the polymer. To improve drug loading capacity and sustained release behaviour, thermosensitive P(NIPAM-co-NVP)/CS semi-IPN hydrogels were developed by Li. G [36]. Hydrogels' LCST was between 35 and 36 °C, which is ideal for applications involving intelligent drug administration. The stability and migration of encapsulated pharmaceuticals in hydrogels were facilitated by the high affinity for anionic medications that CS in semi-IPN networks provided. In contrast to pH 2.2 and pH 5.0, the regulated release of medicines without burst diffusion was achieved at pH 7.4 and 37 °C through the deprotonation of CS and collapsed PNIPAM. The design and development of innovative controlled drug delivery systems with enhanced drug loading capacity and sustained release will greatly benefit from this study.

4 Design and Synthesis of Chitosan-Based IPNs for Drug Delivery

4.1 Crosslinking Strategies in IPN Formation

4.1.1 Physical Crosslinking

Polyacrylamide/Sodium Alginate (PAM/SA) interpenetrating polymer network (IPN) hydrogels were created by Li. Yali [37] and group using covalently crosslinked PAM interpenetrated with ionically crosslinked SA to create robust hydrogels with exceptional formability. The heat resistance of PAM/SA IPN hydrogel is better than that of the individual components, according to TGA findings. According to swelling research, ESR decreases as the amount of chemical crosslinker or ionic crosslinking is increased by adding more N,N′--methylenebisacrylamide (MBA) or SA. Tensile testing indicates that the mechanical strength for hydrogels based on covalent/ionic crosslinking is enhanced when loosely crosslinked PAM and strongly crosslinked SA are combined. Certain polymers with ionic groups (e.g., polyelectrolytes) can undergo crosslinking via ionic bonds or coordination with metal ions. For example, chitosan or alginate can form ionic gels by complexation with divalent cations (such as calcium ions) [38].

4.1.2 Hydrogen Bonding

Hydrogen bonds between particular functional groups, like urethane or amide groups, can be used to crosslink some IPNs. Without the need for long-term chemical bonds, these non-covalent interactions aid in the formation of a network structure. Multiple hydrogen-bonding poly(acrylic acid-co-N-vinyl caprolactam) (PAN) hydrogels can display reversible dynamic responsive behaviours triggered by pH was prepared by Cheng. R and group [39]. The dynamic responsive behaviours are controlled by the hydrogen bonding between the comonomer units when the pH changes. They demonstrate that by varying the acid stimulation strength, the hydrogel crosslinking density, and the composition of the PAN random copolymer, the transparency recovery time, which depends on the nonequilibrium-to-equilibrium structure transition kinetics of the PAN hydrogels, can be controlled.

4.1.3 Crosslinking by Radiation

Crosslinking reactions can result from the generation of free radicals in polymers by high-energy radiation, such as electron beams or gamma rays. When creating crosslinked networks in IPNs, this method is very helpful since it allows one polymer to be pre-crosslinked by radiation prior to the synthesis of the second polymer. Scientist Wang. L and their group develop a double network strategy for epoxy resin that uses two distinct processing techniques to significantly enhance the performance of printed objects. Epoxy molecules crosslink under γ-ray radiation to generate a ray-induced network structure, while epoxy acrylates and polyethylene glycol dimethacrylates crosslink under UV light to form a light-induced network structure. The interpenetrating polymer network (IPN) is formed by the interpenetration of the two types of networks. The printed IPN specimens exhibit high fracture strain (up to 30.4%) and Young's modulus (up to 4.89 GPa) [40].

4.1.4 By Click Chemistry

Crosslinks with great specificity can be produced by employing high-yielding, selective reactions such as azide-alkyne cycloadditions (such as Huisgen's 1,3-dipolar cycloaddition). When creating IPNs that need exact control over network construction, this strategy performs well. For use in biomedical applications, a series of polyethylene glycol (PEG)-based derivatives with many "clickable" groups were created by polycondensation of dihydroxyloligo (ethyleneglycol) with maleic anhydride/thiolmalic acid. Huisgen's 1,3-dipolar cycloaddition of azides with alkynes, catalysed by Cu(I), produced a number of PEG-TMC networks that resulted in fully hydrophilic PEG hydrogels and PEG-poly(TMC) (PTMC) hydrogels with amphiphilic properties [41].

4.1.5 Enzymatic Crosslinking

This method uses an enzyme-catalysed cross-linking process between polymer chains to create an in situ hydrogel. Many enzymes were investigated and documented, including tyrosinase, phosphopantetheinyl transferase, lysyl oxidase, plasma amine oxidase, transglutaminases (TG), peroxidases, and phosphatases.

In a work done by Liu X et al. [42], biopolymer hydrogel fibres made of gelatine and chitosan with an interpenetrating polymer network (IPN) structure were created using a bienzymatic crosslinking technique. Microbial transglutaminase (mTG) catalysed the formation of one network of gelatin and horseradish peroxidase (HRP) in the presence of H_2O_2, which caused another network of chitosan grafted with phloretic acid (chitosan-PA) to intertwine with the first network of gelatin. The IPN biopolymer hydrogel fibres were made by wet spinning. HRP-catalysed injectable tyramine modified hyaluronic acid (HA-Tyr) hydrogels were recently created by Kim et al. (2011) in two phases. First, the carboxyl groups of HA and the amine groups of tyramine formed an amide bond to create the HA-Tyr conjugate. Next, HRP and H_2O_2 were used to create HA-Tyr hydrogels through a radical cross-linking reaction. Dexamethasone was administered intra-articularly using this hydrogel to treat rheumatoid arthritis [43].

4.1.6 Small-Molecule Cross-Linking

In order to create small-molecule cross-linking hydrogels, at least one polymer and a small molecule acting as a cross-linker in the right solvent are needed. Cross-linkers are molecules that enable the creation of bridges between polymeric chains by having at least two reactive functional groups. Drug molecules and bi-functional molecules are the two categories of small cross-linkers. The former occurs when a bi-functional molecule interacts with a polymer to trap a drug molecule inside the hydrogel. Additionally, pharmaceutical molecules with two functional groups that are covalently bound to polymeric chains create hydrogel without the need for cross-linkers [44]. Schiff base is created by the most basic type of cross-linking, which occurs between aldehyde and amino groups. Dialdehydes, such as glyoxal and especially glutaraldehyde [45], for instance, use the Schiff reaction to create covalent imine connections with the amino groups of chitosan. In other instances, hyaluronic acid-tyramine [46] and dextran-tyramine [47] formed covalent bonds utilizing hydrogen peroxide (H_2O_2) and horseradish peroxidase (HRP) as cross-linkers to create hydrogels with more regulated gelation periods that ranged from 5 s to 9 min, depending on the reactant quantities utilized.

4.1.7 Peroxide Linkage

A popular technique that uses peroxides that produce free radicals to crosslink certain polymers, especially those based on unsaturated structures. The polymer chains are crosslinked as a result of the free radicals' induction of covalent bond formation. In a study done by Borova. S et al. [48], they developed the use of peroxide copolymers in their melt to crosslink commercial, unfunctionalized hydrophilic poly(2-ethyl-2-oxazoline). Investigations have been conducted on the effects of temperature, peroxide copolymer concentration, and crosslinking process length. The process makes it possible to manage the mechanical properties of the final materials and produce hydrogels from unfunctionalized polymers in their melt. For many current and future uses of soft materials, including medical applications, the creation of hydrogels with an appropriate mechanical performance is essential.

4.2 Incorporating Natural and Synthetic Polymers with Chitosan

In the field of polymer science, chitosan, a biopolymer made from chitin (usually found in crustacean shells), has drawn a lot of attention because of its special qualities, which include non-toxicity, biocompatibility, and biodegradability, as well as the presence of functional groups (like hydroxyl and amino groups) that are easily modified chemically. Chitosan can provide desired properties to interpenetrating polymer networks (IPNs), particularly for use in the environmental, pharmaceutical, and biomedical industries. The interactions between chitosan and natural and synthetic polymers in the creation of IPNs will be examined in this review, along with the incorporation techniques, advantages of hybrid systems, and their uses [49].

4.2.1 Incorporating Chitosan with Natural Polymers

Interpenetrating polymer network (IPN) systems based on chitosan is showing promise as drug delivery platforms because of their capacity to encapsulate a variety of therapeutic compounds and their biocompatibility and biodegradability. The stability, controlled release, and targeted distribution of medications are improved when chitosan is mixed with natural polymers like gelatin, pectin, or alginate to create an IPN. Because these IPNs may be customized to provide particular release profiles, they are perfect for a range of uses, such as topical, oral, and controlled release systems. A flexible strategy for enhancing patient outcomes and drug delivery effectiveness is provided by the synergy between chitosan and other natural polymers (Table 1).

Table 1 List of some IPN with chitosan and natural polymer for drug delivery system

	Natural polymer	Synthesis method	Drug	Reference
Chitosan	Xanthan-Gum	Simultaneously film	Amoxicillin	[50]
Chitosan	Xanthan Gum	Free radical polymerization	Acyclovir	[51]
Chitosan	Guargum-gt-acrylamide	Water-in-oil (w/o) emulsion cross-linking	5-fluorouracil	[52]
Chitosan	Acrylamide-*grafted*-guar gum	Emulsion cross-linking	ciprofloxacin	[53]
Carboxymethyl chitosan	Starch	Nanocomposite hydrogels	amoxicillin	[54]
Chitosan	Starch +AgNP	Nanocomposite hydrogel	Model drug	[55]
Chitosan	Tamarind seed polysaccharide	Simultaneous blending	Aceclofenac	[56]
Chitosan	Hydroxyethyl cellulose	Blend microspheres	Isoniazid	[57]
Chitosan	Cellulose + halloysite nanotube	Nanocomposite	curcumin	[58]
Chitosan	Starch	Crosslinking with GA	AHA	[59]
Chitosan	Egg albumin with PEG 400	Interpolymeric complexation	Alprazolam	[60]

4.2.2 Incorporating Chitosan with Synthetic Polymers (Table 2)

Table 2 List of some IPNs with chitosan and synthetic polymer for drug delivery system

	Synthetic polymer	Synthesis method	Drug	Reference
Chitosan	Polyacrylic acid	Free radical polymerization	Aspirin	[61]
Chitosan-poly (1-vinyl imidazole)	Chitosan-poly (methacrylic-acid)	Free radical polymerization	Doxorubicin hydrochloride	[62]
Carboxymethyl-chitosan	2-hydroxyethyl methacrylate (HEMA)	Ethylene glycol dimethacrylate as crosslinker	Ciprofloxacin	[63]
Chitosan	Poly(2-hydroxyethyl methacrylate co-itaconic acid)	–	Diclofenac sodium	[64]
Chitosan	Acrylic acid (AA), acrylamide (AAm), polyvinylpyrrolidone (PVP), and montmorillonite (MMT)	Free radical graft copolymerization	Clarithromycin	[65]
Chitosan	Glycine and glutamic acid	Cross linked with glutaraldehyde	Chlorpheniramine maleate	[66]
Chitosan	Acrylic acid	Crosslinking graft copolymerization by MBA	Perindopril	[67]
Chitosan	Poly(meth (methacrylic)) acid	Crosslinking	Nateglinide	[68]

4.3 Fabrication Techniques

Hydrogel: It is made from a mixture of various chemically crosslinked polymers that have the capacity to hold onto water in their structure due to the hydrophilic function of the polymers. Additionally, hydrogels are three-dimensional networks of polymers. After photopolymerization, ionic cross-linking, chemical cross-linking, or in response to an environmental stimulation like temperature, pH, or media ionic strength, a polymer solution can instantly gel. In an effort to improve the mechanical strength of natural polymers and get around the disadvantages of synthetic polymers, an attempt was made to include collagen and hyaluronic acid into IPN hydrogels by Suri. S et al. It was discovered that the hydrogels were robust and stable [69]. In another study, modified gelatin and poly(ethylene glycol) diacrylate of protein-loaded interpenetrating network were discovered to connect with the cell membrane and cause the desired pharmacological reactions [70]. Similarly, two biological polymers, collagen and hyaluronic acid, were combined with two synthetic polymers, poly(vinyl alcohol) and poly(acrylic acid), in order to improve the mechanical strength of natural polymers and get around the biological disadvantages of synthetic polymers. These mixtures were created into hydrogels, films, and sponges, and growth hormone was then added. The findings of this study demonstrated that collagen-poly(vinyl alcohol) hydrogels could release growth hormone in a regulated manner, and that the amount and rate of growth hormone released were primarily influenced by the system's collagen content [71].

Nanocomposite: Nanocomposites are materials that incorporate nanoparticles (e.g., clay, carbon nanotubes, or graphene) into a polymer matrix. These composites can enhance the mechanical, thermal, and electrical properties of IPNs. Materials known as nanocomposite interpenetrating polymer networks (IPNs) have two or more polymer networks interwoven at the nanoscale, with nanoparticles scattered throughout the matrix to improve characteristics like biocompatibility and mechanical strength [72]. Nanoparticles have given polymer hydrogels special physical properties, including reactivity to mechanical, optical, thermal, barrier, acoustic, magnetic, and electric stimulation, among others. Catalysis, separation devices, drug delivery, optics, electronics, sensors, actuators, and microfluidics are just a few of the biotechnological fields that can benefit from these special qualities. Parsa et al. developed a nanocomposite with the incorporation of chitosan to polyvinyl alcohol (PVA), which increases antibacterial action against both gram-positive and gram-negative bacteria as well as cell development in the presence of tetracycline [73]. In another study by Ianchis et al., they developed an IPN system with N, N′-methylenebisacrylamide and Clay, which improved mechanical characteristics, slowed the de-swelling process, and increased swelling degrees of clay depending upon the properties [74].

Microsphere–microspheres are a new medication delivery method based on IPN. These are free-flowing particles, which are small spherical particles with a diameter of 1 to 1000 μm that are composed of either natural or synthetic polymers. The core of a microsphere contains the medicine, while the coating material is an outside layer of polymer. The controlled release of medications is one of the effective uses of interpenetrating networks [75]. In a study conducted by Bhattacharya S. S [76], using the w/o emulsion cross-linking approach, xanthan gum enabled the creation of superabsorbent polymeric microspheres that were effectively manufactured and tested for ciprofloxacin hydrochloride's sustained release. Using X-ray diffraction (XRD) studies and Fourier transform infrared spectroscopy (FTIR), IPN production was verified. A differential scanning calorimetry (DSC) study was performed to investigate the dispersion nature of the medication after encapsulation. To examine the sustained drug delivery from IPN microspheres, a comprehensive evaluation of in vitro drug release studies was conducted.

4.4 *Structural Characterization of Chitosan-Based IPNs*

Understanding the connections between the molecular composition, physical characteristics, and functionality of chitosan-based interpenetrating polymer networks (IPNs) in applications like drug administration, wound healing, and tissue engineering requires a structural characterization of these networks. The shape, chemical structure, crystallinity, and thermal stability of chitosan IPNs are investigated using a variety of analytical techniques, which offer important insights into how they behave in various scenarios [77].

One of the main methods for structural characterization is Fourier Transform Infrared (FTIR) Spectroscopy, which aids in determining the functional groups found in chitosan and the components of its interacting polymers [78]. By identifying certain peaks that correlate to new chemical bonds or shifts in functional groups brought on by crosslinking, FTIR spectra can verify the successful production of IPNs. In the IR spectra, for example, the emergence of a new peak or the shifting of preexisting peaks may signify the interaction between chitosan and other polymers, confirming the network structure's creation [79, 80].

Another essential method for assessing the crystallinity of chitosan-based IPNs is X-ray diffraction (XRD). As a naturally occurring polysaccharide, chitosan usually has some degree of crystallinity; however, the overall crystalline structure may be impacted by the creation of an IPN. The degree of crystallinity and amorphous areas in the IPNs may be determined by XRD analysis, and this information has a direct impact on the IPNs' mechanical qualities, swelling behaviour, and drug release features [81]. When compared to pure chitosan, a shift in the diffraction peaks could indicate successful crosslinking or the development of a more amorphous structure, both of which could be advantageous for specific biomedical applications [82].

An essential tool for analysing the surface morphology and porous structure of chitosan-based IPNs is scanning electron microscopy, or SEM. At the microscopic level, SEM imaging shows the dispersion of the polymer network, surface texture, and porosity [83]. For applications like as drug delivery, where the size and distribution of pores can affect the rate of drug release, the presence of well-defined pores is crucial [84]. In medical applications, the morphology of chitosan IPN, such as their smoothness or roughness, can also affect their bioactivity and tissue compatibility.

A couple of crucial methods for evaluating the thermal stability of chitosan-based IPNs are thermogravimetric analysis (TGA) and differential scanning calorimetry (DSC). By measuring a sample's weight loss while heating, TGA can determine the sample's stability and rate of thermal degradation. The stability and processing conditions of the polymer network are revealed by DSC's evaluation of thermal transitions, such as the melting point and glass transition temperature (Tg) [85, 86]. These methods are essential for figuring out the temperature ranges at which the integrity of the IPNs can be preserved throughout processing or sterilization.

5 Applications of Chitosan-Based IPNs in Drug Delivery

5.1 Oral Drug Delivery

When it comes to drug administration, oral drug delivery is a flexible, economical approach with excellent patient compliance. When using an oral drug delivery method, a medication must pass through a number of obstacles before entering the bloodstream. Digestive enzymes may break down the medication molecules (peptides or nucleic acids) in the gastrointestinal system. Nevertheless, the use of oral methods is limited by the drug's brief intestinal residence period. Because of its mucoadhesive qualities, CS polymer may be an appropriate drug carrier in oral medication delivery methods. The ionic interactions between the cationic amino groups of chitosan and the negative functional groups of sialic acid of the mucus in the gastrointestinal tract may be the cause of its mucoadhesive qualities. Improved medication bioavailability and distribution to the stomach, small intestine, and colon can be achieved with CS-based IPN hydrogels.

Many attempts have been made to use CS-based IPN hydrogels to administer insulin. For oral insulin delivery, Yin and colleagues created a superporous IPN hydrogel using poly(acrylic acid-co-acrylamide)/O-carboxymethyl CS. In Caco-2 cell monolayers and in a rat intestinal model, the developed system's strong enzymatic inhibitory capabilities significantly increased the paracellular permeability of insulin by 4.9- and 4.2-fold, respectively. Insulin administered orally to rats' intestines demonstrated good retentive qualities for over 8 h. In their study, raising the pH from 1.0 to 6.2 increased the rate at which insulin was released. In acidic pH, the carboxylic groups of the polymeric chains were protonated, which leads to shrinkage of the IPN hydrogel [87].

5.2 Transdermal and Topical Drug Delivery

Since transdermal drug delivery avoids the first-pass metabolic effect and enables the distribution of macromolecules and medications at lower doses over a longer duration, it is a preferred method for the delivery of therapeutic substances. It has been demonstrated that CS-based IPN hydrogel membranes are capable of effectively transporting drug molecules through the skin.

Recently, naringenin (NRG) was delivered transdermally to treat atopic dermatitis using a pH-responsive hydrogel based on carboxymethyl cellulose (CMC)/2-hydroxyethyl acrylate. The process of radical polymerization was used to create the hydrogel. According to the findings, the swelling ratio was higher at pH 7.5 and 8.5 than it was at pH 5.5. The ionization of the carboxyl group of CMC was linked to these outcomes. In fact, the carboxyl groups were protonated at pH 5.5, and the resulting strong hydrogen bonds led to a preponderance of polymer-polymer interactions over polymer-water interactions. At pH 5.5 (normal skin pH), pH 7.5 (acne skin pH), and pH 8.5 (atopic skin pH), the drug release from the produced hydrogel was investigated. The way the drug release rate behaved at different pH values indicated that the pH may affect the drug release in a hydrogel system and that basic skin types, such as atopic skin, would have an easier time releasing NRG [88].

5.3 Targeted and Sustained Release Systems

The use of chitosan-based interpenetrating polymer networks (IPNs) in targeted and sustained drug release systems is growing because of their biodegradability, biocompatibility, and capacity to be tailored for certain release patterns. Drug release over long periods of time can be regulated by altering the chitosan network topology, guaranteeing a long-lasting therapeutic impact. Further improving the ability to target drug release to certain areas, such as acidic tumour environments or inflammatory tissues, is the addition of stimuli-responsive components, such as pH or temperature-sensitive polymers. These features make chitosan-based IPNs perfect for enhancing drug delivery applications' effectiveness and minimizing their negative effects.

In a study, using N,N-methylene-bisacrylamide (MBA) as a crosslinking agent and benzoyl peroxide as a catalyst, crosslinked hydrogels based on chitosan (CS) and acrylic acid (AA) were created by free radical polymerization with feed compositions. USP phosphate buffer solutions with pH values of 1.2, 5.5, 6.5, and 7.5 were used for the creation of swelling investigations. Results showed that swelling increased when the amount of AA in the hydrogels' structure increased in solutions with higher pH values. Hydrogel was created by electrostatic contact between cationic ions. Verapamil, a model medication, was added to a few chosen samples. The degree of crosslinking, the medium's pH, and the ratio of CS to AA all affected the release of ventacamil. By fitting the experimental data to model equations and determining the associated parameters, the release mechanisms were investigated. The results showed that non-Fickian diffusion dominated the kinetics of drug release from the hydrogels in both pH 1.2 and pH 7.5 buffer solutions [89].

6 Recent Advances in Chitosan-Based IPNs for Drug Delivery

6.1 Smart Drug Delivery Systems (Responsive IPNs)

The development of a significant strategy to control drug release based on photothermal effects has led to the development of a sustained and controlled drug delivery system based on a PCL–PEG–PCL thermosensitive hydrogel combined with chitosan-multiwalled carbon nanotubes for a near-infrared light-triggered drug delivery. This combination of carbon nanotubes that incorporate hydrogel can enhance the sustained effect of drug delivery by a dual-stage release and allow drug delivery by controlling light irradiation. Using nude mice as models, thermal imaging was used to track this in situ photothermal process, and fluorescence imaging in vivo tracked the controlled drug delivery of doxorubicin in real-time based on the drug's fluorescence ability. The findings imply that the hydrogel's structure may be broken by the photothermal effect of the carbon nanotubes through a gel–sol transition, which would cause the drug to be released from the sustained drug delivery system by NIR irradiation in response to desire [90].

Hydrogel-based stimuli-responsive drug-delivery systems are becoming more and more popular because of their many benefits, which include high drug loading capacity, target-specific delivery, biocompatibility, decreased drug toxicity, and controlled drug release. Using free radical polymerization in the presence of poly (NIPA-co-VSA) nanogels, a pH and temperature-responsive poly(NIPAAm-co-VSA)/Alginate interpenetrating polymer network (IPN) hydrogel was created by Balan et al. [91]. The hydrogels and Ca^{2+} ions were physically cross-linked using a natural polymer called Na-Alginate. The hydrogels' doxorubicin (DOX) loading capabilities, as assessed by the solution impregnation method, ranged from 86 to 161 mg DOX/g polymer. They have examined their DOX release behaviour in relation to temperature, pH, time, and the hydrogel's physical crosslinking degree. Artificial neural network (ANN) modelling was successfully applied to experimentally acquired and very complex DOX release characteristics.

6.2 Nano Formulations and Nanoparticle-Based IPNs

Because chitosan-based nanoparticles are so versatile, they have shown increasing promise for use in drug delivery systems. To improve the efficiency of chitosan-based nanoparticles for stimulus-responsive anticancer drug delivery, free radical polymerization was used to create novel pH-sensitive semi-interpenetrating networks (semi-IPN) of chitosan-poly(1-vinyl imidazole) (CS-PVIm) and chitosan-poly(methacrylic-acid) (CS-PMAA) nanoparticles. To assess the semi-IPN nanoparticles' suitability as drug carriers, an analysis was conducted by Shikuku et al. [62]. Moreover, persistent drug release from doxorubicin-hydrochloride (Dox)-loaded nanoparticles was demonstrated by the protonation and deprotonation of their functional groups in response to pH changes. The encapsulation efficiency (EE) of CS-PVIm-nanoparticles was 57%, whereas that of CS-PMAA-nanoparticles was 52%. This study offers a simple process for creating pH-responsive, intelligent semi-IPN nanoparticles specifically designed for anticancer medication delivery.

In another study, the inclusion of hydrothermally produced carbon microspheres and nanocellulose in the hydrogel network created by SA chelating with Ca^{2+} improved the drug carrier's many characteristics. Using cellulose and sodium alginate, a green biomass-based hydrogel was created as a methylene blue drug carrier model in order to create a stable hydrogel drug carrier with superior biocompatibility, biodegradability, and low toxicity. Furthermore, the produced SA-CNFgelCS0.1 had good compressive and elastic properties and was able to bear a pressure of 8.64 N. In the meantime, its drug loading capacity and encapsulation rate could both reach 19.36% and 95.5%, respectively. In PBS solution, which is similar to the pH of the gastric environment, the drug release rate reached 43.4% at 100 h, demonstrating good pH-responsiveness and long-lasting release ability during the drug-carrying release process [92].

6.3 Combination Therapy and Multi-drug Delivery

The potential of the IPNs' hydrogel films for wound dressing and drug distribution was investigated by Shakouripour et al. [93], using chitosan (CS), 2-hydroxyethyl methacrylate (HEMA), and 2-hydroxypropyl methacrylate (HPMA). At varied drug doses, the release behaviour of produced hydrogel formulations was examined at two distinct pH values for DOX and CUR, which are 5.5 and 7.4, respectively. The formulation of CS/p(HEMA)1 that contained CS/HEMA in a 1:1 ratio demonstrated the highest drug release rates of 75.5% for curcumin and 76.0% for doxorubicin among the IPNs. Drug release profiles in vitro showed a time-dependent release pattern, with all formulations exhibiting a maximum release at 48 h.

7 Challenges and Future Perspectives

Interpenetrating polymer networks (IPNs) based on chitosan have attracted interest due to its potential for use in a number of fields, such as wound healing and drug delivery [10]. However, their use is subject to a number of restrictions

[94]. Furthermore, the IPNs' long-term efficacy may be impacted by the early disintegration caused by chitosan's biodegradability [95, 96]. A further obstacle is the network structure's inability to be uniform, which may have an impact on the stability, drug release, and swelling behaviour of the intended characteristics. Furthermore, the high cost of raw chitosan material and complex processing techniques hinder the widespread commercialization of chitosan-based IPNs in certain industries. Although scalability is still an issue, chitosan-based interpenetrating polymer networks (IPNs) have great potential for industrial applications. Complex processes, including crosslinking and solvent usage, are frequently used in the production of chitosan IPNs, and they might be challenging to scale up for large-scale manufacturing. Furthermore, it might be difficult to achieve consistent quality and uniformity in the finished product because even slight changes in processing parameters can result in performance discrepancies. Notwithstanding these challenges, the biocompatibility and biodegradability of chitosan IPNs make them promising for use in sectors such as food packaging, agriculture, and medicines [97–99]. The scalability and economic viability of chitosan IPNs in a range of applications depend on improvements in manufacturing techniques, cost-effective production, and process optimization in order to realize their industrial potential.

Due to continuous research and technological breakthroughs, chitosan-based interpenetrating polymer networks (IPNs) in drug delivery technologies are expected to see substantial advancements in the future. The creation of chitosan IPNs with improved mechanical characteristics and stability, which addresses present performance constraints, is one significant trend. Researchers want to develop more resilient, long-lasting, and sensitive materials appropriate for regulated and prolonged drug release by combining chitosan with other biopolymers or synthetic polymers and applying sophisticated crosslinking processes [100]. Using nanotechnology to create nanoscale chitosan IPNs for more accurate medication targeting and controlled release at the cellular or molecular level is another exciting field. This could result in less harmful side effects and more effective treatments. Additionally, chitosan IPNs are being investigated for their potential to distribute medications based on metabolic or genetic profiles, among other customized needs, as personalized medicine gains popularity. A developing area of interest is smart drug delivery systems that can release therapeutic medicines just where they are needed. These systems can react to external stimuli such as pH, temperature, or enzymes. It is anticipated that improvements in production methods will make chitosan IPNs more affordable, opening the door for more extensive clinical and commercial uses in tissue regeneration [101, 102], wound healing [103], and drug delivery [10, 104].

8 Conclusion

Because of their intrinsic biocompatibility, biodegradability, and capacity to undergo a variety of drug release processes, chitosan-based interpenetrating polymer networks (IPNs) offer a viable and adaptable platform for drug delivery systems.

Targeted delivery, regulated and prolonged drug release, and reactivity to environmental cues, including pH, temperature, and enzyme activity, are only a few benefits of these systems [10]. The fine-tuning of release profiles made possible by the combination of chitosan and other polymers in IPNs makes them appropriate for a variety of therapeutic uses, including site-specific therapies and oral and injectable drug delivery. The flexibility of these systems is facilitated by important mechanisms such as diffusion, swelling, erosion, pH responsiveness, and enzymatic degradation, which enable customized drug release that can enhance patient compliance and therapeutic efficacy [105, 106].

Acknowledgements This work is financially supported by the DST-SERB funding agency through three sanctioned projects EEQ/2020/000496, EEQ/2016/000712, and ECR/2016/002018. K.S. also acknowledges UGC for the UGC-BSR Research Start-UpGrant (F.30-363/2017(BSR) Dt-08/08/2017) and CU for the UPE-II Nanofabrication project fund (UGC/166/UPE-II, Dt-03/04/2017) for financial support.

References

1. Adepu, S., & Ramakrishna, S. (2021). Controlled drug delivery systems: Current status and future directions. *Molecules, 26*(19), 5905.
2. Hisham, F., Maziati Akmal, M. H., Ahmad, F., Ahmad, K., & Samat, N. (2024). Biopolymer chitosan: Potential sources, extraction methods, and emerging applications. *Ain Shams Engineering Journal, 15*(2), 102424.
3. Opriş, O., Mormile, C., Lung, I., Stegarescu, A., Soran, M.-L., & Soran, A. (2024). An overview of biopolymers for drug delivery applications. *Applied Sciences, 14*(4), 1383.
4. Padhi, S., Behera, A., Hasnain, M. S., & Nayak, A. K. (2022). Chapter 6 – Uses of chitosan in drug delivery. In M. S. Hasnain, S. Beg, & A. K. Nayak (Eds.), *Chitosan in biomedical applications* (pp. 139–162). Academic.
5. Mura, P., Maestrelli, F., Cirri, M., & Mennini, N. (2022). Multiple roles of chitosan in mucosal drug delivery: An updated review. *Marine Drugs, 20*(5), 335.
6. León-Campos, M. I., Mendoza, J. J., Aguayo-Morales, H., Cobos-Puc, L. E., Cabrera-Munguía, D. A., & Claudio-Rizo, J. A. (2024). The biological applications of IPN hydrogels. *Admet DMPK, 12*(4), 581–621.
7. Chatterjee, S., Ghosal, K., Kumar, M., Mahmood, S., & Thomas, S. (2022). A detailed discussion on interpenetrating polymer network (IPN) based drug delivery system for the advancement of health care system. *Journal of Drug Delivery Science and Technology, 79*, 104095.
8. Hasan, M. M., Habib, M. L., Anwaruzzaman, M., Kamruzzaman, M., Khan, M. N., & Rahman, M. M. (2020). Chapter 3 – Processing techniques of chitosan-based interpenetrating polymer networks, gels, blends, composites and nanocomposites. In S. Gopi, S. Thomas, & A. Pius (Eds.), *Handbook of Chitin and Chitosan* (pp. 61–93). Elsevier.
9. Neblea, I. E., Chiriac, A.-L., Zaharia, A., Sarbu, A., Teodorescu, M., Miron, A., Paruch, L., Paruch, A. M., Olaru, A. G., & Iordache, T.-V. (2023). Introducing semi-interpenetrating networks of chitosan and ammonium-quaternary polymers for the effective removal of waterborne pathogens from wastewaters. *Polymers, 15*(5), 1091.
10. Lohani, A., Singh, G., Bhattacharya, S. S., & Verma, A. (2014). Interpenetrating polymer networks as innovative drug delivery systems. *Journal of Drug Delivery, 2014*, 583612.

11. Román-Doval, R., Torres-Arellanes, S. P., Tenorio-Barajas, A. Y., Gómez-Sánchez, A., & Valencia-Lazcano, A. A. (2023). Chitosan: Properties and its application in agriculture in context of molecular weight. *Polymers, 15*(13), 2867.
12. Aranaz, I., Alcántara, A. R., Civera, M. C., Arias, C., Elorza, B., Heras Caballero, A., & Acosta, N. (2021). Chitosan: An overview of its properties and applications. *Polymers, 13*(19), 3256.
13. Aranaz, I., Alcántara, A. R., Civera, M. C., Arias, C., Elorza, B., Heras Caballero, A., & Acosta, N. (2021). Chitosan: An overview of its properties and applications. *Polymers (Basel), 13*(19), 3256.
14. Coquery, C., Negrell, C., Caussé, N., Pébère, N., & David, G. (2019). Synthesis of new high molecular weight phosphorylated chitosans for improving corrosion protection. *Pure and Applied Chemistry, 91*(3), 509–521.
15. Essawy, A. A., Hefni, H., & El-Nggar, A. M. (2019). Biocompatible and biodegradable chitosan composites in wound healing application: In Situ novel photo-induced skin regeneration approach. In Inamuddin, S. Thomas, R. K. Mishra, & A. M. Asiri (Eds.), *Sustainable polymer composites and nanocomposites* (pp. 143–183). Springer.
16. Ortona, O., D'Errico, G., Mangiapia, G., & Ciccarelli, D. (2008). The aggregative behavior of hydrophobically modified chitosans with high substitution degree in aqueous solution. *Carbohydrate Polymers, 74*(1), 16–22.
17. Mohamed, N. A., Sabaa, M. W., El-Ghandour, A. H., Abdel-Aziz, M. M., & Abdel-Gawad, O. F. (2013). Quaternized N-substituted carboxymethyl chitosan derivatives as antimicrobial agents. *International Journal of Biological Macromolecules, 60*, 156–164.
18. Hirano, S., Yamaguchi, Y., & Kamiya, M. (2002). Novel N-saturated-fatty-acyl derivatives of chitosan soluble in water and in aqueous acid and alkaline solutions. *Carbohydrate Polymers, 48*(2), 203–207.
19. Pellá, M. C. G., Lima-Tenório, M. K., Tenório-Neto, E. T., Guilherme, M. R., Muniz, E. C., & Rubira, A. F. (2018). Chitosan-based hydrogels: From preparation to biomedical applications. *Carbohydrate Polymers, 196*, 233–245.
20. Wang, J., Liu, L. G., Jiao, W.-Q., Yang, H., Liu, J., & Liu, D. (2021). Phenylboronic acid-conjugated chitosan nanoparticles for high loading and efficient delivery of curcumin. *Carbohydrate Polymers, 256*, 117497.
21. Olewnik-Kruszkowska, E., Gierszewska, M., Grabska-Zielińska, S., Skopińska-Wiśniewska, J., & Jakubowska, E. (2021). Examining the impact of Squaric acid as a crosslinking agent on the properties of chitosan-based films. *International Journal of Molecular Sciences, 22*(7), 3329.
22. Jiang, H.-L., Kim, Y.-K., Arote, R., Nah, J.-W., Cho, M.-H., Choi, Y.-J., Akaike, T., & Cho, C.-S. (2007). Chitosan-graft-polyethylenimine as a gene carrier. *Journal of Controlled Release, 117*(2), 273–280.
23. Kristiansen, K. A., Potthast, A., & Christensen, B. E. (2010). Periodate oxidation of polysaccharides for modification of chemical and physical properties. *Carbohydrate Research, 345*(10), 1264–1271.
24. Hanafy, M. S., Desoky, W. M., Hussein, E. M., El-Shaer, N. H., Gomaa, M., Gamal, A. A., Esawy, M. A., & Guirguis, O. W. (2021). Biological applications study of bio-nanocomposites based on chitosan/TiO_2 nanoparticles polymeric films modified by oleic acid. *Journal of Biomedical Materials Research, Part A, 109*(2), 232–247.
25. Elsabahy, M., & Hamad, M. A. (2021). Design and preclinical evaluation of chitosan/kaolin nanocomposites with enhanced hemostatic efficiency. *Marine Drugs, 19*(2), 50.
26. Nie, J., Wang, Z., & Hu, Q. (2016). Chitosan hydrogel structure modulated by metal ions. *Scientific Reports, 6*(1), 36005.
27. Patel, S., & Goyal, A. (2017). Chitin and chitinase: Role in pathogenicity, allergenicity and health. *International Journal of Biological Macromolecules, 97*, 331–338.

28. Filion, D., Lavertu, M., & Buschmann, M. D. (2007). Ionization and solubility of chitosan solutions related to thermosensitive chitosan/glycerol-phosphate systems. *Biomacromolecules, 8*(10), 3224–3234.
29. Hajji, S., Younes, I., Ghorbel-Bellaaj, O., Hajji, R., Rinaudo, M., Nasri, M., & Jellouli, K. (2014). Structural differences between chitin and chitosan extracted from three different marine sources. *International Journal of Biological Macromolecules, 65*, 298–306.
30. Szymańska, E., & Winnicka, K. (2015). Stability of chitosan—A challenge for pharmaceutical and biomedical applications. *Marine Drugs, 13*(4), 1819–1846.
31. Yang, Z., Chen, L., McClements, D. J., Qiu, C., Li, C., Zhang, Z., Miao, M., Tian, Y., Zhu, K., & Jin, Z. (2022). Stimulus-responsive hydrogels in food science: A review. *Food Hydrocolloids, 124*, 107218.
32. Vala, A. U., Rane, A. V., Kanny, K., Abitha, V. K., & Thomas, S. (2017). Chapter 11 – ageing behavior of polyurethane based blends and interpenetrating polymer networks. In S. Thomas, J. Datta, J. T. Haponiuk, & A. Reghunadhan (Eds.), *Polyurethane polymers* (pp. 261–281). Elsevier.
33. Merle, G., Wessling, M., & Nijmeijer, K. (2011). Anion exchange membranes for alkaline fuel cells: A review. *Journal of Membrane Science, 377*(1), 1–35.
34. Banerjee, S., Siddiqui, L., Bhattacharya, S. S., Kaity, S., Ghosh, A., Chattopadhyay, P., Pandey, A., & Singh, L. (2012). Interpenetrating polymer network (IPN) hydrogel microspheres for oral controlled release application. *International Journal of Biological Macromolecules, 50*(1), 198–206.
35. Schoener, C. A., Hutson, H. N., Fletcher, G. K., & Peppas, N. A. (2011). Amphiphilic interpenetrating networks for the delivery of hydrophobic, Low Molecular Weight Therapeutic Agents. *Industrial & Engineering Chemistry Research, 50*(22), 12556–12561.
36. Li, G., Guo, L., Chang, X., & Yang, M. (2012). Thermo-sensitive chitosan based semi-IPN hydrogels for high loading and sustained release of anionic drugs. *International Journal of Biological Macromolecules, 50*(4), 899–904.
37. Li, Y., Wang, C., Zhang, W., Yin, Y., & Rao, Q. (2015). Preparation and characterization of PAM/SA tough hydrogels reinforced by IPN technique based on covalent/ionic crosslinking. *Journal of Applied Polymer Science, 132*(4).
38. Wang, Y., Shen, Z., Wang, H., Song, Z., Yu, D., Li, G., Liu, X., & Liu, W. (2024). Progress in research on metal ion crosslinking alginate-based gels. *Gels, 11*(1), 16.
39. Cheng, R., Xu, M., Zhang, X., Jiang, J., Zhang, Q., & Zhao, Y. (2023). Hydrogen bonding enables polymer hydrogels with pH-induced reversible dynamic responsive behaviors. *Angewandte Chemie International Edition, 62*(23), e202302900.
40. Wang, L., Zhang, F., Du, S., & Leng, J. (2022). 4D printing of shape-changing structures based on IPN epoxy composites formed by UV post-curing and γ-ray radiation. *Composites Part A: Applied Science and Manufacturing, 162*, 107146.
41. Bignotti, F., Baldi, F., Grassi, M., Abrami, M., & Spagnoli, G. (2021). Hydrophobically-modified PEG hydrogels with controllable hydrophilic/hydrophobic balance. *Polymers, 13*(9), 1489.
42. Liu, X.-j., Zhang, Y.-m., & Li, X.-s. (2015). Tough biopolymer IPN hydrogel fibers by bienzymatic crosslinking approach. *Chinese Journal of Polymer Science, 33*(12), 1741–1749.
43. Kim, K., Park, S., Yang, J.-A., Jeon, J.-H., Bhang, S., Kim, B.-S., & Hahn, S. (2011). Injectable hyaluronic acid–tyramine hydrogels for the treatment of rheumatoid arthritis. *Acta Biomaterialia, 7*(2), 666–674.
44. Berger, J., Reist, M., Mayer, J., Felt, O., Peppas, N., & Gurny, R. (2004). Structure and interactions in covalently and ionically crosslinked chitosan hydrogels for biomedical applications. *European Journal of Pharmaceutics and Biopharmaceutics, 57*(1), 19–34.
45. Liang, H.-F., Hong, M.-H., Ho, R.-M., Chung, C.-K., Lin, Y.-H., Chen, C.-H., & Sung, H.-W. (2004). Novel method using a temperature-sensitive polymer (methylcellulose) to thermally gel aqueous alginate as a pH-sensitive hydrogel. *Biomacromolecules, 5*(5), 1917–1925.

46. Kurisawa, M., Chung, J. E., Yang, Y. Y., Gao, S. J., & Uyama, H. (2005). Injectable biodegradable hydrogels composed of hyaluronic acid–tyramine conjugates for drug delivery and tissue engineering. *Chemical Communications, 34*, 4312–4314.
47. Jin, R., Hiemstra, C., Zhong, Z., & Feijen, J. (2007). Enzyme-mediated fast in situ formation of hydrogels from dextran–tyramine conjugates. *Biomaterials, 28*(18), 2791–2800.
48. Borova, S., Tokarev, V., Stahlhut, P., & Luxenhofer, R. (2020). Crosslinking of hydrophilic polymers using polyperoxides. *Colloid and Polymer Science, 298*(12), 1699–1713.
49. Thomas, S., Pius, A., & Gopi, S. (2020). *Handbook of Chitin and Chitosan: Volume 3: Chitin- and Chitosan-based polymer materials for various applications*. Elsevier.
50. Thakur, A., Monga, S., & Wanchoo, R. (2014). Sorption and drug release studies from semi-interpenetrating polymer networks of chitosan and xanthan gum. *Chemical and Biochemical Engineering Quarterly, 28*, 105–115.
51. Malik, N. S., Ahmad, M., Minhas, M. U., Tulain, R., Barkat, K., Khalid, I., & Khalid, Q. (2020). Chitosan/xanthan gum based hydrogels as potential carrier for an antiviral drug: Fabrication, characterization, and safety evaluation. *Frontiers in Chemistry, 8*, 50.
52. Sekhar, E. C., Rao, K. K., & Raju, R. R. (2011). Chitosan/guargum-g-acrylamide semi IPN microspheres for controlled release studies of 5-Fluorouracil. *Journal of Applied Pharmaceutical Science, 1*(8), 199–204.
53. Kajjari, P. B., Manjeshwar, L. S., & Aminabhavi, T. M. (2011). Novel interpenetrating polymer network hydrogel microspheres of Chitosan and Poly(acrylamide)-grafted-Guar Gum for controlled release of ciprofloxacin. *Industrial & Engineering Chemistry Research, 50*(23), 13280–13287.
54. Gholamali, I., & Alipour, E. (2020). Carboxymethyl Chitosan/Starch/CuO nanocomposite hydrogels for controlled release of amoxicillin. *Regenerative Engineering and Translational Medicine, 6*(4), 398–406.
55. Nezami, S., & Sadeghi, M. (2020). pH-sensitive free AgNPs composite and nanocomposite beads based on starch as drug delivery systems. *Polymer Bulletin, 77*(3), 1255–1279.
56. Jana, S., Sen, K. K., & Basu, S. K. (2014). In vitro aceclofenac release from IPN matrix tablets composed of chitosan-tamarind seed polysaccharide. *International Journal of Biological Macromolecules, 65*, 241–245.
57. Angadi, S. C., Manjeshwar, L. S., & Aminabhavi, T. M. (2010). Interpenetrating polymer network blend microspheres of chitosan and hydroxyethyl cellulose for controlled release of isoniazid. *International Journal of Biological Macromolecules, 47*(2), 171–179.
58. Doustdar, F., Olad, A., & Ghorbani, M. (2022). Development of a novel reinforced scaffold based on chitosan/cellulose nanocrystals/halloysite nanotubes for curcumin delivery. *Carbohydrate Polymers, 282*, 119127.
59. Viyoch, J., Sudedmark, T., Srema, W., & Suwongkrua, W. (2005). Development of hydrogel patch for controlled release of alpha-hydroxy acid contained in tamarind fruit pulp extract. *International Journal of Cosmetic Science, 27*(2), 89–99.
60. Jana, S., Maji, N., Nayak, A. K., Sen, K. K., & Basu, S. K. (2013). Development of chitosan-based nanoparticles through inter-polymeric complexation for oral drug delivery. *Carbohydrate Polymers, 98*(1), 870–876.
61. Simayi, R., Zhang, X., Sawut, A., Zhu, Z., Wu, T., & Jiang, M. (2024). Structural modulation of chitosan/Polyacrylic acid-based hydrogels for optimized controlled drug release performance. *Chemistry Select, 9*(9), e202303257.
62. Shikuku, R., Hasnat, M. A., Mashrur, S. B. A., Haque, P., Rahman, M. M., & Khan, M. N. (2024). Chitosan-based pH-sensitive semi-interpenetrating network nanoparticles as a sustained release matrix for anticancer drug delivery. *Carbohydrate Polymer Technologies and Applications, 7*, 100515.
63. Ayón, M. A. G., Hoppenstedt, A. M., López, A. Z., Sarabia, J. B., Claverie, A. L., & Urías, M. A. G. (2024). Semi-interpenetrated polymer networks based on PHEMA and modified chitosan as a potential bactericide hydrogel for wound-healing. *Macromolecular Chemistry and Physics, 225*(12), 2400018.

64. Valderruten, N. E., & García, J. (2025). Interpenetrating polymer networks of poly (2-hydroxyethyl methacrylate co-itaconic acid) and chitosan as a controlled release matrix. *Reactive and Functional Polymers, 210*, 106192.
65. Panahi, Y., Gharekhani, A., Hamishehkar, H., Zakeri-Milani, P., & Gharekhani, H. (2019). Stomach-specific drug delivery of clarithromycin using a semi interpenetrating polymeric network hydrogel made of montmorillonite and chitosan: Synthesis, characterization and in vitro drug release study. *Advanced Pharmaceutical Bulletin, 9*(1), 159–173.
66. Rani, M., Agarwal, A., Maharana, T., & Negi, Y. S. (2010). A comparative study for interpenetrating polymeric network (IPN) of chitosan-amino acid beads for controlled drug release. *African Journal of Pharmacy and Pharmacology, 4*(2), 35–54.
67. Ijaz, H., Tulain, U. R., Minhas, M. U., Mahmood, A., Sarfraz, R. M., Erum, A., & Danish, Z. (2022). Design and in vitro evaluation of pH-sensitive crosslinked chitosan-grafted acrylic acid copolymer (CS-co-AA) for targeted drug delivery. *International Journal of Polymeric Materials and Polymeric Biomaterials, 71*(5), 336–348.
68. Kothiya, D., & Vaghani, S. (2023). Fabrication of interpenetrating polymer network-based hydrogel for colon-targeted release of Nateglinide. *Jordan Journal of Pharmaceutical Sciences, 16*(4), 753–769.
69. Suri, S., & Schmidt, C. E. (2009). Photopatterned collagen-hyaluronic acid interpenetrating polymer network hydrogels. *Acta Biomaterialia, 5*(7), 2385–2397.
70. Burmania, J. A., Stevens, K. R., & Kao, W. J. (2003). Cell interaction with protein-loaded interpenetrating networks containing modified gelatin and poly(ethylene glycol) diacrylate. *Biomaterials, 24*(22), 3921–3930.
71. Cascone, M. G., Sim, B., & Downes, S. (1995). Blends of synthetic and natural polymers as drug delivery systems for growth hormone. *Biomaterials, 16*(7), 569–574.
72. Ismail, N. A., & Razali, M. H. (2020). Bio-nanocomposite IPN for Biomedical Application. In S. Jana & S. Jana (Eds.), *Interpenetrating Polymer Network: Biomedical applications* (pp. 313–337). Springer Singapore.
73. Parsa, P., Paydayesh, A., & Davachi, S. M. (2019). Investigating the effect of tetracycline addition on nanocomposite hydrogels based on polyvinyl alcohol and chitosan nanoparticles for specific medical applications. *International Journal of Biological Macromolecules, 121*, 1061–1069.
74. Ianchis, R., Ninciuleanu, C. M., Gifu, I. C., Alexandrescu, E., Somoghi, R., Gabor, A. R., Preda, S., Nistor, C. L., Nitu, S., Petcu, C., Icriverzi, M., Florian, P. E., & Roseanu, A. M. (2017). Novel hydrogel-advanced modified clay nanocomposites as possible vehicles for drug delivery and controlled release. *Nanomaterials (Basel), 7*(12), 443.
75. Banerjee, S., Ray, S., Maiti, S., Sen, K. K., Bhattacharyya, U., Kaity, S., & Ghosh, A. (2010). Interpenetrating polymer network (IPN): A novel biomaterial. *International Journal of Applied Pharmaceutics, 2*(1), 28–34.
76. Bhattacharya, S. S., Mazahir, F., Banerjee, S., Verma, A., & Ghosh, A. (2013). Preparation and in vitro evaluation of xanthan gum facilitated superabsorbent polymeric microspheres. *Carbohydrate Polymers, 98*(1), 64–72.
77. Cona, C., Bailey, K., & Barker, E. (2024). Characterization methods to determine Interpenetrating Polymer Network (IPN) in hydrogels. *Polymers, 16*(14), 2050.
78. Nebhani, L., & Jaisingh, A. (2020). *Chemical analysis of polymers, polymer science and innovative applications* (pp. 69–116). Elsevier.
79. Than-ardna, B., Tamura, H., & Furuike, T. (2022). Preparation, characterization, and properties of Chitosan-based semi-interpenetrating Polymer Networks and Poly(2-hydroxyethyl methacrylate) structure. *Macromolecular Chemistry and Physics, 223*(24), 2200282.
80. Cui, L., Jia, J., Guo, Y., Liu, Y., & Zhu, P. (2014). Preparation and characterization of IPN hydrogels composed of chitosan and gelatin cross-linked by genipin. *Carbohydrate Polymers, 99*, 31–38.
81. Popelka, A., Zavahir, S., & Habib, S. (2020). Morphology analysis. In *Polymer science and innovative applications* (pp. 21–68). Elsevier.

82. M. Nasrollahzadeh, M. Atarod, M. Sajjadi, S.M. Sajadi, Z. Issaabadi, Plant-mediated green synthesis of nanostructures: Mechanisms, characterization, and applications, interface science and technology, Elsevier 2019, pp. 199–322.
83. Suo, H., Zhang, D., Yin, J., Qian, J., Wu, Z. L., & Fu, J. (2018). Interpenetrating polymer network hydrogels composed of chitosan and photocrosslinkable gelatin with enhanced mechanical properties for tissue engineering. *Materials Science and Engineering: C, 92*, 612–620.
84. Falsafi, S. R., Rostamabadi, H., Assadpour, E., & Jafari, S. M. (2020). Morphology and microstructural analysis of bioactive-loaded micro/nanocarriers via microscopy techniques; CLSM/SEM/TEM/AFM. *Advances in Colloid and Interface Science, 280*, 102166.
85. Ratna, D. (2022). Chapter 6—Characterization, performance evaluation and lifetime analysis of thermoset resin. In D. Ratna (Ed.), *Recent advances and applications of thermoset resins* (2nd ed., pp. 503–582).
86. Zoratto, N., & Matricardi, P. (2018). Semi-IPNs and IPN-based hydrogels. In *Polymeric gels* (pp. 91–124).
87. Yin, L., Ding, J., Fei, L., He, M., Cui, F., Tang, C., & Yin, C. (2008). Beneficial properties for insulin absorption using superporous hydrogel containing interpenetrating polymer network as oral delivery vehicles. *International Journal of Pharmaceutics, 350*(1–2), 220–229.
88. Park, S. H., Shin, H. S., & Park, S. N. (2018). A novel pH-responsive hydrogel based on carboxymethyl cellulose/2-hydroxyethyl acrylate for transdermal delivery of naringenin. *Carbohydrate Polymers, 200*, 341–352.
89. Ranjha, N. M., Ayub, G., Naseem, S., & Ansari, M. T. (2010). Preparation and characterization of hybrid pH-sensitive hydrogels of chitosan-co-acrylic acid for controlled release of verapamil. *Journal of Materials Science: Materials in Medicine, 21*, 2805–2816.
90. Dong, X., Wei, C., Liang, J., Liu, T., Kong, D., & Lv, F. (2017). Thermosensitive hydrogel loaded with chitosan-carbon nanotubes for near infrared light triggered drug delivery. *Colloids and Surfaces B: Biointerfaces, 154*, 253–262.
91. Balan, K. E., Boztepe, C., & Künkül, A. (2022). Modeling the effect of physical crosslinking degree of pH and temperature responsive poly(NIPAAm-co-VSA)/alginate IPN hydrogels on drug release behavior. *Journal of Drug Delivery Science and Technology, 75*, 103671.
92. Wang, Y., Zhang, Y., Zhong, H., Guo, M., Chen, X., & Lu, Y. (2025). Construction of a non-toxic interpenetrating network hydrogel drug carrier supported by carbon microspheres and nanocellulose. *Carbohydrate Polymers, 350*, 123035.
93. Shakouripour, F., Olad, A., & Bayramoglu, G. (2025). Preparation of interpenetrating networks from chitosan and poly(hydroxypropyl methacrylate) or p(hydroxyethyl methacrylate) for controlled release of doxorubicin and curcumin: Investigation of potential use in wound dressing. *International Journal of Biological Macromolecules, 301*, 140929.
94. Khodaverdi, K., Naghib, S. M., Mozafari, M. R., & Rahmanian, M. (2025). Chitosan/hydroxyapatite hydrogels for localized drug delivery and tissue engineering: A review. *Carbohydrate Polymer Technologies and Applications, 9*, 100640.
95. Hong, F., Qiu, P., Wang, Y., Ren, P., Liu, J., Zhao, J., & Gou, D. (2024). Chitosan-based hydrogels: From preparation to applications, a review. *Food Chemistry: X, 21*, 101095.
96. Dodero, A., Scarfi, S., Mirata, S., Sionkowska, A., Vicini, S., Alloisio, M., & Castellano, M. (2021). Effect of crosslinking type on the physical-chemical properties and biocompatibility of Chitosan-based electrospun membranes. *Polymers, 13*(5), 831.
97. Jiménez-Gómez, C. P., & Cecilia, J. A. (2020). Chitosan: A natural biopolymer with a wide and varied range of applications. *Molecules, 25*(17), 3981.
98. Jiang, A., Patel, R., Padhan, B., Palimkar, S., Galgali, P., Adhikari, A., Varga, I., & Patel, M. (2023). Chitosan based biodegradable composite for antibacterial food packaging application. *Polymers, 15*(10), 2235.
99. Nandini, T., Sudhalakshmi, C., Sivakumar, K., Parameswari, E., & Thangamani, C. (2025). A review-chitosan nanoparticles towards enhancing nutrient use efficiency in crops. *International Journal of Biological Macromolecules, 306*, 141433.

100. Yarahmadi, A., Dousti, B., Karami-Khorramabadi, M., & Afkhami, H. (2024). Materials based on biodegradable polymers chitosan/gelatin: A review of potential applications. *Frontiers in Bioengineering and Biotechnology, 12*, 1397668.
101. Morello, G., De Iaco, G., Gigli, G., Polini, A., & Gervaso, F. (2023). Chitosan and pectin hydrogels for tissue engineering and in vitro modeling. *Gels, 9*(2), 132.
102. Kim, Y., Zharkinbekov, Z., Raziyeva, K., Tabyldiyeva, L., Berikova, K., Zhumagul, D., Temirkhanova, K., & Saparov, A. (2023). Chitosan-based biomaterials for tissue regeneration. *Pharmaceutics, 15*(3), 807.
103. Fares, M. M., Radaydeh, S. K., & Jabani, Z. H. (2024). IPN based hydrogels for in-vivo wound dressings; catalytic wound healing dynamics and isothermal adsorption models. *Journal of Photochemistry and Photobiology B: Biology, 254*, 112901.
104. Guo, B.-L., & Gao, Q.-Y. (2007). Preparation and properties of a pH/temperature-responsive carboxymethyl chitosan/poly(N-isopropylacrylamide) semi-IPN hydrogel for oral delivery of drugs. *Carbohydrate Research, 342*(16), 2416–2422.
105. Jana, S., Gandhi, A., & Sen, K. (2019). Chitosan-based interpenetrating polymer networks: Drug delivery application. In S. Jana & S. Jana (Eds.), *Functional Chitosan* (pp. 269–295). Springer.
106. Lohani, A., Saxena, R., Khan, S., & Mascarenhas-Melo, F. (2024). pH-responsive IPN beads of carboxymethyl konjac glucomannan and sodium carboxymethyl cellulose as a controlled release carrier for ibuprofen. *International Journal of Biological Macromolecules, 278*, 134676.

Polyelectrolyte Complexes of Chitosan as Potential Drug Delivery Systems

Ljiljana Đekić and Ana Ćirić

Abstract Chitosan-based polyelectrolyte complexes (CH PECs) have attracted significant attention as multifunctional carriers for advanced drug delivery. Formed through non-covalent interactions between chitosan (CH) and anionic polyelectrolytes, these systems offer advantages such as biocompatibility, biodegradability, mucoadhesion, and pH-responsive drug release. This chapter provides an in-depth overview of CH PECs, focusing on their formation mechanisms, physicochemical properties, preparation techniques, and characterization methods. The influence of critical parameters such as the degree of deacetylation, molecular weight, polymer ratios, and pH on complex stability, drug encapsulation efficiency, and release kinetics is thoroughly discussed. Applications across various administration routes, including oral, buccal, ocular, nasal, pulmonary, dermal, transdermal, vaginal, and parenteral delivery, are reviewed, highlighting the ability of CH PECs to enhance drug solubility, prolong therapeutic action, and improve bioavailability. The chapter also addresses translational challenges related to large-scale manufacturing, regulatory requirements, and in vivo performance. Overall, CH PECs represent a promising platform for the development of next-generation drug delivery systems capable of overcoming limitations of conventional formulations and facilitating clinical translation.

Keywords Polyelectrolyte complexes · Chitosan · Anionic polyelectrolytes · Enhanced drug solubility · pH-responsive drug release · Advanced drug delivery

1 Introduction

Drug delivery is a crucial aspect of pharmaceutical research and development with the aim of improving the efficacy, safety, and bioavailability of drugs. Many drugs, especially poorly soluble and highly labile compounds, face significant challenges in conventional formulations, such as rapid metabolism, poor absorption, and

L. Đekić (✉) · A. Ćirić
University of Belgrade – Faculty of Pharmacy, Belgrade, Serbia
e-mail: ljiljana.djekic@pharmacy.bg.ac.rs

S. Jana (ed.), *Chitosan-Based Drug Delivery Systems*, AAPS Introductions in the Pharmaceutical Sciences 16, https://doi.org/10.1007/978-3-032-08347-0_3

non-specific distribution. Inefficient drug delivery can lead to reduced therapeutic efficacy, requiring higher doses and more frequent administration, increasing the risk of systemic side effects and patient non-adherence. Advanced drug delivery systems (DDS) seek to overcome these limitations by incorporating controlled release technologies, site-specific targeting, and improved solubility strategies [27]. Among the new solutions, polyelectrolyte complexes (PECs) have gained prominence due to their ability to encapsulate drugs, improve stability, and provide responsive drug release mechanisms [13].

CH is a natural, cationic polysaccharide and has unique properties that make it particularly suitable for pharmaceutical and biomedical applications [20]. As a biocompatible and biodegradable polymer, CH exhibits low toxicity, making it safe for various routes of drug delivery [41]. One of its most valuable properties is mucoadhesion, which allows it to interact electrostatically with negatively charged mucosal surfaces, thereby prolonging drug retention time and improving absorption [36]. However, the solubility of CH is highly pH-dependent, which limits its independent application in neutral and alkaline environments. To address this limitation, CH is often combined with anionic polyelectrolytes to form stable PECs with improved solubility, enhanced structural integrity, and tunable drug release properties [8].

Polyelectrolytes are charged macromolecules that form self-assembled structures through electrostatic interactions in aqueous solutions. They are divided into cationic polyelectrolytes (e.g., chitosan), anionic polyelectrolytes (e.g., alginate, pectin, hyaluronic acid), and ampholytic polyelectrolytes, which carry both positive and negative charges (e.g., proteins) [6, 10, 33]. When oppositely charged polyelectrolytes interact with each other, they form stable PECs without the need for toxic crosslinkers, ensuring biocompatibility and improved drug delivery [8]. PECs can improve the efficiency of drug encapsulation and allow both hydrophilic and hydrophobic drugs to be incorporated into various types of drug carriers such as nanoparticles, microparticles, nanofibers, hydrogels, and films [6].

Among the various PEC systems, the CH PECs have proven to be particularly valuable for the pharmaceutical sciences. Their pH-responsive nature allows for site-specific and controlled drug release, making them ideal for oral, transdermal, ophthalmic, and injectable formulations [68]. In oral drug delivery, e.g., CH PECs have shown promise in protecting drugs from degradation in the stomach and releasing them in the intestine, where pH-induced swelling facilitates drug diffusion [16, 29]. In addition, these complexes can improve mucosal adhesion and thus drug absorption in nasal, buccal, and ocular DDS [68]. From a broader perspective, CHPEC-based hydrogels have also been explored for tissue engineering and wound healing, providing mechanical support and bioactivity for regenerative medicine applications [9].

This chapter provides a comprehensive overview of CH PECs as DDS, focusing on their formation mechanisms, physicochemical properties, and biomedical applications. The chapter addresses the key properties of CH relevant to drug delivery, mechanisms of PEC formation, characterization techniques, drug loading and release behavior, and their application in various routes of administration. By addressing these aspects, this chapter aims to demonstrate the potential of CH PECs to improve therapeutic efficacy, overcome drug solubility issues, and promote site-specific drug delivery strategies.

2 Properties of Chitosan Relevant to Drug Delivery

CH is a cationic polysaccharide derived from the deacetylation of chitin, a natural biopolymer found in crustacean shells, fungi, and insect exoskeletons. Structurally, it consists of β-(1 → 4)-linked D-glucosamine and N-acetyl-D-glucosamine units with amine ($-NH_2$) and hydroxyl (–OH) functional groups (Fig. 1), which contribute to its biological activity and its ability to form PECs. The degree of deacetylation (DDA) determines solubility, charge density, and mucoadhesive behavior, with a higher DDA value (> 75%) leading to stronger cationic properties and better interactions with anionic molecules. Another key factor influencing the properties of CH is its molecular weight (MW), which significantly affects solubility, viscosity, mechanical strength, and biological activity. These properties make CH a versatile biomaterial for drug delivery [20, 41].

The physicochemical properties of CH (DDA and MW) have a significant impact on drug delivery performance and interactions with biological systems. CH is soluble under acidic conditions (pH <6.5) due to the protonation of its amino groups ($-NH_3^+$) but becomes insoluble at neutral and alkaline pH. This pH-dependent solubility is advantageous for the site-specific release of drugs, especially in gastro-intestinal (GI) and mucosal drug delivery. The solubility of CH is strongly influenced by DDA and MW; higher DDA improves solubility, while low molecular weight (LMW) CH has better water solubility compared to high molecular weight

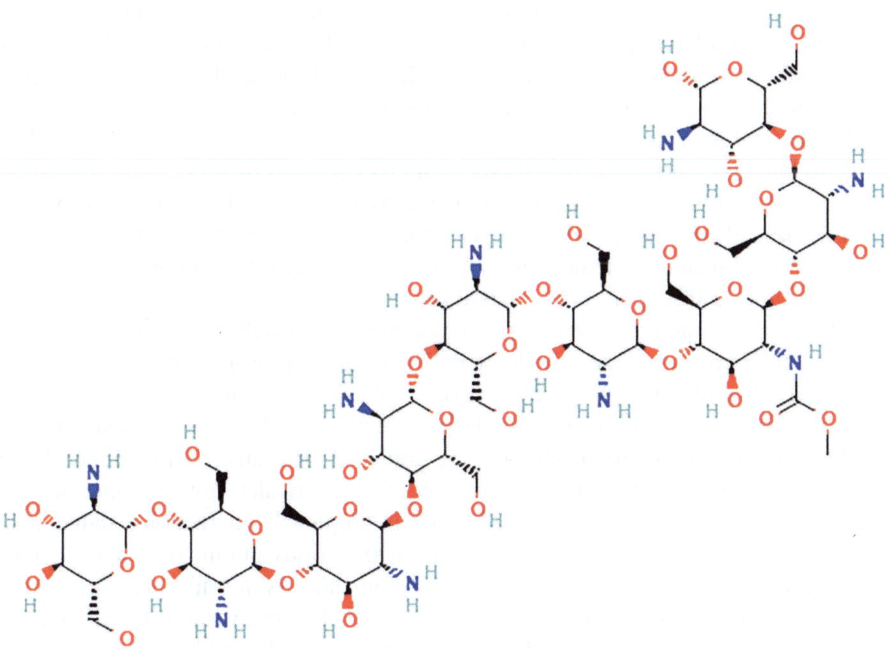

Fig. 1 Chemical structure of CH. (https://pubchem.ncbi.nlm.nih.gov/compound/71853)

(HMW) forms [20]. The cationic charge of CH allows it to interact with negatively charged mucosal surfaces, improving drug retention time and absorption. HMW CH ($>$300,000 gmol^{-1}) exhibits stronger mucoadhesion due to increased chain entanglement, making it ideal for oral, nasal, buccal, and ophthalmic formulations [12, 20]. CH is biodegradable due to enzymatic degradation by lysozymes to non-toxic oligosaccharides, making it suitable for sustained release drug formulations. However, the rate of degradation depends on MW; LMW CH is degraded faster, while HMW CH resists enzymatic degradation and thus allows a longer retention time in vivo [20, 41]. CH modulates tight junctions in epithelial cells and thus improves drug transport across mucosal barriers. This property is also MW-dependent, with LMW CH ($<$50,000 gmol^{-1}) showing better penetration ability due to its smaller size and higher diffusion rates, while HMW CH ($>$150,000 gmol^{-1}) exhibiting stronger bioadhesion but lower permeability [7, 20]. CH is classified as generally recognized as safe (GRAS) by the Food and Drug Administration (FDA), which makes it biocompatible and non-toxic for pharmaceutical applications [34]. Importantly, DDA affects immunogenicity; CH with a higher DDA ($>$85%) has a lower immunogenic potential and is therefore more suitable for systemic and injectable formulations [5]. CH exerts its antimicrobial effect via electrostatic interactions with bacterial membranes, leading to membrane disruption and bacterial death [28]. LMW CH ($<$10,000 gmol^{-1}) penetrates bacterial cells more efficiently, whereas HMW CH ($>$300,000 gmol^{-1}) forms a protective polymer barrier on surfaces, which is useful for wound healing and antimicrobial coatings [20]. CH-based DDS enables controlled, targeted, and sustained drug release, protects encapsulated drugs from degradation and improves bioavailability [39]. Generally, these properties make CH-based formulations highly promising for different routes of administration, including oral, transdermal, ocular, and parenteral drug delivery. However, to overcome short comings of using CH as an individual polymer in the development of DDS, such as pH-dependent solubility, rapid degradation, low mechanical strength, poor compatibility with hydrophobic drugs, rapid degradation, limited mucoadhesion, and potential immunogenicity, chemical modifications or the combination of CH with other polymers, including the formation of PECs, are often undertaken.

The combination of CH with different polymers is a widely researched strategy to improve their properties for drug delivery. Two main approaches are being pursued: chemical crosslinking (copolymerization) and physical crosslinking. Copolymerization leads to the formation of new chemical entities with unknown toxicological profiles, making physical crosslinking the preferred alternative. Three-dimensional polymer networks, such as PECs, are based on non-covalent interactions, such as electrostatic forces, hydrogen bonding, and hydrophobic interactions, rather than irreversible chemical bonds. CH forms PECs with various anionic polymers, including alginate, pectin, carrageenan, xanthan gum, and hyaluronic acid, providing improved drug release control, stability, and bioavailability without the risks associated with toxic crosslinkers, such as formaldehyde and glutaraldehyde [19, 20, 31].

3 Formation of Chitosan-Based Polyelectrolyte Complexes

PECs are formed by non-covalent interactions between oppositely charged polyelectrolytes, leading to the self-assembly of stable, biodegradable, and biocompatible structures. These complexes play a crucial role in drug delivery as they can improve drug encapsulation, control release, and protect sensitive bioactive compounds [31]. Key forces responsible for PEC formation include ionic interactions (the primary mechanism involves electrostatic attraction between positively charged CH and negatively charged polyelectrolyte), hydrogen bonding (secondary interactions contribute to the stabilization of PECs, especially in polysaccharide-based complexes where hydroxyl and carboxyl groups form hydrogen bonds), van der Waals forces and hydrophobic interactions (weak intermolecular forces influence the morphology of PECs and affect drug encapsulation efficiency, swelling behavior and mechanical properties) [11, 19].

CH interacts with a variety of anionic polyelectrolytes and forms PECs with different structural and functional properties [13]. Some of the anionic polymers that can form PECs with CH are alginate [61], pectin [69], carrageenan [21], hyaluronate [26], xanthan gum [15, 16, 43], and carboxymethylcellulose [29].

The formation and stability of PECs depend on several physicochemical factors, including pH, concentration of polyelectrolytes, charge density, ionic strength, mixing ratio of the polymers, order, and rate of polymer mixing. The pH influences the ionization state of the CH and polyanions and thus the stability of the complex and drug release properties. The optimal pH range for the formation of CH PECs is 4–6, where both polymers remain partially charged and crosslinked. Polyelectrolyte concentration can affect the efficiency of complex formation (higher polymer concentrations increase PEC formation but can lead to aggregation or phase separation). Polyelectrolytes with higher charge density (i.e., CH with high DDA) also exhibit stronger ionic interactions, which increase PEC stability and drug loading efficiency. A high ionic strength (e.g., due to added salts) can shield electrostatic interactions, reduce PEC formation, and destabilize already formed complexes. The mixing ratio of the polymers (stoichiometry) can strongly influence the PEC properties. For example, stoichiometric complexes (1:1 charge ratio) tend to form precipitates with high rigidity and low solubility, while non-stoichiometric PECs are often more water-soluble and mucoadhesive. The order and rate of polymer mixing affect the uniformity, particle size, and colloidal stability of PECs. A slow dropwise addition generally favors a controlled assembly [54, 68].

PECs can be prepared using various techniques, which usually involve mixing oppositely charged polymer solutions under controlled physicochemical conditions. One of these techniques is *solvent casting*, in which CH and the anionic polyelectrolyte are dissolved separately, mixed under controlled conditions, and cast onto a substrate to form thin PEC films [57]. During in situ *gel formation*, PECs form spontaneously and generate self-assembled hydrogels [60]. Using the *coacervation technique*, PECs are precipitated from solution at controlled ionic strength and pH (commonly used in nanoparticles, microparticles, and encapsulated drug formulations) [25]. In *layer-by-layer (LbL) assembly*, successive cationic and anionic layers are deposited to build multilayer PECs [43].

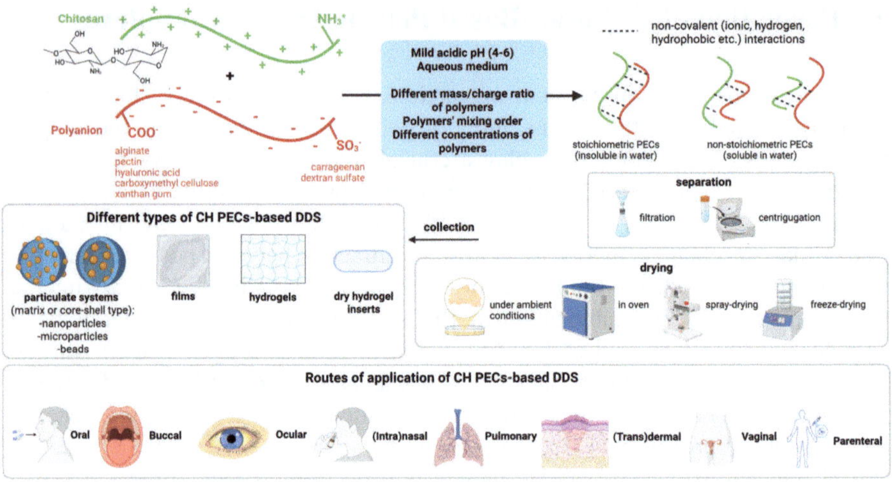

Fig. 2 Formation of CH PECs and the common types of CH PECs-based DDS studied for different routes of application. (Created in https://BioRender.com)

CH PECs-based DDS enable versatile drug delivery platforms, including: hydrogels [21, 49], tablets [29], beads [36], microparticles/nanoparticles [12, 15, 16, 46, 63, 66], films [43, 57], and nanofibers [61].

Figure 2 shows schematically the formation of CH PECs and the common types of CH PECs-based DDS studied for different routes of application.

4 Characterization of Chitosan-Based Polyelectrolyte Complexes

Physicochemical characterization of CH PECs is crucial for determining their stability, drug loading capacity, and release behavior. For CH PECs in the form of nano–/microparticles and beads, the particle size is a key parameter influencing drug diffusion, bioavailability, and cellular uptake. It is usually measured using dynamic light scattering (DLS), which provides hydrodynamic diameter and polydispersity index data to assess the uniformity of size distribution [8]. The zeta potential is used to evaluate the surface charge and colloidal stability of PECs in particulate form. A high absolute zeta potential ($> \pm 30$ mV) usually indicates good dispersion stability, which prevents particle aggregation and improves drug transport in biological fluids [66]. Scanning electron microscopy (SEM) is often used for morphological analysis. SEM provides a detailed surface topography that indicates whether the PECs have a smooth, porous, or rough texture [16]. Fourier transform infrared spectroscopy (FTIR) is one of the most important techniques for detecting interactions between functional groups in PECs. The shift or disappearance of characteristic absorption peaks indicates the formation of ionic bonds between CH and anionic polyelectrolytes [16]. In particular, the peaks of amide I (1650 cm^{-1}) and amide II (1560 cm^{-1}) in CH are shifted or broadened when complexed with carboxyl (–COO^{-}) or sulfate (–SO_3^{-}) groups in anionic polyelectrolytes [12]. Differential scanning calorimetry (DSC) measures thermal transitions such as melting temperature (Tm) and glass transition temperature (Tg), which provide information about complex stability and polymer–polymer interactions. A shift in Tg indicates successful polyelectrolyte complexation, which leads to structural changes in the PEC matrix. Thermogravimetric analysis (TGA) evaluates thermal stability by measuring weight loss as a function of temperature. The multistage degradation pattern observed in the PECs indicates different thermal decomposition phases, confirming the presence of separate polymeric components within the complex [56].

Figure 3 illustrates the importance of physicochemical characterization of CH PECs for evaluating and understanding the influence of different formulation parameters, such as mass and charge mixing ratios of polymers.

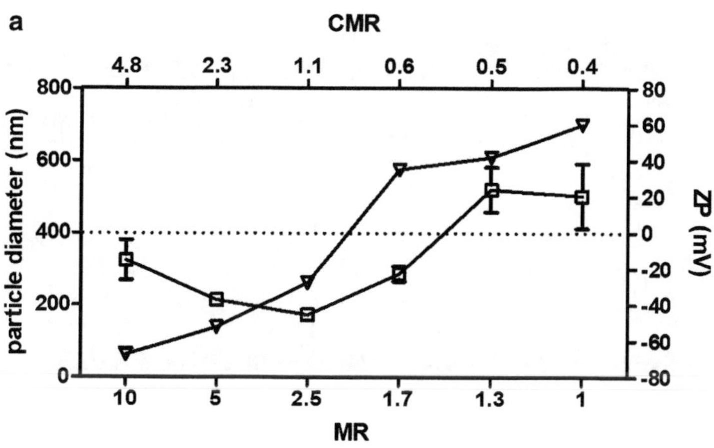

Fig. 3 Examples of physicochemical characterization of CH PECs: (**a**). Particle diameter (open rectangles) and zeta potential (ZP) (open triangles) in relation to mass mixing ratio (MR) and charge mixing ratio (CMR) of the ratio of hyaluronic acid (HA)/chitosan (CS). Data expressed as mean ± SD ($n = 3$). *Reprinted from Sladek* et al., *"An Enteric-Coated Polyelectrolyte Nanocomplex Delivers Insulin in Rat Intestinal Instillations When Combined with a Permeation Enhancer," Pharmaceutics 2020, 12(3), 259, under the terms of the Creative Commons Attribution (CC BY 4.0) license* (https://creativecommons.org/licenses/by/4.0/). © *2020 by the authors*; (**b**). SEM images of CH/pectin (A), CH/carrageenan (B) and CH/polycarbophil (C) PEC-based hydrogel inserts for vaginal drug delivery, 1000 × 10,000 magnification. *Reprinted from Deshkar* et al., *"Polyelectrolyte-Complex-Based Hydrogel Inserts for Vaginal Delivery of Posaconazole and Probiotics," Gels 2023, 9(11), 851, under the terms of the Creative Commons Attribution (CC BY 4.0) license* (https://creativecommons.org/licenses/by/4.0/). © *2023 by the authors*; (**c**). DSC thermograms for pectin (PC), chitosan (CS), clotrimazole (CLO), CLO-loaded PC/CS PEC-based films (F1–F3), placebo films (PF), and the physical mixture of PC and CS at a weight ratio of 2:1. *Reprinted from Potaś* et al., *"Multilayer Films Based on Chitosan/Pectin Polyelectrolyte Complexes as Novel Platforms for Buccal Administration of Clotrimazole," Pharmaceutics 2021, 13(10), 1588, under the terms of the Creative Commons Attribution (CC BY 4.0) license* (https://creativecommons.org/licenses/by/4.0/). © *2021 by the authors*

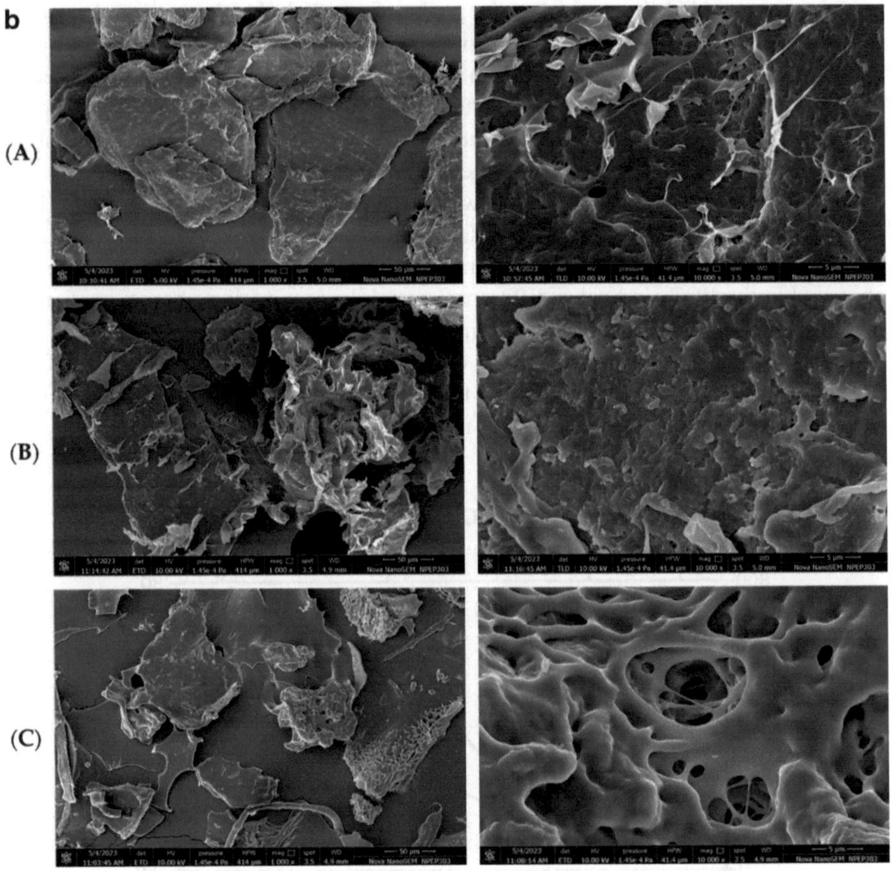

Fig. 3 (continued)

5 Drug Loading and Release Mechanisms

The method by which the drugs are incorporated into PECs has a significant impact on encapsulation efficiency, stability, and release kinetics. There are two main strategies, *pre-loading* and *post-loading* (Fig. 4). In the *pre-loading* method, the drug is incorporated during PEC formation, ensuring homogeneous distribution within the polymer matrix. This approach is widely used for microparticles, nanoparticles, and hydrogels, as it improves the interaction between the drug and

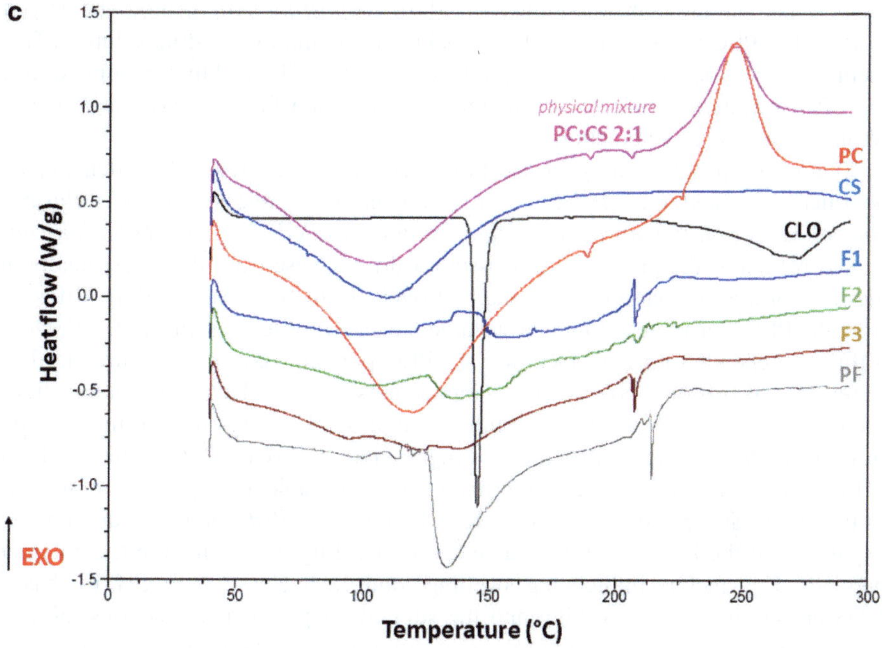

Fig. 3 (continued)

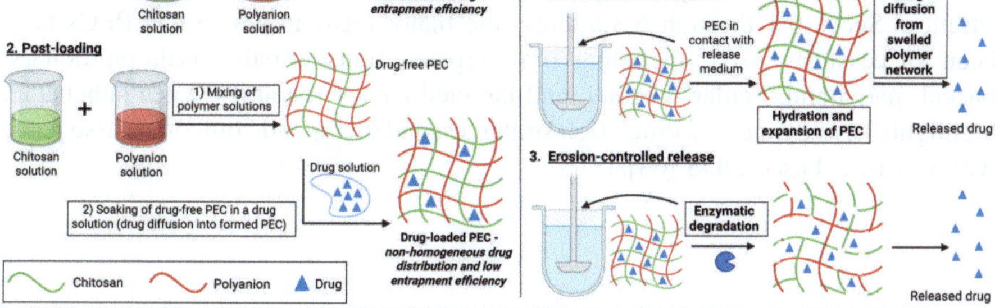

Fig. 4 Schematic representation of the main drug loading and drug release mechanisms from CHPECs-based DDS. (Created in https://BioRender.com)

the polymer and prevents burst release [15]. In the *post-loading* method, the drug is loaded after PEC formation, usually by soaking the complex in a drug solution. This method allows for greater flexibility in loading hydrophilic and hydrophobic drugs, but often results in lower encapsulation efficiency and faster release rates due to weak polymer–drug interactions [24].

The release profile of drug-loaded PECs depends on the crosslinking density, the polymer ratio, and the pH of the environment. Drug release can be *diffusion-controlled, swelling-controlled,* and *erosion-controlled* (Fig. 4). *Diffusion-controlled release* is the primary mechanism for drug release from PECs, especially in the form of micro−/nanoparticles and hydrogels. It involves passive drug diffusion from the PEC matrix, resulting in a sustained release profile over time [14, 29]. Drug diffusion depends on many factors, e.g., PEC porosity (porous structures lead to faster drug diffusion), polymer-drug interactions (strong electrostatic interactions prolong drug release), drug hydrophilicity (hydrophilic drugs diffuse more easily from CH PECs than hydrophobic drugs), and pH (CH PECs often exhibit pH-dependent drug release behavior, making them suitable for targeted formulations in the GI tract) [15, 16, 26, 40, 67]. In *swelling-controlled release*, hydration and expansion of the PEC structure enable controlled drug diffusion from the polymer network. Key factors influencing swelling-controlled release from CHPEC-based DDS are the degree of swelling and the strength of polymer interactions [50]. In *erosion-controlled release*, drug release is controlled by the gradual erosion of the polymer matrix, which may be the result of weakening of PEC bonds or enzymatic degradation (by enzymes such as lysozyme, making them suitable for colon-targeted drug delivery) [30].

6 Applications of Chitosan Polyelectrolyte Complexes in Drug Delivery

CH PECs have gained increasing attention as versatile platforms for a wide range of drug delivery applications. Their biocompatibility, mucoadhesive properties, tunable stability, and ability to provide controlled and site-specific drug release make them suitable candidates for improving therapeutic efficacy via various routes of administration. Sects. 6.1 through 6.8 address the major areas in which CH PECs have been successfully used, including oral, topical, transdermal, nasal, pulmonary, buccal, parenteral, ocular, vaginal, and targeted cancer therapies. Each subchapter highlights key advances, formulation strategies, and therapeutic outcomes associated with CH PEC-based DDS [68].

In Table 1 are summarized the composition, preparation, and drug delivery potential of CH PECs as DDS for different routes of administration.

6.1 Oral Drug Delivery

CH PECs have emerged as promising carriers for oral drug delivery, addressing critical challenges such as low bioavailability, enzymatic degradation, and poor intestinal permeability of drugs [37]. The mucoadhesive properties of CH enable longer retention time of DDS in the GI tract, while its ability to open tight junctions enhances the paracellular transport of macromolecules. In addition, CH PECs can enable pH-sensitive drug release, ensuring the protection of labile drugs under acidic gastric conditions and targeted release in the intestine [37, 64]. As a versatile excipient, CH enables the development of various oral DDS, including hydrogels, microparticles, nanoparticles, and gastroretentive forms suitable for both conventional small-molecule drugs and biologics such as proteins, peptides, and vaccines [13].

An important application of CH PECs is the improvement of the oral bioavailability of drugs. Studies have shown that CH complexes with polyanions such as alginate [61], hyaluronic acid [22], and pectin [69] improve drug solubility and/or stability, leading to improved absorption. In addition, CH/xanthan gum PECs have been shown to improve the controlled release of ibuprofen, with the potential of mitigating GI irritation while maintaining therapeutic efficacy [14, 15].

CH PECs have also been used in gastroretentive DDS. For example, Samanta et al. [60] developed CH/carboxymethyl tamarind gum PEC-based floating capsules for gastroretentive (stomach-specific) delivery of ofloxacin, prolonging gastric residence time (in vitro floating behavior in simulated gastric fluid pH 1.2 over 6 h, without floating lag-time) and achieving sustained drug release (over 10 h).

Nanotechnology-based approaches have further expanded the use of CH PECs in oral drug delivery. The use of electrospun CH-based nanofibers has shown improved solubility and dissolution rate of poorly water-soluble drugs. Schoeller et al. [61] investigated the possibility of surface functionalization of electrospun nanofibers with CH/alginate PEC. Specifically, layers of CH and alginate were sequentially deposited on electrospun poly (lactic-co-glycolic acid) (PLGA) nanofibers to form a polyelectrolyte complex via layer-by-layer assembly and achieve pH sensitivity for the controlled release of ibuprofen. The deposition of CH/alginate PEC-based layers resulted in slowing down the release of ibuprofen in acidic pH (1.0) compared to neutral pH due to the interactions between the drug and the PEC-based coating. The prepared functionalized nanofibers were considered promising to bypass the acidic environment of the GI tract. Recently, nanogels based on CH complexed with hyaluronic acid, carboxymethylpullulan, or their derivatives grafted with aminoguaiacol (with different degrees of substitution) were prepared and evaluated

Table 1 Composition, preparation, route of administration, and drug delivery potential of the selected investigated examples of DDS based on CH PECs

Polyanion	Drug	PEC type/ dosage form	Preparation method	Use/route of administration	Outcome of drug entrapment	Reference
Alginate	Ibuprofen	PEC-based coating of PLGA nanofibers	Layer-by-layer assembly + oven drying (40 °C)	Oral	pH-sensitive and controlled drug release	Schoeller et al., [61]
Pectin/thiolated pectin	5-fluorouracil	Beads	Ionotropic gelation + oven drying (37 °C)	Oral (colon-targeted)	pH-sensitive drug release, cytotoxicity to HT29 colorectal cancer cells	Leonard et al., [36]
Sodium cellulose sulfate	Budesonide	PEC-based coating of solid lipid nanoparticles	Layer-by-layer assembly	Oral (colon-targeted)	pH-sensitive drug release, good anti-inflammatory effect in mice	Zhang et al., [70]
Carboxymethyl tamarind gum	Ofloxacin	Floating capsules	In situ gelling	Oral (gastroretentive)	Prolonged gastric residence (in vitro and in vivo), sustained drug release	Samanta et al., [60]
Xanthan gum	Ibuprofen	Microparticles	Complex coacervation + ambient drying or spray drying	Oral	pH-sensitive drug release, high drug entrapment	Ćirić et al., [15]
Hyaluronic acid (+Eudragit® L-100 coating)	Insulin	Nanoparticles	Complex coacervation + freeze drying	Oral	Increased intestinal permeability (with the addition of permeation enhancer), pH-sensitive release	Sladek et al., [64]
Alginate and poly (lactide-co-glycolide)	MAP antigen	Nanoparticles	Ionic complexation	Oral (vaccine)	Enhanced T-cell response and IgA secretion in mice, pH-sensitive release	Liu et al., [38]
Xanthan gum	Escin	Microparticles	Complex coacervation + ambient drying or spray drying	Oral	Sustained and pH-dependent release, high entrapment	Ćirić et al., [16]

Pectin	Clotrimazole	Multilayer films	Solvent casting + oven drying (@ 30–35 °C)	Buccal	pH-responsive release, good mucoadhesion, antifungal, and safety profile	Potaś et al., [55]
Gum Arabic or Polygalacturonic acid	Nystatin	Films	Solvent casting + oven drying (@ 40 ° C)	Buccal	Sustained release, good mucoadhesion, antifungal activity on *Candida* sp.	Armenta-Rojas et al., [3]
Alginate	Clindamycin phosphate	Films	Solvent casting + ambient drying	Buccal	Good swelling and mucoadhesion, delayed drug release	Kilicarslan et al., [32]
Pectin	Brinzolamide	Nanocapsules	Complex coacervation	Ocular	Sustained reduction of intra-ocular pressure in rabbits, enhanced permeation through the cornea	Dubey et al., [25]
Hyaluronate	Dexamethasone phosphate	Nanoparticles	Ionic gelation + freeze drying	Ocular	Excellent mucoadhesion and prolonged release, good anti-inflammatory effect (in vitro)	Dubashynskaya et al., [23]
κ-Carrageenan	Sumatriptan succinate	Inserts	Ionic complexation + freeze-drying	Intranasal	Controlled release, high mucoadhesion	Alavi and Mortazavi, [1]
Pectin	Levodopa + carbidopa	Nanoparticles	Ionic gelation + freeze-drying	Intranasal	Enhanced mucoadhesion, sustained release	Avhad et al., [4]
Xanthan gum	Rifampicin	Microparticles (dry powder inhaler)	Complex coacerva-tion + spray drying	Pulmonary	pH-responsive release appro-priate for the specific lung uptake of drug	Surini et al., [65]
Alginate, sodium car-boxymethylcellulose or polyvinylpyrrolidone K90	Doxycycline + levofloxacin	Microparticles (dry powder inhaler)	Ionotropic gelation + spray drying	Pulmonary	Sustained co-delivery of drugs, high encapsulation efficiency, no cytotoxicity on H1299 cells, improved delivery to lung	Mishra et al., [45]
κ-Carrageenan	Naproxen	Nanoparticles	Ionotropic gelation + freeze-drying	Transdermal	Sustained and improved per-meation through rat skin, no irritation, reduced inflammation	Noreen et al., [47]

(continued)

Table 1 (continued)

Polyanion	Drug	PEC type/ dosage form	Preparation method	Use/route of administration	Outcome of drug entrapment	Reference
Casein (sodium salt)	Quercetin	Nanoparticles	Ionotropic gelation + freeze-drying	Topical	Prolonged release, suitable for wound healing	Milenkova et al., [44]
Phytic acid	Metronidazole	Nanoparticles	Ionic gelation + freeze-drying	Vaginal	Mucoadhesion, prolonged release and enhanced antibacterial activity against *E. Coli*	Nayak et al., [46]
Carrageenan, pectin, or polycarbophil	Posaconazole + *L. casei*	Hydrogel inserts	Ionic gelation + freeze-drying	Vaginal	Mucoadhesion, sustained drug release, good antifungal and probiotic effect—Reduced *C. Albicans*	Deshkar et al., [21]
Poly(NIPAm-co-AMPS)	Aspirin	Hydrogel	In situ gelation + freeze-drying	Parenteral (injections into tissues)	Thermosensitive (sol-gel) DDS with sustained drug release, no cytotoxicity on L929 cells	Zhao et al., [71]
Pectin	Temozolomide	Nanoparticles	Ionic gelation	Parenteral (brain tumor)	Improved penetration through the blood-brain barrier and tumor affinity (in vitro determined)	Silant'ev et al., [63]
Hyaluronate	Sorafenib	Microparticles	Layer-by-layer deposition (coating) on bovine serum albumin gel core	Parenteral (hepatocellular carcinoma)	Good encapsulation efficiency, sustained drug release, cytotoxicity on HepG2 and mesenchymal stem cells	Paşcalău et al., [52]

as potential DDS for oral administration of ibuprofen to prevent wound infection and reduce tissue inflammation. All PECs were found to be stable at 4 °C and pH 5 (acetate buffer) for 28 days. In phosphate buffer saline (PBS) pH 7.4, better stability was observed at 4 °C for PECs based on carboxymethylpullulan-aminoguaiacol graft derivatives, which was explained by hydrophobic interactions between the aminoguaiacol groups. All investigated PECs exhibited antioxidant activity (detected by the DPPH method). Diclofenac and curcumin were used as model drugs. The release of diclofenac in buffers simulating the GI conditions occurred only at pH 5.8 and 6.6, while the drug was not released at pH 2. These results indicate that the investigated PECs, especially those prepared with carboxymethylpullulan or carboxymethylpullulan-aminoguaiacol-grafted derivatives, can be used for intestinal drug release. The release of curcumin from PECs in PBS pH 7.4/ethanol (60/40 V/V) was found to slowly increase linearly with increasing degree of aminoguaiacol substitution (the presence of aminoguaiacol moieties led to the formation of hydrophobic clusters in the PECs and facilitated the release of curcumin). The expected antibacterial activity against *S. aureus* was difficult to detect, probably due to the low concentration of PECs in the suspensions used in this test to prevent bacterial growth. The authors concluded that the investigated PECs could be used as promising DDS in wound care to prevent wound infections and reduce tissue inflammation [26].

Colon-targeted drug delivery following oral administration has been shown to be an effective strategy for the treatment of colonic diseases, as it allows for more localized drug delivery with fewer systemic side effects. CH PECs have been evaluated for their mucoadhesive properties, controlled drug release, and ability to target the drug to the colorectal region. These formulations prolong mucosal residence time and minimize systemic absorption, making them promising carriers for colorectal cancer therapy and the treatment of inflammatory bowel disease [36, 70]. Colorectal cancer remains an important health problem that necessitates the development of targeted DDS to improve therapeutic efficacy while minimizing toxicity. Leonard et al. [36] developed thiolated pectin/CH composite PEC beads for the targeted delivery of 5-fluorouracil following oral administration to the colon. These mucoadhesive beads were prepared using a triple crosslinking mechanism (calcium pectinate, polyelectrolyte complexation, and disulfide bonds) to improve adhesion to the colonic mucosa and enable controlled drug release for 180 min, with up to 85% of the drug released during this period. The thiolated composite (using thiolated pectin instead of unmodified pectin) exhibited superior mucoadhesion, a crucial factor in increasing drug retention time at the tumor site, and showed selective cytotoxicity against HT29 colorectal cancer cells with minimal effects on normal HEK293 cells. In addition, the study highlighted the advantages of using CH/pectin PECs due to their structural stability and digestibility in the colon, making them ideal candidates for targeted drug delivery to colorectal cancer cells. The formation of PEC between CH and thiolated pectin improved the mechanical integrity of the formulation while ensuring site-specific drug release. The mucoadhesive properties of thiolated pectin prolonged drug retention time and improved therapeutic outcomes by increasing the exposure of cancer tissue to the drug. Inflammatory bowel diseases such as ulcerative colitis require targeted therapeutic strategies to reduce inflammation while minimizing systemic exposure to corticosteroids, which are commonly used in the treatment of these types of diseases. Zhang et al. [70] developed a colon-targeted DDS using solid lipid nanoparticles coated with CH/sodium cellulose sulfate PECs for colon-specific delivery of budesonide after oral administration. This system enabled enzyme-responsive drug release by utilizing cellulase secreted by the gut microbiota as a stimulus for controlled drug release. The solid lipid nanoparticles encapsulating budesonide showed selective drug release at inflamed sites in the colon, reducing systemic absorption and toxicity while enhancing therapeutic efficacy (i.e., anti-inflammatory effect) in a dextran sulfate sodium-induced colitis mouse model. The PEC-coated solid lipid nanoparticles effectively prevented premature drug release and ensured that budesonide reached the colon intact. This study highlighted the potential of CH PECs to optimize the therapeutic index of corticosteroids for the treatment of ulcerative colitis by localizing drug action and minimizing off-target effects.

In the field of protein and peptide therapeutics, CHPECs for oral insulin delivery have gained considerable attention. Conventional insulin delivery by subcutaneous injection poses compliance issues; therefore, oral insulin formulations are highly desirable [37, 66]. Recent studies have reported the development of insulin-loaded CH/poly-γ-glutamic acid PEC nanoparticles crosslinked with procyanidin. The prepared nanoparticles improved insulin encapsulation efficiency (up to 90%), enabled pH-dependent insulin release, and significantly reduced enzymatic degradation in vitro. The transepithelial permeability of orally administered nanoencapsulated insulin has been shown to be 8.1 times higher than that of native insulin. Pharmacokinetic studies showed an oral bioavailability of insulin of $9.81 \pm 1.32\%$ in rats with type I diabetes after intragastric administration of enteric-coated (with Eudragit® L100–55) capsules containing freeze-dried nanoparticles loaded with the drug [37]. In addition, the addition of permeation enhancers such as sucrose laurate to CH PEC nanoparticles has been investigated to further improve insulin bioavailability and efficacy after oral administration [64]. The ability of CH to facilitate the transport of macromolecules via both paracellular and transcellular pathways makes it a strong candidate for the non-invasive delivery of peptide-based therapies.

Another emerging application of CH PECs is the development of oral vaccines. Due to its immunostimulatory properties, CH has been investigated as a vaccine adjuvant to improve mucosal immunity. In a recent study, a ternary PEC of CH/PLGA/alginate/R8 was developed for oral administration of vaccines that showed effective antigen (*Mycobacterium avium subspecies paratuberculosis* (MAP) antigens) protection and immune activation in intestinal lymphoid tissues. The PECs exhibited a crosslinked network structure that prevented antigen degradation under in vitro gastric conditions (simulated gastric fluid) while promoting controlled antigen release in the simulated intestinal fluid, resulting in enhanced IgA secretion and systemic immune responses after oral administration to mice. These results suggested that CH PECs could be an effective alternative to injectable vaccines, especially for large-scale immunization programs [38].

6.2 Topical (Dermal) and Transdermal Drug Delivery

CH PECs have been widely researched as carriers for topical (dermal) and transdermal drug delivery due to their biocompatibility, bioadhesiveness, and ability to form controlled release formulations. These PECs provide enhanced drug penetration into the skin and can be tailored to release therapeutic agents locally or facilitate systemic absorption through transdermal delivery [54]. The applications of CH PECs in these two delivery routes differ in terms of the formulation design, the depth of penetration of the drug, and the intended therapeutic effect.

6.2.1 Topical (Dermal) Drug Delivery

Topical drug delivery aims to deliver therapeutic agents directly to the skin surface to locally treat conditions such as infections, wounds, inflammation, and acne [59]. CH PECs have been shown to improve drug retention at the site of application, increase stability, and enable sustained drug release [54].

Nano- and micro-sized quercetin-loaded casein/CH PEC hydrogels have been developed to provide antioxidant and anti-inflammatory effects in wound healing, with controlled drug release and improved skin permeability. Four types of particles were prepared at pH 6 and casein/CH charge ratios 1:1, 2:1, 4:1, and 6:1. The electrostatic interactions between the oppositely charged polymers were confirmed by FTIR. The yield increased (from 37.5% to 72.5%) with the excess of casein, as did the size of the gel particles (between 377 nm and 5.72 μm). The loading efficiency of quercetin was 95% or more for all hydrogels tested. The release of quercetin in phosphate buffer pH 5.5 (32 °C, 50 rpm) followed zero-order kinetics. The reported quercetin-loaded casein/CH complexes have the potential to be used as an adjunctive treatment in wound healing [44].

Another important area of application for CH PECs is acne treatment. Berberine-loaded CH/pectin PECs were formulated in core-shell microcapsules for controlled (sustained and pH-dependent) drug release and showed significant antibacterial activity on *Cutibacterium acnes* and no cytotoxicity on keratinocyte (HaCaT) cell line. The authors pointed out that further research should focus on specific in vivo studies to evaluate the potential of cutaneous absorption, especially follicular penetration [51].

The development of CH PEC scaffolds using 3D printing technology has also been investigated for tissue engineering and drug delivery. For example, Palamidi et al. [50] developed biocompatible printable inks based on PECs comprising CH, collagen, and gelatin. These inks were 3D printed using a pneumatic-based extrusion printer. The results showed that printing was optimal at an extrusion pressure of 150 ± 40 MPa and G22 or G25 nozzles. Inks with low CH content (< 4% w/v) printed poorly, while inks with 4% w/v CH and 1% w/v gelatin showed satisfactory extrusion and printing quality. The addition of collagen (0.1% w/v) to the optimized ink had no effect on printability. Stabilization after printing with potassium hydroxide resulted in self-supporting scaffolds with consistent morphological integrity. FTIR spectra confirmed the formation of PECs. Swelling tests showed that the constructs neutralized with potassium hydroxide remained stable and had the lowest swelling ratio in different pH environments, indicating their highest structural stability. After optimizing the ink composition, the constructs were loaded with 10% levofloxacin as a model drug. More than 80% of levofloxacin was released from the PECs composed of CH and gelatin within the first hour and reached full release after 24 h, indicating inadequate drug release control. In contrast, PECs composed of CH, gelatin, and collagen initially released levofloxacin more slowly (55% in 15 min), followed by stable release after 4 h, probably due to controlled diffusion of the drug from the expanded constructs. Cell compatibility of the 3D-printed constructs was demonstrated by MTT assay on fibroblasts, while antimicrobial activity of the drug-loaded constructs was demonstrated against *Escherichia coli* and *Staphylococcus aureus*. The authors suggested that by developing hydrogel inks and optimizing the printing parameters, scaffolds suitable for tissue engineering and topical drug applications can be prepared.

6.2.2 Transdermal Drug Delivery

Transdermal drug delivery facilitates the transport of drugs through the skin into the bloodstream and offers an alternative to oral and injectable routes of administration, especially for drugs with extensive first-pass metabolism or GI instability. CH PECs have been developed as transdermal carriers as they improve drug permeation and enable controlled release [49, 53].

For example, Patil et al. [53] developed atorvastatin-loaded CH/chondroitin sulfate PEC nanoparticles and incorporated them into transdermal patches to improve skin permeability and bioavailability of the drug, especially compared to the conventional oral formulations available on the market. The drug-loaded nanoparticles were prepared by the ionic gelation method and optimized (Box-Behnken design). The optimized batch of nanoparticles had an average size of 219.2 ± 5.98 nm, 82.68 ± 2.63% drug entrapment efficiency and 25.41 ± 3.29 mV zeta potential. Moreover, atorvastatin-loaded nanoparticles showed sustained release of the drug and higher skin permeation compared to a suspension of the drug (across human fibroblasts and keratinocytes monolayers (apical to basolateral)). Subsequently, the optimized nanoparticles were incorporated into transdermal patches by solvent evaporation. The in vivo pharmacokinetic study showed that the transdermal patch with the optimized atorvastatin-loaded nanoparticles significantly increased the bioavailability of atorvastatin compared to the marketed oral tablet ($p < 0.05$). The authors concluded that the transdermal patches with drug-loaded PEC-based nanoparticles are a promising DDS for drugs with poor bioavailability. In addition, PEC-based CH/carrageenan nanoparticles loaded with naproxen were prepared and incorporated into Carbopol® 940-based gels and investigated for transdermal drug delivery in the treatment of inflammation associated with rheumatoid arthritis. The percentage of ex vivo permeation patterns of naproxen in the optimized nanoparticle-loaded gel (88.66%) was higher than that of the control gel (36.195%), while an in vivo study in rats showed that the optimized nanoparticle-loaded gel formulation was more effective than the control gel containing naproxen in the treatment of arthritis-associated inflammation. The optimized formulation showed higher drug retention in inflamed tissue with minimal systemic side effects (e.g., GI side effects) compared to oral administration of naproxen, which may lead to improved patient compliance [47].

The potential of pH-sensitive PEC hydrogels for transdermal delivery has also been investigated, especially for drugs that are unstable in the acidic environment of the stomach, which may result in low bioavailability after oral administration. Carboxymethylagarose/CH-PECs were developed as cytocompatible hydrogels for the controlled release of diclofenac sodium. The hydrogels were prepared at different polymer ratios and oven-dried at 34 °C for 12 h to obtain films. The optimal carboxymethylagarose-to-CH ratio was found to be 2:1, due to the highest swelling capacity (25.2 g swelling medium absorbed by 1 g of sample at pH 6.0), high drug encapsulation efficiency (up to 69%), high loading capacity (up to 79%) and sustained release over 72 h (cumulative release of 67% of encapsulated diclofenac sodium) through a Fickian diffusion mechanism. Viability tests with HaCaT cells showed an almost 100% survival rate in the presence of dry hydrogels and diclofenac sodium. With the results presented, the authors demonstrated the potential of PEC-based dry hydrogels (films) as safe DDS for long-term pain management via transdermal application [49].

In an effort to overcome the skin's natural barrier to drug penetration, studies have investigated the use of CH PECs in combination with permeation enhancers. For example, Lefnaoui et al. [35] developed matrix-type CH/alginate PEC transdermal DDS, i.e., films of ketotifen fumarate, an antihistaminic drug. Films with variable PEC compositions were prepared by the film casting method using different CH-to-sodium alginate ratios. Propylene glycol was used as a plasticizer, while polysorbate 80 and sorbitan monolaurate were used as permeation enhancers. All the films prepared exhibited suitable physicochemical properties without significant interactions between ketotifen fumarate and the polymers. The most satisfactory drug release pattern was achieved when polysorbate 80 was used as a permeation enhancer, i.e., 99.88% of ketotifen fumarate was released and 2.121 mg/cm^2 of drug was permeated after 24 h (drug release and permeation through the abdominal skin of rats in a Franz diffusion cell). The optimal film formulation had a permeability coefficient of 14.00 $\pm$ 0.001 cm h^{-1} and a latency time of 0.35 $\pm$ 0.02 h. The in vitro drug release assay showed controlled release of ketotifen fumarate following Korsmeyer-Peppas kinetics. The results obtained suggested that PEC-based controlled-release transdermal formulations in combination with permeation enhancers could be a suitable alternative to the usual forms of antihistaminic drugs.

6.3 Nasal and Pulmonary Drug Delivery

CH PECs have been extensively studied for nasal and pulmonary drug delivery due to their mucoadhesive properties, biocompatibility, and ability to enhance drug permeability. These systems are particularly valuable in overcoming the challenges associated with rapid clearance from the mucosal surface and low bioavailability of certain therapeutic agents. The applications of CH PECs for nasal and pulmonary administration are different. Nasal formulations are aimed at local, systemic, or nose-to-brain drug delivery, while pulmonary formulations focus on respiratory drug administration with or without systemic drug absorption [2, 4, 72].

6.3.1 Nasal Drug Delivery

Nasal drug delivery is a non-invasive route that enables rapid systemic absorption and direct nose-to-brain transport, bypassing the blood-brain barrier. CH PECs are commonly used for this application due to their mucoadhesive properties, which prolong residence time in the nose and improve drug absorption [72]. In one study, lyophilized inserts based on κ-carrageenan/CH PECs were described for intranasal administration of sumatriptan succinate to improve mucoadhesive potential and control drug release in migraine treatment. The formation of PECs between κ-carrageenan and CH was confirmed by FTIR. The results showed that the molar ratio of the polymers plays a crucial role in modulating the properties of the inserts. Inserts with a κ-carrageenan/CH molar ratio of 4:1 had the highest water absorption capacity (approximately 13 g of swelling medium absorbed by 1 g of sample at pH 5.5) and mucoadhesive potential (mucoadhesion force >200 mg/cm^2) and provided the best control of sumatriptan succinate release (approx. 30% of the drug was released in the first 30 min, while further release was continuous for 6 h with a cumulative drug release of ~60%), so they were considered promising for efficient delivery of sumatriptan succinate via the nasal route to terminate migraine attacks [1].

Nasal PEC-based formulations have also been investigated for the treatment of Parkinson's disease. Avhad et al. [4] prepared PEC-based CH/pectin nanoparticles (187.3–210.4 nm) loaded with levodopa (with the addition of carbidopa, a peripheral amino acid decarboxylase inhibitor), which were subsequently lyophilized. The content of levodopa in the freeze-dried nanoparticles ranged from 45.23 mg to 65.12 mg, the drug loading was 78–85% and the encapsulation efficiency was 81.93–89.67%. The main objective of the research was to develop DDS that have the potential to enhance the delivery of levodopa to the brain by facilitating transport from the nose to the brain via the olfactory pathway and trigeminal neurons while preventing peripheral metabolism. The in vitro/ex vivo study of drug diffusion from lyophilized nanoparticles (Franz diffusion cell using excised sheep nasal mucosa at PBS pH 6.6 as receptor medium) revealed that the release of levodopa/carbidopa was mainly influenced by the pH of the preparation of the PECs studied, i.e., higher pH

(10 and 12) led to a significantly slower drug release from all formulations. However, PEC nanoparticles prepared at pH 10 and 12 showed greater drug diffusion compared to PEC nanoparticles prepared at pH 5 and 8, as complexation between polymers and drugs is expected to be greater at a more alkaline pH. According to the results, PECs prepared at pH 10 and 12 were considered suitable for the desired release and diffusion rate of levodopa/carbidopa. Also, lyophilized PEC nanoparticles prepared at pH 10 and 12 were found to have stronger mucoadhesion. The high mucoadhesion can be explained by strong electrostatic interactions between positively charged CH (oriented outside the PEC nanoparticles) and negatively charged structures in the mucin. The authors hypothesize that the bioavailability of levodopa in the brain can be significantly improved by this approach compared to conventional oral formulations.

The potential of CHPECs for peptide drug delivery via the nasal route has also been investigated. For example, insulin and pramlintide are peptide drugs that can be used to regulate blood glucose concentrations in diabetic patients. Interestingly, it has also been reported that impaired glucose metabolism can also occur in the brains of diabetics and non-diabetics with Alzheimer's disease. Therefore, the regulation of glucose concentration in the brain by the administration of insulin and pramlintide has been proposed as a therapeutic approach for the treatment of Alzheimer's disease. However, the clinical use of these peptide drugs is subject to some limitations due to their metabolic instability, undesirable pharmacokinetic properties, and invasive routes of administration. Recently, a group of authors described the development of PECs containing pramlintide and insulin stabilized around nanoemulsion droplets as a new approach for nasal delivery of these peptides. First, dextran sulfate/pramlintide PEC and CH/insulin PEC were prepared. Subsequently, PEC-coated nanoemulsions (consisting of medium-chain triglycerides, sorbitan monooleate, polysorbate 80, glycerol, and water) were prepared. Both the dextran sulfate/pramlintide PEC-coated nanoemulsion and the CH/insulin PEC-coated nanoemulsion showed desirable nanotechnological properties (size ~140 nm, zeta potential between −40 mV and + 20 mV, spherical morphology, drug content up to 1.5 mg/mL and association efficiency up to 99%) before and after incubation in simulated nasal fluid (pH 5.5) at 37 °C and during storage at 8 °C for 30 days, indicating their stability. Negatively charged dextran sulfate/pramlintide PEC-coated nanoemulsion had improved mucodiffusion properties and resulted in sustained permeation of pramlintide through the nasal epithelium over 10 h. In contrast, the positively charged CH/insulin PEC-coated nanoemulsion resulted in faster mucosal permeation of insulin compared to the free drug. The results of this study may support future in vivo studies on the potential use of these new DDS for the treatment of brain disorders by the nasal route [72].

6.3.2 Pulmonary Drug Delivery

Pulmonary DDS using CH PECs has gained increasing interest due to their ability to improve drug delivery and retention in the lung. These formulations are particularly useful for the treatment of respiratory diseases such as tuberculosis, asthma, and lung infections [2, 45, 65].

Inhaled administration of antibiotics in aerosol form may be preferable to conventional methods of drug administration (oral or parenteral administration) in the control of lung infections. Mishra et al. [45] developed spray-dried aerosolized CH PEC-based microparticles containing two antibiotics, doxycycline hyclate and levofloxacin hydrochloride, for the treatment of lung infections. Sodium alginate, sodium carboxymethylcellulose, polyvinylpyrrolidone K90, or Carbopol® 974P were used as anionic components of the PECs. PEC microparticles were mostly spherical and showed a high encapsulation efficiency for both drugs (>85%). The release profiles (rotating paddle apparatus, PBS pH 7.4, 37 °C, 50 rpm) showed an initial "burst" phase (>30% of both drugs were released within the first hour of the release study), followed by a secondary sustained-release phase (complete release of drugs was achieved after 12 h). The good aerosolization behavior was evidenced by high values of recovered dose (~96–98%), emitted dose (~75–85%), and fine particle fraction (~60–75%). The cytotoxicity of the investigated PEC-based microparticles was not detected in the alveolar cell line H1299. More than eight-fold higher C_{max} values were obtained in bronchial alveolar lavage fluid with microparticles compared to intravenously administered solutions of doxycycline hyclate and levofloxacin hydrochloride. The results of this study suggest that CHPEC-based microparticles as a dry powder inhaler may be an efficient DDS for the sustained and effective co-administration of two antibiotics commonly used in the treatment of lung infections.

CH/xanthan gum-based PECs have also been formulated in dry powder inhalation systems for tuberculosis treatment [65]. An optimized formulation (out of 14 initial formulations) of rifampicin dry powder inhalation with CH/xanthan gum PEC was prepared by spray drying using the response surface method (Box–Behnken method). The Design Expert predicted that the optimized formulation was the one obtained with a 0.5% CH solution, a rifampicin-to-CH ratio 1:1, and a spray dryer inlet temperature of 143 °C. This formulation resulted in rifampicin dry powder inhalation with an entrapment efficiency of 108.9%, a particle size of 1.177 μm, and a rifampicin release of 14.49% (at pH 7.4) and of 37.97% (at pH 4.5). The optimized formulation has a great potential to improve the uptake of the drug by pulmonary macrophages and thus reduce systemic side effects.

Alnaief et al. [2] prepared hybrid CH/alginate aerogel microparticles using the emulsifying-gelation method as a potential pulmonary DDS. The effects of the mixing order of polymer (CH, alginate, or CH/alginate PEC), paraffin oil, and surfactant (sorbitan monooleate or sorbitan trioleate) on the properties of the final aerogel microparticles were investigated. The results of the zeta potential measurements showed that the surface charge of the particles depended mainly on the surfactant type, while the order of polymer mixing had less influence on the surface charge. The amphiphilic properties of CH altered the hydrophilic-lipophilic balance of the surfactants. The specific surface area of the prepared microparticles was in the range of 29.36–86.20 m^2g^{-1} with a mesoporous pore size of 12.48–13.38 nm and a pore volume of 0.09–0.29 cm^3g^{-1}. The calculated aerodynamic diameter of the prepared microparticles was in the range of 0.17–2.29 μm. Toxicity studies in adult male Sprague–Dawley rats showed that the prepared hybrid CH/alginate aerogel microparticles caused mild lung inflammation, but with some acute kidney and liver toxicities (after a single administration of the microparticles).

6.4 *Buccal Drug Delivery*

CH PECs have been extensively researched for buccal formulations due to their mucoadhesive properties and biocompatibility. These formulations are mostly intended for local drug administration [3, 55].

Buccal films have been extensively studied as a delivery system for the local treatment of infections such as oral candidiasis and periodontal disease. CH/pectin PEC-based multilayer films were formulated for the controlled delivery of clotrimazole, which showed strong mucoadhesion and antifungal activity against *Candida* species, which was partly attributed to the antimicrobial properties of CH [55]. These films provided a prolonged and pH-dependent drug release (higher amounts of clotrimazole were released in a simulated saliva solution pH 4.8 than in a phosphate buffer pH 6.8), which could help to maintain therapeutic concentrations on the oral mucosa over a longer period of time and reduce the frequency of administration. The investigated films also showed a satisfactory safety profile in the cytotoxicity test on human gingival fibroblasts.

Another study focused on the development of nystatin-loaded CH PEC films for the treatment of oropharyngeal candidiasis [3]. The solvent casting method was used to prepare mucoadhesive PEC-based films consisting of CH/gum arabic and CH/polygalacturonic acid. The formation of PECs was detected by FTIR and DSC-TGA analysis. The release of nystatin from PEC-based films was rapid in the initial phase (approximately 55% of the drug was released in the second hour from both CH/gum arabic and CH/polygalacturonic acid PEC-based films). The rapid initial release was then followed by a slower release (84% of nystatin was released from both formulations after 24 h). The films were also shown to adhere well to the oral mucosa. The films showed significant antifungal activity against *Candida albicans*, suggesting that they are an alternative to conventional oral suspensions that suffer from poor retention.

6.5 *Parenteral Drug Delivery*

Parenteral administration of drugs provides direct systemic administration that bypasses first-pass metabolism and increases bioavailability. CH PECs have been extensively studied for parenteral applications due to their biocompatibility, biodegradability, and ability to encapsulate hydrophobic drugs and biomolecules. These systems enable controlled drug release and can be used in various formulations, including nanoparticles and injectable hydrogels [17, 71].

CH PECs have been investigated for immunotropic applications. In a recent study [17], the immunomodulatory and anti-inflammatory potential of CH/κ-carrageenan PEC nanoparticles administered by intraperitoneal injection was investigated. Two formulations of soluble PECs were prepared with an excess of CH (mass ratio CH:-κ-carrageenan 10:1) and with an excess of κ-carrageenan (mass ratio CH:-κ-carrageenan 1:10). The results showed significant activation of immune cells, including macrophages and neutrophils, and increased production of cytokines such as TNF-α and IL-10. The ability of PECs to scavenge NO has been shown to be dependent on the content of κ-carrageenan (significantly higher at a CH:-κ-carrageenan mass ratio of 1:10). These results suggest that the investigated nanoparticles based on CH/κ-carrageenan PEC could serve as novel immunostimulatory agents for vaccine adjuvants or immunotherapies.

Injectable hydrogels composed of CH PECs have gained attention as minimally invasive DDS due to their biocompatibility and biodegradability. Zhao et al. [71] developed thermosensitive, dual-ionic crosslinked PEC-based CH hydrogels for controlled drug release. Poly-N-isopropylacrylamide-2-acrylamide-2-methylpropanulonic acid (poly (NIPAm-co-AMPS)) and sodium bicarbonate were used as the anionic component of the PECs and the crosslinking agent, respectively. The results showed that the hydrogel with a volume ratio of CH/poly(NIPAm-co-AMPS)/sodium bicarbonate of 1/0.2/0.5 (i.e., the optimized PEC hydrogel) exhibited a sol–gel transition at 37 °C. This hydrogel also had good mechanical strength and a gelation time of 1–3 min at 37 °C. The potential of the optimized PEC hydrogel as a DDS was investigated by evaluating the in vitro release profile of aspirin as a model drug. The hydrogel showed sustained release of aspirin for 24 h (~87%), with no significant burst release. In addition, L929 cells showed good survival and proliferation rates when mixed with the optimized PEC hydrogel after 24 h, indicating excellent biocompatibility. This injectable thermoresponsive hydrogel showed great potential for controlled drug release applications.

6.6 Ocular Drug Delivery

CH PECs have been investigated for ocular drug delivery due to their mucoadhesive properties, biocompatibility, and ability to enhance drug permeation through the cornea. Conventional ocular drug formulations, such as eye drops and ointments, suffer from rapid precorneal elimination, low bioavailability, and limited penetration into intraocular tissues. CH PECs offer an effective alternative by improving drug retention at the site of action, facilitating controlled drug release, and increasing corneal permeability [23, 25].

Dubashynskaya et al. [23] developed zinc sulfate crosslinked PEC nanoparticles consisting of hyaluronic acid and a water-soluble cationic derivative diethylaminoethyl CH to improve the delivery of dexamethasone phosphate to the eye by increasing local bioavailability and reducing systemic toxicity. The investigated PEC nanoparticles had a hydrodynamic diameter of 244 nm and a zeta potential of +24.4 mV, indicating their colloidal stability. These PECs exhibited high encapsulation efficiency (75.6%), excellent mucoadhesion, sustained drug release (i.e., prolonged drug release with approximately 70% of the drug released within 10 h), and enhanced anti-inflammatory activity compared to free dexamethasone phosphate solutions. The presence of zinc ions additionally increased the stability of the nanoparticles, prolonged drug release, and maintained therapeutic concentrations for up to 10 h.

A study by Dubey et al. [25] described the development of CH PEC hydrogels for sustained ocular drug delivery for the treatment of glaucoma. The aim of this study was to prepare brinzolamide-loaded CH/pectin PEC-based mucoadhesive nanocapsules by a complex coacervation method. The prepared nanocapsules ranged in size from 217.01 ± 0.21 nm to 240.05 ± 0.08 nm and had appropriate physicochemical properties (pH 4.29–4.57, refractive index of about 1.3, and viscosity around 7.5 cP) and good mucoadhesive strength (7.45–10.34 dyne/cm^2). The in vitro release mechanism of brinzolamide from the nanocapsules in simulated tear fluid (pH 7.4) was a combination of their erosion, which ensured an immediate therapeutic effect, and the diffusion of the drug during 8 h, demonstrating its prolonged release. An ex vivo corneal permeation study using freshly excised goat eyes showed that the brinzolamide-loaded PEC-based nanoparticles penetrated the cornea to a greater extent than the marketed product (Azopt® ophthalmic suspension containing 1% brinzolamide). In a pharmacodynamic study in albino rabbits, a greater reduction in intraocular pressure was achieved after administration of drug-loaded PEC nanoparticles compared to the marketed product. The authors concluded that brinzolamide-loaded mucoadhesive nanocapsules based on CH/pectin-PEC can be considered as an alternative to conventional eye drops, as they improve the ocular availability of brinzolamide by a longer retention time in the cornea and the ability to sustain the release of the drug.

6.7 Vaginal Drug Delivery

CH PECs have also been investigated for the vaginal administration of drugs, as they have mucoadhesive properties, allow controlled drug release, and improve drug retention at the site of action. The vaginal route is an attractive alternative for drug delivery, especially for localized treatments such as antimicrobial therapies, antifungal therapies, and probiotic treatments. However, challenges such as rapid clearance by vaginal secretions necessitate advanced DDS such as CH PECs [21, 46].

Nayak et al. [46] developed metronidazole-loaded nanoparticles based on CH/phytic acid PECs for the treatment of bacterial vaginosis. Morphological examination revealed that the nanoparticles prepared with 0.535% w/v CH and 0.112% w/v phytic acid were non-spherical, with a zeta potential ranging from 25.00 ± 0.45 to 39 ± 0.7 mV. The DSC and X-ray diffraction results showed no changes in the physical state of metronidazole after encapsulation. The investigated nanoparticles exhibited good mucoadhesive properties, which may lead to prolonged drug retention on the vaginal mucosa. In addition, a cumulative metronidazole release of approximately 98 ± 1.5% within 8 h was demonstrated. Antimicrobial in vitro studies showed that the optimized formulation had enhanced efficacy against *Escherichia coli* with an MIC of 0.9 ± 0.1 μgmL^{-1}. Compared to the non-encapsulated metronidazole, the nanoparticles tested showed a fourfold reduction in the *Escherichia coli* count in vivo in female Swiss albino mice. Based on the results obtained, the authors concluded that CH/phytic acid PEC-based nanoparticles loaded with metronidazole can be used as a potential alternative for the treatment of bacterial vaginosis with fewer systemic side effects compared to oral metronidazole formulations due to their excellent physicochemical and antibacterial properties.

Beyond antimicrobial applications, CH PECs have been developed for multifunctional vaginal therapy containing both antifungal agents and probiotics. In a recent study by Deshkar et al. [21], lyophilized hydrogel inserts based on CH/pectin, CH/carrageenan, and CH/polycarbophil PECs were described for the co-administration of the antifungal agent posaconazole and the probiotic *Lactobacillus casei*. PEC formation was confirmed by SEM, FTIR and DSC. The probiotic, *Lactobacillus casei*, was added to the PECs during the freeze-drying process to avoid a possible negative influence of the PEC preparation conditions on cell viability. The PECs were then compressed with the addition of posaconazole to form inserts, which were then optimized. The CH/pectin-based insert showed the best mucoadhesion (1.25 N), sustained drug release (88.2 ± 2.4% in 8 h), and a swelling index of 154.7%. The in vitro efficacy study of the prepared inserts on a co-culture of *Lactobacillus casei* and *Candida albicans* showed an increase in *Lactobacillus casei* count and a significant decrease in viable *Candida albicans* count (fourfold reduction in 24 h), indicating the efficacy of the investigated CH PEC-based inserts in alleviating fungal infection. The results of this study demonstrate the potential of PEC-based hydrogel inserts, especially CH/pectin-based, to prevent vaginal infections and restore a normal vaginal microbiota.

6.8 Targeted Drug Delivery for Cancer Treatment

CH PECs have emerged as promising carriers for targeted delivery of anticancer drugs due to their biocompatibility, biodegradability, and ability to achieve tumor-specific drug delivery and improve their retention in tumors. CH PECs can facilitate the controlled release of chemotherapeutics while reducing systemic toxicity. These CH PEC-based DDS can also improve the pharmacokinetics of anticancer drugs by increasing their solubility, stability, and tumor selectivity [42, 52, 63].

Malviya et al. [42] prepared CH/tamarind gum polysaccharide PEC-stabilized cubic nanoparticles for the targeted delivery of simvastatin to breast cancer cells using the antisolvent precipitation technique. These nanoparticles had high encapsulation efficiency (up to 83.36%) and small size (53.3–383.1 nm). Also, controlled drug release was demonstrated according to the Baker-Lonsdale model. In addition, stability studies were performed for 45 days, and only minor variations in particle size and a slight decrease in drug entrapment efficiency were observed. In vitro cytotoxicity assays against MCF-7 breast cancer cells showed enhanced inhibition of tumor cells by simvastatin-loaded PEC-based nanoparticles compared to free simvastatin. The study highlights the potential of CH PECs to improve the solubility and bioavailability of lipophilic anticancer drugs while minimizing their side effects.

One of the greatest challenges in the treatment of brain tumors is overcoming the blood-brain barrier in order to effectively administer chemotherapeutic drugs. Silant'ev et al. [63] developed pectin/CH PEC nanoparticles for the targeted delivery of temozolomide in the treatment of glioblastoma. The establishment of ionic interactions between the functional groups of the oppositely charged polymers was demonstrated by IR spectroscopy and dynamic rheology. These nanoparticles had a particle size of 30–330 nm and a zeta potential of −29 to 73 mV, indicating good colloidal stability. A comprehensive evaluation of the sorption properties and sorption kinetics of the investigated PEC-based nanoparticles with temozolomide was performed. Of all the nanoparticles prepared, the PEC-based vehicle based on an equal pectin/CH ratio was found to be the most potentially effective for crossing the blood-brain barrier, immobilizing temozolomide and delivering it to brain tumors, while minimizing systemic toxicity, as the nanoparticles exhibited strong affinity for glioblastoma cells and promoted local accumulation of the drug at the tumor site.

In a study by Paşcalău et al. [52], microparticles containing a bovine serum albumin gel core and a multilayer CH/hyaluronic acid PEC-based shell were developed for the targeted delivery of sorafenib, a tyrosine kinase inhibitor used in the treatment of hepatocellular carcinoma. Confocal laser scanning microscopy, scanning electron microscopy and atomic force microscopy revealed the spherical shape of the microcapsules with a size of about 10 μm as well as their core-shell structure and proved the successful encapsulation of the drug. FTIR confirmed only physical interactions between sorafenib and the polymers, while X-ray diffraction patterns and DSC thermograms showed the amorphous structure of the microcapsules. These microcapsules had an encapsulation efficiency of up to 45.64 ± 2.70% and showed sustained drug release for 60 h in both release media tested, i.e., the buffer solution pH 5 (simulating tumoral acidic environment) and pH 7.4 (physiological conditions). The authors suggest that the drug release pattern shown could ensure site-specific action and drug delivery to the tumor tissue. The low cytotoxicity shown in vitro in HepG2 and mesenchymal stem cells makes sorafenib-loaded PEC-based microcapsules a potentially suitable DDS for further in vivo studies on targeted drug delivery in hepatocellular carcinoma therapy.

7 Challenges and Future Prospects

CH PECs have shown great promise for drug delivery due to their biocompatibility, biodegradability, and controlled release capability. However, their clinical implementation is hindered by several issues, including formulation stability, drug loading efficiency, mucoadhesion limitations, and regulatory considerations. Ensuring the stability of PEC formulations remains a major concern, as factors such as pH, ionic strength, and storage conditions can affect their physicochemical properties and lead to aggregation or premature drug release. Variability in CH molecular weight and the degree of deacetylation further complicates reproducibility, making large-scale production difficult. For this reason, researchers have explored crosslinking strategies and surface modifications to improve the long-term stability of PEC-based drug carriers [68]. Despite extensive preclinical studies, regulatory challenges remain a significant obstacle to the commercialization of CHPECs-based DDS. The variability of raw material sources, lack of standardized manufacturing protocols, and concerns regarding immunogenicity and degradation products need to be thoroughly investigated [48]. Establishing regulatory guidelines for quality control and safety assessment will be critical to facilitating clinical approval. Future research should focus on optimizing scalable production techniques, such as microfluidic PEC synthesis [62] and spray drying approaches [18], to ensure batch-to-batch consistency and improve commercial feasibility.

8 Conclusions

CH PECs have shown great potential for drug delivery as they offer advantages such as biocompatibility, biodegradability, mucoadhesive properties, and controlled drug release. Their applications span a wide range of routes of administration, including oral, transdermal, ocular, nasal, pulmonary, buccal, vaginal, and parenteral delivery. These systems have also shown promise for targeted drug delivery in cancer therapy, where their ability to increase drug solubility, improve site-specific accumulation, and reduce systemic toxicity has led to significant advances. In addition, the use of CH PECs in protein and vaccine delivery underscores their versatility in modern pharmaceutical applications. The ability of these complexes to protect drugs from enzymatic degradation while enabling improved absorption has made them an attractive platform for innovative drug formulations.

Despite their numerous advantages, there are still challenges in the clinical implementation of CH PECs. Issues such as formulation stability, limited drug loading, penetration through biological barriers, and regulatory hurdles need to be addressed. However, recent research has focused on improving these formulations through functionalization strategies, hybrid PEC systems, and stimuli-responsive mechanisms. Advances in nanotechnology and polymer chemistry are helping to further refine PEC structures and optimize their pharmacokinetic profiles and therapeutic efficacy. The ongoing development of large-scale production methods and the standardization of regulations will play a crucial role in bringing these systems closer to widespread clinical application.

With advances in this field, CH PECs are poised to play a transformative role in next-generation DDS. Their adaptability, tunability, and biocompatibility offer immense potential for personalized medicine and targeted therapies. With continued research and innovation, the use of CH PECs-based DDS is expected to improve pharmaceutical formulations and thus treatment outcomes and patient adherence in various medical specialties.

References

1. Alavi, S., & Mortazavi, S. A. (2018). Freeze-dried K-Carrageenan/Chitosan polyelectrolyte complex-based insert: A novel intranasal delivery system for Sumatriptan succinate. *The Iranian Journal of Pharmaceutical Research (IJPR), 17*, 1172–1181.
2. Alnaief, M., Obaidat, R. M., & Alsmadi, M. M. (2020). Preparation of hybrid alginate-chitosan aerogel as potential carriers for pulmonary drug delivery. *Polymers, 12*, 2223. https://doi.org/10.3390/polym12102223
3. Armenta-Rojas, E., Cornejo-Bravo, J. M., Serrano-Medina, A., López-Maldonado, E. A., Olivas-Sarabia, A., Castillo-Martínez, N. A., Villarreal-Gómez, L. J., & Hurtado-Ayala, L. A. (2022). Nystatin-loaded polyelectrolyte complex films as a mucoadhesive drug delivery system for potential buccal application. *Biointerface Research In Applied Chemistry, 12*, 4384–4398. https://doi.org/10.33263/BRIAC124.43844398
4. Avhad, P. S., Gupta, R. A., Firke, S. N., Roge, A. B., Sarje, S. K., Raut, A. S., & Devaliya, R. (2022). Formulation and evaluation of chitosan based polyelectrolyte complex of levodopa for nasal drug delivery. *International Journal of Health Sciences, 6*, 1316–1327. https://doi.org/10.53730/ijhs.v6nS3.5555
5. Barbosa, J. N., Amaral, I. F., Águas, A. P., & Barbosa, M. A. (2010). Evaluation of the effect of the degree of acetylation on the inflammatory response to 3D porous chitosan scaffolds. *Journal of Biomedical Materials Research. Part A, 93A*, 20–28. https://doi.org/10.1002/jbm.a.32499
6. Bediako, J. K., Mouele, E. S. M., El Ouardi, Y., & Repo, E. (2023). Saloplastics and the polyelectrolyte complex continuum: Advances, challenges and prospects. *Chemical Engineering Journal, 462*, 142322. https://doi.org/10.1016/j.cej.2023.142322
7. Bernkop-Schnürch, A., & Dünnhaupt, S. (2012). Chitosan-based drug delivery systems. *European Journal of Pharmaceutics and Biopharmaceutics, 81*, 463–469. https://doi.org/10.1016/j.ejpb.2012.04.007
8. Blagodatskikh, I. V., Vyshivannaya, O. V., Tishchenko, N. A., Orlov, V. N., Tikhonov, V. E., Bezrodnykh, E. A., Ezernitskaya, M. A., & Khokhlov, A. R. (2023). Complexation between chitosan and carboxymethyl cellulose in weakly acidic, neutral, and weakly alcaline media. *International Journal of Biological Macromolecules, 243*, 125277. https://doi.org/10.1016/j.ijbiomac.2023.125277
9. Bombaldi de Souza, F. C., Bombaldi de Souza, R. F., Drouin, B., Mantovani, D., & Moraes, Â. M. (2019). Comparative study on complexes formed by chitosan and different polyanions: Potential of chitosan-pectin biomaterials as scaffolds in tissue engineering. *International Journal of Biological Macromolecules, 132*, 178–189. https://doi.org/10.1016/j.ijbiomac.2019.03.187
10. Bravo-Anaya, L. M., Fernández-Solís, K. G., Rosselgong, J., Nano-Rodríguez, J. L. E., Carvajal, F., & Rinaudo, M. (2019). Chitosan-DNA polyelectrolyte complex: Influence of chitosan characteristics and mechanism of complex formation. *International Journal of Biological Macromolecules, 126*, 1037–1049. https://doi.org/10.1016/j.ijbiomac.2019.01.008
11. Buriuli, M., & Verma, D. (2017). Polyelectrolyte complexes (PECs) for biomedical applications. In A. Tripathi & J. S. Melo (Eds.), *Advances in biomaterials for biomedical applications, advanced structured materials 66*. Springer. https://doi.org/10.1007/978-981-10-3328-5_2
12. Cardoso, V. M. d. O., de Brito, N. A. P., Ferreira, N. N., Boni, F. I., Ferreira, L. M. B., Carvalho, S. G., & Gremião, M. P. D. (2021). Design of mucoadhesive gellan gum and chitosan nanoparticles intended for colon-specific delivery of peptide drugs. *Colloids And Surfaces.*

Physicochemical And Engineering Aspects, 628, 127321. https://doi.org/10.1016/j.colsurfa.2021.127321

13. Ćirić, A., Krajišnik, D., Čalija, B., & Đekić, L. (2020). Biocompatible non-covalent complexes of chitosan and different polymers: Characteristics and application in drug delivery. *Arhiv za farmaciju, 70*, 173–197. https://doi.org/10.5937/arhfarm2004173q
14. Ćirić, A., Medarević, Đ., Čalija, B., Dobričić, V., Rmandić, M., Barudžija, T., Malenović, A., & Djekic, L. (2021). Effect of ibuprofen entrapment procedure on physicochemical and controlled drug release performances of chitosan/xanthan gum polyelectrolyte complexes. *International Journal of Biological Macromolecules, 167*, 547–558. https://doi.org/10.1016/j.ijbiomac.2020.11.201
15. Ćirić, A., Milinković Budinčić, J., Medarević, Đ., Dobričić, V., Rmandić, M., Barudžija, T., Malenović, A., Petrović, L., & Đekić, L. (2022a). Influence of spray-drying process on properties of chitosan/xanthan gum polyelectrolyte complexes as carriers for oral delivery of ibuprofen. *Arhiv za farmaciju, 72*, 36–60. https://doi.org/10.5937/arhfarm72-35133
16. Ćirić, A., Milinković Budinčić, J., Medarević, Đ., Dobričić, V., Rmandić, M., Barudžija, T., Malenović, A., Petrović, L., & Djekic, L. (2022b). Evaluation of chitosan/xanthan gum polyelectrolyte complexes potential for pH-dependent oral delivery of escin. *International Journal of Biological Macromolecules, 221*, 48–60. https://doi.org/10.1016/j.ijbiomac.2022.08.190
17. Davydova, V. N., Sorokina, I. V., Volod'ko, A. V., Sokolova, E. V., Borisova, M. S., & Yermak, I. M. (2020). The comparative immunotropic activity of carrageenan, chitosan and their complexes. *Marine Drugs, 18*, 458. https://doi.org/10.3390/md18090458
18. de la Paz, N., Fernández, M., López, O., Garcia, C., Nogueira, A., Torres, L., Turiño, W., & Heinämäki, J. (2021). Spray drying of chitosan acid salts: Process development, scaling up and physicochemical material characterization. *Marine Drugs, 19*, 329. https://doi.org/10.3390/md19060329
19. De Robertis, S., Bonferoni, M. C., Elviri, L., Sandri, G., Caramella, C., & Bettini, R. (2015). Advances in oral controlled drug delivery: The role of drug-polymer and interpolymer non-covalent interactions. *Expert Opinion on Drug Delivery, 12*, 441–453. https://doi.org/10.1517/17425247.2015.966685
20. Desai, N., Rana, D., Salave, S., Gupta, R., Patel, P., Karunakaran, B., Sharma, A., Giri, J., Benival, D., & Kommineni, N. (2023). Chitosan: A potential biopolymer in drug delivery and biomedical applications. *Pharmaceutics, 15*, 1313. https://doi.org/10.3390/pharmaceutics15041313
21. Deshkar, S., Yeole, P., Mahore, J., Shinde, A., & Giram, P. (2023). Polyelectrolyte-complex-based hydrogel inserts for vaginal delivery of Posaconazole and probiotics. *Gels Basel Switz, 9*, 851. https://doi.org/10.3390/gels9110851
22. Dubashynskaya, N. V., Bokatyi, A. N., Sall, T. S., Egorova, T. S., Demyanova, E. V., Dubrovskii, Y. A., Murashko, E. A., Anufrikov, Y. A., Shasherina, A. Y., Vlasova, E. N., & Skorik, Y. A. (2024). Hyaluronan/B12-chitosan polyelectrolyte complex for oral colistin administration. *International Journal of Biological Macromolecules, 263*, 130177. https://doi.org/10.1016/j.ijbiomac.2024.130177
23. Dubashynskaya, N. V., Bokatyi, A. N., Trulioff, A. S., Rubinstein, A. A., Kudryavtsev, I. V., & Skorik, Y. A. (2023). Development and bioactivity of zinc sulfate cross-linked polysaccharide delivery system of dexamethasone phosphate. *Pharmaceutics, 15*, 2396. https://doi.org/10.3390/pharmaceutics15102396
24. Dubashynskaya, N. V., Petrova, V. A., Romanov, D. P., & Skorik, Y. A. (2022). pH-sensitive drug delivery system based on chitin nanowhiskers–Sodium Alginate Polyelectrolyte Complex. *Materials, 15*, 5860. https://doi.org/10.3390/ma15175860
25. Dubey, V., Mohan, P., Dangi, J. S., & Kesavan, K. (2020). Brinzolamide loaded chitosan-pectin mucoadhesive nanocapsules for management of glaucoma: Formulation, characterization and pharmacodynamic study. *International Journal of Biological Macromolecules, 152*, 1224–1232. https://doi.org/10.1016/j.ijbiomac.2019.10.219

26. Dulong, V., Thebault, P., Karakasyan, C., Picton, L., & le Cerf, D. (2024). Polyelectrolyte complexes of chitosan and hyaluronic acid or carboxymethylpullulan and their aminoguaiacol derivatives with biological activities as potential drug delivery systems. *Carbohydrate Polymers, 341*, 122330. https://doi.org/10.1016/j.carbpol.2024.122330
27. Ezike, T. C., Okpala, U. S., Onoja, U. L., Nwike, C. P., Ezeako, E. C., Okpara, O. J., Okoroafor, C. C., Eze, S. C., Kalu, O. L., Odoh, E. C., Nwadike, U. G., Ogbodo, J. O., Umeh, B. U., Ossai, E. C., & Nwanguma, B. C. (2023). Advances in drug delivery systems, challenges and future directions. *Heliyon, 9*, e17488. https://doi.org/10.1016/j.heliyon.2023.e17488
28. Goñi, M. G., Tomadoni, B., Roura, S. I., & Moreira, M. d. R. (2017). Lactic acid as potential substitute of acetic acid for dissolution of chitosan: Preharvest application to Butterhead lettuce. *Journal of Food Science and Technology, 54*, 620–626. https://doi.org/10.1007/s13197-016-2484-5
29. Guarnizo-Herrero, V., Torrado-Salmerón, C., Pabón, N. S. T., Durán, G. T., Morales, J., & Torrado-Santiago, S. (2021). Study of different chitosan/sodium carboxymethyl cellulose proportions in the development of polyelectrolyte complexes for the sustained release of clarithromycin from matrix tablets. *Polymers, 13*, 2813. https://doi.org/10.3390/polym13162813
30. Hamman, J. H. (2010). Chitosan based polyelectrolyte complexes as potential carrier materials in drug delivery systems. *Marine Drugs, 8*, 1305–1322. https://doi.org/10.3390/md8041305
31. Ishihara, M., Kishimoto, S., Nakamura, S., Sato, Y., & Hattori, H. (2019). Polyelectrolyte complexes of natural polymers and their biomedical applications. *Polymers, 11*, 672. https://doi.org/10.3390/polym11040672
32. Kilicarslan, M., Ilhan, M., Inal, O., & Orhan, K. (2018). Preparation and evaluation of clindamycin phosphate loaded chitosan/alginate polyelectrolyte complex film as mucoadhesive drug delivery system for periodontal therapy. *European Journal of Pharmaceutical Sciences, 123*, 441–451. https://doi.org/10.1016/j.ejps.2018.08.007
33. Koetz, J., & Kosmella, S. (2007). *Polyelectrolytes and nanoparticles*. Springer.
34. Kulkarni, N., Jain, P., Shindikar, A., Suryawanshi, P., & Thorat, N. (2022). Advances in the colon-targeted chitosan based multiunit drug delivery systems for the treatment of inflammatory bowel disease. *Carbohydrate Polymers, 288*, 119351. https://doi.org/10.1016/j.carbpol.2022.119351
35. Lefnaoui, S., Moulai-Mostefa, N., Yahoum, M. M., & Gasmi, S. N. (2018). Design of antihistaminic transdermal films based on alginate–chitosan polyelectrolyte complexes: Characterization and permeation studies. *Drug Development and Industrial Pharmacy, 44*, 432–443. https://doi.org/10.1080/03639045.2017.1395461
36. Leonard, T. E., Liko, A. F., Gustiananda, M., Putra, A. B. N., Juanssilfero, A. B., & Hartrianti, P. (2023). Thiolated pectin-chitosan composites: Potential mucoadhesive drug delivery system with selective cytotoxicity towards colorectal cancer. *International Journal of Biological Macromolecules, 225*, 1–12. https://doi.org/10.1016/j.ijbiomac.2022.12.012
37. Liu, K., Chen, Y., Yang, Z., & Jin, J. (2023). Preparation and characterization of CS/γ-PGA/PC complex nanoparticles for insulin oral delivery. *Colloid & Polymer Science, 301*, 481–490. https://doi.org/10.1007/S00396-023-05078-1
38. Liu, Y., Long, M., Wang, Y., Liang, Z., Dong, Y., Qu, M., Ge, X., Nan, Y., Chen, Y., & Zhou, X. (2024). Chitosan-alginate/R8 ternary polyelectrolyte complex as an oral protein-based vaccine candidate induce effective mucosal immune responses. *International Journal of Biological Macromolecules, 275*, 133671. https://doi.org/10.1016/j.ijbiomac.2024.133671
39. Liu, Z., Wang, K., Peng, X., & Zhang, L. (2022). Chitosan-based drug delivery systems: Current strategic design and potential application in human hard tissue repair. *European Polymer Journal, 166*, 110979. https://doi.org/10.1016/j.eurpolymj.2021.110979
40. Maleki, R., Khedri, M., Rezvantalab, S., & Beheshtizadeh, N. (2024). Investigation of pH-dependent paclitaxel delivery mechanism employing chitosan-Eudragit bioresponsive nanocarriers: A molecular dynamics simulation. *Journal of Biological Engineering, 18*, 49. https://doi.org/10.1186/s13036-024-00445-0

41. Maliki, S., Sharma, G., Kumar, A., Moral-Zamorano, M., Moradi, O., Baselga, J., Stadler, F. J., & García-Peñas, A. (2022). Chitosan as a tool for sustainable development: A mini review. *Polymers, 14*, 1475. https://doi.org/10.3390/polym14071475
42. Malviya, R., Raj, S., Fuloria, S., Subramaniyan, V., Sathasivam, K., Kumari, U., Meenakshi, D. U., Porwal, O., Kumar, D. H., Singh, A., Chakravarthi, S., & Fuloria, N. K. (2021). Evaluation of antitumor efficacy of chitosan-tamarind gum polysaccharide polyelectrolyte complex stabilized nanoparticles of simvastatin. *International Journal of Nanomedicine, 16*, 2533–2553. https://doi.org/10.2147/IJN.S300991
43. Martín-Illana, A., Chinarro, E., Cazorla-Luna, R., Notario-Perez, F., Veiga-Ochoa, M. D., Rubio, J., & Tamayo, A. (2022). Optimized hydration dynamics in mucoadhesive xanthan-based trilayer vaginal films for the controlled release of tenofovir. *Carbohydrate Polymers, 278*, 118958. https://doi.org/10.1016/j.carbpol.2021.118958
44. Milenkova, S., Zahariev, N., Ambrus, R., Pilicheva, B., & Marudova, M. (2023). A study on the stoichiometry of casein/chitosan gel complexes as a delivery system for quercetin. *Applied Sciences, 13*, 10868. https://doi.org/10.3390/app131910868
45. Mishra, B., Mishra, M., & Yadav, S. K. (2017). Antibacterial loaded spray dried chitosan polyelectrolyte complexes as dry powder aerosol for the treatment of lung infections. *The Iranian Journal of Pharmaceutical Research (IJPR), 16*, 74–92. https://doi.org/10.22037/ijpr.2017.1953
46. Nayak, R., Halder, J., Rajwar, T. K., Pradhan, D., Rai, V. K., Dubey, D., Kar, B., Ghosh, G., & Rath, G. (2024). Metronidazole loaded chitosan–phytic acid polyelectrolyte complex nanoparticles as mucoadhesive vaginal delivery system for bacterial vaginosis. *International Journal of Biological Macromolecules, 255*, 128212. https://doi.org/10.1016/j.ijbiomac.2023.128212
47. Noreen, S., Pervaiz, F., Ashames, A., Buabeid, M., Fahelelbom, K., Shoukat, H., Maqbool, I., & Murtaza, G. (2021). Optimization of novel naproxen-loaded chitosan/carrageenan Nanocarrier-based gel for topical delivery: Ex vivo, histopathological, and in vivo evaluation. *Pharmaceuticals, 14*, 557. https://doi.org/10.3390/ph14060557
48. Nyamweya, N. N. (2021). Applications of polymer blends in drug delivery. *Future Journal of Pharmaceutical Sciences, 7*, 18. https://doi.org/10.1186/S43094-020-00167-2
49. Ortiz, J. A., Sepúlveda, F. A., Panadero-Medianero, C., Murgas, P., Ahumada, M., Palza, H., Matsuhiro, B., & Zapata, P. A. (2022). Cytocompatible drug delivery hydrogels based on carboxymethylagarose/chitosan pH-responsive polyelectrolyte complexes. *International Journal of Biological Macromolecules, 199*, 96–107. https://doi.org/10.1016/j.ijbiomac.2021.12.093
50. Palamidi, A., Koumentakou, I., Michopoulou, A., Bikiaris, D. N., & Terzopoulou, Z. (2024). Optimization of chitosan-gelatin-based 3D-printed scaffolds for tissue engineering and drug delivery applications. *International Journal of Pharmaceutics, 666*, 124776. https://doi.org/10.1016/j.ijpharm.2024.124776
51. Paşcalău, V., Bogdan, C., Pall, E., Matroş, L., Pandrea, S. L., Suciu, M., Borodi, G., Iuga, C. A., Ştiufiuc, R., Topală, T., Pavel, C., Popa, C., & Moldovan, M. L. (2020a). Development of BSA gel/Pectin/Chitosan polyelectrolyte complex microcapsules for Berberine delivery and evaluation of their inhibitory effect on Cutibacterium acnes. *Reactive and Functional Polymers, 147*, 104457. https://doi.org/10.1016/j.reactfunctpolym.2019.104457
52. Paşcalău, V., Tertis, M., Pall, E., Suciu, M., Marinca, T., Pustan, M., Merie, V., Rus, I., Moldovan, C., Topala, T., Pavel, C., & Popa, C. (2020b). Bovine serum albumin gel/polyelectrolyte complex of hyaluronic acid and chitosan based microcarriers for Sorafenib targeted delivery. *Journal of Applied Polymer Science, 137*, e49002. https://doi.org/10.1002/app.49002
53. Patil, P., Vankani, A., & Sawant, K. (2024). Design, optimization and characterization of atorvastatin loaded chitosan-based polyelectrolyte complex nanoparticles based transdermal patch. *International Journal of Biological Macromolecules, 274*, 133219. https://doi.org/10.1016/j.ijbiomac.2024.133219

54. Potaś, J., Szymańska, E., & Winnicka, K. (2020). Challenges in developing of chitosan—Based polyelectrolyte complexes as a platform for mucosal and skin drug delivery. *European Polymer Journal, 140*, 110020. https://doi.org/10.1016/j.eurpolymj.2020.110020
55. Potaś, J., Szymańska, E., Wróblewska, M., Kurowska, I., Maciejczyk, M., Basa, A., Wolska, E., Wilczewska, A. Z., & Winnicka, K. (2021). Multilayer films based on chitosan/pectin polyelectrolyte complexes as novel platforms for buccal Administration of Clotrimazole. *Pharmaceutics, 13*, 1588. https://doi.org/10.3390/pharmaceutics13101588
56. Potaś, J., Wach, R. A., Rokita, B., Wróblewska, M., & Winnicka, K. (2024). Evaluation of the impact of tragacanth/xanthan gum interpolymer complexation with chitosan on pharmaceutical performance of gels with secnidazole as potential periodontal treatment. *European Journal of Pharmaceutical Sciences, 192*, 106657. https://doi.org/10.1016/j.ejps.2023.106657
57. Potaś, J., Wilczewska, A. Z., Misiak, P., Basa, A., & Winnicka, K. (2022). Optimization of multilayer films composed of chitosan and low-Methoxy Amidated pectin as multifunctional biomaterials for drug delivery. *International Journal of Molecular Sciences, 23*, 8092. https://doi.org/10.3390/ijms23158092
58. PubChem, Chitosan. https://pubchem.ncbi.nlm.nih.gov/compound/71853. Accessed 14 Apr 25.
59. Raina, N., Rani, R., Thakur, V. K., & Gupta, M. (2023). New insights in topical drug delivery for skin disorders: From a nanotechnological perspective. *ACS Omega, 8*, 19145–19167. https://doi.org/10.1021/acsomega.2c08016
60. Samanta, R., Nayak, S., Das, B., & Nayak, A. K. (2023). Chitosan-carboxymethyl tamarind gum in situ polyelectrolyte complex-based floating capsules of ofloxacin: In vitro-in vivo studies. *International Journal of Biological Macromolecules, 253*, 127507. https://doi.org/10.1016/j.ijbiomac.2023.127507
61. Schoeller, J., Itel, F., Wuertz-Kozak, K., Gaiser, S., Luisier, N., Hegemann, D., Ferguson, S. J., Fortunato, G., & Rossi, R. M. (2021). pH-responsive chitosan/alginate polyelectrolyte complexes on electrospun PLGA nanofibers for controlled drug release. *Nanomaterials Basel Switz, 11*. https://doi.org/10.3390/nano11071850
62. Siavashy, S., Soltani, M., Ghorbani-Bidkorbeh, F., Fallah, N., Farnam, G., Mortazavi, S. A., Shirazi, F. H., Tehrani, M. H. H., & Hamedi, M. H. (2021). Microfluidic platform for synthesis and optimization of chitosan-coated magnetic nanoparticles in cisplatin delivery. *Carbohydrate Polymers, 265*, 118027. https://doi.org/10.1016/j.carbpol.2021.118027
63. Silant'ev, V. E., Belousov, A. S., Trukhin, F. O., Struppul, N. E., Shmelev, M. E., Patlay, A. A., Shatilov, R. A., & Kumeiko, V. V. (2024). Rational design of Pectin-Chitosan Polyelectrolyte Nanoparticles for enhanced Temozolomide delivery in brain tumor therapy. *Biomedicine, 12*, 1393. https://doi.org/10.3390/biomedicines12071393
64. Sladek, S., McCartney, F., Eskander, M., Dunne, D. J., Santos-Martinez, M. J., Benetti, F., Tajber, L., & Brayden, D. J. (2020). An enteric-coated polyelectrolyte nanocomplex delivers insulin in rat intestinal instillations when combined with a permeation enhancer. *Pharmaceutics, 12*, 259. https://doi.org/10.3390/pharmaceutics12030259
65. Surini, S., Providya, R., & Putri, K. S. S. (2019). Formula optimization of rifampicin dry powder inhalation with chitosan-xanthan carrier using response surface methodology. *Journal of Applied Pharmaceutical Science, 9*, 33–41. https://doi.org/10.7324/japs.2019.90106
66. Vasconcelos Silva, E. d. L., Oliveira, A. C. d. J., Moreira, L. M. C. d. C., Silva-Filho, E. C., Wanderley, A. G., Soares, M. F. d. L. R., & Soares-Sobrinho, J. L. (2023). Insulin-loaded nanoparticles based on acetylated cashew gum/chitosan complexes for oral administration and diabetes treatment. *International Journal Of Biological Macromolecules, 242*, 124737. https://doi.org/10.1016/j.ijbiomac.2023.124737
67. Wang, W.-Y., Hui, P. C. L., Wat, E., Ng, F. S. F., Kan, C.-W., Lau, C. B. S., & Leung, P.-C. (2016). Enhanced transdermal permeability via constructing the porous structure of Poloxamer-based hydrogel. *Polymers, 8*, 406. https://doi.org/10.3390/polym8110406
68. Wu, D., Zhu, L., Li, Y., Zhang, X., Xu, S., Yang, G., & Delair, T. (2020). Chitosan-based colloidal polyelectrolyte complexes for drug delivery: A review. *Carbohydrate Polymers, 238*, 116126. https://doi.org/10.1016/j.carbpol.2020.116126

69. Wu, M., Ping, H., Wang, K., Ding, H., Zhang, M., Yang, Z., & Du, Q. (2023). Oral delivery of pectin-chitosan hydrogels entrapping macrophage-targeted curcumin-loaded liposomes for the treatment of ulcerative colitis. *International Journal of Pharmaceutics, 647*, 123510. https://doi.org/10.1016/j.ijpharm.2023.123510
70. Zhang, Y., Wang, L., Wang, Z.-D., Zhou, Q., Zhou, X., Zhou, T., Guan, Y.-X., & Liu, X. (2023). Surface-anchored microbial enzyme-responsive solid lipid nanoparticles enabling colonic budesonide release for ulcerative colitis treatment. *Journal of Nanobiotechnology, 21*, 145. https://doi.org/10.1186/s12951-023-01889-0
71. Zhao, J., Cui, L., Wang, X., & Deng, C. (2022). Dual ionically crosslinked chitosan–based injectable hydrogel as drug delivery system. *Colloid & Polymer Science, 300*, 1075–1086. https://doi.org/10.1007/S00396-022-05003-Y
72. Zuglianello, C., Martins Silva, N. G., & Lemos-Senna, E. (2023). Polysaccharide-peptide complexes stabilized around nanoemulsion droplets: A new approach for nasal delivering of pramlintide and insulin. *Journal of Drug Delivery Science and Technology, 85*, 104527. https://doi.org/10.1016/j.jddst.2023.104527

Chitosan-Based Nanogels in Drug Delivery Applications

Kavipriya Murugaiyan, Sasvat Sayee Ram Ramesh, and Aravind Kumar Rengan

Abstract Chitosan-based nanogels represent a significant advancement in the field of biomedical applications, particularly in drug delivery and regenerative medicine. These nanogels leverage the unique properties of chitosan, a natural polysaccharide derived from chitin, which is known for its biodegradability, biocompatibility, and non-toxicity. The integration of chitosan into nanogels enhances their functionality, making them suitable for a variety of applications, including targeted drug delivery, antibacterial treatments, and tissue engineering. This chapter delves into the synthesis methods of chitosan nanogels and also highlights their potential as drug delivery carriers, such as in anti-tumour drugs and proteins, for bioimaging, bone, and dental tissue engineering. The properties of the chitosan nanogels are also compared with other polymeric nanogels, such as PEG and PLGA-based nanogels. While chitosan-based nanogels offer numerous advantages, challenges remain in optimizing their synthesis and ensuring consistent performance across different applications. The potential for greater interactions between nanomaterials and biological systems necessitates thorough biocompatibility assessments. Future research is likely to focus on enhancing the functionalization of chitosan nanogels and expanding their applications in diverse biomedical fields.

Keywords Nanogels · Synthesis · Drug delivery · Tissue engineering

1 Introduction

Biopolymers are prepared from renewable agricultural or marine food sources to meet the ecological demands. Polysaccharide-based biopolymers are known for their bioactivity and are mainly derived from agricultural sources and crustaceans [1]. They have a specific structure, physico-chemical properties, chemical stability, high reactivity, and excellent selectivity towards aromatic compounds and metals

K. Murugaiyan · S. S. R. Ramesh · A. K. Rengan (✉)
Department of Biomedical Engineering, Indian Institute of Technology-Hyderabad, Sangareddy, Telangana, India
e-mail: aravind@bme.iith.ac.in

S. Jana (ed.), *Chitosan-Based Drug Delivery Systems*, AAPS Introductions in the Pharmaceutical Sciences 16, https://doi.org/10.1007/978-3-032-08347-0_4

[2]. Major polysaccharide polymers include cellulose and chitin. Cellulose is a polymer formed by thousands of β-linked-D-glucose monosaccharide units. A glycosidic bond is formed by the condensation of monosaccharide units. Chitosan is a copolymer consisting of N-acetyl-2-amino-2-deoxy-β-D-glucopyranose and 2-amino-2-deoxy-D-glucopyranose linked (1 → 4)-β glycosidic bonds. Similar to cellulose but has a hydroxyl group at the C-2 position converted to an aceto-amido group. Starch is also one of the abundant polymers, present as an energy-storing material in plants. Chitosan came into existence when Rouget discussed the deacetylated forms of chitin in 1859. He observed that deacetylated forms were able to dissolve in acidic conditions. Numerous works have been done demonstrating the substantial role of chitosan in biomedical applications. Formulations derived from chitosan have been explored for potential in drug delivery applications. Its precursor, chitin, is derived from natural sources such as the exoskeleton of insects, crustaceans, fungi, etc., and chitin is the second most abundant polysaccharide present [3, 4].

The pharmacokinetics and pharmacodynamics of a drug play a significant role in the design of the drug delivery vehicles [5]. Drug delivery vehicles are explored and established, but still limited by their poor biocompatibility, stability, non-degradability, and off-target effects [6]. Hydrogels are highly favored in this regard due to their high biocompatibility, stability, and good degradability. Hydrogels are hydrophilic, polymeric, three-dimensional structures that have water-absorbing properties. The crosslinking of the hydrogels is formed by following either chemical or physical methods. Physical crosslinking involves hydrogen bonding, ionic bonding, or hydrophobic forces. These gels can be made environmentally responsive to conditions such as pH, temperature, and ionic strength of the solution. The hydrogels can be charged or non-charged based on their functional groups [7, 8]. Nanogels are gels of any shape but in the nanometer range, according to the IUPAC definition [9]. They can retain more water content compared to the conventional hydrogels, which makes them excellent carriers having good biocompatibility. These polymeric networks are most suitably used as drug delivery carriers. They have good stability, increased loading capacity, responsive to environmental factors such as pH, temperature, and ionic strength [8, 10]. This chapter discusses the overview and specific role of chitosan nanogels for drug delivery applications.

2 Chitosan and Its Origin

Chitin ranks as the second most prevalent natural polysaccharide after cellulose, primarily located in the shells of crustaceans or the cell walls of fungi, crustaceans, molluscs, arthropods, and certain types of seaweed [11, 12]. Other sources include cockroaches, mosquitoes, silkworms, honeybees, Drosophila melanogaster, and microorganisms such as fungus, yeast, algae, and septa of Ascomycetes, Zygomycetes, Basidiomycetes, and Deuteromycetes, worms, mushrooms, etc. [13]. Despite its abundance, chitin is not widely utilized in industrial applications due to its insolubility in many solvents and the challenges associated with extracting it in a pure form from natural sources and processing it consistently under economically viable conditions [14]. The insolubility is attributed to the presence of numerous intra- and intermolecular hydrogen bonds in most solvents, including acids [15]. Typically, the conventional methods prepare chitosan using three fundamental stages: deproteinization, demineralization and deacetylation. In industries, the preparation of chitosan involves four processing stages: deproteinization, demineralization, decolourization, and deacetylation. Deproteinization is the process of eliminating proteins using a strong base, whereas demineralization involves the removal of calcium carbonates and calcium phosphates with a strong acid. Additionally, decolourization and deacetylation refer to the removal of pigments and the transformation of chitin into chitosan, respectively. Decolourization can be achieved with an alcohol like ethanol, while the deacetylation process is carried out using a strong alkali such as NaOH [13, 14].

3 Preparation of Nanogels

The preparation of nanogels involves the formation of crosslinked polymeric networks at the nanoscale, using either physical or chemical crosslinking methods. Various synthetic and natural polymers can be used for the desired properties and intended application of the nanogels [16, 17]. Nanotechnology also allows for the incorporation of stimuli-responsive elements, enabling drug release in response to environmental triggers like pH, temperature, or enzymes, thus enhancing the specificity and efficacy of drug delivery [18]. Generally, Nanogels can be prepared by the following methods.

(a) *Radical crosslinking polymerization*

By using mono-vinylic monomers and di- or multi-functional crosslinkers, these nanogels can be synthesized. This method involves techniques such as solution polymerization, heterogeneous polymerization and controlled or living polymerization [19, 20].

(b) *Ring-opening polymerization*

This involves mechanisms of ring-opening polymerization of cyclic monomers that include covalent, coordination, ionic, enzymatic, etc. Mainly involves the use of carboxy anhydrides obtained from α-amino acids by using a difunctional crosslinker and PEG-containing macroinitiator. The developed core-shell nano-gel comprises a synthetic polypeptide covering an outer-PEG shell [21–23].

(c) *Cross-linking of prepolymers*

This can be achieved by aggregation of hydrophobic polysaccharides due to critical micellar concentration, resulting in physically crosslinked nanogels [24]. Inverse nanoprecipitation or solvent evaporation method can be used to prepare in situ physically or chemically crosslinked prepolymers [25]. Self-assembly or crosslinking of certain co-polymers can be utilised to form nanogels [26].

(d) *Click chemistry approaches*

This method includes copper-catalysed and copper-free methodologies such as azide cyclo-additions and thiol-ene reactions [27].

(e) *Thiol-disulfide chemistry*

It manipulates the reversible nature of disulfide bonds. Polymeric chains having pendant thiols and disulfides can be crosslinked to form nanogels [28, 29].

4 Chitosan Derivatives

The major drawback of chitosan is its inability to dissolve in neutral aqueous conditions. The solubility of chitosan is mainly determined by its molecular weight. Chemical modification allows chitosan to overcome these drawbacks and also to serve the intended purpose. For instance, glycated chitosan, formed by the reaction between galacturonic acid and chitosan, is water soluble. Similarly, trimethyl chitosan is soluble in neutral and alkaline conditions [30, 31]. The -OH and -NH2 active groups of the chitosan molecule facilitate versatile chemical modifications [32, 33]. Chemically modified chitosan derivatives have better physical and chemical properties. They have increased biocompatibility, bioactivity, degradability and nontoxicity [34–36].

5 Methods of Preparation of Chitosan Nanogels

Chitosan nanogels are versatile materials with significant potential in drug delivery and biomedical applications due to their biocompatibility, biodegradability, and ability to respond to stimuli. Various mechanisms for the formation of chitosan nanogels have been explored, each offering unique advantages and characteristics. These mechanisms include ionic gelation, covalent and ionic crosslinking, host-guest interactions, and physical gelation. Each method influences the properties and potential applications of the resulting nanogels.

5.1 Ionic Gelation

Ionic gelation is a widely used method for synthesising chitosan nanoparticles and nanogels. It involves the interaction of chitosan with polyanions, such as tripolyphosphate, leading to the formation of nanoparticles through complexation and chain scission processes. This most common method relies on the use of sodium tripolyphosphate (TPP), which was first presented by Calvo et al. in 1997 [37]. This method is advantageous due to its simplicity and the ability to control particle size and distribution. The ionic gelation process can be further optimised by adjusting the molecular interactions and cross-linking configurations, such as H-link, T-link, and M- link, which affect the stability and interaction energies of the formed nanogels [38–40].

5.2 Covalent and Ionic Crosslinking

Chitosan nanogels can also be formed through covalent and ionic crosslinking, which enhances their stability and drug encapsulation capabilities. Crosslinkers like genipin are used to create biocompatible nanogels with increased hydrophobicity and reduced charge density, which are crucial for uniform size distribution and controlled drug release [41, 42]. The reaction kinetics of genipin crosslinking show a size-dependent, first-order reaction, leading to branched nanogels with varying molar masses and consistent radii of gyration [42].

5.3 Host-Guest Interactions

Supramolecular chitosan nanogels can be synthesised using host-guest chemistry, where chitosan is crosslinked through interactions with molecules like cucurbit [8] uril. This method allows for stimulus-responsive drug release, making it suitable for targeted cancer therapy. The encapsulated drugs are released in response to specific stimuli, such as endogenous spermine or exogenous amantadine, which disrupt the host-guest interactions [43, 44].

5.4 Physical Gelation

Physical gelation methods, such as ammonia-induced gelation in reverse-phase emulsions, result in colloidal chitosan nanogels with controlled particle size and morphology. These nanogels exhibit electrosteric stabilization, which is influenced by the conformation of chitosan chains at the gel-liquid interface. This method is beneficial for creating nanogels with a core-shell structure and partial crystallinity [45–48].

5.5 *pH-Responsive and Injectable Hydrogels*

Chitosan nanogels can also be synthesised to be pH-responsive and injectable, which is advantageous for biomedical applications. These nanogels are formed through chemical and physical crosslinking with polymers like poly methacrylic acid, resulting in stable, biocompatible nanogels with high thermal stability and low cytotoxicity [49, 50]. While each method of chitosan nanogel formation offers distinct advantages, the choice of mechanism depends on the intended application and desired properties of the nanogels. For instance, ionic gelation is favoured for its simplicity and control over particle size, while host-guest interactions provide targeted drug delivery capabilities. The development of chitosan nanogels continues to evolve, with ongoing research focusing on optimising these mechanisms to enhance their functionality and application in various fields.

6 Chitosan Nanogels in Drug Delivery

The modern concept of Drug delivery aims at site-specific delivery of drugs to avoid the off-target effects [6]. Nano-formulations offer targeted delivery that causes effective localization of the drug in the targeted biological area [51]. Many nano-formulations have been proved to reduce the dosage, enhance the stability and biocompatibility, increase efficacy and cost effectiveness compared to drugs available as conventional medicines [52]. Apart from having good wound healing capacity, chitosan nanoformulations can serve as ideal drug carriers [53]. They have good loading, releasing and targeting functions serving as effective drug delivery systems [54, 55]. Chitosan nanogels have gained significant attention in many biomedical areas because of their distinctive structural and functional characteristics (Fig. 1). Their most prominent applications are in targeted drug delivery, tissue engineering, cell culture, bioimaging, and therapeutics. Monodisperse chitosan nanogels were developed with sizes ranging from 268–382 nm. They showed biocompatibility in vitro and zebrafish embryos up to 100 μg/mL [56]. The study focused on the formulation and assessment of chitosan-based nanogels for the sustained delivery of madecassoside. pH- responsive nanogels using a free radical polymerization involving cross-linking of chitosan with methacrylic acid in the presence of a cross-linker, N′, N′-methylene bis(acrylamide). The characterization of the nanogels showed a decrease in polymer crystallinity after polymerization. The swelling behaviour of the nanogels was significantly influenced by pH levels. High swelling was observed at pH 7.4, while lower swelling occurred at pH 1.2 and 4.6, demonstrating the excellent pH-sensitive nature of the engineered nanogels. The release of the drug madecassoside from the nanogels was also pH-dependent. Maximum drug release occurred at pH 7.4, indicating that the nanogels can effectively sustain the release of the drug in a physiological environment [57]. Chitosan nanogels loaded with the drug sorafenib tosylate were

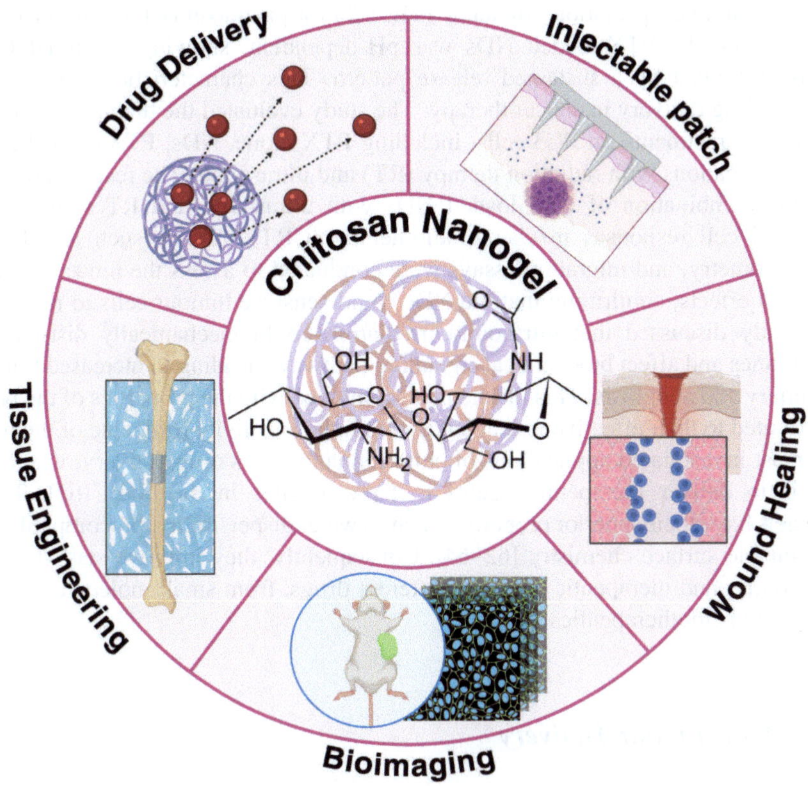

Fig. 1 Chitosan nanogel in drug delivery

successfully created using a modified solvent evaporation technique. A total of nine batches were formulated by varying the concentration of chitosan (12, 14, 16 mg/ml) and the sonication time (5, 10, 15 min) to optimise the formulation. The optimised batch, referred to as F7, exhibited a particle size of 182 nm and a zeta potential of 64.5 mV, with an impressive %EE of 98%. The drug release profile indicated that the release of Sorafenib Tosylate from the F7 nanogel was significantly higher at pH 6 (97.556%) compared to pH 7.4 (45.13%). The stability tests conducted over 6 months showed that the nanogel maintained its integrity, with no colour change, phase separation, or opacity observed, highlighting the promising nature of the developed nanogel for targeted cancer therapy [58]. Nanogels prepared using chitosan hydrochloride- carboxymethyl starch behaved as viscoelastic solids in rheological experiments. These nanogels showed spherical morphology and higher encapsulation efficiency of curcumin [59]. Nanodroplets(NDs) were successfully prepared using a homogenization/emulsion method, resulting in round, spherical shapes with relatively smooth surfaces and core-shell structures. The sizes of the NDs were uniform, measuring less than 300 nm with a polydispersity index (PDI) of less than 0.3 at pH levels of 6.0 or higher. The NDs exhibited high loading efficiency

(20.35%) and encapsulation efficiency (91.58%) for paclitaxel (PTX). In vitro drug release from the PTX-loaded NDs was pH-dependent, showing an initial burst release followed by a sustained release pattern. This characteristic is crucial for targeted drug delivery in cancer therapy. The study evaluated the therapeutic effects of various treatments on PC3 cells, including PTX alone, NDs, PTX-loaded NDs, and combinations with radiation therapy (RT) and ultrasound. The results indicated that the combination of PTX-loaded NDs with ultrasound and RT significantly enhanced cell responses in chemoradiotherapy (CRT). Assays such as CCK-8, flow cytometry, and migration assays were conducted to assess the bioactivity and radiation effects, confirming that the NDs could sensitise tumour cells to radiation. The study discussed that ultrasound irradiation could mechanically disrupt cell membranes and affect biochemical signalling pathways, leading to increased tumour cell injury and apoptosis. This mechanism, combined with the properties of the NDs, contributed to their effectiveness in enhancing radiosensitivity [60]. One of the main and most investigated applications of chitosan nanogels is drug delivery, where the aim is to deliver therapeutic agents to targeted sites in the body [61]. These nanogels have their superior properties such as water dispersibility, biocompatibility, and tunable surface chemistry [62–64]. Consequently, they improve stability, bioavailability, and therapeutic effect of different drugs, from small molecules to proteins and chemotherapeutics.

6.1 Anti-tumour Delivery

Tumour is defined as a neoplasm of infiltrating local tissue cells formed due to local tissue factors [65]. Most of the anti-cancer agents are cytotoxic, hence, chitosan is explored for drug delivery applications. Chitosan nanogels have been extensively investigated for the treatment of cancer, particularly for drug delivery.

6.1.1 Doxorubicin

The most commonly used chemotherapeutic agent is extracted from Streptomyces spp. It acts by intercalating the DNA and inducing apoptosis [66]. Chitosan nanogels permit controlled release, site-specific delivery to tumour cells such as lung cancer, colorectal cancer, [67, 68] A pH-responsive CSNPs/P(NIPAm-co-AAc) (poly-N-Isopropylacrylamide) nanogel loaded with doxorubicin was developed, which showed enhanced anti-cancer activity against Ehrlich ascites carcinoma–bearing mice. The developed nanogel was given intraperitoneally at doses equivalent to 1.5 and 3 mg/Kg of Doxorubicin for 14 days, which exhibited superior activity against free drug [69]. Carboxymethyl chitosan was modified with glycidyl methacrylate and folic acid to obtain nanogels (GCMCS-FA-DOX). The nanogels

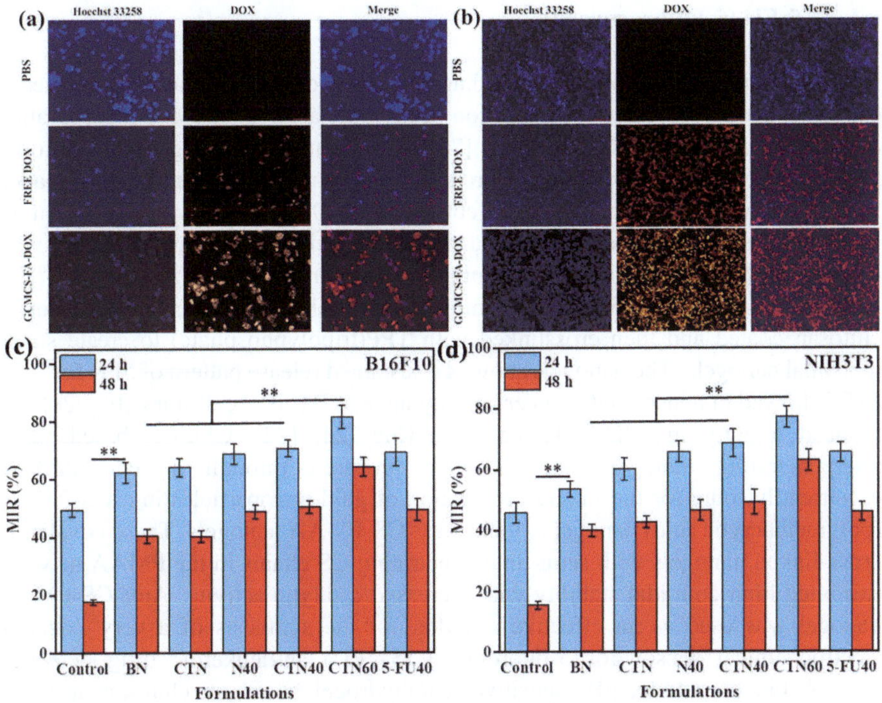

Fig. 2 Fluorescence images showing higher DOX uptake from GCMCS-FA-DOX nanogels than free DOX, in (**a**) HepG2 and especially more in (**b**) HCT-116 cells, due to folate receptor-mediated targeting [70]. Maximum internalization ratio (MIR) of the formulation was higher in (**c**) B16F10 cells than (**d**) NIH3T3 cells, likely due to enhanced uptake via the EPR effect and greater drug release in the acidic tumour microenvironment [72]. (Figure (c, d) is reproduced from an open-access article distributed under the terms of the Creative Commons Attribution 4.0 (CC BY-4.0) license)

showed an encapsulation efficiency of 94.7% and a loading efficiency of 15.6%. The nanogels were responsive to high glutathione and pH. The particle size of the nanoparticles was around 220 ± 38 nm. Enhanced drug internalization and higher toxicity in HCT-116 cells were observed with GCMCS-FA-DOX compared to HepG2 cells (Fig. 2a, b) [70]. Nanoparticle-nanogel system was developed using multi-core iron oxide and manganese ferrite nanoparticles, which were further incorporated into chitosan- alginate nanogels to create a multi-modal drug delivery vehicle. The magnetic nanogels were loaded with doxorubicin, which showed an encapsulation efficiency of 70%. Surface modification with citrate and bovine serum albumin improved colloidal stability and changed surface properties. BS-functionalized nanogels showed enhanced release stability at physiological pH, with release profiles being controlled by diffusion-controlled kinetics [71].

6.1.2 5-FU (5-Fluorocuracil)

It is an antimetabolite drug that is used against many cancers, particularly colorectal cancer. It causes its effects by inhibition of thymidylate synthase and incorporation of its metabolites into RNA or DNA [73]. A pH-sensitive biodegradable chitosan nanogel(FCNGL) was synthesised through the ionotropic gelation method, loaded with 5-Fluorouracil at a low drug concentration(0.2%). In-vivo assessment of DMBA-croton oil induced melanoma model in Swiss albino mice coupled with gamma scintigraphy showed substantial tumour-targeted localization of ^{99m}Tc-labelled nanogels [74]. 5FU loaded niosomal particles were effectively prepared, chitosan coated and then crosslinked with TPP(tripolyphosphate) to create stable niosomal nanogels. The nanogels showed a sustained release pattern of 5FU for 72 h. MTT showed enhanced anti-cancer activity against B16F10 cell lines (Fig. 2c) and reduced toxicity in NIH-3 T3 cell line (Fig. 2d) [72]. Chitosan-based semi-interpenetrating nanogel catalysts were prepared through a reductant-free, eco-friendly route for the in situ preparation of gold nanoparticles in a CS-PMAA (polymethacrylic acid) nanogel to yield Au@CS-PMAA nanogels. These covalently cross-linked nanogels with semi- interpenetrating CS chains in the PMAA network exhibited high colloidal stability with increased catalytic activity. Au@CS-PMAA nanogel was used as an effective modifier for the synthesis of a very selective electrochemical sensor for 5-fluorouracil (5- FU) antileukemic drug detection [75]. A biocompatible, pH- sensitive nanohydrogel based on chitosan, agarose, and graphene oxide (CS/AG/GO) was prepared by glyoxal as a cross-linking agent. Later, 5-FU-loaded nanocomposites (NCs) were prepared through a water-in-oil-in-water (W/O/W) emulsification process. The NCs exhibited significantly high drug loading (57%) and encapsulation efficiency (92%), superior to those of previously reported carriers. At pH 5.4, sustained drug release was obtained with almost complete release of 5-FU in 48 h. In addition, cytotoxicity studies against MCF-7 breast cancer cells indicated significant inhibition of cell viability (~23%) after treatment with CS/AG/GO/5-FU (Fig. 3a), suggesting its potential as an efficient pH-sensitive nanocarrier for targeted breast cancer therapy [76].

6.1.3 Methotrexate

The hydrophilic anticancer drug methotrexate is used as a folate antagonist that inhibits the enzyme dihydrofolate reductase, which helps in the synthesis of DNA OR RNA [77]. Methotrexate-loaded hydrogel nanoparticles were prepared by the ionic gelation method and characterised in vitro, and their capacity to cross the blood–brain barrier (BBB) in intact animal models was assessed in vivo. Surface modification with polysorbate 80 was utilised to improve brain delivery. After intravenous injection with a dose of 25 mg/kg, both the surface-modified and unmodified nanogels delivered significantly increased levels of methotrexate to the brain, levels greater than tenfold those of the free drug in some instances (Fig. 3b, c) [78].

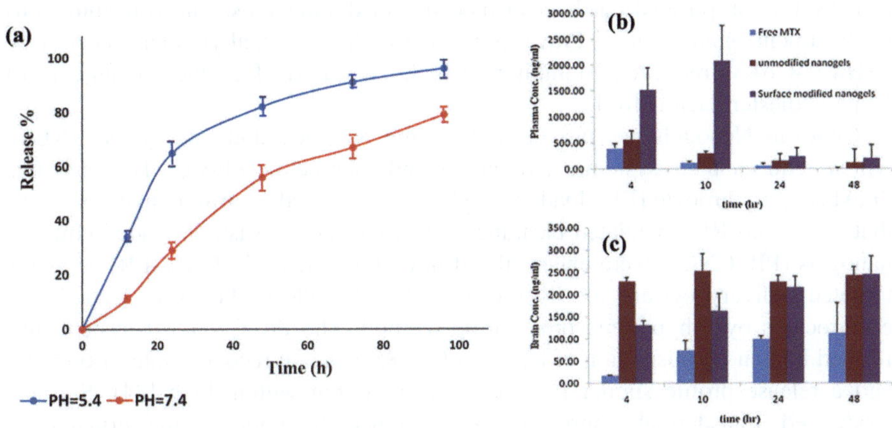

Fig. 3 (**a**) pH-sensitive release profile of 5-FU from nanocarriers, showing ~65% release at pH 5.4 in 24 h and nearly complete release by 48 h, while only ~50% release occurred at pH 7.4 [76]. Plasma (**b**) and brain (**c**) methotrexate levels over time in rats after 25 mg/kg IV of free drug, unmodified, and surface-modified nanogels [78]. (Figure (a) is reproduced from an open-access article distributed under the terms of the Creative Commons Attribution 4.0 (CC BY-4.0) license).

6.1.4 Paclitaxel

Paclitaxel is an anti-microtubule agent that stabilizes the existing microtubules and prevents their disassembly, leading to cell cycle arrest [79]. Chitosan–Pluronic P123- based nanogel systems were developed with different folate ligand conjugation schemes for targeted delivery of paclitaxel (PTX). In one system, the Conjugation of folic acid (FA) to chitosan directly (FA-Cs-P123) was done, whereas FA was conjugated to the P123 segment in the other system (Cs-P123-FA). Quantitative analysis indicated Cs-P123-FA contained considerably more FA (0.71 wt%) than FA- Cs-P123 (0.31 wt.%). The drug release studies showed a sustained, pH-responsive release profile. The Cs-P123-FA nanogels displayed better cytotoxic activity against MCF-7 cells due to their higher FA density, which improved targeting efficiency, showing highly promising targeted drug delivery systems for breast cancer treatment [80]. Self-assembled nanogels were synthesised based on poly (ethylene glycol) methyl ether (mPEG) and chitosan for the co-delivery of chemotherapeutic agents, paclitaxel (PTX) and 5-fluorouracil (5-FU). For improved aqueous solubility of chitosan and enhancing the potential of drug delivery systems (DDDS), different molar ratios of mPEG were used. The entrapment efficiencies were about 50% for PTX and 10% for 5-FU. Drug release experiments showed that both PTX and 5-FU release profiles were affected by the hydrophilic nature of mPEG, which resulted in sustained and controlled release [81]. Core-shell nanogels were developed using gold and chitosan(gold/CS) loaded with curcumin, which showed good cellular uptake and cytotoxicity in MCF-7 cells compared to Huh7 cells [82]. Repurposed drugs have also been loaded into chitosan nano-formulations for treatment against cancer. Pravastatin is a 3-hydroxy-3-methylglutaryl-coenzyme A(HMG-CoA) reductase inhibitor which is used for the treatment of Hypercholesterolaemia [83].

Chitosan Nanogels are fast cleared by the reticuloendothelial system (RES), which reduces their systemic circulation and therapeutic effects. To solve this drawback, erythrocyte (ER)-loaded CNGs were explored as drug delivery systems that can avoid RES-mediated clearance. In this work, pravastatin-loaded chitosan nanogels (PR–CNGs) were encapsulated in erythrocytes (PR–CNG–ER) as a new targeted delivery system for liver cancer therapy. PR–CNGs were loaded into erythrocytes by a hypotonic preswelling method. The PR–CNG–ER system thus formed had an encapsulation efficiency of 36.85%, a cell recovery rate of 66.82%, and a release profile similar to that of haemoglobin within 48 h [84]. Surface-modulated transdermal approach was designed to improve the efficacy of Temozolomide by a Poly(lactic-co-glycolic acid)–chitosan double-walled nanogel (PCNGL). The nanogel was prepared by a dual-approach method, and the haemolytic ratio was around 3.37 ± 0.19, affirming the biological safety of the formulation for transdermal delivery.

In vitro drug release studies showed 85% drug release at slightly acidic pH conditions simulating the microenvironment of the skin and 80% drug incorporation with porcine epidermal tissue in ex vivo studies [85]. Magnetic chitosan nanogels were prepared by crosslinking epichlorohydrin with O-carboxymethyl chitosan, further embedded with iron oxide nanoparticles. The developed hydrogel showed a ferro- ferrimagnetic behaviour and sustained release of the maleimide compounds under an external magnetic field [86]. The developed nanoformulations showed high encapsulation of the bioactive compounds, drugs with increased biocompatibility and efficacy.

6.2 Protein Delivery

Proteins are the fundamental building blocks to create basic structures such as collagen and the cytoskeleton [87]. Chitosan/TPP microgels and nanogels were evaluated for their protein uptake, which was found to be higher depending on the yield, so that more protein binding sites are available. The formed chitosan/TPP particles were also competed by dissolved chitosan for the protein uptake [88]. An injectable hybrid composite nanogel(nPCGs) system was developed through physical and chemical crosslinking with bioactive materials such as protein cytokines, recombinant human interleukin-2, recombinant human interferon-Gamma and doxorubicin, which exhibited simultaneous delivery of proteins and doxorubicin for cancer treatment [67]. Chitosan nanogels with positive charge were prepared using hexanoyl chitosan using the ionotropic gelation method, and with negative charge by the calcium-induced coacervation method. Protein complexation was found to be preferential with oppositely charged proteins, and reduced particle size suggested strong electrostatic interactions, confirming the significant role played by electrostatics in mediating nanogel-biomolecule interactions [89]. Using the reverse emulsion method, chitosan nanogels were prepared in a triglyceride mixture. The obtained nanogels formed polyelectrolyte complexes with chondroitin sulphate without altering the size of the previous dispersion. HIV-1 P24, a capsid protein, showed multiple interactions with the hydrogels, with the limited amount suggesting a surface association. In the case of IgG, the negative hydrogels showed higher loading capacity compared to the positive ones [48]. Recombinant NcPDI from *E. coli* was successfully loaded in the chitosan/alginate and chitosan/alginate-mannose nanogels. Vaccination of the nanogels was done in mice with Neospora caninum tachyzoites infection. Intranasal delivery of the alginate-coated nanogels with recNcPDI protected 80% of the mice [90].

6.3 Bioimaging

Bioimaging has become significant in research and also in the clinic, with the capacity to study and analyse biological processes from the subcellular grade to the whole organism. As a result of its capability to automatically discover, measure, and profile phenotypic variation, bioimaging has been an effective method to study intricate details of biological phenomena [91, 92]. Chitosan-based hybrid nanogels were prepared using chitosan and poly (methacrylic acid) immobilised with CdSe quantum dots. The covalently crosslinked nanogels showed lower toxicity and helped in optical sensing of B16F10 cells, pH sensing of the environment and regulated the release of the anticancer drug according to the pH [93]. A multifunctional chitosan-quantum dot composite nanogel probe was designed using cyanogen bromide(CnBr) coupling. The probe showed very limited toxicity and was able to bind normal cells, cancer cells and macrophages. Modification with folic acid increased its targetability in human ovarian cancer cells(OVCAR3) [94]. Self- assembling nanogels of silver nanoclusters and chitosan were developed, which showed enhanced luminescence via strong matrix-ligand coordination between thiolate ligands and chitosan that ultimately strengthens the molecular structure on the surface of silver nanoclusters (Ag NCs). Nanogel showed excellent antibacterial which was ten-fold higher than the control AgNCs (Fig. 4a) [95]. Fluorescent carbon dots were prepared using a hydrothermal reaction from grapeseed extracts. These carbon dots were then simultaneously loaded into folic acid-chitosan conjugate along with antimiR- 21 for the formation of the nanogels. The developed nanoparticles showed uniform distribution in the hydrodynamic diameter (104.9 ± 30.68 nm) with good stability that allowed the transfer of antimiR-21 sequence to cancer cells with good entrapment efficiency (80–90%). They demonstrated higher cellular uptake in SKOV3(ovarian cancer cells), which expressed increased folate receptors compared to folate-negative NIH3T3 fibroblasts (Fig. 4b–d) [96]. Hybrid PEG-chitosan and graphitic carbon dots were prepared for two-photon fluorescence bioimaging, pH and near-infrared light (NIR)-

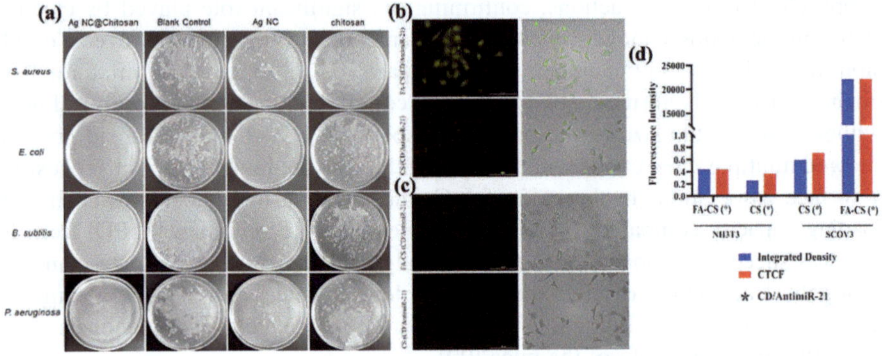

Fig. 4 (**a**) Antibacterial activity of Ag NC@Chitosan, Ag NCs (1 mM Ag), and chitosan (1 mg/mL) was tested against *S. aureus*, *E. coli*, *B. subtilis*, and *P. aeruginosa* after 24 h incubation; blank LB agar served as a control. Colony differences indicate effective bacterial inhibition [95]. Fluorescence microscopy of (**b**) SKOV-3 (FR^+) and (**c**) NIH3T3 (FR^-) cells treated with FA-CS (CD/antimiR-21) and CS(CD/antimiR-21); intracellular fluorescence (**d**) quantified using ImageJ [96]

responsive drug release and bioimaging. They performed as excellent confocal and TPF imaging contrast agent and fluorescent pH-sensing probes with enhanced loading of hydrophobic anti-cancer drugs. The combined chemotherapeutic and photothermal therapy increased the therapeutic efficacy, promoting synergism [97]. Gold nanoclusters were loaded in chitosan nanogels (AuNCs@CS) with increased 2.4- fold electrochemiluminescence (ECL) intensity. An ultrasensitive immunosensor was developed further with fluorine-doped tin oxide (FTO) electrode for the detection of aflatoxin B1(AFB1) with a detection limit of $9.3*10^{-15}$ g/mL [98]. Deep eutectic solvents (DES) were used as reaction media for the oxidation of hydroxytyrosol (HT), favouring oligomer formation. These oligomers were grafted along with chitosan to form nanogels with an average size of around 300 nm. It demonstrated enhanced antioxidant activity and anti-microbial activity against Gram-positive bacteria [99].

6.4 *In Wound Healing and Antibacterial Activity*

Chitosan nanogels were prepared from the probiotic supernatant of three bacteria (*Lactobacillus reuteri, Lactobacillus fermentum, Bacillus subtilis*). The probiotics were aimed at increasing the antibacterial, anti-inflammatory, immunomodulatory and angiogenic activity for the promotion of wound healing. The prepared nanogels had uniform size distribution (10–50 nm) with good stability. In vivo studies on Sprague Dawley rats showed that the group receiving *Bacillus subtillus* natto had better wound healing quality compared to other groups [100]. Trinitroglycerin-loaded chitosan nanogels (TN-Ngs) were prepared, which showed a size distribution around 96 ± 18 nm with loading capacity and encapsulation efficiency being 70.2%

and 2.1% respectively. The TN-Ngs200 showed good wound healing with complete wound closure ratio, reduced scar width, increased formation of granulation tissue and collagen deposition, thereby reducing inflammation and acting as an effective topical agent [101]. Multifunctional sprayable chitosan nanogel with Ag nanoparticles and a nitric oxide donor(S-nitroso-glutathione, GSNO) was developed using TPP as a crosslinker. In a chronic diabetic wound, the Ag nanoparticles undergo reduction in an acidic environment to form Ag^+, which further enhances the NO release from the donor, enhancing anti-bacterial activity and eradicating biofilm formation. The sprayable nanogel causes the transformation of M1 to M2 macrophages, decreasing the TNF-α release and potentiating the tissue remodelling process in chronic diabetic wounds [102]. Imidazolidinyl hydrazide crosslinked chitosan nanogels were developed, which could adsorb Ag^+ ions and facilitate their reduction into nanocrystalline nanogels (Agn@CSAI) in a confined nanoenvironment. The formed nanocrystallines promote reactive oxygen species (ROS) generation, which are potent anti-microbial agents [103]. Silver sulphadiazine loaded nanosuspension based chitosan nanogel was developed using nanoprecipitation and ultrasonication. The prepared nanosupension- based nanogel had good spreading ability, stability and showed increased therapeutic efficacy in ex vivo skin permeation and in vivo burn wound studies [104]. Nisin-loaded chitosan nanogel was developed, and its potential was evaluated in an acetic acid-induced colitis model in rats. It showed improved stability, bioavailability and efficacy against antibiotic-resistant bacteria [105]. Sprayable zwitterion-modified chitosan- based Schiff base crosslinked nanogels (ZC-CSNG) were developed by glutaraldehyde crosslinking with incorporated sulfobetaine and p-coumaric acid. The designed formulation helped in increasing solubility, sprayability, biocompatibility, ROS- scavenging activity and good anti-bacterial activity [106]. On-demand release chitosan nanogels were prepared from Schiff base reaction between oxidised hyaluronic acid and chitosan oligosaccharide encapsulating enrofloxacin (ENR). ENR nanogels had pH-and hyaluronidase(HA) responsiveness with greater anti-bacterial activity and accelerated healing of infected wounds [107]. Dhvar5, a potent antimicrobial peptide, was grafted onto chitosan using click chemistry to form chitosan nanogels (Dhvar5-NG). Synergistic activity was seen between Dhvar5-NG with oxacillin and piperacillin [108]. Chitosan nanogels were prepared and studied against treating *C.albicans* infections in Nile tilapia. Treatment with chitosan nanogels improved antioxidant activity and also modulated immune-related and apoptotic genes [109]. Free radical polymerization technique was employed for the crosslinking of chitosan with methacrylic acid to form nanogels loaded with madecassoside. The system showed good swelling and drug release behaviour with excellent pH-responsiveness [57]. Synthesis of core-shell alginate-coated chitosan (ACC) and chitosan-coated alginate (CCA) nanogels loaded with mupirocin was done using ionic crosslinking with both TPP and calcium chloride. CCA showed more uniform dispersion in size than ACC. Both the nanogels showed good anti-bacterial activity against both gram-positive and gram-negative bacteria. 96% of Mupirocin was released from CCA, whereas only 70% was released from ACC nanogels [110].

6.5 In Bone and Dental Tissue Engineering

Hollow mesoporous MnO2 nanoparticles were developed by using the core-shell method and then loaded with dexamethasone. This was further loaded into Alginate methacrylate hydrogel, and following electrostatic self-assembly using chitosan coating to form nanogels (Alg-MD@CS). In-vitro experiments indicated the ROS-scavenging potential, decreased inflammation and restored mitochondrial health (Fig. 5). In vivo studies revealed an anti-inflammatory microenvironment and promoted bone regeneration [111]. Engineered cell-membrane-coated nanogels capable of scavenging receptor activator of nuclear factor-κB ligand (RANL) were developed, which promoted the release of PTH 1-34 for treating postmenopausal osteoporosis. Dose frequency was reduced, prolonged circulation time and targeted delivery were achieved using the chitosan-based biomimetic nanogel [112].

Chitosan nanogels were loaded with *Mentha piperita* essential oil (MPEO) to serve as an anti-biofilm agent at 50 and 400 μg/mL against Streptococcus mutans. The release of MPEO is around 50% and the bacteria exhibited high sensitivity towards the nanoformulation [113]. Chlorogenic acid (CGA) was loaded into porous calcium phosphate-chitosan nanogel (CaPNP@Chi) using the ionic gelation method in the ratio of 1.25:1 (calcium phosphate nanoparticles to chitosan). The zone of inhibition against *S.aureus* was found to be 18.7 mm ± 0.6 mm and 68% biofilm degradation. The composite (CaPNP@Chi@CGA) has a 70% encapsulation efficiency and has potential applications in restorative dentistry [114]. Triclosan was encapsulated into poly(caprolactone) nanoparticles, which were loaded into chitosan hydrogel along with flurbiprofen, forming a nanogel system. The dual-responsive hydrogel was pH sensitive and temperature dependent. In vivo studies on

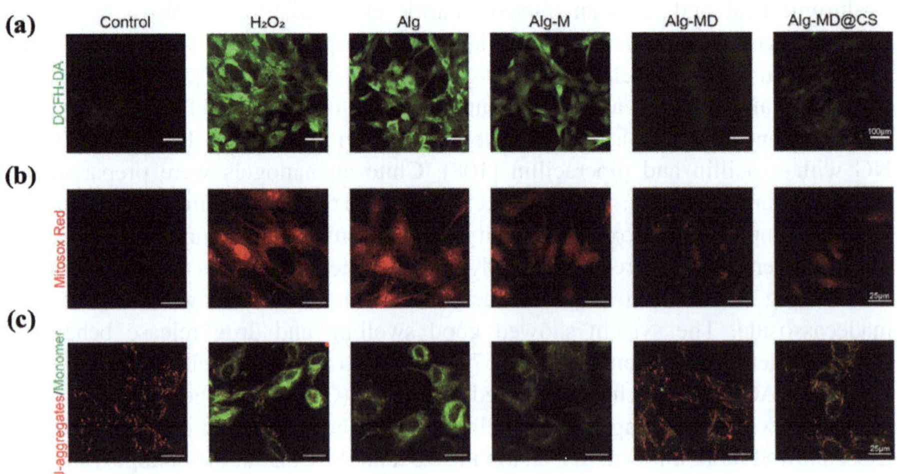

Fig. 5 Nanogels protect BMSC mitochondrial function- (**a**) DCFH-DA, (**b**) Mitosox Red, and (**c**) JC-1 assays post-H2O2 [111]

experimental periodontitis rats showed good anti-inflammatory and anti-bacterial effects [115]. Carboxymethyl chitosan and lysozyme (CMC/LYZ) nanogels encapsulated with amorphous calcium phosphate (ACP) were developed to serve as a potential nanoformulation to reduce dentinal sensitivity by occluding the dentinal tubules. SEM images showed the thickness of the coating to be around 1-2 μm with a penetrating depth of 4–8 μm approximately. XRD and nanoindentation methods confirmed the similarity of the mineralised substance with natural dentin [116]. ACP from CMC/LYZ-ACP nanogels underwent controllable dissociation and converted to hydroxyapatite (HAP) to form an aprismatic enamel-like layer on the demineralized enamel de novo. This could aid in the remineralization of early enamel caries [117].

6.6 Other Disease Conditions

Naringenin-β cyclodextrin (NAR/β-CD) complexes were loaded with chondroitin sulfate and low molecular weight chitosan to form nanogels for the treatment of early diabetic retinopathy. It showed sustained release, radical scavenging activity and better biocompatibility [118]. Glycol chitosan/fucoidan nanogels encapsulating an anti- inflammatory peptide KAFAK (GC/Fu@KAFAK NGs) were synthesised using genipin crosslinking and electrostatic interaction.

In the osteoarthritis model of rats, intra-articular injection of GC/Fu@KAFAK NGs decreased the inflammatory cytokines and increased the formation of Type-2 collagen and chondrogenic marker Sox9 [119]. Functionalized chitosan nanogels with B1 bradykinin receptor antagonist (BQ-123-CH1) showed promising results on equine articular chondrocytes. Thus, functionalization of chitosan nanogels can help in reducing inflammation and degradation processes that occur during osteoarthritis [120, 121]. Chitosan-methyl cyclodextrin (MCD) nanogel, along with antibiotics and adjuvants, was used to target infections of the gastrointestinal tract. In vivo studies on sturgeon fish exhibited effective antibacterial activity up to 11 days [122].

7 Regulatory Approval and Clinical Translation

Another significant challenge is the regulatory approval and clinical translation of chitosan-based nanogels. The development of nanogel-based drug delivery systems requires rigorous testing to ensure their safety and efficacy, which can be a lengthy and costly process [123, 124]. Additionally, there is a need for standardized protocols for the fabrication and characterization of chitosan-based nanogels to facilitate their translation from the laboratory to the clinic. The chitosan nanogels when compared to polymeric PEG, PLGA nanogels showed superior biocompatibility with increased drug loading capacity, stimuli-responsiveness, muco-adhesiveness and targeting ability highlighting their remarkable versatility for biomedical applications (Table 1).

Table 1 Comparison of Chitosan Nanogels with Other Nanogels

Property	Chitosan-based Nanogels	PEG-based Nanogels	PLGA-based Nanogels
Biocompatibility	Highly biocompatible and biodegradable [125, 126]	Excellent biocompatibility but less biodegradable than chitosan- based nanogels [127]	Biocompatible and Biodegradable [128–130]
Drug loading capacity	High drug loading capacity due to three dimensional structure [125]	Moderate drug loading capacity compared to chitosan-based nanogels [127]	Moderate drug loading capacity [129, 130]
Stimuli-responsiveness	Can be engineered to respond to pH, temperature and other stimuli [131, 132]	Limited stimuli- responsive behaviour compared to chitosan-based nanogels [127]	Limited stimuli-responsive behaviour [129, 130]
Muco-adhesive properties	Excellent mucoadhesive properties, enhancing drug absorption in mucosal tissues [133, 124]	Lack mucoadhesive properties [127]	Lack mucoadhesive properties [129, 130]
Targeting ability	Can be functionalized with targeting ligands for specific tissue targeting [124, 127]	Can be functionalized with targeting ligands [118]	Can be functionalized with targeting ligands [129, 130]

8 Conclusion and Future Perspectives

Reactive groups of chitosan allow us to chemically modify and form versatile drug delivery vehicles. It is widely used in biomedical applications including drug delivery, bioimaging, wound healing, bone and dental tissue engineering etc., This is possible due to its properties such as biocompatibility, biodegradability, flexibility, bio-adhesion, stimuli responsiveness, antibacterial and antifungal properties. Modern formulations involve a smart delivery system for multimodal therapies, and chitosan is the biomaterial of choice owing to its versatility. Focus should be made on translating the chitosan material for industrial applications to fully exploit its potential.

To overcome these challenges, future research should focus on developing more stable and scalable fabrication methods for chitosan-based nanogels. Additionally, there is a need for further studies on the biocompatibility and toxicity of these nanogels, as well as their performance in preclinical and clinical trials. Collaboration between researchers, manufacturers, and regulatory agencies will be crucial to address these challenges and ensure the successful translation of chitosan-based nanogels into clinical practice.

References

1. Sahoo, D., Sahoo, S., Mohanty, P., Sasmal, S., & Nayak, P. L. (2009 Jan). Chitosan: A new versatile bio-polymer for various applications. *Designed Monomers and Polymers., 12*(5), 377–404.
2. Crini, G. (2005 Jan). Recent developments in polysaccharide-based materials used as adsorbents in wastewater treatment. *Progress in Polymer Science., 30*(1), 38–70.
3. Dash, M., Chiellini, F., Ottenbrite, R. M., & Chiellini, E. (2011 Aug). Chitosan—A versatile semi- synthetic polymer in biomedical applications. *Progress in Polymer Science., 36*(8), 981–1014.
4. Xiong Chang, X., Mujawar Mubarak, N., Ali Mazari, S., Sattar Jatoi, A., Ahmad, A., Khalid, M., et al. (2021 Dec). A review on the properties and applications of chitosan, cellulose and deep eutectic solvent in green chemistry. *Journal of Industrial and Engineering Chemistry., 104*, 362–380.
5. Glassman, P. M., & Muzykantov, V. R. (2019 Sep). Pharmacokinetic and Pharmacodynamic properties of drug delivery systems. *The Journal of Pharmacology and Experimental Therapeutics, 370*(3), 570–580.
6. Ezike, T. C., Okpala, U. S., Onoja, U. L., Nwike, C. P., Ezeako, E. C., Okpara, O. J., et al. (2023). Advances in drug delivery systems, challenges and future directions. *Heliyon., 9*(6), e17488.
7. Trinadha Rao, M., Phanindra, C. H. V. S., Yamini, M., & Prasad, C. H. (2021). Hydrogels the three dimensional networks: A review. *International Journal of Current Pharmaceutical Sciences, 13*, 12–17.
8. Sivaram, A. J., Rajitha, P., Maya, S., Jayakumar, R., & Sabitha, M. (2015 Jul). Nanogels for delivery, imaging and therapy. *WIREs Nanomedicine and Nanobiotechnology, 7*(4), 509–533.
9. Alemán, J. V., Chadwick, A. V., He, J., Hess, M., Horie, K., Jones, R. G., et al. (2007). Definitions of terms relating to the structure and processing of sols, gels, networks, and inorganic-organic hybrid materials (IUPAC recommendations 2007). *Pure and Applied Chemistry., 79*(10), 1801–1829.
10. Raemdonck, K., Demeester, J., & De Smedt, S. (2009). Advanced nanogel engineering for drug delivery. *Soft Matter, 5*(4), 707–715.
11. Muzzarelli, R. A. A. (2011). Chitin nanostructures in living organisms. In N. S. Gupta (Ed.), *Chitin [Internet]* (pp. 1–34). Springer. [cited 2025 Apr 17]. (Topics in Geobiology; vol. 34). Available from: http://link.springer.com/10.1007/978-90-481-9684-5_1
12. Lichtfouse, E., Morin-Crini, N., Fourmentin, M., Zemmouri, H., Nascimento, I. O. D. C., Queiroz, L. M., et al. (2019). Chitosan for direct bioflocculation processes. In G. Crini & E. Lichtfouse (Eds.), *Sustainable agriculture reviews 36 [Internet]* (pp. 335–380). Springer International Publishing. [cited 2025 Apr 17]. (Sustainable agriculture reviews; vol. 36). Available from: http://link.springer.com/10.1007/978-3-030-16581-9_9
13. Issahaku, I., Tetteh, I. K., & Tetteh, A. Y. (2023 Apr 1). Chitosan and chitosan derivatives: Recent advancements in production and applications in environmental remediation. *Environmental Advances., 11*, 100351.
14. Ruiz, G. A. M., & Corrales, H. F. Z. (2017). Chitosan, Chitosan derivatives and their biomedical applications. In E. A. Shalaby (Ed.), *Biological activities and application of marine polysaccharides [Internet]*. InTech. [cited 2025 Apr 17]. Available from: http://www.intechopen.com/books/biological-activities-and-application-of- marine-polysaccharides/chitosan-chitosan-derivatives-and-their-biomedical-applications.
15. Rouhani Shirvan, A., Shakeri, M., & Bashari, A. (2019). Recent advances in application of chitosan and its derivatives in functional finishing of textiles. In *The impact and prospects of green chemistry for textile technology [Internet]* (pp. 107–133). Elsevier. [cited 2025 Apr 17]. Available from: https://linkinghub.elsevier.com/retrieve/pii/B9780081024911000058

16. Soni, K. S., Desale, S. S., & Bronich, T. K. (2016). Nanogels: An overview of properties, biomedical applications and obstacles to clinical translation. *Journal of Controlled Release, 240*, 109–126.
17. Rolland, J. P., Maynor, B. W., Euliss, L. E., Exner, A. E., Denison, G. M., & DeSimone, J. M. (2005). Direct fabrication and harvesting of monodisperse, shape-specific nanobiomaterials. *Journal of the American Chemical Society, 127*(28), 10096–10100.
18. Rahim, M. A., Jan, N., Khan, S., Shah, H., Madni, A., Khan, A., et al. (2021). Recent advancements in stimuli responsive drug delivery platforms for active and passive cancer targeting. *Cancers, 13*(4), 670.
19. Sanson, N., & Rieger, J. (2010). Synthesis of nanogels/microgels by conventional and controlled radical crosslinking copolymerization. *Polymer Chemistry, 1*(7), 965.
20. Gao, H., & Matyjaszewski, K. (2009 Apr). Synthesis of functional polymers with controlled architecture by CRP of monomers in the presence of cross-linkers: From stars to gels. *Progress in Polymer Science., 34*(4), 317–350.
21. Duda, A., & Kowalski, A. (2009). Thermodynamics and kinetics of ring-opening polymerization. In P. Dubois, O. Coulembier, & J. Raquez (Eds.), *Handbook of ring-opening polymerization [Internet]* (1st ed., pp. 1–51). Wiley. [cited 2025 Apr 17]. Available from: https://onlinelibrary.wiley.com/doi/10.1002/9783527628407.ch1
22. Mazo, A. R., Allison-Logan, S., Karimi, F., Chan, N. J. A., Qiu, W., Duan, W., et al. (2020). Ring opening polymerization of α-amino acids: Advances in synthesis, architecture and applications of polypeptides and their hybrids. *Chemical Society Reviews, 49*(14), 4737–4834.
23. Raftery, E. D., Gharkhanian, E. G., Ricapito, N. G., McNamara, J., & Deming, T. J. (2018). Influence of sulfur-containing Diamino acid structure on covalently crosslinked copolypeptide hydrogels. *Chemistry An Asian Journal., 13*(22), 3547–3553.
24. Akiyoshi, K., Deguchi, S., Moriguchi, N., Yamaguchi, S., & Sunamoto, J. (1993 Jun). Self-aggregates of hydrophobized polysaccharides in water. Formation and characteristics of nanoparticles. *Macromolecules., 26*(12), 3062–3068.
25. Moya-Ortega, M. D., Alvarez-Lorenzo, C., Sigurdsson, H. H., Concheiro, A., & Loftsson, T. (2012 Feb). Cross-linked hydroxypropyl-β-cyclodextrin and γ-cyclodextrin nanogels for drug delivery: Physicochemical and loading/release properties. *Carbohydrate Polymers., 87*(3), 2344–2351.
26. Lee, W., Li, Y., & Chu, I. (2006). Amphiphilic Poly(D, L -lactic acid)/Poly(ethylene glycol)/Poly(D, L -lactic acid) Nanogels for controlled release of hydrophobic drugs. *Macromolecular Bioscience., 6*(10), 846–854.
27. Lallana, E., Riguera, R., & Fernandez-Megia, E. (2011). Reliable and efficient procedures for the conjugation of biomolecules through Huisgen Azide–Alkyne Cycloadditions. *Angewandte Chemie, International Edition, 50*(38), 8794–8804.
28. Sun, W., An, Z., & Wu, P. (2017). Revealing the distinct thermal transition behavior between PEGA-based linear polymers and their disulfide cross-linked nanogels. *Physical Chemistry Chemical Physics, 19*(37), 25746–25753.
29. Kumar, P., Liu, B., & Behl, G. (2019 Aug). A comprehensive outlook of synthetic strategies and applications of redox-responsive Nanogels in drug delivery. *Macromolecular Bioscience., 19*(8), 1900071.
30. Moramkar, N., & Bhatt, P. (2021 Jul). Insight into chitosan derived nanotherapeutics for anticancer drug delivery and imaging. *European Polymer Journal., 154*, 110540.
31. Liang, X., Mu, M., Fan, R., Zou, B., & Guo, G. (2022 Aug). Functionalized chitosan as a promising platform for cancer immunotherapy: A review. *Carbohydrate Polymers., 290*, 119452.
32. Razmi, F. A., Ngadi, N., Wong, S., Inuwa, I. M., & Opotu, L. A. (2019 Sep). Kinetics, thermodynamics, isotherm and regeneration analysis of chitosan modified pandan adsorbent. *Journal of Cleaner Production., 231*, 98–109.

33. Ren, L., Xu, J., Zhang, Y., Zhou, J., Chen, D., & Chang, Z. (2019 Aug). Preparation and characterization of porous chitosan microspheres and adsorption performance for hexavalent chromium. *International Journal of Biological Macromolecules., 135*, 898–906.
34. Christou, C., Philippou, K., Krasia-Christoforou, T., & Pashalidis, I. (2019). Uranium adsorption by polyvinylpyrrolidone/chitosan blended nanofibers. *Carbohydrate Polymers., 219*, 298–305.
35. Wang, W., Meng, Q., Li, Q., Liu, J., Zhou, M., Jin, Z., et al. (2020 Jan). Chitosan derivatives and their application in biomedicine. *International Journal of Molecular Sciences., 21*(2), 487.
36. Kaczmarek, B., Owczarek, A., Nadolna, K., & Sionkowska, A. (2019). The film-forming properties of chitosan with tannic acid addition. *Materials Letters., 245*, 22–24.
37. Lai, P., Daear, W., Löbenberg, R., & Prenner, E. J. (2014). Overview of the preparation of organic polymeric nanoparticles for drug delivery based on gelatine, chitosan, poly(d,l-lactide-co-glycolic acid) and polyalkylcyanoacrylate. *Colloids and Surfaces B: Biointerfaces., 118*, 154–163.
38. Solomevich, S. O., Dmitruk, E. I., Bychkovsky, P. M., Salamevich, D. A., Kuchuk, S. V., & Yurkshtovich, T. L. (2021 Feb). Biodegradable polyelectrolyte complexes of chitosan and partially crosslinked dextran phosphate with potential for biomedical applications. *International Journal of Biological Macromolecules., 1*(169), 500–512.
39. Kunjachan, S., Jose, S., & Lammers, T. (2010). Understanding the mechanism of ionic gelation for synthesis of chitosan nanoparticles using qualitative techniques. *Asian Journal of Pharmaceutics (AJP) [Internet], 4*(2) [cited 2025 Jun 22]; Available from:https://www.asiapharmaceutics.info/index.php/ajp/article/view/220
40. Koukaras, E. N., Papadimitriou, S. A., Bikiaris, D. N., & Froudakis, G. E. (2012). Insight on the formation of chitosan nanoparticles through ionotropic gelation with Tripolyphosphate. *Molecular Pharmaceutics, 9*(10), 2856–2862.
41. Pérez-Álvarez, L., Laza, J. M., & Álvarez-Bautista, A. (2016). Covalently and ionically crosslinked chitosan nanogels for drug delivery. *Current Pharmaceutical Design, 22*(22), 3380–3398.
42. Zhao, X., Tang, J., Liu, Y., Hu, B., Chen, Q., & Liu, Y. (2023 Nov 8). Reaction kinetics of chitosan nanogels crosslinked by genipin. *Journal of Chromatography A., 1710*, 464427.
43. Ding, Y. F., Wei, J., Li, S., Pan, Y. T., Wang, L. H., & Wang, R. (2019 Aug 14). Host–guest interactions initiated supramolecular chitosan nanogels for selective intracellular drug delivery. *ACS Applied Materials & Interfaces, 11*(32), 28665–28670.
44. Speight, J. G. (2011). Monomers, polymers, and plastics. In *Handbook of industrial hydrocarbon processes [Internet]* (pp. 499–537). Elsevier. [cited 2025 Apr 21]. Available from: https://linkinghub.elsevier.com/retrieve/pii/B9780750686327100143
45. Brunel, F., Véron, L., Ladavière, C., David, L., Domard, A., & Delair, T. (2009). Synthesis and structural characterization of chitosan Nanogels. *Langmuir, 25*(16), 8935–8943.
46. Puig, J. E., & Rabelero, M. (2016). Semicontinuous microemulsion polymerization. *Current Opinion in Colloid & Interface Science., 25*, 83–88.
47. Brunel, F., El Gueddari, N. E., & Moerschbacher, B. M. (2013 Feb). Complexation of copper (II) with chitosan nanogels: Toward control of microbial growth. *Carbohydrate Polymers., 92*(2), 1348–1356.
48. Brunel, F., Véron, L., David, L., Domard, A., Verrier, B., & Delair, T. (2010). Self-assemblies on Chitosan nanohydrogels. *Macromolecular Bioscience., 10*(4), 424–432.
49. Bhattacharjee, S., Goswami, S., Das, S., Bhattacharjee, S., & Bhaladhare, S. (2023). pH-responsive, stable, and biocompatible functional nanogels based on chitosan (CS)/poly methacrylic acid (PMAA) polymers: Synthesis and characterization. *Materials Today Communications., 36*, 106541.
50. Wang, K., Buschle-Diller, G., & Misra, R. D. K. (2015). Chitosan-based injectable hydrogels for biomedical applications. *Materials Technology, 30*(sup8), B198–B205.
51. Salunke, P. A., Wagh, R. S., Patil, S. S., Barhate, D. S. D. An overview: Site specific drug delivery system. 2016;

52. Petrovic, S., Bita, B., & Barbinta-Patrascu, M. E. (2024 Jan). Nanoformulations in pharmaceutical and biomedical applications: Green perspectives. *International Journal of Molecular Sciences., 25*(11), 5842.
53. Loo, H. L., Goh, B. H., Lee, L. H., & Chuah, L. H. (2022). Application of chitosan-based nanoparticles in skin wound healing. *Asian Journal of Pharmaceutical Sciences., 17*(3), 299–332.
54. Grewal, A. K., & Salar, R. K. (2024). Chitosan nanoparticle delivery systems: An effective approach to enhancing efficacy and safety of anticancer drugs. *Nano TransMed., 3*, 100040.
55. Herdiana, Y., Febrina, E., Nurhasanah, S., Gozali, D., Elamin, K. M., & Wathoni, N. (2024). Drug loading in Chitosan-based nanoparticles. *Pharmaceutics., 16*(8), 1043.
56. Manivong, S., Garcia Ac, A., Patten, S. A., Fernandes, J. C., Benderdour, M., Banquy, X., et al. (2022 Jan). Chitosan-based Nanogels: Synthesis and toxicity profile for drug delivery to articular joints. *Nanomaterials (Basel), 12*(8), 1337.
57. Suhail, M., Chiu, I. H., Ullah, A., Khan, A., Ullah, H., Al-Sowayan, N. S., et al. (2024). Formulation and in vitro assessment of polymeric pH-responsive Nanogels of chitosan for sustained delivery of Madecassoside. *ACS Omega, 9*(17), 19345–19352.
58. Mali, K. K., Gavhane, Y. N., & Chakole, R. D. (2024). Natural polymer-based Nanogel for pH-responsive delivery of Sorafenib Tosylate in Hemangiosarcoma. *AAPS PharmSciTech, 25*(4), 83.
59. Li, X. M., Wu, Z. Z., Zhang, B., Pan, Y., Meng, R., & Chen, H. Q. (2019 Sep). Fabrication of chitosan hydrochloride and carboxymethyl starch complex nanogels as potential delivery vehicles for curcumin. *Food Chemistry., 293*, 197–203.
60. Shang, M., Sun, X., Guo, L., Shi, D., Liang, P., Meng, D., et al. (2020). pH- and ultrasound-responsive paclitaxel-loaded Carboxymethyl Chitosan Nanodroplets for combined imaging and synergistic Chemoradiotherapy. *International Journal of Nanomedicine, 15*, 537–552.
61. Herdiana, Y., Wathoni, N., Shamsuddin, S., & Muchtaridi, M. (2022). Drug release study of the chitosan-based nanoparticles. *Heliyon., 8*(1), e08674.
62. Huang, Y., & Lapitsky, Y. (2011). Monovalent salt enhances colloidal stability during the formation of chitosan/Tripolyphosphate microgels. *Langmuir, 27*(17), 10392–10399.
63. Pereira, P., Correia, A., & Gama, F. M. (2016). In vivo imaging of glycol chitosan-based Nanogel biodistribution. *Macromolecular Bioscience., 16*(3), 432–440.
64. Goswami, N., Lin, F., Liu, Y., Leong, D. T., & Xie, J. (2016). Highly luminescent Thiolated gold nanoclusters impregnated in Nanogel. *Chemistry of Materials, 28*(11), 4009–4016.
65. Baba, A. I., & Câtoi, C. (2007). Tumor cell morphology. In *Comparative oncology [Internet]*. The Publishing House of the Romanian Academy. [cited 2025 Apr 21]. Available from: https://www.ncbi.nlm.nih.gov/books/NBK9553/
66. Venkatesh, P., & Kasi, A. (2025). Anthracyclines. In *StatPearls [Internet]*. StatPearls Publishing. [cited 2025 Apr 23]. Available from: http://www.ncbi.nlm.nih.gov/books/NBK538187/
67. Wu, X., He, C., Wu, Y., Chen, X., & Cheng, J. (2015 Nov). Nanogel-incorporated physical and chemical hybrid gels for highly effective chemo–protein combination therapy. *Advanced Functional Materials, 25*(43), 6744–6755.
68. Feng, C., Li, J., Kong, M., Liu, Y., Cheng, X. J., Li, Y., et al. (2015). Surface charge effect on mucoadhesion of chitosan based nanogels for local anti-colorectal cancer drug delivery. *Colloids and Surfaces B: Biointerfaces., 128*, 439–447.
69. Abo-ser, M. M., Toson, E. S. A., El-Bindary, A. A., Schlatter, G., & Shoueir, K. R. (2024 May). Smart chitosan nanogel for targeted doxorubicin delivery, ensuring precise release, and minimizing side effects in Ehrlich ascites carcinoma-bearing mice. *International Journal of Biological Macromolecules., 267*, 131390.
70. Yang, L., Ling, J., Wang, N., Jiang, Y., Lu, Y., Yang, L. Y., et al. (2022). Delivery of doxorubicin by dual responsive carboxymethyl chitosan based nanogel and in vitro performance. *Materials Today Communications., 31*, 103781.

71. Veloso, S. R. S., Marta, E. S., Rodrigues, P. V., Moura, C., Amorim, C. O., Amaral, V. S., et al. (2023). Chitosan/alginate Nanogels containing multicore magnetic nanoparticles for delivery of doxorubicin. *Pharmaceutics., 15*(9), 2194.
72. Ahmed, M. M., Ameen, M. S. M., Abazari, M., Badeleh, S. M., Rostamizadeh, K., & Mohammed, S. S. (2023 Aug). Chitosan-decorated and tripolyphosphate-crosslinked pH-sensitive niosomal nanogels for controlled release of fluoropyrimidine 5-fluorouracil. *Biomedicine & Pharmacotherapy., 164*, 114943.
73. Longley, D. B., Harkin, D. P., & Johnston, P. G. (2003 May). 5-fluorouracil: Mechanisms of action and clinical strategies. *Nature Reviews. Cancer, 3*(5), 330–338.
74. Sahu, P., Kashaw, S. K., Sau, S., Kushwah, V., Jain, S., Agrawal, R. K., et al. (2019 Feb 1). pH responsive 5-fluorouracil loaded biocompatible Nanogels for topical chemotherapy of aggressive melanoma. *Colloids and Surfaces B: Biointerfaces., 174*, 232–245.
75. Deepak Vishnu, S. K., Ranganathan, P., Rwei, S. P., Pattamaprom, C., Kavitha, T., & Sarojini, P. (2020). New reductant-free synthesis of gold nanoparticles-doped chitosan-based semi- IPN nanogel: A robust nanoreactor for exclusively sensitive 5-fluorouracil sensor. *International Journal of Biological Macromolecules., 148*, 79–88.
76. Rajaei, M., Rashedi, H., Yazdian, F., Navaei-Nigjeh, M., Rahdar, A., & Díez-Pascual, A. M. (2023). Chitosan/agarose/graphene oxide nanohydrogel as drug delivery system of 5- fluorouracil in breast cancer therapy. *Journal of Drug Delivery Science and Technology., 82*, 104307.
77. Hanoodi, M., & Mittal, M. (2025). Methotrexate. In *StatPearls [Internet]*. StatPearls Publishing. [cited 2025 Apr 23]. Available from: http://www.ncbi.nlm.nih.gov/books/NBK556114/
78. Azadi, A., Hamidi, M., & Rouini, M. R. (2013). Methotrexate-loaded chitosan nanogels as 'Trojan horses' for drug delivery to brain: Preparation and in vitro/in vivo characterization. *International Journal of Biological Macromolecules., 62*, 523–530.
79. Awosika, A. O., Farrar, M. C., & Jacobs, T. F. (2025). Paclitaxel. In *StatPearls [Internet]*. StatPearls Publishing. [cited 2025 Apr 23]. Available from: http://www.ncbi.nlm.nih.gov/books/NBK536917/
80. Nguyen, V. T., Doan, P., Nguyen, D. T., Doan, V.-D., Dao, T. P., Plavskii, V., et al. (2022). Effect of targeting ligand designation of self-assembly chitosan-poloxamer nanogels loaded Paclitacel on inhibiting MCF-7 cancer cell growth. *Journal of Biomaterials Science, Polymer Edition, 33*(4), 426–442.
81. Tran, T. V., Phuong, T. H. D., Tran, N. Q., Nguyen, C. K., & Nguyen, D. H. (2018 Jan). Polymeric chitosan based nanogels as a potential platform for dual targeted drug delivery in cancer therapy. *International Journal of Nanotechnology, 15*(1–3), 188–198.
82. Amanlou, N., Parsa, M., Rostamizadeh, K., Sadighian, S., & Moghaddam, F. (2019 Mar). Enhanced cytotoxic activity of curcumin on cancer cell lines by incorporating into gold/chitosan nanogels. *Materials Chemistry and Physics., 226*, 151–157.
83. Hatanaka, T. (2000). Clinical pharmacokinetics of pravastatin. *Clinical Pharmacokinetics, 39*(6), 397–412.
84. Harisa, G. I., Badran, M. M., AlQahtani, S. A., Alanazi, F. K., & Attia, S. M. (2016 Jan). Pravastatin chitosan nanogels-loaded erythrocytes as a new delivery strategy for targeting liver cancer. *Saudi Pharmaceutical Journal, 24*(1), 74–81.
85. Sahu, P., Kashaw, S. K., Kashaw, V., Shabaaz, J. P., & Dahiya, R. (2021). Synthesis and ex vivo evaluation of PLGA chitosan surface modulated double walled transdermal Pluronic nanogel for the controlled delivery of Temozolomide. *International Journal of Biological Macromolecules., 187*, 742–754.
86. Demarchi, C. A., Debrassi, A., de Campos Buzzi, F., Corrêa, R., Filho, V. C., Rodrigues, C. A., et al. (2014). A magnetic nanogel based on O-carboxymethylchitosan for antitumor drug delivery: Synthesis, characterization and in vitro drug release. *Soft Matter, 10*(19), 3441–3450.
87. Shen, Y., Levin, A., Kamada, A., Toprakcioglu, Z., Rodriguez-Garcia, M., Xu, Y., et al. (2021). From protein building blocks to functional materials. *ACS Nano, 15*(4), 5819–5837.

88. Cai, Y., & Lapitsky, Y. (2017). Analysis of chitosan/tripolyphosphate micro- and nanogel yields is key to understanding their protein uptake performance. *Journal of Colloid and Interface Science., 494*, 242–254.
89. Zubareva, A., Ilyina, A., Prokhorov, A., Kurek, D., Efremov, M., Varlamov, V., et al. (2013 Jul). Characterization of protein and peptide binding to Nanogels formed by differently charged chitosan derivatives. *Molecules, 18*(7), 7848–7864.
90. Debache, K., Kropf, C., Schütz, C. A., Harwood, L. J., Käuper, P., Monney, T., et al. (2011). Vaccination of mice with chitosan nanogel-associated recombinant NcPDI against challenge infection with Neospora caninum tachyzoites. *Parasite Immunology., 33*(2), 81–94.
91. Pang, Y. X., Li, X., Zhang, X., Yeoh, J. X., Wong, C., Manickam, S., et al. (2022). The synthesis of carbon-based quantum dots: A supercritical fluid approach and perspective. *Materials Today Physics., 27*, 100752.
92. Juvatkar, P. V., & Kale, M. K. Bioimaging: A powerful tool for diagnosis: A review. 2021. [cited 2025 Apr 25]; Available from: https://zenodo.org/record/5118358
93. Wu, W., Shen, J., Banerjee, P., & Zhou, S. (2010). Chitosan-based responsive hybrid nanogels for integration of optical pH-sensing, tumor cell imaging and controlled drug delivery. *Biomaterials, 31*(32), 8371–8381.
94. Maxwell, T., Banu, T., Price, E., Tharkur, J., Campos, M. G. N., Gesquiere, A., et al. (2015 Dec). Non-cytotoxic quantum dot–chitosan nanogel biosensing probe for potential cancer targeting agent. *Nanomaterials, 5*(4), 2359–2379.
95. Liu, J., Liu, L., Li, S., Kang, Q., Zhang, R., & Zhu, Z. (2021). Self-assembled nanogels of luminescent thiolated silver nanoclusters and chitosan as bactericidal agent and bacterial sensor. *Materials Science and Engineering C, Materials for Biological Applications, 118*, 111520.
96. Radnia, F., Mohajeri, N., Hashemi, F., Imani, M., & Zarghami, N. (2021). Design and development of folate-chitosan/CD nanogel: An efficient fluorescent platform for cancer-specific delivery of AntimiR-21. *Reactive and Functional Polymers., 160*, 104814.
97. Wang, H., Di, J., Sun, Y., Fu, J., Wei, Z., Matsui, H., et al. (2015). Biocompatible PEG-chitosan@carbon dots hybrid Nanogels for two-photon fluorescence imaging, near-infrared light/pH dual-responsive drug carrier, and synergistic therapy. *Advanced Functional Materials., 25*(34), 5537–5547.
98. Wang, M., Li, H., Wang, Y., Hu, X., Fang, S., Ma, H., et al. (2023). Spatial confinement-enhanced electrochemiluminescence of gold nanoclusters in chitosan nanogels and its application for ultrasensitive detection of aflatoxin B1. *Journal Of Electroanalytical Chemistry, 943*, 117595.
99. Bellou, M. G., Skonta, A., Chatzikonstantinou, A. V., Polydera, A. C., Katapodis, P., Voutsas, E., et al. (2025 Feb). Enzymatic oxidation of Hydroxytyrosol in deep eutectic solvents for chitosan functionalization and preparation of bioactive Nanogels. *Catalysts, 15*(2), 180.
100. Ashoori, Y., Mohkam, M., Heidari, R., Abootalebi, S. N., Mousavi, S. M., Hashemi, S. A., et al. (2020). Development and in vivo characterization of probiotic lysate-treated chitosan Nanogel as a novel biocompatible formulation for wound healing. *BioMed Research International., 2020*(1), 8868618.
101. Asadi, K., Azarpira, N., Heidari, R., Hamidi, M., Yousefzadeh-Chabok, S., Nemati, M. M., et al. (2024). Trinitroglycerin-loaded chitosan nanogels accelerate angiogenesis in wound healing process. *International Journal of Biological Macromolecules., 278*, 134937.
102. Huang, Q., Yang, Z., Tao, X., Ma, C., Cao, P., Wei, P., et al. (2024). Sprayable chitosan nanogel with nitric oxide to accelerate diabetic wound healing through bacteria inhibition, biofilm eradication and macrophage polarization. *International Journal of Biological Macromolecules., 254*, 127806.
103. Chen, H., Liu, J., Qiao, Y., Bian, D., He, L., Hu, Y., et al. (2025). Imidazolyl crosslinked chitosan nanogel for nanoconfined synthesis of silver nanocrystallines as antimicrobials. *International Journal Of Biological Macromolecules, 305*, 140838.

104. Barkat, H., Barkat, M. A., Ali, R., & Hadi, H. (2025). Nanotechnological advances in burn wound care: Silver sulfadiazine-loaded Nanosuspension-based chitosan-incorporated Nanogel for partial thickness burns. *The International Journal of Lower Extremity Wounds.*, 15347346241309425.
105. Lorestanizadeh, G., Arasteh, J., Ardani, G. S., & Mehrjo, H. (2025). Immunomodulatory effect of nisin loaded with chitosan nanogel in ulcerative colitis rat model. *JoBAZ., 86*(1), 20.
106. Basak, S., Singh, P., Weller, A., Kadumudi, F. B., Kempen, P. J., Mijakovic, I., et al. (2025). Cost- effective and eco-friendly sprayable nanogels (ZC-CSNG) for multifunctional wound dressing applications. *Chemical Engineering Journal., 503*, 158312.
107. Luo, W., Jiang, Y., Liu, J., Ju, M., Algharib, S. A., & Dawood, A. S. (2023 Dec 31). On-demand release of enrofloxacin-loaded chitosan oligosaccharide-oxidized hyaluronic acid composite nanogels for infected wound healing. *International Journal of Biological Macromolecules., 253*, 127248.
108. Costa, B., Alves, P. M., Fonseca, D. R., Campos, F., Monteiro, A. C., Shahrour, H., et al. (2024). Dhvar5-chitosan nanogels and their potential to improve antibiotics activity. *International Journal of Biological Macromolecules., 277*, 134059.
109. Mahboub, H. H., Eltanahy, A., Omran, A., Mansour, A. T., Safhi, F. A., Alwutayd, K. M., et al. (2024). Chitosan nanogel aqueous treatment improved blood biochemicals, antioxidant capacity, immune response, immune-related gene expression and infection resistance of Nile tilapia. *Comparative Biochemistry and Physiology Part B: Biochemistry and Molecular Biology., 269*, 110876.
110. Hesan, M., Gholipour-Kanani, A., Lotfi, M., & Shafiee, M. (2023 Jan). The synthesis and characterization of core-shell nanogels based on alginate and chitosan for the controlled delivery of mupirocin. *Biochemical Engineering Journal., 1*(190), 108742.
111. Huang, Z., Jiang, X., Zhang, L., Wang, W., Li, Z., Huang, Y., et al. (2025). Multifunctional manganese-based nanogels catalyze immune energy metabolism to promote bone repair. *Composites Part B: Engineering., 291*, 112005.
112. Cui, Y., Lv, B., Li, Z., Ma, C., Gui, Z., Geng, Y., et al. (2024 Feb). Bone-targeted biomimetic Nanogels re-establish osteoblast/osteoclast balance to treat postmenopausal osteoporosis. *Small, 20*(6), 2303494.
113. Ashrafi, B., Rashidipour, M., Marzban, A., Soroush, S., Azadpour, M., Delfani, S., et al. (2019 May). Mentha piperita essential oils loaded in a chitosan nanogel with inhibitory effect on biofilm formation against S. Mutans on the dental surface. *Carbohydrate Polymers., 212*, 142–149.
114. Palaniraj, S., Murugesan, R., & Narayan, S. (2019). Chlorogenic acid- loaded calcium phosphate chitosan nanogel as biofilm degradative materials. *The International Journal of Biochemistry & Cell Biology., 114*, 105566.
115. Aminu, N., Chan, S. Y., Yam, M. F., & Toh, S. M. (2019). A dual-action chitosan-based nanogel system of triclosan and flurbiprofen for localised treatment of periodontitis. *International Journal of Pharmaceutics, 570*, 118659.
116. Song, J., Wang, H., Yang, Y., Xiao, Z., Lin, H., Jin, L., et al. (2018 Jun). Nanogels of carboxymethyl chitosan and lysozyme encapsulated amorphous calcium phosphate to occlude dentinal tubules. *Journal of Materials Science: Materials in Medicine, 29*(6), 84.
117. Song, J., Li, T., Gao, J., Li, C., Jiang, S., & Zhang, X. (2021). Building an aprismatic enamel-like layer on a demineralized enamel surface by using carboxymethyl chitosan and lysozyme-encapsulated amorphous calcium phosphate nanogels. *Journal Of Dentistry, 107*, 103599.
118. Zucca, G., Vigani, B., Valentino, C., Ruggeri, M., Marchesi, N., Pascale, A., et al. (2025). Chondroitin Sulphate-chitosan based Nanogels loaded with Naringenin-β-Cyclodextrin complex as potential Tool for the treatment of diabetic retinopathy: A formulation study. *International Journal of Nanomedicine, 20*, 907–932.
119. Li, T., Yang, J., Weng, C., Liu, P., Huang, Y., Meng, S., et al. (2021). Intra-articular injection of anti- inflammatory peptide-loaded glycol chitosan/fucoidan nanogels to inhibit

inflammation and attenuate osteoarthritis progression. *International Journal of Biological Macromolecules., 170*, 469–478.
120. Cullier, A., Cassé, F., Manivong, S., Contentin, R., Legendre, F., Garcia Ac, A., et al. (2022 Jan). Functionalized Nanogels with Endothelin-1 and bradykinin receptor antagonist peptides decrease inflammatory and cartilage degradation markers of osteoarthritis in a horse organoid model of cartilage. *International Journal of Molecular Sciences., 23*(16), 8949.
121. Terlinden, A., Jacquet, S., Manivong, S., Cullier, A., Cassé, F., Legendre, F., et al. (2024). Double- blinded, randomized tolerance study of a biologically enhanced Nanogel with endothelin-1 and bradykinin receptor antagonist peptides via intra-articular injection for osteoarthritis treatment in horses. *Bmc Veterinary Research, 20*(1), 1–15.
122. Zlotnikov, I. D., Belogurova, N. G., Poddubnaya, I. V., & Kudryashova, E. V. (2023 Sep). Mucosal adhesive chitosan Nanogel formulations of antibiotics and adjuvants (Terpenoids, flavonoids, etc.) and their potential for the treatment of infectious diseases of the gastrointestinal tract. *Pharmaceutics., 15*(9), 2353.
123. Ferreira, L. M. B., & Zucolotto, V. (2024 Jun 1). Chitosan-based nanomedicines: A review of the main challenges for translating the science of polyelectrolyte complexation into innovative pharmaceutical products. *Carbohydrate Polymer Technologies and Applications., 7*, 100441.
124. Yadav, S., Singh, A., Palei, N. N., Pathak, P., Verma, A., & Yadav, J. P. (2024). Chitosan-based Nanoformulations: Preclinical investigations, Theranostic advancements, and clinical trial prospects for targeting diverse pathologies. *AAPS PharmSciTech, 25*(8), 263.
125. Wang, Y., Lou, X., Yang, L., & Hou, Y. Application of Chitosan-based Nanogel in Cancer Nanomedicine. *Current Pharmaceutical Design., 31*(16), 1247–1258.
126. Passos Gibson, V., Nunes, J. B. B., Falcao, D. Q., Roullin, V. G., & Leblond, C. J. (2021). Lipid coating of chitosan Nanogels for improved colloidal stability and in vitro biocompatibility. *AAPS PharmSciTech, 22*(5), 159.
127. Al-Shadidi, J. R., Al-Shammari, S., Al-Mutairi, D., Alkhudhair, D., Thu, H. E., & Hussain, Z. (2024). Chitosan nanoparticles for targeted cancer therapy: A review of stimuli-responsive, passive, and active targeting strategies. *IJN., 19*, 8373–8400.
128. Sadeghi, S., Mohammadnejad, J., Eidi, A., & Jafary, H. (2024). Biocompatible PLGA-PCL nanobeads for efficient delivery of curcumin to lung cancer. *Cancer Nanotechnology., 15*(1), 34.
129. Vashist, A., Alvarez, G. P., Camargo, V. A., Raymond, A. D., Arias, A. Y., Kolishetti, N., et al. (2024). Recent advances in nanogels for drug delivery and biomedical applications. *Biomaterials Science, 12*(23), 6006–6018.
130. Chellathurai, M. S., Chung, L. Y., Hilles, A. R., Sofian, Z. M., Singha, S., Ghosal, K., et al. (2024). Pharmaceutical chitosan hydrogels: A review on its design and applications. *International Journal of Biological Macromolecules., 280*, 135775.
131. Pérez-Pacheco, Y., Tylkowski, B., & García-Valls, R. (2025 Jan). Chitosan micro/Nanocapsules in action: Linking design, production, and therapeutic application. *Molecules, 30*(2), 252.
132. Salahuddin, A., Ashraf, A., Ahmad, K., & Hou, H. (2024). Recent advances in chitosan-based smart hydrogel for drug delivery systems. *International Journal of Biological Macromolecules., 280*, 135803.
133. Haider, A., Khan, S., Iqbal, D. N., Shrahili, M., Haider, S., Mohammad, K., et al. (2024). Advances in chitosan-based drug delivery systems: A comprehensive review for therapeutic applications. *European Polymer Journal., 210*, 112983.

Chitosan-Based Micelles in Drug Delivery and Bioimaging

Disha Ghosh and Kishor Sarkar

Abstract Chitosan-based micelles have gained significant attention in the fields of drug delivery and bioimaging due to their versatile properties, including biocompatibility, biodegradability, and the ability to encapsulate a wide range of therapeutic agents. This chapter provides a comprehensive overview of the design, synthesis, and application of chitosan-based micelles, focusing on their dual role as carriers for drug delivery and agents for bioimaging.

The chapter begins with a discussion of chitosan's unique chemical structure and its role in the formation of micelles through self-assembly. Key factors affecting micelle stability, drug encapsulation efficiency, and release kinetics are explored in detail. Various preparation techniques and functionalization strategies for chitosan-based micelles are outlined, emphasizing their adaptability for different drug delivery platforms, including oral, intravenous, and targeted therapies. In addition to drug delivery, the chapter delves into the applications of chitosan-based micelles in bioimaging, including fluorescence and magnetic resonance imaging (MRI). The potential of these micelles in theranostics—combining drug delivery with real-time imaging capabilities—is also examined, providing insight into their role in personalized medicine.

The chapter concludes by addressing recent advancements in stimuli-responsive micelles and hybrid systems, along with challenges related to scalability and clinical translation. This comprehensive overview positions chitosan-based micelles as a promising tool in the development of next-generation drug delivery and imaging technologies.

Keywords Chitosan micelles · Drug delivery systems · Bioimaging · Self-assembly;Targeted therapy · Theranostics

D. Ghosh · K. Sarkar (✉)
Gene Therapy and Tissue Engineering Lab, Department of Polymer Science and Technology, University of Calcutta, Kolkata, West Bengal, India
e-mail: kspoly@caluniv.ac.in; http://www.kishorgttl.com/

S. Jana (ed.), *Chitosan-Based Drug Delivery Systems*, AAPS Introductions in the Pharmaceutical Sciences 16, https://doi.org/10.1007/978-3-032-08347-0_5

1 Introduction

Micelles are amphiphilic molecules that self-assemble into nanostructures with a hydrophilic shell and a hydrophobic core. Because these structures may solubilize hydrophobic pharmaceuticals, increasing their stability and bioavailability, they are especially useful in drug delivery systems [1]. A wide range of medicinal substances, such as peptides, proteins, nucleic acids, and chemotherapeutic medications, can be encapsulated by micelles to enable targeted and regulated release. Furthermore, micelles can be designed to react to particular stimuli, such pH, temperature, or enzymes, enabling medication release at specified sites [2]. In bioimaging, micelles can include imaging agents such as quantum dots or superparamagnetic iron oxide nanoparticles (SPIONs) [3], allowing for multimodal imaging methods including fluorescence and magnetic resonance imaging (MRI) [4]. This combination of therapeutic and diagnostic capabilities, known as theranostics, offers the potential for personalized medicine by allowing real-time monitoring of drug distribution and therapeutic efficacy.

Because of its biocompatibility, biodegradability, and easily modifiable amino groups, chitosan—a naturally occurring polymer produced from chitin—has attracted a lot of interest in nanomedicine [5]. Chitosan is a perfect choice for the creation of nanocarriers because of these characteristics. Chitosan improves the stability, mucoadhesive qualities, and possibility for controlled drug release of micellar systems. A variety of therapeutic agents, including anticancer medications like doxorubicin [6] and paclitaxel [7], have been delivered via chitosan-based micelles, which have shown enhanced drug solubility and therapeutic efficacy. Chitosan's cationic properties also improve the transport of proteins and nucleic acids by facilitating endosomal escape and cellular absorption. Furthermore, chitosan can be functionalized with targeting ligands, such as folic acid or antibodies, to achieve site-specific delivery, thereby reducing systemic side effects and improving treatment outcomes.

Micelles are essential for improving the effectiveness and delivery of medicinal substances. They can deliver poorly water-soluble chemicals, which are frequently limited in their clinical applicability due to poor bioavailability, thanks to their ability to solubilize hydrophobic medicines. These medications' solubility and stability are increased by encapsulating them in micelles, which boosts therapeutic efficacy. Furthermore, controlled and targeted drug delivery is possible by engineering micelles to release their payload in response to particular stimuli, such as pH changes or the presence of particular enzymes. By limiting the medicinal agent's accessibility to healthy tissues, this targeted delivery lowers side effects and enhances patient results.

Moreover, micelles present a promising platform for the administration of a variety of medicinal and imaging agents, especially those based on chitosan. They are positioned as useful instruments in the development of personalized medicine because of their capacity to improve solubility, stability, and targeted distribution, as well as their potential for controlled release and multimodal imaging. It is anticipated that future studies and technical developments would improve micelle-based systems even more, resolving current issues and broadening their therapeutic uses.

2 Fundamentals of Chitosan-Based Micelles

2.1 Chemical Structure and Properties of Chitosan

One of the most often used polymers in biomedical science is chitosan (CS), a naturally occurring amino polysaccharide that is the second most prevalent after cellulose. It is made by partially deacetylating chitin, which is found in the exoskeleton of insects, marine aquatic creatures, and microorganisms, including yeast, fungus, and microalgae, in an alkaline environment. With 1,4-glycosidic linkages connecting them, it is a copolymer of N-acetyl-D-glucosamine (acetylated unit) and D-glucosamine (deacetylated unit) [8]. By N-deacetylating the chitin molecule, chitosan, a linear polymer, is produced. Amino groups, primary hydroxyl groups, and secondary hydroxyl groups are the three rings and functional groups that make up the molecule. Chitosan is a polycationic natural polymer that has reactive amino and hydroxyl groups. Chitosan's solubility, viscosity, and reactivity are all influenced by the degree of deacetylation, which is normally between 60 and 95% [9]. Its reactivity and cationic character are enhanced by a larger DD, which indicates more free amino groups. Chitosan's degree of deacetylation, which typically ranges from 60% to 95%, affects its solubility, viscosity, and reactivity. A greater DD, which denotes more free amino groups, improves its reactivity and cationic nature. Chitosan is a powder that ranges from white to off-white. The protonation of the -NH_2 at the C-1 position of the D-glucosamine repeat unit makes it more soluble than chitin in acidic water environments. It is insoluble in water, organic solvents, and neutral or alkaline conditions but soluble in acidic solutions because amino groups have been protonated. The molecular weight ranges from 10 kDa to over 1000 kDa, and it can affect viscosity and functionality. Reactive groups found in chitosan include a primary amino group (C2) and primary and secondary hydroxyl groups (C6, C3). Along with functional groups, the acetamide group and glycosidic connections are taken into account. These functional groups allow for a wide range of modifications, producing polymers with unique properties and behaviours. In order to improve chitosan's solubility or biodegradability or to add new characteristics or functions, derivatives of the substance have been introduced. To improve solubility in aqueous circumstances, for instance, deacetylation, de-polymerization, or quaternization have been employed, among other techniques [10, 11].

Chitosan is a non-toxic, biocompatible, and biodegradable substance. Other noteworthy biological characteristics include its ability to heal wounds and its antibacterial and haemostatic effects. It can be processed into fibres, gels, microspheres-microcapsules, and micro/nanoparticles. It is also a great film forming [12]. Additionally, due to the presence of free -OH and -NH_2 groups in its structure, it can be chemically modified to enhance some of its qualities for specific uses.

Because of its exceptional physical, chemical, and biological qualities, CS is a great choice for use in the food, pharmaceutical, cosmetics, and medical industries. For many of these uses, chitosan is preferred over other cationic polymers such polylysine, polyarginine, or polyethyleneimine due to its relatively reduced toxicity [13, 14].

2.2 Micelle Formation: Self-Assembly Mechanism

The amphiphilic characteristic of chemically modified chitosan molecules drives the highly ordered, spontaneous self-assembly mechanism of chitosan-based micelles. Deacetylation of chitin produces chitosan, a cationic polymer with several amino and hydroxyl groups that enable a variety of chemical changes. Native chitosan, on the other hand, lacks the amphiphilic nature necessary for micelle formation and is naturally hydrophilic [15, 16]. Alkyl chains, fatty acids (such as palmitic or stearic acid), cholesterol, or hydrophobic polymers like polycaprolactone (PCL) or polylactic acid (PLA) are chemically added to chitosan to cause micellization. These amphiphilic derivatives replicate the behaviour of surfactants by displaying separate polar (hydrophilic) and non-polar (hydrophobic) regions [17, 18]. These modified chitosan molecules go through a spontaneous self-assembly process when they are distributed in an aqueous environment above a particular concentration known as the critical micelle concentration (CMC). The hydrophilic chitosan segments orient outward to interact favourably with the surrounding water, providing a stabilizing shell, while the hydrophobic segments combine to minimize their exposure to water due to hydrophobic interactions, forming the micelle's core [19]. This thermodynamically stable core–shell structure aids in the solubilization of hydrophobic medications or substances within the micellar core while maintaining the micelle's dispersion in water because of the hydrophilic shell. Hydrophobic contacts, van der Waals forces, hydrogen bonds, and occasionally electrostatic interactions—particularly significant when ionic species or charged medicines are involved—are the forces that propel this self-assembly [20]. Crucially, a number of variables, including the degree of hydrophobic group substitution, the molecular weight of chitosan, the medium's pH, and the solution's ionic strength, affect the chitosan micelles'size, stability, and ability to load drugs. Furthermore, micelle behaviour under physiological settings is significantly influenced by the pH-sensitive behaviour of chitosan [21]. Chitosan's amino groups become protonated at acidic pH, which improves its solubility and mucoadhesiveness. At neutral or basic pH, however, the reduced protonation may cause the micelle to destabilize or disassemble, which is an effect that is used for targeted and controlled drug release [22]. To further control micelle behaviour and drug delivery kinetics, the functional groups on chitosan can be further modified to include targeting ligands, stimuli-responsive elements like pH-, temperature-, or redox-sensitive linkages, or polyethylene glycol (PEG) for stealth properties [23]. Chitosan-based micelles can encapsulate a wide range of hydrophobic medicinal agents due to their dynamic and reversible self-assembly, which improves their solubility, bioavailability, and circulation time while lowering systemic toxicity. These micelles can also passively accumulate in tumour tissues through the increased permeability and retention (EPR) effect because of their nanoscale size (10–200 nm). Chitosan-based micelles' self-assembly mechanism thus highlights their considerable promise in pharmaceutical formulations and nanomedicine by offering a flexible and adjustable nanoplatform for effective drug encapsulation, distribution, and targeted therapy [24, 25].

2.3 *Stability, Size, and Surface Characteristics of Chitosan Micelles*

One important factor that affects chitosan micelle performance and shelf life in biological settings is stability. A number of variables, including the medium's ionic strength, pH, degree of hydrophobic group substitution, and critical micelle concentration (CMC), affect micellar stability. When the micelles are diluted in bodily fluids, a low CMC value guarantees that they stay intact. Adding hydrophobic components to chitosan, such as cholesterol, fatty acids, or alkyl chains, increases micelle stability and core hydrophobicity. The micelle shell or core can also be crosslinked to improve structural integrity and increase resistance to disintegration in physiological settings [26].

Chitosan micelles generally measure between 20 and 500 nm, influenced by the preparation method and the extent of hydrophobic modification. Amphiphilic chitosan derivatives, for example, can produce micelles that are appropriate for drug delivery applications and have an average size of 140 nm. For better bioavailability and increased cellular absorption, smaller micelles are frequently chosen [27].

Chitosan micelles normally possess a positively charged surface. The micelles' stability is improved by increasing the degree of amino substitution since it lowers the critical micelle concentration (CMC) and raises the zeta potential. The zeta potentials of stearic acid-grafted chitosan (CS-SA) micelles, for example, varied from +22.9 to +48.4 mV, depending on the degree of substitution, while their CMC ranged from 0.16 to 0.25 mg/mL [28]. A key factor in the stability of micelles is the zeta potential, which stands for the surface charge. Increased zeta potentials inhibit aggregation and improve dispersion stability by increasing the electrostatic repulsion between particles.

3 Synthesis of Chitosan-Based Micelles

3.1 *Methods of Chitosan Micelle Preparation*

Chitosan micelles can be prepared using methods like emulsion, solvent evaporation, and precipitation-based techniques. By dispersing chitosan in a solvent and then removing the solvent or altering the circumstances, these techniques allow chitosan to self-assemble into micelle structures.

3.1.1 Emulsion Solvent Diffusion Method

This method creates an internal phase by dissolving a polymer in an organic solvent that is water soluble, such as acetonitrile or ethyl acetate. After that, this phase is emulsified to create an oil-in-water (O/W) emulsion in an aqueous phase with a stabilizer. When diluted with water, the solvent diffuses out, causing the polymer to precipitate and micelles or nanoparticles to form.

3.1.2 Ionic-Gelation Method

In ionic gelation, positively charged chitosan is dissolved in a room-temperature acetic acid solution while being magnetically stirred for an hour. Polyanion tripolyphosphate (TPP) dissolved in deionized water (DI) is used to create a second solution. TPP is typically employed as an ionic cross-linker. In order to create a complex coacervate aqueous phase, an aqueous solution containing chitosan and another containing TPP are mixed during the ionic gelation synthesis process. It is necessary to magnetically stir the TPP–chitosan mixture when it is at room temperature. The solution undergoes three distinct phases based on the stage of the process: clear (chitosan solution), opalescent, or milky (after adding TPP to the chitosan solution), and aggregated (after adding more TPP to a milky solution). The milky appearance is a sign that CNPs are forming. The extracted CNPs are excellent for in vivo or in vitro drug delivery applications [29].

3.1.3 Microemulsion Method

A microemulsion, which is stabilized by a surfactant and frequently a cosurfactant, is a clear, transparent, and thermodynamically stable mixture of two immiscible liquids, typically water and oil. These systems are single-phase and optically isotropic, which means that the water and oil phases do not appear to be separated. Stearic acid (SA) and chitosan oligosaccharide (CSO) were coupled by Hu et al. [30] using 1-ethyl-3-(3-dimethylaminopropyl) carbodiimide (EDC), which caused the two substances to form micelle-like nanoparticles. To assess the effectiveness of the CSO-SA micelle-type nanoparticles as a gene delivery device, plasmid DNA was loaded onto them. To create CNPs cross-linked with glutaraldehyde, Manchanda and Nimesh [31] employed the same microemulsion technique as Banerjee et al. [32]. To investigate their release in vitro under various pH levels, synthetic oligonucleotides (ODNs) were loaded onto the CNPs. The authors came to the conclusion that ODNs were released more readily in basic environments as opposed to neutral or acidic ones.

3.1.4 Emulsification Solvent Diffusion Method

An organic phase (including a dissolved polymer and drug) and an aqueous phase are combined to form a stable emulsion, which is the process by which the emulsification solvent diffusion method produces nanoparticles. When water is added to dilute the emulsion, the solvent diffuses out of the droplets, causing precipitation and the creation of nanoparticles. When encapsulating hydrophobic medications in polymeric nanoparticles, this technique is especially helpful. By adding an organic phase to a chitosan-containing solution that contains a stabilizing agent (such as poloxamer), an emulsion is produced through emulsification solvent diffusion. The entire combination must then be mechanically agitated and homogenized under pressure. Niwa et al. [33] exploited a D, L-lactide/glycolide copolymer (PLGA) in a modified emulsification technique that was initially described by El-Shabouri [34]. This process involves diluting the emulsion with large amounts of water, which leads to the creation of nanoparticles and polymeric precipitation. The emulsification solvent diffusion method's regulating mechanism is the organic solvent's diffusion in the aqueous medium. This method's drawbacks include the significant shear forces that arise during the production of nanoparticles and the use of organic solvents.

3.1.5 Polyelectrolyte Complex Method

The polyelectrolyte complex (PEC) method is a technique that includes combining solutions of oppositely charged polyelectrolytes to build complexes between them. Drug delivery, water treatment, and material science are just a few of the uses for these complexes, which are created through electrostatic interactions. The PEC self-assembles as a result of the charge neutralization, which significantly lowers its hydrophilicity. The size of the micelle ranges from 50 to 700 nm. Drugs, proteins, peptides, and plasmid DNA have all been delivered via these PEC-complexed CNPs. Liu and colleagues [35] also prepared CNPs loaded with heparin using the PEC approach. They took into account how the yield size of nanoparticles was impacted by the solution's pH, molecular weight (MW), and constituent concentration.

3.2 Functionalization and Modification of Chitosan for Improved Performance

3.2.1 Covalent Functionalization

Common methods for CS covalent modification involve partial oxidative breakage of the backbone or chemical derivatization of the major amine and alcohol functions [36].

3.2.2 Amine Functionalization

Alkylation

The most traditional method of functionalizing CS amines involves using a coupling agent to activate the carboxylic acid derivative of interest, followed by acylation of the amines of CS with the reactive intermediate that is created. N-(3-Dimethylaminopropyl)-N′-ethyl carbodiimide chloride (EDC.HCl) is the preferred coupling agent because of its advantageous solubility in water. It is frequently recommended to employ it in polar solvents like DMF or N-methyl pyrrolidone (NMP) when feasible due to its low stability in aqueous solutions.

Nucleophilic substitution (SN) is used in a number of publications to graft tiny aliphatic halides onto CS amines, creating secondary amine bonds. Among the most popular quaternary ammonium CS salts is TMC. Even though TMC can be purchased commercially, multiple processes have shown that it can be produced at 60 ° C with extra MeI and NaOH. To guarantee full CS N-trimethylation and a greater yield of the quaternized polymer, a two-step process was specifically chosen.

Phthaloylation

The preferred technique for protecting CS amines and improving the solubility of the resultant intermediate is phthaloylation. Phthalic anhydride is typically used in DMF at 120–130 °C. By adding water to the reaction solvent, the selectivity toward N-phthaloylation over protection of the hydroxyl groups can be improved [37].

Acylation

It is frequently reported that urea bonds form on CS amines in polar solvents or water when 1,1′-carbonyldiimidazole (CDI) is present as an activating agent. The majority of approaches use a one-pot method in order to prevent the sensitive acyl imidazole intermediate from being isolated. When CS was conjugated with BPEI-cystamine in the presence of CDI, bioreducible copolymers for gene delivery were created [38] (Fig. 1).

3.2.3 Epoxide and Aziridine Ring Openings

The functionalization of CS amines by SN is likewise governed by the epoxide ring opening of glycidol derivatives. The resultant derivatives are frequently employed for drug/gene administration and have been reported to produce micelles, hydrogels, or membranes. Because of the resulting CS derivatives' improved solubility in organic and aqueous environments, the remaining free amines can be further derivatized using activated polymers such as isocyanate terminated polycaprolactone [39]. Another glycidol derivative that has been documented for CS quaternization is glycidyl trimethylammonium chloride (2,3-epoxypropyltrimethylammonium chloride, EPTAC).

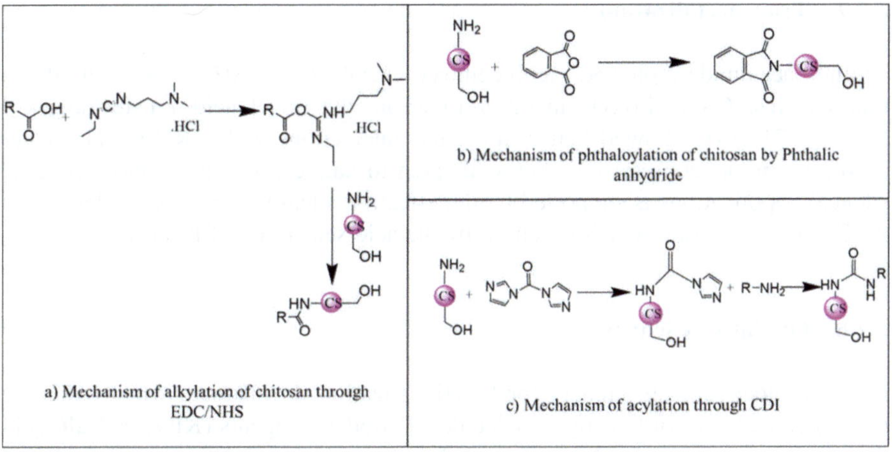

Fig. 1 Possible mechanism of chitosan modification through amine functionalizing

3.2.4 Hydroxyl Functionalization

Both CS chemical derivatization may occur for primary alcohol in the C6 position and secondary alcohol in the C3 position. Because CS amines are more nucleophilic than alcohols, selective hydroxyl functionalization necessitates N-protection or functionalization.

3.2.5 Acylation Through Carbamate

Carbamate bonds guarantee dependable, covalent, and long-lasting connections to CS alcohols, just like urea bonds do for amine alterations. Both amino-terminated PEG derivatives and aliphatic amines were grafted to CS hydroxyl groups in the presence of CDI, an activating agent, in a dry solvent. After this process, CS-based self-assembled micelles for gene delivery were produced by conjugation to dodecyl amine [40].

3.2.6 Acylation Through Ester

In the conjugation to CS alcohols, ester linkages were rarely reported because of their vulnerability to hydrolysis under physiological circumstances. In order to create functionalized membranes for metal affinity chromatography applications, Zhu et al. [41] explained how to O-succinylate cross-linked CS.

3.2.7 Transacetalization

Trifluoroacetamide-protected aminoethoxy ketal (TFA-AE-k) was used to transacetalize CS hydroxyls in order to create the novel acid-transforming CS (ATC) [37]. ATC showed better molecular interactions with siRNA and greater absorption in aqueous media when compared to native CS. This method for gene silencing applications is supported by the effective release of siRNA condensed by ATC into the cytosol, which is caused by the acid sensitivity of ketal linkages.

3.2.8 Oxidative Cleavage

Reactive aldehydes are created for functionalization downstream when oxidative cleavage breaks down CS sugar molecules. Periodate reagents (KIO_4 or $NaIO_4$) in degassed acetate buffer at 4 °C were used in the majority of documented oxidation processes. Higher degrees of deacetylation of the initial polymer might restrict the amount of ammonia released, which causes substantial depolymerization under these conditions [42].

4 Drug Loading and Release Mechanisms in Chitosan Micelles

4.1 Encapsulation Efficiency and Drug Loading Capacity

Encapsulation efficiency (EE) is the percentage of the drug that is successfully contained within the nanoparticle or other drug carrier after being introduced to the system. Which is calculated by—(weight of drug encapsulated/total weight of drug added) * 100%.

Drug loading capacity (DLC) is a percentage of the total weight of the nanoparticle that indicates how much medicine may be loaded into the nanoparticle carrier. Which is calculated by—(weight of drug encapsulated/weight of the nanoparticle) * 100% [43–45].

Although they are connected, EE and DLC are not comparable. While a high DLC does not always imply a high EE, a high EE might help produce a high DLC. While EE depends simply on the amount of drug encapsulated, the DLC depends on both the EE and the amount of drug added.

4.2 Mechanisms of Drug Release

Chitosan micelles'amphiphilic nature allows for a high drug loading capacity and regulated release behaviour, making them an effective and adaptable nanocarrier system for drug delivery. Usually, chemical conjugation, physical encapsulation, or electrostatic interactions are used to load drugs into chitosan micelles. Hydrophobic medications are contained within the micellar core created by hydrophobically modified chitosan (such as grafting oleic or stearic acid), whereas hydrophilic medications may engage in ionic or hydrogen bonding interactions with the hydrophilic shell. The stability and effectiveness of drug encapsulation are greatly influenced by the chitosan's degree of substitution, molecular weight, and pH sensitivity.

The main mechanisms controlling drug release from chitosan micelles include diffusion, micelle destabilization, or polymer breakdown. In acidic environments, such as tumour tissues or intracellular compartments, where protonation of amino groups causes micelle enlargement or disintegration, chitosan's pH sensitivity enables selective release. Furthermore, persistent release may be triggered by lysozymes' enzymatic breakdown of chitosan [46, 47]. Release kinetics can be precisely altered for therapeutic uses by adjusting the micelle's crosslinking density and chemical makeup. Chitosan micelles are ideal for delivering proteins, nucleic acids, and chemotherapeutics due to their regulated and responsive release profile, which also improves drug bioavailability, stability, and site-specific administration in sophisticated drug delivery systems [48].

5 Chitosan-Based Micelles in Drug Delivery

5.1 Oral and Intravenous Drug Delivery Applications

Polymeric micelles, which are nano systems made up of a hydrophilic shell and a hydrophobic core, are frequently used to encapsulate hydrophobic medications. These systems are effective oral delivery methods because they can shield the medication from the harsh GIT environment while also enhancing drug stability [49, 50]. Another benefit of polymeric micelles' high drug encapsulation capacity is their ability to quickly reach a therapeutic dose while avoiding the epithelium efflux pumps and limiting undesirable side effects. Chitosan is frequently used to create polymeric micelles because it is easily transformed into an amphiphilic polymer with self-assembly properties. Chitosan's mucoadhesive qualities and capacity to momentarily relax the epithelium's tight connections enhance drug absorption in the intestines [51, 52].

In 2020, Tu et al. [53] created a new chitosan derivative called gallic acid-Chitosan-D-α-tocopherol polyethylene glycol 1000 succinate (GA-CS-TPGS copolymer) and paclitaxel micelles called PTX-Micelles. Their aim was to increase the bioavailability and anti-tumour efficacy of PTX (by inhibiting P-gp efflux and drug metabolism in the liver) by improving the permeability and solubility of the micelles. PTX-Micelles were more effective against lung cancers and had about 3.80-fold higher bioavailability than Taxol® (Bristol-Myers Squibb, Princeton, NJ, USA).

In 2023, Yang and colleagues [54] introduced a novel delivery method based on linolenic acid (LA)-linked chitosan (CS) polymeric micelles to improve the oral absorption of doxorubicin (DOX). The transporter ligand LA was grafted onto CS to form CS-LA micelles, which encapsulate DOX. The micelles had a critical micelle concentration (CMC) of 23.6 μg/mL and were completely spherical. The micelles had a drug loading of 8.77 ± 0.02% and a particle size of 117.1 ± 1.6 nm. Pharmacokinetic tests showed a 166% relative bioavailability when compared to DOX•HCl. The intestinal absorption process of micelles mediated by fatty acid transport protein 4 was validated by the penetration test, intracellular uptake, and immunofluorescence imaging.

Chitosan-based polymeric micelles loaded with camptothecin (CPT) were coupled into 3D printing systems (printfills) sealed with an enteric layer by Almeida and colleagues [55], in order to shield nanosystems from the harsh GIT environment. At a pH equal to that of the stomach's simulated gastric juice, the printfills remained stable and only released their micelles in the colon. The intestinal absorption was subsequently replicated using the dissolving media, and in a 3D intestinal cell-based model, the chitosan micelles considerably enhanced the CPT permeability in comparison to the free pharmaceuticals, with an apparent permeability coefficient (Papp) of almost 9 × 10–6 cm/s. By employing 3D printing and nanotechnology, polymeric micelles can be released specifically into the colon, improving intestinal absorption and avoiding systemic or drug-specific breakdown throughout the tract.

5.2 Targeted and Sustained Release Systems

Chitosan-based micelles, which combine the natural biocompatibility of chitosan with sophisticated formulation techniques, have become a viable platform for targeted and sustained drug administration. These micelles, which are usually 20–500 nm in size, are created by adding hydrophobic substances like cholesterol or stearic acid to chitosan. This allows hydrophobic medications like doxorubicin and paclitaxel to be encapsulated. Because these micelles are amphiphilic, they can self-assemble in aquatic settings to form core-shell structures that improve the stability and solubility of drugs. Chitosan micelles can be functionalized with ligands like folic acid or antibodies to allow targeted distribution. These ligands can identify particular receptors that are overexpressed on target cells and facilitate receptor-mediated endocytosis. To improve cellular absorption and circulation time, the micelles' surface charge and size can also be adjusted. Modifications such as pH-sensitive groups or redox-sensitive linkers (such as disulfide bonds) can be added for sustained release, guaranteeing that medication release mostly takes place in the intracellular compartments or tumour microenvironment and reducing systemic side effects.

Yuting Yuan and colleagues [56] created a novel amphiphilic low molecular weight chitosan-graft-nicotinic acid bearing decyl groups (LCND) in an effort to improve the water solubility and photostability of α-tocopherol. The LCND was created by a two-step reaction and spontaneously assembled into a cationic micelle using an ultrasonication technique. Results indicated that when the feeding ratio of LCND to α-tocopherol reached 10: 3, the cationic micelle demonstrated the strongest loading capacity with a loading content of 18.50% and a strong ability to self-assemble with nanoscale spherical morphology. In the meantime, α-tocopherol's significantly improved water solubility, photostability, and prolonged release behaviour in cationic micelles were noted. While free α-tocopherol was fully released in 10 h, the cumulative release of α-tocopherol in a cationic micelle increased to 82.18% in 96 h.

Xue Zhang and his team [57] developed a micelle system for hepatocellular carcinoma therapy. For the in vivo administration of celastrol (Cela), a carboxymethyl chitosan-thioketal-rhein (GCTR) conjugate modified with glycyrrhetinic acid was created and produced. In order to create Cela/GCTR PMs, which have a tiny particle size and a high drug loading capacity, Cela was encapsulated inside polymeric micelles (PMs) created by GCTR conjugates self-assembling in water. In the blood environment, Cela/GCTR PMs had a sustained-release characteristic; in the tumour microenvironment, they exhibited a rapid-release feature. At low concentrations, Cela/GCTR PMs had a negligible effect on L-02 cells but a strong proliferation-inhibitory effect on HepG2 and BEL-7402 cells. Cela/GCTR PMs had the ability to respond to ROS both in cells and in vitro, increase Cela's bioavailability, and have exceptional hepatoma-targeting capabilities. Cela/GCTR PMs may also successfully stop the growth of tumors without causing any harm to other organs.

In another work [58], pH-sensitive for colon-targeted drug administration, curcumin (CUR) has been incorporated into N-naphthyl-N, O-succinyl chitosan (NSCS) and N-octyl-N,O-succinyl chitosan (OSCS) polymeric micelle carriers. Dropping, O/W emulsion, cosolvent evaporation, and dialysis were the physical entrapment techniques used. The dialysis process produced CUR-loaded micelles, which had the maximum loading capacity. The polymeric micelles' loading capacity increased when the initial CUR content was increased from 5 to 40 weight percent to the polymer. There were no notable variations in the loading capacity of CUR-loaded micelles among the hydrophobic cores. Every micelle loaded with CUR had particle diameters between 120 and 338 nm. The micelles' ability to respond to pH was confirmed when their shape changed in response to interaction with varying pH values.

5.3 Micelles for Delivery of Hydrophobic and Hydrophilic Drugs

In order to overcome issues with solubility and bioavailability, chitosan micelles have become effective nanocarriers for the administration of both hydrophilic and hydrophobic medications. Usually, hydrophobic moieties like oleic or stearic acid are added to chitosan to make these micelles, which are amphiphilic structures that self-assemble in aqueous solutions. While the hydrophilic shell of the micelles guarantees biocompatibility and makes interactions with biological systems easier, the hydrophobic core contains lipophilic medicines, improving their solubility and stability. For instance, through the catalytic action of CDI, we were able to successfully prepare three amphiphilic conjugates with varying hydrophobic chain lengths (LC-Cys-OA, LC-Cys-DA, and LC-Cys-CA). These conjugates could then self-assemble into core-shell structural micelles to optimize the therapeutic efficacy of Dox by controlling the length of hydrophobic segments. Of the three Dox-loaded micelles, Dox-loaded LC-Cys-CA exhibited the highest redox sensitivity to tumour cells. When the disulfide bond was broken and the entire micelle dissociated, the in vitro release experiment showed that the three Dox-loaded micelles exhibited accelerated Dox release behaviour in the redox environment of tumour cells [59].

Chitosan micelles have the ability to transport hydrophilic medications in addition to hydrophobic ones. These micelles can encapsulate water-soluble medications, increasing their bioavailability and therapeutic efficacy, by altering chitosan to increase its stability and solubility. Furthermore, targeting ligands like folic acid or antibodies can be functionalized onto the surface of chitosan micelles to accomplish site-specific distribution, which lowers systemic side effects and enhances therapeutic results. The Stenzel team conducted a number of intriguing research on the development of BSA-based drug delivery systems for cancer treatment. The study where BSA was treated with PCL and the resulting micelles were loaded with albendazole, with a DLE of 16.4% is noteworthy among all of these findings. With diameters of about 100 nm, these drug-loaded micelles demonstrated significant cytotoxicity and strong growth inhibition effects on MCTS, suggesting that the system may be employed to treat pancreatic cancer. In a different work [60], BSA-PCL copolymer and poly(oligo(ethylene glycol) methyl ether acrylate)-poly (ε-caprolactone) (POEGMEA-PCL) copolymer were co-assembled to create Cur-loaded drug delivery systems (DLE about 40%).

6 Chitosan-Based Micelles for Bioimaging

6.1 Principles of Bioimaging and Role of Micelles

The topic of bioimaging is one that is developing quickly and includes a variety of methods for visualizing biological processes and structures in living things. The creation of contrast, spatial and temporal resolution, sensitivity, specificity, and non-invasiveness are among the fundamentals of bioimaging. Imaging modalities like magnetic resonance imaging (MRI) [61], computed tomography (CT) [62], positron emission tomography (PET) [63], fluorescence imaging [64], and ultrasound are chosen and optimized according to these principles. To produce photographs of internal biological features, each method uses a different physical interaction, such as magnetic fields, X-rays, sound waves, or light absorption/emission. In order to improve signal-to-noise ratios and make it easy to distinguish between healthy and sick tissues, contrast agents and probes are essential. Targeted imaging, which seeks to selectively visualize particular tissues or molecular markers linked to disease, such as malignant tumours, inflammatory areas, or infection sites, is a key focus of contemporary bioimaging. This has been made possible by the development of nanoparticles and nanocarriers that can deliver imaging agents to particular locations with improved biocompatibility and specificity. Among these, chitosan micelles have drawn a lot of interest because of their distinct biological and physicochemical characteristics. Because it is non-toxic, biodegradable, and biocompatible, chitosan—a naturally occurring cationic polymer derived from chitin—is perfect for use in biomedical applications. Chitosan can self-assemble into micelles, which are spherical, nanoscale structures with a hydrophilic shell and a hydrophobic core, when it is modified with hydrophobic chains. To prevent deterioration and improve their solubility and durability in biological settings, these micelles can enclose hydrophobic imaging agents (such fluorescent dyes or MRI contrast materials) within their core. Furthermore, to guarantee receptor-mediated distribution to certain cells or tissues, the surface of chitosan micelles can be readily functionalized with targeted ligands like folic acid, antibodies, or peptides. Because of this, chitosan micelles are very useful as intelligent nanocarriers for focused bioimaging. Additionally, by interacting electrostatically with negatively charged cell membranes, their cationic composition enhances cellular absorption. Numerous imaging modalities have made use of chitosan micelles. For instance, they can transport radionuclides for PET and SPECT imaging, fluorescent dyes for visual imaging, or gadolinium or iron oxide nanoparticles for MRI. Additionally, they can be included in multipurpose systems for theragnostic applications, which combine therapy and diagnosis on a single platform. Notwithstanding their benefits, there are still issues with repeatability, large-scale production, and possible immunological reactions that need to be investigated and improved. However, a potential new area in the creation of multipurpose, target-specific, and biocompatible imaging systems is represented by chitosan micelles. Their efficiency and versatility provide enormous promise for improving the accuracy of disease diagnosis, tracking the results of treatment, and supporting the more general objectives of non-invasive and personalized care.

6.2 Micelles for Fluorescence Imaging

Optical imaging offers quick, real-time imaging and is both safe and extremely sensitive. Its primary limitations include light scattering, autofluorescence, and absorption by bodily lipids, water, and neighbouring tissues. Advanced methods like optical coherence tomography, fluorescence, bioluminescence, and diffuse optical tomography have recently improved these limitations. NIRF imaging is very frequently used for small animal imaging in vivo. By using the NIR window (700–900 nm), the NIRF imaging system minimizes the problems of autofluorescence or excessive absorption while identifying the basic activities at the subcellular levels. Red blood cell haemoglobin has significant absorbance and autofluorescence in the visible spectrum only, and water and lipids absorb infrared light more often than near-infrared light, which makes NIRF imaging feasible [65].

There have been numerous reports of imaging nanoagents based on chitosan that employ different NIR fluorophores. Zhu et al. [66] showed that α-hederin chitosan nanoparticles, which had an average particle size of 148.23 nm, were linked with a CD147 antibody. Internalization and lysosomal escape were facilitated by the CD147 antibody through a caveolae-dependent mechanism. The α-hederin caused cancer cells to undergo apoptosis. The nanoparticles contained cyanine 7 (Cy7) fluorophore labels for in vivo imaging. In tumour tissues treated with α-Hed-CS-CD147-NPs, the fluorescence intensity was brighter, indicating a higher targeting effectiveness than that of α-Hed-CS-NPs, which led to increased antitumor efficacy. Quantum dots, which are artificial fluorescent semiconductor nanocrystals, are superior to fluorophores in a few ways, including their outstanding photostability without bleaching, limited emission wavelength, and strong photoluminescence. There have been reports of chitosan nanoparticles encasing quantum dots. Alexandra et al. [67], for instance, reported tripeptide-modified chitosan organic shells with quantum dot cores. RGD peptides were used to target these nanoparticles, while quantum dots were used for imaging. After being coupled with chitosan, the RGD peptide was applied to quantum dots. The amphiphilic polymer mPEG-CS-Bio-Hz-Q was developed and produced for minimal protein adsorption, active tumour targeting, pH-responsive drug release, and near-infrared (NIR) fluorescence imaging. The polymer was used to administer pH-triggered paclitaxel (PTX) release via a micellar carrier, investigated by the group of Cheng et al. [68]. In another study by Shi et al. [69], TPE-bi(SS-CS-Bio) is a new polymer that was designed and manufactured using aggregation-induced emission (AIE) fluorogen, biotin, and disulfide bond modified chitosan, able to self-assemble into micelles and greater encapsulation efficiency of paclitaxel (PTX), which had a strong blue fluorescence emission and a good AIE feature, which allowed tracking of their distribution of drug throughout the cells.

6.3 Chitosan Micelles for Magnetic Resonance Imaging (MRI)

One of the most effective in vivo imaging techniques for obtaining anatomical, physiological, and molecular data is magnetic resonance imaging (MRI). Small animals can have a spatial resolution of 10–100 μm with the magnetic field powers that are now available. However, because of MRI's limited sensitivity, exogenous contrast agents like gadolinium or iron oxide must be used. Using a Larmor frequency and a net magnetic moment, MRI detects the relaxation processes of hydrogen protons during gyroscopic precession. When the net magnetic precession is flipped by a resonant radio frequency, the net magnetic moment gradually returns to its initial state once the radio frequency is removed; this process is known as relaxing. Both longitudinal (T1) and transverse (T2) relaxation can be used to quantify relaxation processes. By reducing either the T1 or T2 relaxation period, MRI contrast agents alter the strength of the surrounding MRI signal [70].

Chitosan nanoparticle-based MRI contrast agents have been documented for MRI imaging of cancer. T1 contrast agents were described by Nwe et al. [71]. employing water-soluble glycol chitosan (GC) polymer and 1,4,7,10-tetraazacyclododecane-1,4,7,10-tetraacetic acid 1-(2,5-dioxo-1-pyrrolidinyl) ester (DOTA-NHS), followed by Gd chelation. Superparamagnetic iron oxide (SPIO) nanoparticles are mostly utilized for T2 contrast effects because of their low magnetic aggregation characteristics. Hydrophobic chitosan cores have been used to encapsulate SPIO nanoparticles [72].

6.4 Applications in Theragnostic: Combined Drug Delivery and Imaging

Targeting moieties, imaging agents, or therapeutic agents were added to CS derivatives to entrap drug/imaging agents and obtain multimodal imaging and combined therapy. The entrapped drug/imaging agents inside the CS-based nano systems may have better imaging and therapeutic effects thanks to these functionalized CS derivatives.

By connecting photosensitizers with CS, a theragnostic can be easily constructed. Jia et al. [73] conjugated protoporphyrin IX (PpIX) and PEG-modified glycol CS together (GC-PEG-PpIX) to create CS-based plasma membrane activatable self-assembled NPs. To prevent fluorescence self-quenching, PpIX was retained inside the NPs' core by strong π–π stacking in this nano system. When PpIX reached the cell membrane, its strong affinity for the plasma membrane encouraged its exposure, which improved PDT and fluorescence imaging.

Apart from acting as a drug carrier, CS may also contribute to the creation of inorganic nanoparticles and provide them with functional groups for additional modification. In the detection and therapy of cancer, these CS-functionalized organic–inorganic NPs show outstanding therapeutic/imaging benefits. CS derivatives have been applied to the surface of some inorganic NPs, such as rare-earth-based UCNPs, AuNPs, molybdenum nanosheets, silver-neodymium bimetallic NPs, and SPIONs, to improve their biocompatibility and prevent aggregation in solution. This has allowed the drug carriers to perform additional functions. Additionally, a site for integrating many components into a single nanoplatform was made possible by the coating of the CS derivatives (Table 1).

7 Recent Advances in Chitosan-Based Micelles

7.1 *Smart Micelles for Stimuli-responsive Drug Release*

7.1.1 pH-Responsiveness

The ability of smart biomaterials to react to changes in environmental parameters can facilitate the local delivery of treatments in situations where the pH of the environment is either basic or acidic. It can also enable a release in certain organ systems or pathological circumstances where the pH fluctuates. The systemic concentrations of very hazardous chemotherapeutics can be limited via pH-controlled release mechanisms [83]. Solid tumours' acidic extracellular environment allows for the development of pH-sensitive drug delivery systems that, when exposed to acidic pH, change chemically, physically, or both. Smart delivery methods investigate conditions like diabetes and stomach ulcers in addition to cancer treatment. Weak acidic or basic groups are typically found in the polymer backbone of polyelectrolyte polymeric systems. Proton acceptance or release is triggered by changes in the pH of the surrounding environment, which promotes bond breakage, solubility, and/or structure.

Gao and their team [84] developed a system using epoxy chemicals as cross-linking agents; carboxymethyl chitosan (CMC) was created as a pH-sensitive hydrogel. The cumulative release of doxorubicin, according to release experiments, was 29.9%. In 144 h, 49.3% and 65% of the gel were removed at pH values of 7.4, 6.5, and 5.0, respectively. Studies conducted in vivo confirmed that the implantation of drug-loaded hydrogels may break down continuously in tumour tissues, improving the accumulation of doxorubicin in tumour locations and lowering the tumour volume in the mice at the end of the seven-day trial. The aforementioned experimental study demonstrated that the hydrogels enhanced the drug's concentration in the tumoral area and markedly extended the release of doxorubicin.

7.1.2 Photo-Responsiveness

Since light can be controlled externally, it is especially well-suited for creating cutting-edge biomedical platforms that enhance the dynamic properties of hydrogel

Table 1 Summary of recent chitosan-based theranostic nanoplatforms for cancer treatment, highlighting their composition, target tumour type, imaging agents, imaging modalities, therapeutic agents, theranostic modes, and relevant references

Composition	Tumour type	Imaging agent	Mode of imaging	Therapeutic agent	Mode of theragnostic	Ref.
PS@CS@Au–Fe_3O_4–FA	HeLa	SPIONs/ICG/AuNPs	MRI/CT/OI	ICG	PDT	[74]
PCL-SS-CMC-GA	HepG2	DOX	OI	DOX/PHA	Chemo/PDT/PTT	[64]
CS-HA-pA20–36	A20	SPIONs/ATTO680	MRI/OI	pA20–36	Immunotherapy	[75]
CFMQ-CS-PLGA	C6/U87-MG	Cy5	OI	Temozolomide	Chemotherapy	[76]
CMC/DSPE-PEG	22RV1/PC-3/16HBE	Perfluorohexane	USI	RNA	Gene therapy	[77]
CS@FA	MCF-7	Curcumin	OI	Curcumin	Chemotherapy	[78]
M-MSN/alginate/CS	MCF-7/EMT-6	Fe_3O_4-AuNPs	CT/MRI	shRNA (P-gp)/ DOX/Ce6	Gene/chemo/PDT	[79]
CS@FA	MCF-7	Rhodamine B	OI	Curcumin	Chemotherapy	[80]
CS@FA	HeLa	UCNPs (NaYF4: Yb, Er@NaYF4: Yb, Nd)	CT/OI	DOX/POM	Chemo/PTT	[81]
DOTA-Ppa-GC	4 T1	99mTC	SPECT	177Lu/Ppa	Radio/PDT	[82]

materials. Exposure to light waves with the right wavelength and intensity can modify photo-stimulated responsive hydrogels. Common light sources include ultraviolet (UV), visible (Vis), and near-infrared (NIR) light.

Ryplida et al. [85] created a film hydrogel with a regulated hydrophobic-hydrophilic transition when exposed to UV–visible light by preparing TiO_2 polydopamine perfluoro silicon carbon-dot linked chitosan polyvinyl alcohol loaded tannic acid. In particular, when exposed to light stimulation, the hydrogel displayed up to 278% mechanical strain. Furthermore, following UV and visible light irradiation, the film hydrogel's electrical conductivity increased to 1.096 × 10–3 and 1.026 × 10–3 S/cm, respectively.

7.1.3 Glucose-Responsiveness

The primary energy source for the body's regular physiological processes is glucose, and its level is often strictly regulated via feedback. The human pancreas can release endogenous insulin by monitoring glucose levels as a glucose reaction platform.

The blood glucose level is out of balance when this balanced feedback regulatory system is disrupted, as in diabetes. Patients must administer insulin on a frequent basis to lower and maintain their blood sugar levels because of the erroneous insulin release. It was discovered that when the produced glucose stimulated hydrogel was exposed to glucose molecules in the outside world, its structure and characteristics would alter. Despite being thought to be promising, glucose-responsive hydrogels (like insulin delivery systems) react too slowly to return to their initial state after reacting to variations in glucose concentration.

Abureesh et al. [86] created chitosan/polyvinyl alcohol hydrogels using boric acid as a cross-linker, and they found that freeze-thaw cycling methods improved the hydrogels' capacity to swell and their sensitivity to glucose. The test findings demonstrated that pH could have an impact on the hydrogels' swelling, which showed a notable sensitivity to glucose in physiological settings. 86.45% of bovine serum albumin (BSA) was released under ideal circumstances (pH = 7.4, glucose, 37 °C).

7.1.4 Temperature-Responsiveness

One stimulus response factor that is easier to manage is temperature. According to research, thermosensitive responsive hydrogels can alter their size in response to temperature changes. A denser network structure results from an increase in the hydrophobic contact between non-polar groups in the hydrogel polymer chain at ambient temperatures above LCST. Conversely, a more open network structure is formed as the ambient temperature drops below LCST because the hydrophobic contact between the hydrogel polymer chains becomes less strong. As a result, when heated, hydrogels with LCST can develop, and when cooled, hydrogels with UCST can.

In vivo drug release at precise places is made possible by thermally responsive injectable chitosan hydrogels. For loxoprofen, Ahmad et al. [87] created injectable hydrogels using chitosan copolymer (N-isopropyl-acrylamide). Maximum swelling (4.5-fold) was seen in swelling studies, and at pH = 1.2 and 25 °C, there was a quick release of the medication. The produced hydrogel was shown to have a regulated release in rabbits (Cmax = 10 μg/mL; Tmax = 36 h; AUCtot = 646 μg/h·mL) and to be safe using the MTT assay (survival rate of around 80%).

7.2 *Hybrid Micelles Incorporating Nanoparticles for Enhanced Bioimaging*

Nanoparticle-containing hybrid chitosan micelles have become cutting-edge platforms for improved bioimaging uses. These nanostructures combine the high imaging sensitivity of inorganic nanoparticles like iron oxide [88], gold [89], or quantum dots [90] with the biocompatibility and functional diversity of chitosan. Amphiphilic chitosan derivatives self-assemble to create chitosan micelles, which have a hydrophilic shell and a hydrophobic core that are ideal for encasing imaging agents. When nanoparticles are incorporated into or on top of the micellar structure, signal contrast is improved and multimodal imaging capabilities, including MRI, fluorescence, and photoacoustic imaging, can be used in a single system. Additionally, site-specific distribution is made possible by surface functionalization with targeting ligands, which reduces systemic toxicity while increasing imaging precision [91]. Theranostic uses are made possible by the hybrid design, which also permits regulated release and possible integration with therapeutic drugs. These multifunctional systems represent a promising approach for non-invasive, targeted, and high-resolution visualization of pathological conditions in real time.

Gholami et al. [92] investigated the use of chitosan-based hybrid platforms loaded with Dox and SPIONs to treat glioblastoma. Dox was released quickly in drug-release assays at a pH of 5.5, which is similar to the pH of the tumour microenvironment and suggests pH-dependent drug-release capabilities. Furthermore, fluorescence microscopy was utilized to verify the hybrid platform's cellular internalization. Overall, the study showed that this formulation has potential for glioblastoma diagnosis and therapy. Gd(III)-encapsulated glycol chitosan nanoparticles have been shown by Na et al. [93], to image malignancies using T1-weighted MR imaging. They produced particles with an average size of roughly 280 nm by chelating Gd(III) using DOTA-modified chitosan nanoparticles. They chemically bonded Cy5.5 onto Gd(III)-chitosan nanoparticles for MR-optical multimodal imaging. A liver tumour model was used to demonstrate both NIRF and MR imaging. Through EPR, the chitosan nanoparticles helped prevent the growth of liver tumours.

7.3 Advances in Targeted Micelle Systems for Personalized Medicine

Polymer micelle carriers for targeted drug delivery have emerged as a key tactic for improving therapeutic efficacy while reducing the systemic toxicity linked to traditional chemotherapy. Usually between 10 and 100 nm in size, polymer micelles are ideal for taking advantage of the EPR effect, which is a phenomenon where macromolecules preferentially concentrate in tumour tissues because of the special characteristics of the tumour vasculature [94–96]. Large endothelial gaps, impaired artery integrity, and inadequate lymphatic drainage are characteristics of the broad yet aberrant vascular network that tumours typically form. Higher concentrations of the therapeutic substance at the target site result from these characteristics, which enable micelles to passively collect within tumour tissues more efficiently than they do in healthy tissues.

In order to target monocytes for the purpose of delivering drugs to specific tumors, chitosan-stearic acid micelles were employed by Yang et.al [97]. To create the chitosan-stearic acid polymer, 1-ethyl-3-(3-dimethyl-aminopropyl) carbodiimide chemistry was used. The prepared micelles were 85.93 ± 0.68 nm in size and exhibited a zeta potential of 23.17 ± 0.90 mV. According to the study, the monocytes selectively absorbed the micelles. Additionally, a high rate of micelle internalization by tumour macrophages was noted. In vitro release tests were conducted using the micelles in phosphate-buffered saline (pH 7.4) following doxorubicin loading, in addition to the previously stated investigations. Despite being steady for 72 h, the micelles released less than 25% of the medication.

8 Challenges and Future Perspectives

In order to overcome issues with solubility and bioavailability, chitosan micelles have become effective nanocarriers for the administration of both hydrophilic and hydrophobic medications. Usually, hydrophobic moieties like oleic or stearic acid are added to chitosan to make these micelles, which are amphiphilic structures that self-assemble in aqueous solutions. While the hydrophilic shell of the micelles guarantees biocompatibility and makes interactions with biological systems easier, the hydrophobic core contains lipophilic medicines, improving their solubility and stability. For example, the hydrophobic anticancer drug camptothecin has been effectively encapsulated in chitosan micelles, exhibiting enhanced solubility and regulated release patterns [98].

There are many obstacles to overcome when moving from laboratory-scale synthesis to large-scale output. Small batch sizes, high manufacturing costs, and a lack of reproducibility and standardization of techniques make standardizing and certifying chitosan-based nanomedicines a major challenge. Using cross-flow filtration techniques, recent research has shown that it is feasible to scale up the manufacture of chitosan derivatives, including methoxypolyethylene glycol-chitosan (mPEG-CS) and methoxypolyethylene glycol-chitosan-oleic acid (mPEG-CS-OA). By decreasing production time, improving yield, and maintaining the micelles' physicochemical characteristics, these techniques have demonstrated promise in speeding the transition from bench to bedside [99].

Future developments in chitosan-based micelle systems are expected to concentrate on improving their usefulness and suitability for imaging and medication delivery. The development of stimuli-responsive systems, including pH-sensitive, redox-sensitive, and ultrasound-triggered micelles, has the potential to improve therapeutic outcomes by enabling targeted and regulated drug release. Real-time monitoring of drug distribution and treatment efficacy is also made possible by the combination of chitosan micelles with imaging agents, such as superparamagnetic iron oxide nanoparticles, which provide multimodal imaging and theragnostic applications [100]. The creation of thiomers, or thiolate chitosan derivatives, has also demonstrated promise in improving mucoadhesion and penetration through mucosal barriers, hence increasing the number of possible delivery routes.

9 Conclusion

A promising platform in nanomedicine, chitosan-based micelles provide improved solubility, stability, and regulated release of medicinal chemicals. They are good candidates for a number of drug delivery applications, such as gene delivery, anticancer therapy, and the treatment of infectious disorders, due to their biocompatibility and biodegradability. Adding chitosan to micellar systems enhances drug administration and makes bioimaging applications easier. These systems can be used for both therapeutic administration and diagnostic imaging by encapsulating imaging agents like quantum dots or superparamagnetic iron oxide nanoparticles (SPIONs) in chitosan micelles. Personalized treatment plans are made possible by this combination, which is referred to as theragnostic, enabling real-time monitoring of drug distribution and therapy response. Despite its benefits, chitosan-based micelles still face difficulties in clinical translation. It is necessary to address problems such as restricted scalability, batch-to-batch variability, and possible immunogenicity. These difficulties may be overcome with the help of recent developments in the synthesis and functionalization of chitosan micelles, such as the creation of stimuli-responsive systems and surface alterations for targeted distribution. To fully realize their promise in drug delivery and bioimaging applications, future research must concentrate on improving the physicochemical characteristics of chitosan-based micelles and carrying out thorough preclinical and clinical trials. In conclusion, chitosan-based micelles are a flexible and successful strategy in the realm of nanomedicine. They are useful instruments in the development of personalized medicine because of their capacity to encapsulate a broad variety of therapeutic substances and their potential for dual functionality in drug delivery and bioimaging. Chitosan-based micelles have the potential to greatly influence the treatment of numerous illnesses, providing better therapeutic results and medical attention, provided further research and development are conducted.

Acknowledgements This work is financially supported by the DST-SERB funding agency through three sanctioned projects EEQ/2020/000496, EEQ/2016/000712, and ECR/2016/002018. K.S. also acknowledge UGC for the UGC-BSR Research Start-Up Grant (F.30-363/2017(BSR) Dt-08/08/2017) and CU for the UPE-II Nanofabrication project fund (UGC/166/UPE-II, Dt-03/04/2017) for financial support.

References

1. Ghezzi, M., Pescina, S., Padula, C., Santi, P., Del Favero, E., Cantù, L., & Nicoli, S. (2021). Polymeric micelles in drug delivery: An insight of the techniques for their characterization and assessment in biorelevant conditions. *Journal of Controlled Release, 332*, 312–336.
2. Sun, L., Liu, H., Ye, Y., Lei, Y., Islam, R., Tan, S., Tong, R., Miao, Y. B., & Cai, L. (2023). Smart nanoparticles for cancer therapy. *Signal Transduction and Targeted Therapy, 8*(1), 418.
3. Dulińska-Litewka, J., Łazarczyk, A., Hałubiec, P., Szafrański, O., Karnas, K., & Karewicz, A. (2019). Superparamagnetic iron oxide nanoparticles-current and prospective medical applications. *Materials (Basel), 12*(4), 617.
4. Agrawal, P., Strijkers, G. J., & Nicolay, K. (2010). Chitosan-based systems for molecular imaging. *Advanced Drug Delivery Reviews, 62*(1), 42–58.
5. Shukla, S. K., Mishra, A. K., Arotiba, O. A., & Mamba, B. B. (2013). Chitosan-based nanomaterials: A state-of-the-art review. *International Journal of Biological Macromolecules, 59*, 46–58.
6. Kong, F., Tang, C., & Yin, C. (2020). Benzylguanidine and galactose double-conjugated chitosan nanoparticles with reduction responsiveness for targeted delivery of doxorubicin to CXCR 4 positive tumors. *Bioconjugate Chemistry, 31*(10), 2446–2455.
7. Wang, F., Li, J., Tang, X., Huang, K., & Chen, L. (2020). Polyelectrolyte three layer nanoparticles of chitosan/dextran sulfate/chitosan for dual drug delivery. *Colloids and Surfaces B: Biointerfaces, 190*, 110925.
8. Huq, T., Khan, A., Brown, D., Dhayagude, N., He, Z., & Ni, Y. (2022). Sources, production and commercial applications of fungal chitosan: A review. *Journal of Bioresources and Bioproducts, 7*(2), 85–98.
9. Aranaz, I., Alcántara, A. R., Civera, M. C., Arias, C., Elorza, B., Heras Caballero, A., & Acosta, N. (2021). Chitosan: An overview of its properties and applications. *Polymers (Basel), 13*(19), 3256.
10. Kou, S., Peters, L., & Mucalo, M. (2022). Chitosan: A review of molecular structure, bioactivities and interactions with the human body and micro-organisms. *Carbohydrate Polymers, 282*, 119132.
11. Thadathil, N., & Velappan, S. P. (2014). Recent developments in chitosanase research and its biotechnological applications: A review. *Food Chemistry, 150*, 392–399.
12. Rajinikanth, B. S., Rajkumar, D. S. R., K, K., & Vijayaragavan, V. (2024). Chitosan-based biomaterial in wound healing: A review. *Cureus, 16*(2), e55193.
13. Shete, A., Chavan, A., Potekar, P., Yadav, G., & Shah, N. (2024). Modification of physicochemical properties of chitosan to improve its pharmaceutical and agrochemical potential applications. *International Journal of Biological Macromolecules, 267*, 131404.
14. Mawazi, S. M., Kumar, M., Ahmad, N., Ge, Y., & Mahmood, S. (2024). Recent applications of chitosan and its derivatives in antibacterial, anticancer, wound healing, and tissue engineering fields. *Polymers (Basel), 16*(10), 1351.

15. Aranaz, I., Harris, R., & Heras, A. (2010). Chitosan amphiphilic derivatives. *Chemistry and applications, Current Organic Chemistry, 14*(3), 308–330.
16. Patel, M. P., Patel, R. R., & Patel, J. K. (2010). Chitosan mediated targeted drug delivery system: A review. *Journal of Pharmacy & Pharmaceutical Sciences, 13*(4), 536–557.
17. Esmaeili, J., Jalise, S. Z., Pisani, S., Rochefort, G. Y., Ghobadinezhad, F., Mirzaei, Z., Mohammed, R. U. R., Fathi, M., Tebyani, A., & Nejad, Z. M. (2024). Development and characterization of Polycaprolactone/chitosan-based scaffolds for tissue engineering of various organs: A review. *International Journal Of Biological Macromolecules, 272*, 132941.
18. Liu, K., Jiang, X., & Hunziker, P. (2016). Carbohydrate-based amphiphilic nano delivery systems for cancer therapy. *Nanoscale, 8*(36), 16091–16156.
19. Chiappisi, L., & Gradzielski, M. (2015). Co-assembly in chitosan–surfactant mixtures: Thermodynamics, structures, interfacial properties and applications. *Advances in Colloid and Interface Science, 220*, 92–107.
20. Yang, Y., Wang, S., Wang, Y., Wang, X., Wang, Q., & Chen, M. (2014). Advances in self-assembled chitosan nanomaterials for drug delivery. *Biotechnology Advances, 32*(7), 1301–1316.
21. Woraphatphadung, T., Sajomsang, W., Gonil, P., Treetong, A., Akkaramongkolporn, P., Ngawhirunpat, T., & Opanasopit, P. (2016). pH-responsive polymeric micelles based on amphiphilic chitosan derivatives: Effect of hydrophobic cores on oral meloxicam delivery. *International Journal of Pharmaceutics, 497*(1–2), 150–160.
22. Casettari, L., Vllasaliu, D., Lam, J., Soliman, M., & Illum, L. (2012). Biomedical applications of amino acid-modified chitosans: A review. *Biomaterials, 33*, 7565–7583.
23. Chawla, S., Kanatt, S., & Sharma, A. K. (2015). *Chitosan Polysaccharides* (pp. 219–246). Springer.
24. Jafernik, K., Ładniak, A., Blicharska, E., Czarnek, K., Ekiert, H., Wiącek, A. E., & Szopa, A. (2023). Chitosan-based nanoparticles as effective drug delivery systems—A review. *Molecules, 28*(4), 1963.
25. Poshina, D. N., Rakshina, A. D., & Skorik, Y. A. (2025). Hydrophobic chitosan derivatives for gene and drug delivery in cancer therapies. *Polysaccharides, 6*(1), 11.
26. Ul-Islam, M., Alabbosh, K. F., Manan, S., Khan, S., Ahmad, F., & Ullah, M. W. (2024). Chitosan-based nanostructured biomaterials: Synthesis, properties, and biomedical applications. *Advanced Industrial and Engineering Polymer Research, 7*(1), 79–99.
27. Bu, X., Ji, N., Dai, L., Dong, X., Chen, M., Xiong, L., & Sun, Q. (2021). Self-assembled micelles based on amphiphilic biopolymers for delivery of functional ingredients. *Trends in Food Science & Technology, 114*, 386.
28. Yadav, H. K. S., Almokdad, A. A., shaluf, S. I. M., & Debe, M. S. (2019). Chapter 17 – Polymer-based nanomaterials for drug-delivery carriers. In S. S. Mohapatra, S. Ranjan, N. Dasgupta, R. K. Mishra, & S. Thomas (Eds.), *Nanocarriers for drug delivery* (pp. 531–556). Elsevier.
29. Hembram, K. C., Prabha, S., Chandra, R., Ahmed, B., & Nimesh, S. (2016). Advances in preparation and characterization of chitosan nanoparticles for therapeutics. *Artificial Cells, Nanomedicine, and Biotechnology, 44*(1), 305–314.
30. Hu, F.-Q., Zhao, M.-D., Yuan, H., You, J., Du, Y.-Z., & Zeng, S. (2006). A novel chitosan oligosaccharide–stearic acid micelles for gene delivery: Properties and in vitro transfection studies. *International Journal of Pharmaceutics, 315*(1), 158–166.
31. Manchanda, R., & Nimesh, S. (2010). Controlled size chitosan nanoparticles as an efficient, biocompatible oligonucleotides delivery system. *Journal of Applied Polymer Science, 118*, 2071–2077.
32. Banerjee, T., Mitra, S., Kumar Singh, A., Kumar Sharma, R., & Maitra, A. (2002). Preparation, characterization and biodistribution of ultrafine chitosan nanoparticles. *International Journal of Pharmaceutics, 243*(1), 93–105.
33. Niwa, T., Takeuchi, H., Hino, T., Kunou, N., & Kawashima, Y. (1993). Preparations of biodegradable nanospheres of water-soluble and insoluble drugs with D,L-lactide/glycolide

copolymer by a novel spontaneous emulsification solvent diffusion method, and the drug release behavior. *Journal of Controlled Release, 25*(1), 89–98.
34. El-Shabouri, M. H. (2002). Positively charged nanoparticles for improving the oral bioavailability of cyclosporin-a. *International Journal of Pharmaceutics, 249*(1), 101–108.
35. Liu, Z., Jiao, Y., Liu, F., & Zhang, Z. (2007). Heparin/chitosan nanoparticle carriers prepared by polyelectrolyte complexation. *Journal of Biomedical Materials Research Part A: An Official Journal of The Society for Biomaterials, The Japanese Society for Biomaterials, and The Australian Society for Biomaterials and the Korean Society for Biomaterials, 83*(3), 806–812.
36. Jhaveri, J., Raichura, Z., Khan, T., Momin, M., & Omri, A. (2021). Chitosan nanoparticles-insight into properties, functionalization and applications in drug delivery and Theranostics. *Molecules, 26*(2), 272.
37. Edson, J. A., Ingato, D., Wu, S., Lee, B., & Kwon, Y. J. (2018). Aqueous-soluble, acid-transforming chitosan for efficient and stimuli-responsive gene silencing. *Biomacromolecules, 19*(5), 1508–1516.
38. Lu, H., Dai, Y., Lv, L., & Zhao, H. (2014). Chitosan-graft-polyethylenimine/DNA nanoparticles as novel non-viral gene delivery vectors targeting osteoarthritis. *PLoS One, 9*(1), e84703.
39. Herzberger, J., Niederer, K., Pohlit, H., Seiwert, J., Worm, M., Wurm, F. R., & Frey, H. (2016). Polymerization of ethylene oxide, propylene oxide, and other Alkylene oxides: Synthesis, novel polymer architectures, and bioconjugation. *Chemical Reviews, 116*(4), 2170–2243.
40. Nicolle, L., Casper, J., Willimann, M., Journot, C. M., Detampel, P., Einfalt, T., Grisch-Chan, H. M., Thöny, B., Gerber-Lemaire, S., & Huwyler, J. (2021). Development of covalent chitosan-polyethylenimine derivatives as gene delivery vehicle: Synthesis, characterization, and evaluation. *International Journal of Molecular Sciences, 22*(8), 3828.
41. Zhu, C., Wu, J., Jin, X., Yan, Y., Ding, C.-F., Tang, K., & Zhang, D. (2021). Post-synthesis of biomimetic chitosan with honeycomb-like structure for sensitive recognition of phosphorylated peptides. *Journal of Chromatography A, 1643*, 462072.
42. Vold, I. M., & Christensen, B. E. (2005). Periodate oxidation of chitosans with different chemical compositions. *Carbohydrate Research, 340*(4), 679–684.
43. Liu, M., Dasgupta, A., Koczera, P., Schipper, S., Rommel, D., Shi, Y., Kiessling, F., & Lammers, T. (2020). Drug loading in poly(butyl cyanoacrylate)-based polymeric microbubbles. *Molecular Pharmaceutics, 17*(8), 2840–2848.
44. Massella, D., Celasco, E., Salaün, F., Ferri, A., & Barresi, A. A. (2018). Overcoming the limits of flash nanoprecipitation: Effective loading of hydrophilic drug into polymeric nanoparticles with controlled structure. *Polymers, 10*(10), 1092.
45. Çalış, S., Öztürk Atar, K., Arslan, F. B., Eroğlu, H., & Çapan, Y. (2019). Chapter 4 – Nanopharmaceuticals as drug-delivery systems: for, against, and current applications. In S. S. Mohapatra, S. Ranjan, N. Dasgupta, R. K. Mishra, & S. Thomas (Eds.), *Nanocarriers for drug delivery* (pp. 133–154). Elsevier.
46. Kamaly, N., Yameen, B., Wu, J., & Farokhzad, O. C. (2016). Degradable controlled-release polymers and polymeric nanoparticles: Mechanisms of controlling drug release. *Chemical Reviews, 116*(4), 2602–2663.
47. Han, Y., Pan, J., Liang, N., Gong, X., & Sun, S. (2021). A pH-sensitive polymeric micellar system based on chitosan derivative for efficient delivery of paclitaxel. *International Journal of Molecular Sciences, 22*(13), 6659.
48. Fentahun Darge, H., Yibru Hanurry, E., Simegniew Birhan, Y., Worku Mekonnen, T., Tizazu Andrgie, A., Chou, H.-Y., Lai, J.-Y., & Tsai, H.-C. (2021). Multifunctional drug-loaded micelles encapsulated in thermo-sensitive hydrogel for in vivo local cancer treatment: Synergistic effects of anti-vascular and immuno-chemotherapy. *Chemical Engineering Journal, 406*, 126879.

49. Xu, W., Ling, P., & Zhang, T. (2013). Polymeric micelles, a promising drug delivery system to enhance bioavailability of poorly water-soluble drugs. *Journal of Drug Delivery, 2013*, 340315.
50. Miyata, K., Christie, R., & Kataoka, K. (2010). Polymeric micelles for nano-scale drug delivery. *Reactive and Functional Polymers, 17*, 227.
51. Desai, N., Rana, D., Salave, S., Gupta, R., Patel, P., Karunakaran, B., Sharma, A., Giri, J., Benival, D., & Kommineni, N. (2023). Chitosan: A potential biopolymer in drug delivery and biomedical applications. *Pharmaceutics, 15*(4), 1313.
52. Kumar, R., Sirvi, A., Kaur, S., Samal, S. K., Roy, S., & Sangamwar, A. T. (2020). Polymeric micelles based on amphiphilic oleic acid modified carboxymethyl chitosan for oral drug delivery of bcs class iv compound: Intestinal permeability and pharmacokinetic evaluation. *European Journal of Pharmaceutical Sciences, 153*, 105466.
53. Chen, T., Tu, L., Wang, G., Qi, N., Wu, W., Zhang, W., & Feng, J. (2020). Multi-functional chitosan polymeric micelles as oral paclitaxel delivery systems for enhanced bioavailability and anti-tumor efficacy. *International Journal of Pharmaceutics, 578*, 119105.
54. Yang, Y., Chen, Y., Li, D., Lin, S., Chen, H., Wu, W., & Zhang, W. (2023). Linolenic acid conjugated chitosan micelles for improving the oral absorption of doxorubicin via fatty acid transporter. *Carbohydrate Polymers, 300*, 120233.
55. Almeida, A., Linares, V., Mora-Castaño, G., Casas, M., Caraballo, I., & Sarmento, B. (2021). 3D printed systems for colon-specific delivery of camptothecin-loaded chitosan micelles. *European Journal of Pharmaceutics and Biopharmaceutics, 167*, 48–56.
56. Yuan, Y., Wang, Z., Su, S., Lin, C., Mi, Y., Tan, W., & Guo, Z. (2023). Self-assembled low molecular weight chitosan-based cationic micelle for improved water solubility, stability and sustained release of α-tocopherol. *Food Chemistry, 429*, 136886.
57. Zhang, X., Xu, X., Wang, X., Lin, Y., Zheng, Y., Xu, W., Liu, J., & Xu, W. (2023). Hepatoma-targeting and reactive oxygen species-responsive chitosan-based polymeric micelles for delivery of celastrol. *Carbohydrate Polymers, 303*, 120439.
58. Woraphatphadung, T., Sajomsang, W., Rojanarata, T., Ngawhirunpat, T., Tonglairoum, P., & Opanasopit, P. (2018). Development of chitosan-based pH-sensitive polymeric micelles containing curcumin for colon-targeted drug delivery. *AAPS PharmSciTech, 19*(3), 991–1000.
59. Yuan, Y., Tan, W., Mi, Y., Wang, L., Qi, Z., & Guo, Z. (2024). Effect of hydrophobic chain length in amphiphilic chitosan conjugates on intracellular drug delivery and smart drug release of redox-responsive micelle. *Marine Drugs, 22*(1), 18.
60. Jiang, Y., Wong, S., Chen, F., Chang, T., Lu, H., & Stenzel, M. H. (2017). Influencing selectivity to cancer cells with mixed nanoparticles prepared from albumin–polymer conjugates and block copolymers. *Bioconjugate Chemistry, 28*(4), 979–985.
61. Xiao, Y., Lin, Z. T., Chen, Y., Wang, H., Deng, Y. L., Le, D. E., Bin, J., Li, M., Liao, Y., & Liu, Y. (2015). High molecular weight chitosan derivative polymeric micelles encapsulating superparamagnetic iron oxide for tumor-targeted magnetic resonance imaging. *International Journal of Nanomedicine, 10*, 1155–1172.
62. Lin, W., Zhang, X., Qian, L., Yao, N., Pan, Y., & Zhang, L. (2017). Doxorubicin-loaded unimolecular micelle-stabilized gold nanoparticles as a theranostic nanoplatform for tumor-targeted chemotherapy and computed tomography imaging. *Biomacromolecules, 18*(12), 3869–3880.
63. Zia, N., Iqbal, Z., Raza, A., Zia, A., Shafique, R., Andleeb, S., & Walker, G. C. (2022). Glycol-chitosan-based Technetium-99m-loaded multifunctional nanomicelles: Synthesis, evaluation, and in vivo biodistribution. *Nanomaterials, 12*(13), 2198.
64. Yan, T., Hui, W., Zhu, S., He, J., Liu, Z., & Cheng, J. (2021). Carboxymethyl chitosan based redox-responsive micelle for near-infrared fluorescence image-guided photo-chemotherapy of liver cancer. *Carbohydrate Polymers, 253*, 117284.
65. Weissleder, R. (2001). A clearer vision for in vivo imaging. *Nature Biotechnology, 19*(4), 316–317.

66. Zhu, R., Zhang, C.-g., Liu, Y., Yuan, Z.-q., Chen, W.-l., Yang, S.-d., Li, J.-z., Zhu, W.-j., Zhou, X.-f., & You, B.-g. (2015). CD147 monoclonal antibody mediated by chitosan nanoparticles loaded with α-hederin enhances antineoplastic activity and cellular uptake in liver cancer cells. *Scientific Reports, 5*(1), 17904.
67. Mansur, A. A., de Carvalho, S. M., & Mansur, H. S. (2016). Bioengineered quantum dot/chitosan-tripeptide nanoconjugates for targeting the receptors of cancer cells. *International Journal of Biological Macromolecules, 82*, 780–789.
68. Cheng, P., Liang, N., Zhao, W., Gong, X., Wang, W., & Sun, S. (2023). Chitosan-based near-infrared fluorescent micelles for controlled drug delivery and bioimaging in cancer therapy. *European Polymer Journal, 189*, 111974.
69. Shi, H., Liang, N., Liu, J., Li, S., Gong, X., Yan, P., & Sun, S. (2021). AIE-active polymeric micelles based on modified chitosan for bioimaging-guided targeted delivery and controlled release of paclitaxel. *Carbohydrate Polymers, 269*, 118327.
70. Key, J., & Leary, J. F. (2014). Nanoparticles for multimodal in vivo imaging in nanomedicine. *International Journal of Nanomedicine, 9*, 711–726.
71. Nwe, K., Huang, C.-H., & Tsourkas, A. (2013). Gd-labeled glycol chitosan as a pH-responsive magnetic resonance imaging agent for detecting acidic tumor microenvironments. *Journal of Medicinal Chemistry, 56*(20), 7862–7869.
72. Javid, A., Ahmadian, S., Saboury, A. A., Kalantar, S. M., & Rezaei-Zarchi, S. (2013). Chitosan-coated superparamagnetic iron oxide nanoparticles for doxorubicin delivery: Synthesis and anticancer effect against human ovarian cancer cells. *Chemical Biology & Drug Design, 82*(3), 296–306.
73. Jia, H. R., Jiang, Y. W., Zhu, Y. X., Li, Y. H., Wang, H. Y., Han, X., Yu, Z. W., Gu, N., Liu, P., Chen, Z., & Wu, F. G. (2017). Plasma membrane activatable polymeric nanotheranostics with self-enhanced light-triggered photosensitizer cellular influx for photodynamic cancer therapy. *Journal of Controlled Release, 255*, 231–241.
74. Wang, Y., Liu, X., Deng, G., Wang, Q., Zhang, L., Wang, Q., & Lu, J. (2017). Multifunctional PS@CS@Au–Fe3O4–FA nanocomposites for CT, MR and fluorescence imaging guided targeted-photothermal therapy of cancer cells. *Journal of Materials Chemistry B, 5*(22), 4221–4232.
75. Torino, E., Auletta, L., Vecchione, D., Orlandella, F. M., Salvatore, G., Iaccino, E., Fiorenza, D., Grimaldi, A. M., Sandomenico, A., Albanese, S., Sarnataro, D., Gramanzini, M., Palmieri, C., Scala, G., Quinto, I., Netti, P. A., Salvatore, M., & Greco, A. (2018). Multimodal imaging for a theranostic approach in a murine model of B-cell lymphoma with engineered nanoparticles. *Nanomedicine, 14*(2), 483–491.
76. Schmitt, R. R., Mahajan, S. D., Pliss, A., & Prasad, P. N. (2022). Small molecule based EGFR targeting of biodegradable nanoparticles containing temozolomide and Cy5 dye for greatly enhanced image-guided glioblastoma therapy. *Nanomedicine: Nanotechnology, Biology, and Medicine, 41*, 102513.
77. Guo, L., Shi, D., Shang, M., Sun, X., Meng, D., Liu, X., Zhou, X., & Li, J. (2022). Utilizing RNA nanotechnology to construct negatively charged and ultrasound-responsive nanodroplets for targeted delivery of siRNA. *Drug Delivery, 29*(1), 316–327.
78. Yu, B., Li, X., Zheng, W., Feng, Y., Wong, Y.-S., & Chen, T. (2014). pH-responsive cancer-targeted selenium nanoparticles: A transformable drug carrier with enhanced theranostic effects. *Journal of Materials Chemistry B, 2*(33), 5409–5418.
79. Yang, H., Chen, Y., Chen, Z., Geng, Y., Xie, X., Shen, X., Li, T., Li, S., Wu, C., & Liu, Y. (2017). Chemo-photodynamic combined gene therapy and dual-modal cancer imaging achieved by pH-responsive alginate/chitosan multilayer-modified magnetic mesoporous silica nanocomposites. *Biomaterials Science, 5*(5), 1001–1013.
80. Hanafy, N. A. N. (2021). Optimally designed theranostic system based folic acids and chitosan as a promising mucoadhesive delivery system for encapsulating curcumin LbL nano-template against invasiveness of breast cancer. *International Journal of Biological Macromolecules, 182*, 1981–1993.

81. Liu, S., Li, W., Gai, S., Yang, G., Zhong, C., Dai, Y., He, F., Yang, P., & Suh, Y. D. (2019). A smart tumor microenvironment responsive nanoplatform based on upconversion nanoparticles for efficient multimodal imaging guided therapy. *Biomaterials Science, 7*(3), 951–962.
82. Zhang, Y., Wang, G., Li, Q., Jiang, Y., Chen, W., Zhao, M., Liang, G., & Miao, Q. (2022). Acidity-activated charge conversion of 177Lu-labeled Nanoagent for the enhanced photodynamic radionuclide therapy of cancer. *ACS Applied Materials & Interfaces, 14*(3), 3875–3884.
83. Wells, C. M., Harris, M., Choi, L., Murali, V. P., Guerra, F. D., & Jennings, J. A. (2019). Stimuli-responsive drug release from smart polymers. *Journal of Functional Biomaterials, 10*(3), 34.
84. Gao, J., Xu, Y., Zheng, Y., Wang, X., Li, S., Yan, G., Wang, J., & Tang, R. (2019). pH-sensitive carboxymethyl chitosan hydrogels via acid-labile ortho ester linkage as an implantable drug delivery system. *Carbohydrate Polymers, 225*, 115237.
85. Ryplida, B., Lee, K., In, I., & Park, S. (2019). Light-induced swelling-responsive conductive, adhesive, and stretchable wireless film hydrogel as electronic artificial skin. *Advanced Functional Materials, 29*, 1903209.
86. Abureesh, M. A., Oladipo, A. A., & Gazi, M. (2016). Facile synthesis of glucose-sensitive chitosan–poly(vinyl alcohol) hydrogel: Drug release optimization and swelling properties. *International Journal of Biological Macromolecules, 90*, 75–80.
87. Ahmad, U., Sohail, M., Ahmad, M., Minhas, M. U., Khan, S., Hussain, Z., Kousar, M., Mohsin, S., Abbasi, M., Shah, S. A., & Rashid, H. (2019). Chitosan based thermosensitive injectable hydrogels for controlled delivery of loxoprofen: Development, characterization and in-vivo evaluation. *International Journal of Biological Macromolecules, 129*, 233–245.
88. Khmara, I., Strbak, O., Zavisova, V., Koneracka, M., Kubovcikova, M., Antal, I., Kavecansky, V., Lucanska, D., Dobrota, D., & Kopcansky, P. (2019). Chitosan-stabilized iron oxide nanoparticles for magnetic resonance imaging. *Journal of Magnetism and Magnetic Materials, 474*, 319–325.
89. Liu, Z., Zhou, D., Yan, X., Xiao, L., Wang, P., Wei, J., & Liao, L. (2022). Gold nanoparticle-incorporated chitosan Nanogels as a Theranostic Nanoplatform for CT imaging and tumour chemotherapy. *International Journal of Nanomedicine, 17*, 4757–4772.
90. Hassani, S., Gharehaghaji, N., & Divband, B. (2022). Chitosan-coated iron oxide/graphene quantum dots as a potential multifunctional nanohybrid for bimodal magnetic resonance/fluorescence imaging and 5-fluorouracil delivery. *Materials Today Communications, 31*, 103589.
91. Han, X., Xu, K., Taratula, O., & Farsad, K. (2019). Applications of nanoparticles in biomedical imaging. *Nanoscale, 11*(3), 799–819.
92. Gholami, L., Tafaghodi, M., Abbasi, B., Daroudi, M., & Kazemi Oskuee, R. (2019). Preparation of superparamagnetic iron oxide/doxorubicin loaded chitosan nanoparticles as a promising glioblastoma theranostic tool. *Journal of Cellular Physiology, 234*(2), 1547–1559.
93. Na, J. H., Lee, S., Koo, H., Han, H., Lee, K. E., Han, S. J., Choi, S. H., Kim, H., Lee, S., & Kwon, I. C. (2016). T 1-weighted mr imaging of liver tumor by gadolinium-encapsulated glycol chitosan nanoparticles without non-specific toxicity in normal tissues. *Nanoscale, 8*(18), 9736–9745.
94. Farhoudi, L., Maryam Hosseinikhah, S., Vahdat-Lasemi, F., Sukhorukov, V. N., Kesharwani, P., & Sahebkar, A. (2024). Polymeric micelles paving the way: Recent breakthroughs in camptothecin delivery for enhanced chemotherapy. *International Journal of Pharmaceutics, 659*, 124292.
95. Zhang, Y., Huang, Y., & Li, S. (2014). Polymeric micelles: Nanocarriers for cancer-targeted drug delivery. *AAPS PharmSciTech, 15*(4), 862–871.
96. Wan, D., Li, C., & Pan, J. (2020). Polymeric micelles with reduction-responsive function for targeted cancer chemotherapy. *ACS Applied Bio Materials, 3*(2), 1139–1146.
97. Yang, X., Lian, K., Tan, Y., Zhu, Y., Liu, X., Zeng, Y., Yu, T., Meng, T., Yuan, H., & Hu, F. (2020). Selective uptake of chitosan polymeric micelles by circulating monocytes for enhanced tumor targeting. *Carbohydrate Polymers, 229*, 115435.

98. Almeida, A., Castro, F., Resende, C., Lúcio, M., Schwartz, S., Jr., & Sarmento, B. (2022). Oral delivery of camptothecin-loaded multifunctional chitosan-based micelles is effective in reduce colorectal cancer. *Journal of Controlled Release, 349*, 731–743.
99. Almeida, A., Günday-Türeli, N., & Sarmento, B. (2021). A scale-up strategy for the synthesis of chitosan derivatives used in micellar nanomedicines. *International Journal of Pharmaceutics, 609*, 121151.
100. Rahman, M. (2023). Magnetic resonance imaging and iron-oxide nanoparticles in the era of personalized medicine. *Nano, 7*(4), 424.

Advances of Chitosan-Based Transdermal Drug Delivery Systems

Arindam Maity, Prajna Gupta, Sougata Jana, and Sreejan Manna

Abstract Transdermal drug delivery has emerged as a promising alternative to conventional drug delivery systems. The use of biopolymers in transdermal drug delivery systems have shown favourable results with the required safety profile. Chitosan is a naturally occurring polysaccharide originated from marine sources. It offers excellent biocompatible and biodegradable nature as pharmaceutical excipient. It is also widely employed in the biomedical field due to its favourable properties. The hydrogel forming ability in combination with controlled release of incorporated drugs have widened the scope of chitosan as transdermal drug carriers. The antimicrobial and wound healing ability of chitosan has also aided in its use as a transdermally applied material. Chitosan possesses easy fabrication techniques, compatibility with most of the pharmaceutical excipients and therapeutic agents have also promoted its use. It can be easily fabricated into different transdermal dosage forms, including gels, patches, microneedles etc. The better understanding of its chemical modifications and physicochemical properties can help to enhance the efficacy of chitosan-based transdermal formulations.

Keywords Transdermal drug delivery · Chitosan · Transdermal patches · Hydrogels · Microneedles · Wound healing

A. Maity
Belarani Institute of Medical Science, Bankura, West Bengal, India

P. Gupta
BCDA College of Pharmacy & Technology, Hridaypur, Barasat, Kolkata, West Bengal, India

S. Jana
Department of Health and Family Welfare, Directorate of Health Services, Kolkata, West Bengal, India

S. Manna (✉)
Department of Pharmaceutical Technology, Brainware University, Barasat, Kolkata, West Bengal, India
e-mail: sm.pt@brainwareuniversity.ac.in; manna.sreejan@gmail.com

S. Jana (ed.), *Chitosan-Based Drug Delivery Systems*, AAPS Introductions in the Pharmaceutical Sciences 16, https://doi.org/10.1007/978-3-032-08347-0_6

1 Introduction

Naturally derived polymers have emerged as promising materials for drug delivery systems due to their biocompatibility, ease of extraction, and widespread availability. These characteristics make them attractive alternatives to synthetic polymers [72]. Polysaccharides represent a major class of polymers widely employed in the pharmaceutical and food industries [66]. Many polysaccharides are also reported to be obtained from marine sources [43]. Chitin is one naturally occurring polysaccharide majorly originating from marine sources [65].It is synthesized by a high number of biotic organisms [81]. Generally, chitin is created by some marine creatures like crustaceans, insects, molluscs and fungi [65]. After cellulose, chitin is mostly found in natural polymers in the environment [56, 81]. However, chitin has remarkable properties such as biocompatibility, biodegradability, massive mechanical strength and biologically active. But chitin has been used restricted because of low solubility [45]. Chitosan is a biopolymer which is obtained from chitin [5, 83]. The degree of deacetylation ≥75% of chitin is granted as chitosan [23]. Chitosan has found significant application in several kinds of drug delivery and biomedical fields because it has shown, low toxicity, mucoadhesive properties, antimicrobial activity [27, 37], good biocompatibility [38], biodegradability, non-antigenicity [45], chelating capability [75]. Currently, chitosan derivatives are extensively employed in biomedicine and medical substances. In addition, chitosan derivatives are being developed as nanoparticles, micelles, hydrogel and microspheres. Furthermore, it has the potential to utilize chitosan derivatives as vehicles of targeted drug delivery systems and delivery carriers of vaccines [76]. Chitosan and its derivatives have been applied by researchers employing a variety of technical methods to various transdermal dosage forms, including emulsions, nanocapsules, transdermal microneedles, hydrogels, transdermal membranes, transdermal patches, nanoscaffolds, and liposomes to facilitate the drugs penetrate the skin [27, 38]. Modified chitosan has been shown to be an excellent choice for an anticancer drug delivery; thus, it can deliver the anticancer agents like photodynamic agents, chemical drugs, and nucleic acids. Also, the native chitosan is an excellent alternative for wound healing applications since it has several characteristics that make it appropriate for used as a wound dressing [72]. Moreover, chitosan microspheres have significant uses in controlled release of protein and peptide drugs due to their extraordinary muco-adhesiveness and permeability across the biological surfaces [74]. However, chitosan-based delivery approaches may integrate two strategies to establish the multifunctional system with controlled release of medication and targeted delivery along with minimal side effects [20].

Transdermal drug delivery systems (TDDS) have been a popular drug delivery route. It began with the Scopolamine patch which was introduced to the market in 1979 [49, 78]. TDDS is another approach to drug delivery through the skin layer. In these delivery systems, the drug enters the bloodstream via the skin and circulates throughout the body before arriving at its target location [63, 77]. TDDS may provide convenient and efficient drug delivery locally and/or systemically over an extended period of time [32, 44].

Transdermal drug delivery systems have several advantages. TDDS provides avoidance of hepatic first pass metabolism [28, 58] and avoidance of enzymatic degradation of the drug [45], improving bioavailability as compared to oral drug delivery [50, 78] and reducing the gastrointestinal adverse effects [13]. Importantly, TDDS can provide painless medication delivery over injectable. Generally, TDDS formulations can be given by healthcare professionals or taken by the patient themselves with minimal training [32]. TDDS has been remarkably influenced by the delivery of several therapeutic agents, hormonal therapy, central nervous system-related diseases, pain management, and cardiovascular disease treatment [28]. Additionally, it is utilized for the delivery of drugs in motion sickness and smoking cessation [77]. Other advantages, like the frequency of medicine administration, may be decreased by TDDS. Protein drugs that are not able to be delivered orally, but it can be delivered by TDDS [36]. This drug delivery system showed a highly consistent rate of absorption of drugs and maintained a stable drug concentration in the blood [16, 78]. Furthermore, vaccination via transdermal drug delivery is a substitute for conventional vaccination routes because of the abundance of dendritic cells within the skin [21]. Moreover, the TDDS approach has been utilized for the delivery of hydrophilic and hydrophobic drugs [58]. Therefore, transdermal drug delivery systems are patient favourable, and it is widely used drug delivery technique for the treatment of various diseases. This review focused on recent advances and development of chitosan and its derivatives integrated transdermal preparations.

2 Sources of Chitosan

Chitosan is obtained from the exoskeleton of shrimps, lobsters, and crabs [45]. A traditional source of chitosan is crustaceans' shell waste, generally obtained from the seafood industry [26]. Chitosan is produced when chitin undergoes partial deacetylation [33, 67, 82]. One of the sources of chitosan is fungal biomass [25]. Some fungi contain chitosan, which is primarily located in their cell walls. Notably, important sources of chitosan include fungi from the class Zygomycetes, such as the genera Absidia, Cunninghamella, Rhizopus, Mucor, Benjaminiella, and Gongronella [12, 34]. Another source of chitosan is the exoskeleton of beetles. Macro-mushrooms are one of the natural sources of chitosan. Chitosan derived from insects has similar properties like crustacean derivatives [26].

3 Extraction of Chitosan

Chitin and chitosan are often extracted by using chemical, biological methods and microwave irradiation methods [22]. There are many taxonomic groups that contain chitin. Still, marine crustaceans are typically the source of commercial chitin. Crustacean shells are composed of 20–30% chitin, 30–40% proteins, and 30–50%

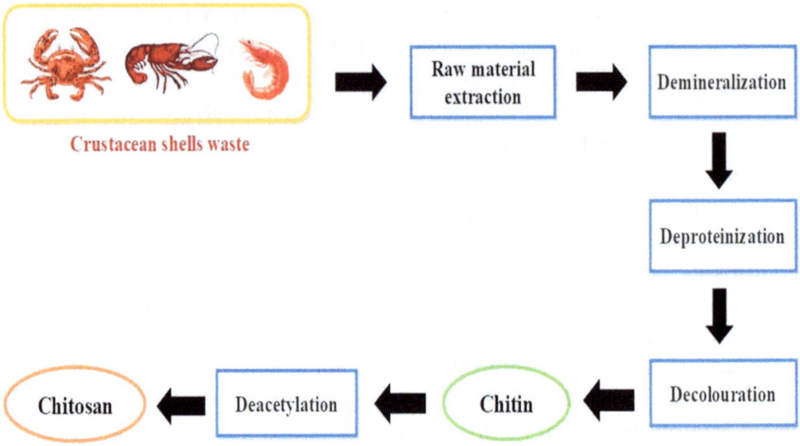

Fig. 1 Steps of chitosan extraction and production from crustacean shells waste

calcium carbonate. Lipidic pigments like carotenoids are also present in crustacean shells [9]. The conventional chitin extraction method is carried out by following steps, i.e. raw material washing and conditioning, demineralization, deproteinization, discoloration or bleaching [54, 55]. The deacetylation step is the chitosan formation stage [22]. When the protein separated from chitin is intended for use as a food ingredient in animal food, then the deproteinization step can be executed before the demineralization step [55]. However, the deproteinization step is needed for fungal extraction, but the demineralization step is not required [54]. The general extraction procedure of chitosan from marine sources was schematically represented in Fig. 1.

3.1 Chemical Method

Raw Material Washing and Conditioning
Crustacean shells are thoroughly rinsed with water to eliminate the organic matter that adheres to their surface [55].

Demineralization
Demineralization step is accomplished by acid treatment with hydrochloric, nitric, sulfuric, oxalic, acetic, and formic acids [54]. Hydrochloric acid is most frequently used for eliminating the mineral component [12]. Generally, dilute hydrochloric acid (HCl) solution is used to remove the calcium carbonate and calcium chloride, which are the primary inorganic compounds of exoskeleton of the crustaceans [23]. The emission of CO_2 during the digestion reaction is an essential measure of mineral content. After that, the resultant materials are filtered and washed with distilled water and kept to dry in an oven overnight [22].

Deproteinization

The deproteinization step is carried out by using a dilute sodium hydroxide (NaOH) solution for protein elimination [23, 45]. To eliminate excess sodium hydroxide, the mixture is filtered and rinsed continuously with deionized water after it has dried overnight in an oven. Finally, the purified chitin is obtained [22].

Discoloration or Bleaching

Crustacean shells contain pigments such as canthaxanthin, lutein, astaxanthin, astacene, β-carotene. In this step, mild oxidizing agents have been used, such as hydrogen peroxide (0.5–3%) and sodium hypochlorite (0.32%) [55] or extraction with acetone, chloroform, and ethanol [12]. However, they may damage the free amino groups, and polymer modification could happen. Strongly coloured shells are treated with the acetone and sodium hypochlorite mixture at room temperature [55].

Deacetylation

The deacetylation step involves converting chitin to chitosan by the removal of chitin-containing acetyl groups [46]. Generally, chitosan is prepared by treatment with the concentrated solution of sodium hydroxide or potassium hydroxide at high temperature (110–115 °C) without oxygen for several hours. When the degree of deacetylation exceeds 50%, chitosan is produced [22, 23].

3.2 Biological Method

Biological treatments provide an ecologically safe alternative to acid and alkali treatments for chitin and chitosan extraction. Bacteria that produce lactic acid have been utilized for the demineralization step. However, protease enzymes from bacteria utilized for the deproteinization step. Deacetylation of chitin is done by the utilization of chitin deacetylase [22].

Different scientific studies of the extraction of chitin and chitosan by the use of chemical and biological methods from various sources are briefly described below.

Teli and Sheik reported the extraction process of chitosan from the shrimp shell waste. The power of the shrimp shell was immersed in 1 M sodium hydroxide solution for 24 h, after that, washed then dried. Demineralization was done with the treatment of 1 M hydrochloric acid solution. In deproteinization step was executed by use of 1 M sodium hydroxide solution, and the discoloured step was completed by the utilization of oxalic acid and $KMnO_4$. Then resultant chitin was deacetylated by use of 50% sodium hydroxide solution, and finally, the chitosan [70].

Kumari and Rath reported the extraction of chitin and chitosan from Fish Scales, *Labeo rohita*, by the use of 1% HCl solution for 36 h for the demineralization. Then, deproteinization was done by the utilization of 0.5 N NaOH solutions at room temperature for 18 h. After that, 50% of NaOH solution was treated with chitin at 80 °C for 2 h with an oil bath [35].

Soon et al. conveyed that chitin and chitosan were extracted from *Zophobas morio* larvae. Here, the demineralization step was accomplished by utilizing 1 M of HCl solution for 30 min at 35 °C temperature. During deproteinization, 0.5 M, 1 M, and 2 M concentrations was used at 80 °C for 20 h. After that deacetylation process was done by extracting chitin with treatment of 50% NaOH in a water bath maintained at 90 °C temperature for 30 h [19].

Ma et al. reported the preparation and characterization of chitosan from the *Catharsius molossus* residue. Demineralization and deproteinization steps were carried out by the utilization of 1.3 M of hydrochloric acid solution at 80 °C temperature for 30 min and 4 M of sodium hydroxide solution at 90 °C temperature for 6 h, respectively. After that extracted chitin was immersed in 18 M of sodium hydroxide solution for 24 h at room temperature, then at 90 °C temperature, the sample was heated for 7 h, and the alkali solution was changed 3, 5, and 7 h accordingly [39].

Arancibia et al. reported that chitin and chitosan were extracted from the shrimp *Litopenaeus vannamei* wastes. Demineralization was accomplished by the use of lactic acid with the sample material at 21 °C temperature for 36 h. Viscozyme and Alcalase were used in an enzymatic hydrolysis to remove the protein. The remaining solid material was deacetylated for 4 h at 110 °C using 40% sodium hydroxide solution [10].

3.3 Microwave Irradiation Method

Conventionally, the extraction of chitosan using chemical and biological techniques is a time-consuming process. For these reasons, researchers have found an alternative method for chitin and chitosan extraction, i.e. microwave or radiofrequency synthesis method, which is a faster, efficient, and eco-friendly process. This novel approach heats the reaction mass in three dimensions, replacing traditional heating methods. Microwave irradiation activates molecules with dipole rotation or ionic conductivity, causing a rapid increase in temperature [22].

Samar et al. conveyed that the chitosan extraction process accomplished by the use of the microwave technique, specifically, in the deacetylation step [41].

4 Chemistry of Chitosan

Chitosan is a naturally obtained polycationic polymer [48]. The chemical structure of chitosan is represented in Fig. 2. Chitosan is constituted of the repeated D-glucosamine and the N-acetyl-D-glucosamine units, and these units are attached by β-[1-4] glycosidic bonds [37]. The reactive groups that present in the structure of chitosan are primary amino group ($-NH_2$) at second carbon atom (C_2) position [8]. The primary hydroxyl group ($-OH$) present at sixth carbon atom (C_6) position and secondary hydroxyl group ($-OH$) present at third carbon atom (C_3) position [8, 56]. Chitosan containing acetamide group and the glycosidic bonds

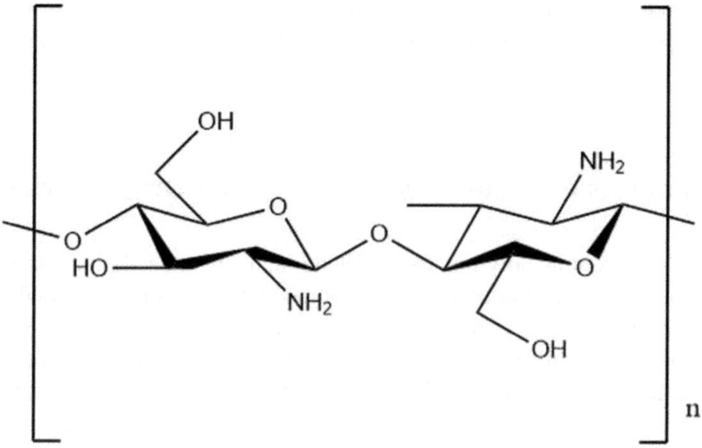

Fig. 2 Chemical structure of chitosan

can be considered as functional groups. All of these functional groups are facilitated an extensive number of modifications, resulting in polymers with unique properties and functions [8]. The proportion of deacetylated units to total units or degree of deacetylation is an important structural factor that affecting the chitosan's physicochemical properties. The chitosan can recognize when degree of deacetylation is 50% above. Additionally, the degree of deacetylation influences chitosan's physicochemical properties like solubility, crystallinity, and hydrophilicity as well as biological properties like cell response, biodegradability, and biocompatibility [22, 47]. The higher degree of deacetylation of chitosan possesses a significant role as a biomaterial in tissue engineering because of its polycationic property that interacts with the cell membrane which negatively charged [19]. The degree of deacetylation is determined by several analytical techniques, including potentiometric titration, near-infrared spectroscopy (NIR), magnetic resonance (NMR), UV spectroscopy, infrared spectroscopy (FTIR), etc. [31]. Chitosan's physical properties in aqueous medium are dependent upon the degree of deacetylation as well as the distribution of acetyl groups in the chitosan polymer chain [17]. The higher number of protonated $-NH_2$ groups present in the chitosan structure shows solubility in acidic aqueous solutions. Chitosan becomes soluble when approximately 50% of all $-NH_2$ groups are protonated [8]. The amino groups are protonated at pH below its pKa value (<6.5) [67] and act as cationic polyelectrolytes because amino groups possess a positive charge [30]. Chitosan is insoluble in water at neutral pH [23, 67], as well as insoluble in alkaline solutions [23]. The presence of more acetylated groups in chitosan promotes hydrogen bonding [67]. Besides the degree of deacetylation, the chitosan's molecular weight has played a significant role in its viscosity as well as solubility. Chitosan's molecular weight depends on the sources and preparation process. Chitosan with a higher molecular weight of greater than 250 kDa, medium molecular weight has a range between 50 and 250 kDa and lower molecular weight has less than 50 kDa. Chitosan possessing a high molecular weight produces higher viscosity. Low molecular weight chitosan exhibits greater solubility

[67]. The changes or modifications of physicochemical properties, as well as biological properties of chitosan, make it suitable for various pharmaceutical and biomedical applications.

5 Advantages of Chitosan as Transdermal Drug Carrier

Chitosan is a potential polymer for application in the transdermal drug delivery system, generally utilized in transdermal patches [52, 80]. Chitosan-based coating as a drug delivery approach shows specific advantages such as process stability, increased biocompatibility, drug targeting at specific sites, and promotes the high therapeutic index of the drug [7]. Chitosan and its derivatives are utilized to improve skin penetrability. Specifically, chitosan, which has a low molecular weight, is frequently used in transdermal drug delivery to increase permeability. However, chitosan, as well as its derivatives, can be utilized as a coating of transdermal liposomes, along with increasing their durability. N-succinyl chitosan has the potential to provide a controlled-release coating and its response to alter the skin pH and release content accordingly [52].

Chitosan-based transdermal patches are better than other materials, facilitating the wound healing. Both antibacterial and anti-inflammatory combined properties of chitosan may be synergistic with the administered drugs. Interestingly, Chitosan can promote fibroblast and granulation proliferation, stimulate neutrophils and IgM, increase analgesia and haemostasis, interact with phospholipid membrane, active macrophages, improve antibacterial activity, and produce extracellular matrix [38].

Particularly, adhesive chitosan extends the drug accommodation time on the surface of skin and eases the skin's hydration and swelling. Importantly, chitosan can open the compact junctions between the keratinocytes, also weakening the stratum corneum barrier layer [40]. Chitosan gives the scaffolds better biological qualities and promotes the effective adherence and proliferation of cells because of its high biocompatibility. It is a favourable biomaterial for several tissue engineering applications due to its properties like flexibility, nontoxic, biodegradability, and antimicrobial properties. Additionally, chitosan has a strong affinity for biological membranes that make it an excellent choice for a drug delivery system [62]. However, microneedles composed of chitosan and its derivatives do not need to be removed after drug administration because it can degrade safely in vivo [38]. Chi et al. developed a chitosan integrated microneedle patch which was utilized for wound healing of the skin [18].

6 Chitosan-Based Transdermal Drug Delivery Systems

There are several types of transdermal formulations made up with chitosan such as transdermal patches, microneedles, hydrogels, etc. Details have been discussed below. The application of chitosan in transdermal delivery is summarized in

Table 1 Application of different types of chitosan-containing transdermal formulations

Sl. no	Type of formulation	Formulating polymers	Incorporated drugs	Study objective	References
1.	Transdermal Patches	Chitosan and tamarind gum polysaccharide	Protein (bovine serum albumin)	Transdermal delivery of protein	[42]
		Chitosan	Lornoxicam	Prolonged anti- inflammatory effect	[79]
		Chitosan	Glibenclamide nanocrystal	Hyperglycemic effect through stable nanosized drug	[6]
		Chitosan	Ondansetron	Enhancement bioavailability for nausea and vomiting treatment	[11]
2.	Microneedles	Chitosan	Meloxicam	Pain management in cattle	[14]
		Thiolated chitosan	Tacrolimus	Enhanced bioavailability of immunosuppressant tacrolimus	[2]
		Chitosan hydrogel and PVP	*Centella asiatica*, asiatic acid	Wound healing and inflammatory effect	[61]
		Chitosan	Growth hormone releasing hormone	Prolonged release of luteinizing hormone-releasing hormone analogues for cancer treatment	[15]
3.	Hydrogels	Chitosan	Terbinafine hydrochloride (T-HCl)	Topical antifungal activity	[51]
		Chitosan	Nanocurcumin	Enhancement of curcumin bioavailability for stress, cancer etc.	[24]
4.	Wound dressing hydrogels	HMW chitosan-co poly AMPS	Mupirocin	Non-irritant anti-bacterial wound healing hydrogel preparation	[1]
		Chitosan in lactic acid solution	Fucoidan	Treatment of dermal burns on rabbit	[64]

Table 1. A diagrammatic representation of chitosan-based transdermal application is also given in Fig. 3.

6.1 Transdermal Patches

Amongst the various transdermal drug delivery approaches, patches are the most common. Backing membrane, matrix, drug reservoir, and adhesive layer are the

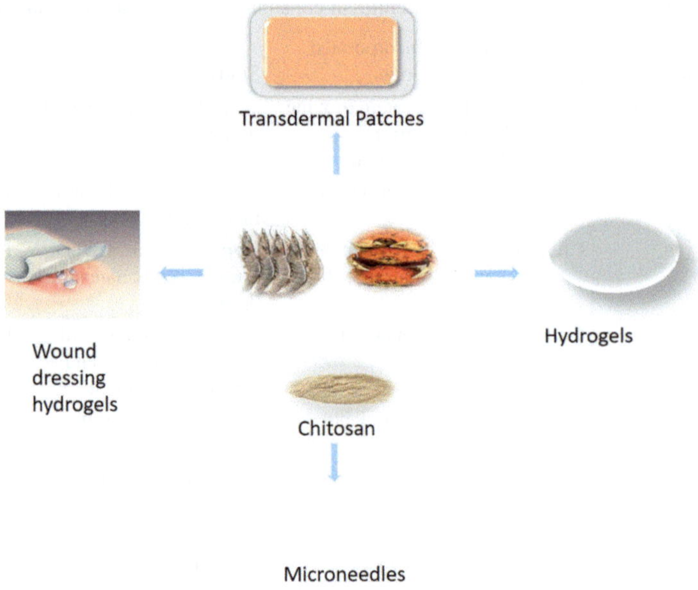

Fig. 3 Different types of transdermal formulations of chitosan

basic components of patches. A variety of permeation enhancer materials can be used to modify transdermal drug delivery systems, allowing for predictable control of the drug absorption profile. Different transdermal drug delivery systems, including vapour patches, membrane-moderated transdermal systems, matrix systems with drug-in-adhesive or matrix-dispersion systems, and single or multilayer drugs in the adhesive system, have different mechanisms to regulate the drug release rate. Matrix can be formulated with several types of polymers, chitosan is one of the widely used polymers in TDDS patch formulation. Various methods are available to fabricate chitosan patch, here we have discussed few of them [4, 38].

Malviya et al. designed chitosan composite-based films for the delivery of protein. Composite-based films had been developed by varying the proportions of chitosan (50, 100, 150 mg) and tamarind gum polysaccharide (100, 150, 200 mg) with the model compound. The solution was made by mixing different volumes of gum and chitosan and stirring them for 10 min at 40 °C. After 24 h, the polysaccharide composite solution was treated with bovine serum albumin. After being stirred for 15 min at 45 degrees Celsius, it was transferred to a mould and allowed to dry at 40 degrees Celsius [42].

Kim et al. fabricated a porous patch of chitosan, which was hybridized with silver. In a solution of 1 % v/v acetic acid and 10 % v/v n-butanol, chitosan was dissolved to 1 wt%. To the chitosan solution, several concentrations of silver nitrate (0, 0.5, 1, 3, 6 % w/v) were added and stirred. After that, 0.003 wt% sodium borohydride was added to the silver nitrate/chitosan solutions. The solutions were vigorously stirred before being injected into the mould, frozen for 24 h at −70 °C, and then lyophilized for another 24 h. The chitosan patch obtained from the mould was washed with ethanol (100, 85, 70 and 50% v/v) stagewise for 2 h each. The scaffold was then lyophilized for 24 h and frozen at −70 degrees Celsius for 24 h following a 48-h wash in distilled water. Silver nitrate/chitosan/sodium borohydride solutions ($AgNO_3$ content: 0, 0.5, 1, 3, 6 % w/v) were cast onto flat glass slips for silver hybridized porous chitosan films, and the solvent was allowed to evaporate for 3 weeks at room temperature. After that, the produced films were completely dried for 24 h at 30 °C in a vacuum oven. The resultant films, which were utilized for the hydrophilic ratio and crystal intensity experiments, were roughly 0.5 mm thick [29].

Yadav et al. created chitosan-based matrix-type TDDS patch for prolonged delivery of Lornoxicam using a mercury substrate and the solvent casting technique. The patches were cast into 4-cm-diameter glass rings and set on a mercury surface inside a Petri plate. A solution of 1% glacial acetic acid was used to dissolve 1.5% chitosan with an 80% deacetylation degree. Different quantities of Tween 20 (1,1.5,2%) were utilized as a penetration enhancer, while 2% propylene glycol was utilized as a plasticizer. There was enough medication added to the solution for each patch to contain 8 mg of Lornoxicam. The mercury surface, which has a glass ring with a diameter of 4 cm, was covered with the drug solution. A funnel that was inverted was placed over these spots. For 24 h, the solvent was left to evaporate under normal circumstances. The patches were placed in desiccators and wrapped in aluminium foil once they had dried [79].

Glibenclamide (GBD) nanocrystals were successfully incorporated into a chitosan transdermal patch. Nano-crystallized GBD had better drug release than micro GBD containing patch. The nanocrystals were physico-chemically stable in the chitosan patch. The drug was able to reduce blood glucose levels efficiently after skin penetration [6]. Propranolol hydrochloride was included in the chitosan membrane. The chitosan membrane was crosslinked with glutaraldehyde. The drug permeation was successfully controlled depending on the degree of crosslinking [71]. A chitosan patch of ondansetron was developed, where 2-(2-ethoxy-ethoxy) ethanol was used as a plasticizer. In this formulation, eucalyptol as permeation enhancer had greater permeability than terpenes the penetration enhancer [11].

6.2 Microneedles

Most medicines are unable to penetrate the skin at effective rates, which has significantly hindered the efficiency of transdermal drug delivery. It has been suggested that using micron-scale needles to increase skin permeability can significantly improve transdermal delivery, particularly for macromolecules. Microneedles of various sizes, forms, and materials have been created using the instruments of the microelectronics industry. Solid microneedles, which have been demonstrated to enhance skin permeability to nanoparticles and a wide variety of molecules in vitro, have been the focus of the majority of drug delivery investigations. Oligonucleotide delivery, insulin-induced blood glucose reduction, and protein and DNA vaccine-induced immune response stimulation have all been shown in in vivo investigations [57].

Chitosan microneedles have been designed for a wide variety of applications; here few of the preparation techniques have been elaborated.

Castilla-Casadiego et al. developed a microneedle patch of chitosan for the delivery of meloxicam. Prior to microneedle fabrication, solutions of 10% (w/v) chitosan in acetic acid–water (10, 50 & 90% w/v) were prepared, and it was heated for 3 h at 70 °C. Then, meloxicam (50 mg) was dissolved in chitosan (1 ml) solution. The chitosan/meloxicam solution was put into a PDMS mould to create the microneedle patch. The first step of the chitosan/meloxicam microneedle patch fabrication procedure involved applying 500 mg of the casting solution to the mould to form a monolayer. The mould was then placed inside a 50 mL flat-bottomed centrifuge tube and centrifuged for 30 min at 3000 RPM (step 2). For a total of 120 min of centrifugation, steps 1 and 2 were carried out four times. The microneedles with meloxicam were created as a result of this technique. The microneedles and a chitosan foundation make up the entire patch. A second layer of the chitosan solution devoid of meloxicam was added in order to create the chitosan base. The second layer is on the surface of the skin, thus not penetrating the skin to allow drug release. The second layer was prepared by adding 5 mL of chitosan solution (0.1 g/mL in acetic acid at 10% (v/v), 50% (v/v), or 90% (v/v)) into the 50 mL tube containing the mould loaded with the chitosan/meloxicam solution and centrifuged at 3000 RPM for 60 min. Lastly, the mould in the 50 mL tube was placed in an oven to dry without the cap at 28 °C for 3 days. The microneedle patch was gently removed from the tube with tweezers [14].

Suzuki et al. fabricated hollow biodegradable microneedles by the dip coating method. At first, a chitosan acetate solution was prepared by dissolving chitosan in an acetic acid solution. This solution was made bubble-free by stirring in a centrifugal mixer. A film can be deposited using the dip-coating technique on a substrate, which is typically cylindrical or flat. The substrate was dipped in the liquid phase raw material during the dip-coating process and then removed. After being soaked in the chitosan-acetate solution, an Al wire of 100 μm in diameter was removed. The wire coated with chitosan acetate was dried for 3 min at 200 °C in a clean oven. Four repetitions of the dip-coating and drying procedures were conducted. The solution-coated Al wire was then submerged in a 4% sodium hydroxide solution for 24 h in order to dissolve and extract the Al wire. Concurrently, a neutralization reaction with sodium hydroxide transformed chitosan-acetate back into chitosan; that is, the fine pipe of chitosan was produced by submerging the wire coated in chitosan-acetate in sodium hydroxide solution. The chitosan pipes were then roasted at 200 °C in a sterile oven [68].

Polymeric microneedles offer a better alternative route of administration and can be tailored to optimize the pharmacokinetics of tacrolimus (TM). Ahmad et al., prepared the thiolated chitosan-based microneedle patch of TM (MNP-TM), an alternate delivery method that enhances bioavailability by bypassing intestinal P-gp efflux and hepatic first-pass metabolism.

A number of in vitro, ex vivo, and in vivo characteristics were used to characterize the MNP-TM. When compared to two other manufactured MNPs, MNP-TM made with 2% thiolated chitosan showed relatively superior characteristics in terms of loading efficiency, formulation development and physicochemical parameters. Better conveyance against the skin was made possible by the patch's successful skin piercing. Improved control over drug release for 48 h was demonstrated by the TM release, suggesting the creation of an extended-release delivery system. Therefore, in comparison to the currently available ointment-based dosage form, the thiolated chitosan-based microneedle patch could be further characterized for its other in vitro and in vivo parameters and developed as an effective delivery system for transdermal administration of tacrolimus for greater patient compliance and therapeutic outcomes [2]. The active ingredient of *Centella asiatica*, asiatic acid (AA), decreases inflammation and promotes the formation of extracellular matrix, however it is unable to reach deeper layers of the skin through the stratum corneum. Two kinds of AA-loaded microneedle arrays were developed in order to circumvent the stratum corneum. Ryall et al., created, characterized, and improved a hydrogel array using chitosan and PVP, as well as a dissolving array using chitosan and PVA. Strong and sufficiently long, both needles could penetrate the epidermis without shattering. Both demonstrated a sustained drug release over 48 h and were biocompatible with fibroblasts and keratinocytes (>75% viability at 100% dose). More AA (52. 2%) was released by the hydrogel microneedle than by the dissolving formulation (26.4%) [61]. Chen et al. used a chitosan microneedle method to create a transdermal delivery device for Luteinizing Hormone-Releasing Hormone (LHRH). Mice in such a system may get castrated following an LH surge. It is a viable, practical, and hopeful way for androgen-deprivation therapy (ADT) to go in the future [15].

6.3 Hydrogels

In brief, hydrogels are three-dimensional network structures made of particular synthetic and natural polymers. Because of their porous structure, they can absorb and transport a significant amount of water. Both synthetic and natural polymers can be used to create hydrogels. Excellent permeability, biocompatibility, flexibility, and viscoelasticity are the hallmarks of the perfect hydrogel. Transdermal medication delivery is one from the various biological and pharmacological uses for hydrogels. Because of its special qualities, the development of hydrogel-based patches has recently garnered significant attention. The ability of hydrogels, which are mostly made of hydrophilic polymers, to retain a significant amount of water, between 10 and 20%, can reach 1000 times their dry weight. Hydrogels' significant water content contributes to the suppleness and moisturization of the skin, making them an appropriate topical dose form [3]. Chitosan is an abundantly available natural polymer which have high swellability, making it an excellent choice for hydrogel formulation [53, 60].

Özcanet al. formulated chitosan hydrogels containing terbinafine hydrochloride (T-HCl). In a diluted lactic acid solution (1%), various types of chitosan with varying molecular weights were used to create the gel formulations at a 3% (w/v) concentration. To achieve a smooth dispersion, the mixture was prepared by stirring it at about 600 rpm with the use of a mechanical stirrer. T-HCl was added to the formulations at a concentration of 1% (w/v) [51].

Fariset al. created chitosan-coated nanocurcumin-loaded hydrogel for transdermal delivery of curcumin. Different batches of curcumin were prepared with varying stabilizer concentrations. A modified anti-solvent nano precipitation approach was used to create nanocurcumin formulations. Using a syringe and a volume ratio of 1: 1, the selected nanocurcumin dispersion was injected dropwise to a chitosan solution (CS,1%acetic acid solution) on a magnetic stirrer for 60 min. It was dropping at a rate of 2.5 ml/min. The final chitosan concentrations (w/v) upon mixing with the nanocurcumin dispersion were 0.01%, 0.03%, 0.05%, and 0.07%. Next, measurements of the zeta potential and particle size were made [24].

Tsai et al., utilized chitosan hydrogel as a base for Berberine ointment preparation. In double-distilled water, chitosan powder was dispersed, and in a stirring condition, lactic acid was added to it in some batches. Other batches were prepared using EDTA or an absorption enhancer. Berberine was then added to this preparation to make a 0.1% concentration [73].

The antibacterial effectiveness of a transdermal chitosan-based polymeric hydrogel for the transdermal therapy of bacterial-resistant strains was evaluated by Tariq et al. Ipenem-cilastatin (IMI-CIL) containing nanostructured lipid carriers (NLCs) was created and optimized using the Box–Behnken design model and the microemulsion method. Compritol 888 ATO was used as the lipid core, while Tween 80 was used as a stabilizer. Both the hydrogel and the NLCs loaded with IMI-CIL demonstrated long-lasting in vitro drug release. Furthermore, a noteworthy 15.77-fold rise in drug permeability was shown by the hydrogel enhanced with permeation enhancers. Additionally, as compared to IMI-CIL solution, which showed little effect against non-resistant bacterial species and no activity against resistant isolates, the IMI-CIL-loaded NLCs showed much higher antibacterial activity. The chitosan hydrogel formulation of NLCs loaded with IMI-CIL played a crucial part in providing prolonged drug release and enhancing the antibacterial activity of carbapenem antibiotics. This indicated potential as a possible remedy for bacterial illnesses that are resistant to medications [69].

6.4 *Wound Dressing Hydrogel*

HMW chitosan-co-poly AMPS hydrogel patch membranes were synthesized using an optimized modified free radical polymerization process. A weighed amount of HMW (high molecular weight) chitosan was continuously stirred while being dissolved in a 1% acetic acid solution. It was purged with nitrogen gas to produce a homogeneous solution devoid of oxygen. At room temperature, AMPS (2-acrylamido-2-methylpropane sulfonic acid) was dissolved in a predetermined volume of water. Both the initiators (APS and SHS) were introduced to the solution containing dissolved AMPS while it was being continuously stirred. The monomer and initiator combination was introduced drop by drop to the polymer-containing beaker while shaking constantly. With constant stirring, the cross-linker was gradually added to the chitosan-AMPS solution. The solution was transferred onto petri dishes that had been labelled with a monogram. Petri plates were kept in a water bath at 40 °C for 2 h, 50 °C for 2 h, and 60 °C for 20 h. Following a successful process, the petri dishes were taken out of the water bath, and the unreactive ingredients were removed from the formed patches by washing them with a water–ethanol (70:30) solution. After being cleaned, the hydrogel patches were dried for 2 days at 40 °C in a vacuum oven. Using the swelling diffusion approach, mupirocin was added to the chitosan-co-poly (AMPS) hydrogels. At room temperature, 1% mupirocin solution was made using a water–ethanol (70:30) solution. The patches were weighed and left in the medication solution for 24 h at room temperature prior to the drug being loaded into the dry hydrogel patches. The patches were taken out of the solution after the allotted amount of time, cleaned with distilled water to get rid of any leftover surface solution, and then dried in a hot air oven set at 40 °C. To determine the loaded medicine, they were weighed once more after drying [1].

Films of fucoidan and chitosan were made for wound healing. After dissolving chitosan in a lactic acid solution with fucoidan and stirring it for the entire night, 2.5% propylene glycol was added as a plasticizer. After raising the temperature to 30 °C, the resultant solution was sonicated to eliminate air bubbles, then transferred onto a 3.0 mL Petri dish and allowed to dry. The films were dried, then peeled off and kept at room temperature in an airtight container until they could be examined further. To optimize the film qualities, a number of variables were examined. A digital micrometre was used to measure the films' thickness at five distinct points (the centre and four corners), and the mean thickness was computed [64].

In the study by Ribeiro et al., the cytotoxicity of a chitosan hydrogel as a wound dressing was assessed using fibroblast cells isolated from rat skin. The hydrogel was found to be able to boost cell adhesion and proliferation, and cell viability studies demonstrated that the hydrogel and its breakdown by-products are noncytotoxic. The treatment of dermal burns in Wistar rats was evaluated by inducing full-thickness transcutaneous dermal wounds. The animals those received chitosan hydrogel had wound beds that were significantly smaller than the controls. The absence of pathological anomalies in the organs acquired by necropsy and the lack of a reactive or granulomatous inflammatory reaction in skin lesions with chitosan hydrogel were revealed by histological analysis, supporting the biomaterial's local and systemic histocompatibility. According to the findings, this biomaterial might help restore the structure of the skin [59].

7 Conclusion

Transdermal route is becoming popular due to its several advantages as discussed. Choice of a proper polymer according to the formulation needs is one of the vital steps for any drug delivery system. Chitosan is cationic polymer obtained from natural source which is biodegradable and biocompatible. The major advantage of chitosan is it can be chemically modified and its derivatives possess enhanced properties over the native form. Chitosan have high swellability, which makes it a good choice for hydrogel preparation; it also has wound healing properties. Chitosan can be extensively used as a polymer for transdermal patch. Microneedle patches and microneedle arrays are also widely formulated with chitosan. Various active substances can be coated with chitosan, or nano-formulations are made using chitosan, then it can be incorporated into transdermal formulations. A variety of drugs can be administered via chitosan TDDS. Its biosafety profile is also high. Not only at the laboratory level, continuous research will establish chitosan as an efficient polymer of transdermal formulation at the clinical level also.

References

1. Ahmad, S., Minhas, M. U., Ahmad, M., Sohail, M., Abdullah, O., & Badshah, S. F. (2018). Preparation and evaluation of skin wound healing Chitosan-based hydrogel membranes. *AAPS PharmSciTech, 19*(7), 3199–3209. https://doi.org/10.1208/s12249-018-1131-z
2. Ahmad, Z., Khan, M. I., Siddique, M. I., Sarwar, H. S., Shahnaz, G., Hussain, S. Z., Bukhari, N. I., Hussain, I., & Sohail, M. F. (2020). Fabrication and characterization of Thiolated Chitosan microneedle patch for transdermal delivery of Tacrolimus. *AAPS PharmSciTech, 21*(2). https://doi.org/10.1208/s12249-019-1611-9
3. Ahsan, A., Tian, W.-X., Farooq, M. A., & Khan, D. H. (2020). An overview of hydrogels and their role in transdermal drug delivery. *International Journal of Polymeric Materials and Polymeric Biomaterials, 70*(8), 1–11. https://doi.org/10.1080/00914037.2020.1740989
4. Al Hanbali, O. A., Khan, H. M. S., Sarfraz, M., Arafat, M., Ijaz, S., & Hameed, A. (2019). Transdermal patches: Design and current approaches to painless drug delivery. *Acta Pharmaceutica, 69*(2), 197–215. https://doi.org/10.2478/acph-2019-0016
5. Alemu, D., Getachew, E., & Mondal, A. K. (2023). Study on the physicochemical properties of Chitosan and their applications in the biomedical sector. *International Journal of Polymer Science, 2023*, e5025341. https://doi.org/10.1155/2023/5025341
6. Ali, H. S. M., & Hanafy, A. F. (2017). Glibenclamide nanocrystals in a biodegradable Chitosan patch for transdermal delivery: Engineering, formulation, and evaluation. *Journal of Pharmaceutical Sciences, 106*(1), 402–410. https://doi.org/10.1016/j.xphs.2016.10.010
7. Visan, A. I., & Cristescu, R. (2023). Polysaccharide-based coatings as drug delivery systems. *Pharmaceutics (Basel), 15*(9), 2227–2227. https://doi.org/10.3390/pharmaceutics15092227
8. Aranaz, I., Alcántara, A. R., Civera, M. C., Arias, C., Elorza, B., Heras Caballero, A., & Acosta, N. (2021). Chitosan: An overview of its properties and applications. *Polymers, 13*(19), 3256. https://doi.org/10.3390/polym13193256
9. Aranaz, I., Mengibar, M., Harris, R., Panos, I., Miralles, B., Acosta, N., Galed, G., & Heras, A. (2009). Functional characterization of Chitin and Chitosan. *Current Chemical Biology, 3*(2), 203–230. https://doi.org/10.2174/187231309788166415

10. Arancibia, M., Alemán, A., Calvo, M., López-Caballero, M. E., Montero, P., & Carmen del. (2014). Antimicrobial and antioxidant chitosan solutions enriched with active shrimp (Litopenaeusvannamei) waste materials. *Food Hydrocolloids, 35*, 710–717. https://doi.org/10.1016/j.foodhyd.2013.08.026
11. Can, A. S., Erdal, M. S., Güngör, S., & Özsoy, Y. (2013). Optimization and characterization of Chitosan films for transdermal delivery of Ondansetron. *Molecules, 18*(5), 5455–5471. https://doi.org/10.3390/molecules18055455
12. Bastiaens, L., Soetemans, L., D'Hondt, E., & Elst, K. (2019). Sources of Chitin and Chitosan and their isolation. In *Chitin and Chitosan: Properties and applications* (pp. 1–34). Willey.
13. Bird, D., & Ravindra, N. M. (2020). Transdermal drug delivery and patches—An overview. *Medical Devices & Sensors, 3*(6). https://doi.org/10.1002/mds3.10069
14. Castilla-Casadiego, D. A., Carlton, H., Gonzalez-Nino, D., Miranda-Muñoz, K. A., Daneshpour, R., Huitink, D., Prinz, G., Powell, J., Greenlee, L., & Almodovar, J. (2021). Design, characterization, and modeling of a chitosan microneedle patch for transdermal delivery of meloxicam as a pain management strategy for use in cattle. *Materials Science and Engineering: C, 118*, 111544. https://doi.org/10.1016/j.msec.2020.111544
15. Chen, M.-Y., Chen, Y.-Y., Tsai, H.-T., Tzai, T.-S., Chen, M.-C., & Tsai, Y.-S. (2017). Transdermal delivery of luteinizing hormone-releasing hormone with Chitosan microneedles: A promising tool for Androgen Deprivation therapy. *Anticancer Research, 37*(12), 10.21873/anticanres.12139.
16. Cheng, T., Tai, Z., Shen, M., Liu, Y., Yu, J., Wang, J., Zhu, Q., & Chen, Z. (2023). Advance and challenges in the treatment of skin diseases with the transdermal drug delivery system. *Pharmaceutics, 15*(8), 2165–2165. https://doi.org/10.3390/pharmaceutics15082165
17. Cheung, R. C. F., Ng, T. B., Wong, J. H., & Chan, W. Y. (2015). Chitosan: An update on potential biomedical and pharmaceutical applications. *Marine Drugs, 13*(8), 5156–5186. https://doi.org/10.3390/md13085156
18. Chi, J., Zhang, X., Chen, C., Shao, C., Zhao, Y., & Wang, Y. (2020). Antibacterial and angiogenic chitosan microneedle array patch for promoting wound healing. *Bioactive Materials, 5*(2), 253–259. https://doi.org/10.1016/j.bioactmat.2020.02.004
19. Soon, C. Y., Tee, Y. K., Tan, C. P., Rosnita, A. T., & Abdan, K. (2018). Extraction and physicochemical characterization of chitin and chitosan from Zophobasmorio larvae in varying sodium hydroxide concentration. *International Journal of Biological Macromolecules, 108*, 135–142. https://doi.org/10.1016/j.ijbiomac.2017.11.138
20. dos Santos Rodrigues, B., Sushant Lakkadwala, Sharma, D., & Singh, J. (2019). Chitosan for gene, DNA vaccines, and drug delivery. In *Materials for biomedical engineering* (pp. 515–550). Elsevier BV.
21. Economidou, S. N., Lamprou, D. A., & Douroumis, D. (2018). 3D printing applications for transdermal drug delivery. *International Journal of Pharmaceutics, 544*(2), 415–424. https://doi.org/10.1016/j.ijpharm.2018.01.031
22. El Knidri, H., Belaabed, R., Addaou, A., Laajeb, A., & Lahsini, A. (2018). Extraction, chemical modification and characterization of chitin and chitosan. *International Journal of Biological Macromolecules, 120*, 1181–1189. https://doi.org/10.1016/j.ijbiomac.2018.08.139
23. ElgadirMAbd, U. M. S., Ferdosh, S., Adam, A., Chowdhury, A. J. K., & MdZaidulI, S. (2015). Impact of chitosan composites and chitosan nanoparticle composites on various drug delivery systems: A review. *Journal of Food and Drug Analysis, 23*(4), 619–629. https://doi.org/10.1016/j.jfda.2014.10.008
24. Faris, T. M., Ahmad, A. M., Mohammed, H. A., Abdullah Alamoudi, J., Alsunbul, M., Alrashidi, A., Abdullah, O., Altwaijry, N., & Hassan, A. S. (2023). Preparation and evaluation of transdermal hydrogel of chitosan coated nanocurcumin for enhanced stability and skin permeability. *Arabian Journal of Chemistry, 16*(12), 105302. https://doi.org/10.1016/j.arabjc.2023.105302
25. Hazmi, A. T., Ahmad, F. B., Maziati Akmal, M. H., Md Ralib, A. A., & Binti Ali, F. (2022). Fungal chitosan for potential application in piezoelectric energy harvesting: Review on

experimental procedure of chitosan extraction. *Alexandria Engineering Journal, 67*, 105–116. https://doi.org/10.1016/j.aej.2022.08.020
26. Huq, T., Khan, A., Brown, D., Dhayagude, N., He, Z., & Ni, Y. (2022). Sources, production and commercial applications of fungal chitosan: A review. *Journal of Bioresources and Bioproducts, 7*(2), 85–98. https://doi.org/10.1016/j.jobab.2022.01.002
27. Jana, S., Manna, S., Nayak, A. K., Sen, K. K., & Basu, S. K. (2014). Carbopol gel containing chitosan-egg albumin nanoparticles for transdermal aceclofenac delivery. *Colloids and Surfaces B: Biointerfaces, 114*, 36–44. https://doi.org/10.1016/j.colsurfb.2013.09.045
28. Jeong, W. Y., Kwon, M., Choi, H. E., & Kim, K. S. (2021). Recent advances in transdermal drug delivery systems: A review. *Biomaterials Research, 25*(1). https://doi.org/10.1186/s40824-021-00226-6
29. Kim, J. H., Kim, S. I., Kwon, I.-B., Kim, M. H., & Lim, J. I. (2013). Simple fabrication of silver hybridized porous chitosan-based patch for transdermal drug-delivery system. *Materials Letters, 95*, 48–51. https://doi.org/10.1016/j.matlet.2012.12.078
30. Joseph, S. M., Krishnamoorthy, S., Paranthaman, R., Moses, J. A., & Anandharamakrishnan, C. (2021). A review on source-specific chemistry, functionality, and applications of chitin and chitosan. *Carbohydrate Polymer Technologies and Applications, 2*, 100036. https://doi.org/10.1016/j.carpta.2021.100036
31. Gzyra-Jagieła, K., Pęczek, B., Wiśniewska-Wrona, M., & Gutowska, N. (2019). Physicochemical properties of Chitosan and its degradation products. In *Chitin and Chitosan: Properties and applications* (pp. 61–80). Willey.
32. Karve, T., Dandekar, A., Agrahari, V., Peet, M. M., Banga, A. K., & Doncel, G. F. (2024). Long-acting transdermal drug delivery formulations: Current developments and innovative pharmaceutical approaches. *Advanced Drug Delivery Reviews, 210*. https://doi.org/10.1016/j.addr.2024.115326
33. Kozma, M., Acharya, B., & Bissessur, R. (2022). Chitin, Chitosan, and Nanochitin: Extraction, synthesis, and applications. *Polymers, 14*(19), 3989. https://doi.org/10.3390/polym14193989
34. Kumari, S., Rath, P., Sri Hari Kumar, A., & Tiwari, T. N. (2015). Extraction and characterization of chitin and chitosan from fishery waste by chemical method. *Environmental Technology & Innovation, 3*, 77–85. https://doi.org/10.1016/j.eti.2015.01.002
35. Kumari, S., & Rath, P. K. (2014). Extraction and characterization of Chitin and Chitosan from (Labeorohit) fish scales. *Procedia Materials Science, 6*, 482–489. https://doi.org/10.1016/j.mspro.2014.07.062
36. Li, Z., Fang, X., & Yu, D. (2021). Transdermal drug delivery systems and their use in obesity treatment. *International Journal of Molecular Sciences, 22*(23), 12754. https://doi.org/10.3390/ijms222312754
37. Ways, M. T., Lau, W., & Khutoryanskiy, V. (2018). Chitosan and its derivatives for application in Mucoadhesive drug delivery systems. *Polymers, 10*(3), 267. https://doi.org/10.3390/polym10030267
38. Ma, J., Wang, Y., & Lu, R. (2022). Mechanism and application of Chitosan and its derivatives in promoting permeation in transdermal drug delivery systems: A review. *Pharmaceuticals, 15*(4), 459. https://doi.org/10.3390/ph15040459
39. Ma, J., Xin, C., & Tan, C. (2015). Preparation, physicochemical and pharmaceutical characterization of chitosan from Catharsius molossus residue. *International Journal of Biological Macromolecules, 80*, 547–556. https://doi.org/10.1016/j.ijbiomac.2015.07.027
40. Ma, Q., Gao, Y., Sun, W., Cao, J., Liang, Y., Han, S., Wang, X., & Sun, Y. (2020). Self-assembled chitosan/phospholipid nanoparticles: From fundamentals to preparation for advanced drug delivery. *Drug Delivery, 27*(1), 200–215. https://doi.org/10.1080/10717544.2020.1716878
41. Mahdy Samar, M., El-Kalyoubi, M. H., Khalaf, M. M., & Abd El-Razik, M. M. (2013). Physicochemical, functional, antioxidant and antibacterial properties of chitosan extracted from shrimp wastes by microwave technique. *Annals of Agricultural Sciences, 58*(1), 33–41. https://doi.org/10.1016/j.aoas.2013.01.006

42. Malviya, R., Tyagi, A., Fuloria, S., Subramaniyan, V., Sathasivam, K., Sundram, S., Karupiah, S., Chakravarthi, S., Meenakshi, D. U., Gupta, N., Sekar, M., Sudhakar, K., & Fuloria, N. K. (2021). Fabrication and characterization of Chitosan—Tamarind seed polysaccharide composite film for transdermal delivery of protein/peptide. *Polymers, 13*(9), 1531. https://doi.org/10.3390/polym13091531
43. Manna, S., & Jana, S. (2022). Marine polysaccharides in Tailor-made drug delivery. *Current Pharmaceutical Design, 28*(13), 1046–1066. https://doi.org/10.2174/1381612828666220328122539
44. Manna, S., Nandi, G., & Jana, S. (2022). Alginate-based carriers for transdermal drug delivery. In *Polysaccharide-based biomaterials: Delivery of therapeutics and biomedical applications* (pp. 69–89). The Royal Society of Chemistry.
45. Manna, S., Seth, A., Gupta, P., Nandi, G., Dutta, R., Jana, S., & Jana, S. (2023). Chitosan derivatives as carriers for drug delivery and biomedical applications. *ACS Biomaterials Science & Engineering, 09*(5). https://doi.org/10.1021/acsbiomaterials.2c01297
46. Mati-Baouche, N., Elchinger, P.-H., de Baynast, H., Pierre, G., Delattre, C., & Michaud, P. (2014). Chitosan as an adhesive. *European Polymer Journal, 60*, 198–212. https://doi.org/10.1016/j.eurpolymj.2014.09.008
47. MeryiemDerraz, A. E., ChouaibEnnawaoui, A. M., Rjafallah, A., Laadissi, E. M., Khallok, H., Hatim, Z., & Hajjaji, A. (2023). Extraction and physicochemical characterization of an environmentally friendly biopolymer: Chitosan for composite matrix application. *Journal Of Composites Science, 7*(6), 260–260. https://doi.org/10.3390/jcs7060260
48. Morris, G. A., Kök, S. M., Harding, S. E., & Adams, G. G. (2010). Polysaccharide drug delivery systems based on pectin and chitosan. *Biotechnology and Genetic Engineering Reviews, 27*(1), 257–284. https://doi.org/10.1080/02648725.2010.10648153
49. Matharoo, N., Mohd, H., & Michniak-Kohn, B. (2023). Transferosomes as a transdermal drug delivery system: Dermal kinetics and recent developments. *Wiley Interdisciplinary Reviews-Nanomedicine and Nanobiotechnology, 16*(1). https://doi.org/10.1002/wnan.1918
50. Ng, L. C., & Gupta, M. (2020). Transdermal drug delivery systems in diabetes management: A review. *Asian Journal of Pharmaceutical Sciences, 15*(1), 13–25. https://doi.org/10.1016/j.ajps.2019.04.006
51. Özcan, İ., Abacı, Ö., Uztan, A. H., Aksu, B., Boyacıoğlu, H., Güneri, T., & Özer, Ö. (2009). Enhanced topical delivery of Terbinafine hydrochloride with Chitosan hydrogels. *AAPS PharmSciTech, 10*(3). https://doi.org/10.1208/s12249-009-9299-x
52. Pal, R., Pandey, P., Rai, B., Koli, M., Chakrabarti, M., Thakur, P., Rizwan, M., & Saxena, A. (2023). Chitosan: As highly potential biopolymer obtainable in several advance drug delivery systems including biomedical applications. *Environmental Science, 12*(Special issue (10)), 322–344. https://doi.org/10.48047/ecb/2023.12.si10.0036
53. Park, H., Park, K., & Kim, D. (2006). *Journal of Biomedical Materials Research, 76A*(1). https://doi.org/10.1002/(issn)1552-4965
54. Pellis, A., Guebitz, G. M., & Nyanhongo, G. S. (2022). Chitosan: Sources, processing and modification techniques. *Gels, 8*(7), 393. https://doi.org/10.3390/gels8070393
55. Peniche, C., Argüelles-Monal, W., & Goycoolea, F. M. (2008). Chitin and Chitosan: Major sources, properties and applications. *Monomers, Polymers and Composites from Renewable Resources, 1*, 517–542. https://doi.org/10.1016/b978-0-08-045316-3.00025-9
56. Petroni, S., Tagliaro, I., Antonini, C., Massimiliano D'Arienzo, Sara Fernanda Orsini, Mano, J. F., Brancato, V., Borges, J., & Cipolla, L. (2023). Chitosan-based biomaterials: Insights into chemistry, properties, devices, and their biomedical applications. *Marine Drugs, 21*(3), 147–147. https://doi.org/10.3390/md21030147
57. Prausnitz, M. R. (2004). Microneedles for transdermal drug delivery. *Advanced Drug Delivery Reviews, 56*(5), 581–587. https://doi.org/10.1016/j.addr.2003.10.023
58. Ramadon, D., McCrudden, M. T. C., Courtenay, A. J., & Donnelly, R. F. (2021). Enhancement strategies for transdermal drug delivery systems: Current trends and applications. *Drug Delivery and Translational Research, 12*(4). https://doi.org/10.1007/s13346-021-00909-6

59. Ribeiro, M. P., Espiga, A., Silva, D., Baptista, P., Henriques, J., Ferreira, C., Silva, J. C., Borges, J. P., Pires, E., Chaves, P., & Correia, I. J. (2009). Development of a new chitosan hydrogel for wound dressing. *Wound Repair and Regeneration, 17*(6), 817–824. https://doi.org/10.1111/j.1524-475x.2009.00538.x
60. Rohindra, D. R., Nand, A. V., & Khurma, J. R. (2004). Swelling properties of chitosan hydrogels. *The South Pacific Journal of Natural and Applied Sciences, 22*(1), 32. https://doi.org/10.1071/sp04005
61. Ryall, C., Chen, S., Duarah, S., & Wen, J. (2022). Chitosan-based microneedle arrays for dermal delivery of Centella asiatica. *International Journal of Pharmaceutics, 627*, 122221–122221. https://doi.org/10.1016/j.ijpharm.2022.122221
62. Saylam, E., Akkaya, Y., Ilhan, E., Cesur, S., Guler, E., Sahin, A., Cam, M. E., Ekren, N., Oktar, F. N., Gunduz, O., Ficai, D., & Ficai, A. (2021). Levodopa-loaded 3D-printed poly (lactic) acid/Chitosan neural tissue Scaffold as a promising drug delivery system for the treatment of Parkinson's disease. *Applied Sciences, 11*(22), 10727. https://doi.org/10.3390/app112210727
63. Sen, O., Poddar, P., Sarkar, P., Das, S., & Manna, S. (2025). Current advancements in microneedle technology for therapeutic and biomedical applications. *Sensors International, 6*, 100325. https://doi.org/10.1016/j.sintl.2024.100325
64. Sezer, A. D., Hatipoglu, F., Cevher, E., Oğurtan, Z., Bas, A. L., & Akbuğa, J. (2007). Chitosan film containing fucoidan as a wound dressing for dermal burn healing: Preparation and in vitro/in vivo evaluation. *AAPS PharmSciTech, 8*(2), E94–E101. https://doi.org/10.1208/pt0802039
65. Shafabakhsh, R., Yousefi, B., Asemi, Z., Nikfar, B., Mansournia, M. A., & Hallajzadeh, J. (2020). Chitosan: A compound for drug delivery system in gastric cancer-a review. *Carbohydrate Polymers, 242*, 116403. https://doi.org/10.1016/j.carbpol.2020.116403
66. Shariatinia, Z. (2019). Pharmaceutical applications of natural polysaccharides. In M. S. Hasnain & A. K. Nayak (Eds.), *Natural polysaccharides in drug delivery and biomedical applications* (pp. 15–57). Elsevier.
67. Sharkawy, A., Barreiro, M. F., & Rodrigues, A. E. (2020). Chitosan-based Pickering emulsions and their applications: A review. *Carbohydrate Polymers, 250*, 116885. https://doi.org/10.1016/j.carbpol.2020.116885
68. Suzuki, M., Takahashi, T., & Aoyagi, S. (2020). Fabrication and characterization of a biodegradable hollow microneedle from Chitosan. *Journal of Robotics and Mechatronics, 32*(2), 401–407. https://doi.org/10.20965/jrm.2020.p0401
69. Tariq, A., Kazim, T., Khan, A., Usman, M., & Khan, A. (2024). Reshaping antibiotic delivery: Chitosan-based polymeric hydrogel for transdermal treatment of drug resistant bacteria. *Polymers for Advanced Technologies, 35*(3). https://doi.org/10.1002/pat.6331
70. Teli, M. D., & Sheikh, J. (2012). Extraction of chitosan from shrimp shells waste and application in antibacterial finishing of bamboo rayon. *International Journal of Biological Macromolecules, 50*(5), 1195–1200. https://doi.org/10.1016/j.ijbiomac.2012.04.003
71. Thacharodi, D., & Panduranga, R. K. (1995). Development and in vitro evaluation of chitosan-based transdermal drug delivery systems for the controlled delivery of propranolol hydrochloride. *Biomaterials, 16*(2), 145–148. https://doi.org/10.1016/0142-9612(95)98278-m
72. Tong, X., Pan, W., Su, T., Zhang, M., Dong, W., & Qi, X. (2020). Recent advances in natural polymer-based drug delivery systems. *Reactive and Functional Polymers, 148*, 104501. https://doi.org/10.1016/j.reactfunctpolym.2020.104501
73. Tsai, C. J., Hsu, L. R., Fang, J. Y., & Lin, H. H. (1999). Chitosan hydrogel as a base for transdermal delivery of Berberine and its evaluation in rat skin. *Biological and Pharmaceutical Bulletin, 22*(4), 397–401. https://doi.org/10.1248/bpb.22.397
74. Varshosaz, J. (2007). The promise of chitosan microspheres in drug delivery systems. *Expert Opinion on Drug Delivery, 4*(3), 263–273. https://doi.org/10.1517/17425247.4.3.263
75. Wang, H., Qian, J., & Ding, F. (2018). Emerging Chitosan-based films for food packaging applications. *Journal of Agricultural and Food Chemistry, 66*(2), 395–413. https://doi.org/10.1021/acs.jafc.7b04528

76. Wang, W., Meng, Q., Li, Q., Liu, J., Zhou, M., Jin, Z., & Zhao, K. (2020). Chitosan derivatives and their application in biomedicine. *International Journal of Molecular Sciences, 21*(2). https://doi.org/10.3390/ijms21020487
77. Wong, W. F., Ang, K. P., Sethi, G., & Looi, C. Y. (2023). Recent advancement of medical patch for transdermal drug delivery. *Medicina (Kaunas, Lithuania), 59*(4), 778–778. https://doi.org/10.3390/medicina59040778
78. Xu, Y., Zhao, M., Cao, J., Fang, T., Zhang, J., Zhen, Y., Wu, F., Yu, X., Liu, Y., Li, J., & Wang, D. (2023). Applications and recent advances in transdermal drug delivery systems for the treatment of rheumatoid arthritis. *Acta Pharmaceutica Sinica B, 13*(11). https://doi.org/10.1016/j.apsb.2023.05.025
79. Yadav, A. V., & Urade, M. N. (2019). Formulation and evaluation of Chitosan based transdermal patches of Lornoxicam for prolonged drug release and to study the effect of permeation enhancer. *Indian Journal Of Pharmaceutical Education And Research, 53*(1), 88–96. https://doi.org/10.5530/ijper.53.1.12
80. Yewale, C., Tandel, H., Patel, A., & Misra, A. (2021). Polymers in transdermal drug delivery. In *Applications of polymers in drug delivery* (pp. 131–158). Elsevier.
81. Younes, I., & Rinaudo, M. (2015). Chitin and Chitosan preparation from marine sources. Structure, properties and applications. *Marine Drugs, 13*(3), 1133–1174. https://doi.org/10.3390/md13031133
82. Zhao, D., Yu, S., Sun, B., Gao, S., Guo, S., & Zhao, K. (2018). Biomedical applications of chitosan and its derivative nanoparticles. *Polymers, 10*(4). https://doi.org/10.3390/polym10040462
83. Zhu, Y., Guo, S., Ravichandran, D., Ramanathan, A., Sobczak, M. T., Sacco, A. F., Patil, D., Sri Vaishnavi Thummalapalli, Pulido, T. V., Lancaster, J. N., Yi, J., Cornella, J. L., Lott, D. G., Chen, X., Mei, X., Zhang, Y. S., Wang, L., Wang, X., Zhao, Y., & Hassan, M. K. (2024). 3D-printed polymeric biomaterials for health applications. *Advanced Healthcare Materials, 14*(1). https://doi.org/10.1002/adhm.202402571

pH-Responsive Chitosan-Based System for Targeted Drug Delivery

Sayed Maeen Badshah, Ndumiso Vukile Mdlovu, and Kuen-Song Lin

Abstract Chitosan-based nanomaterials have emerged as a highly promising platform for pH-responsive targeted drug delivery in cancer therapy, leveraging their inherent biocompatibility, biodegradability, and pH-sensitive properties. This chapter explores the structural and functional attributes of chitosan, focusing on the protonation-deprotonation dynamics of its amino groups, which enable precise pH-triggered drug release in acidic tumor microenvironments. Key advancements in nanocarrier design, including crosslinking strategies, ligand functionalization (e.g., glycyrrhizin for hepatocellular carcinoma, bisphosphonates for osteosarcoma), and integration with porous materials (mesoporous silica, MOFs), are discussed, highlighting enhanced tumor targeting and reduced systemic toxicity. Applications in liver and bone cancers demonstrate improved therapeutic efficacy through controlled drug encapsulation and stimuli-responsive release mechanisms. Characterization techniques (DLS, FTIR, TEM/SEM) and toxicological considerations are reviewed, alongside the future prospect for developing multi-targeted, multi-stimuli-responsive platforms activated by internal (pH, redox, enzymes) and optional external (light, heat) cues. By addressing current challenges and opportunities, this work underscores the potential of chitosan-based nano systems to revolutionize precision oncology.

Keywords Chitosan · pH-responsive drug delivery · Targeted cancer therapy · Glycyrrhizin · Bisphosphonates

---

S. M. Badshah · N. V. Mdlovu · K.-S. Lin (✉)
Department of Chemical Engineering and Materials Science, Yuan Ze University, Chung-Li District, Taoyuan, Taiwan
e-mail: kslin@saturn.yzu.edu.tw

S. Jana (ed.), *Chitosan-Based Drug Delivery Systems*, AAPS Introductions in the Pharmaceutical Sciences 16, https://doi.org/10.1007/978-3-032-08347-0_7

1 Introduction of Chitosan-Based Nanomaterials

1.1 Overview of Chitosan: Structure, Sources, and Properties

Chitosan is a naturally derived polysaccharide obtained through the partial deacetylation of chitin, the second most abundant biopolymer in nature after cellulose [1]. Chitin is primarily found in the exoskeletons of crustaceans such as crabs, shrimp, and lobsters, as well as in the cell walls of fungi and insects [2]. The conversion of chitin to chitosan involves a deacetylation process, typically under alkaline conditions, which results in a polymer predominantly composed of β-(1→4)-linked D-glucosamine and N-acetyl-D-glucosamine units [3, 4].

The degree of deacetylation (DDA), molecular weight, and pattern of acetylation critically influence the physicochemical properties of chitosan, such as solubility, charge density, and viscosity [5]. Chitosan is a cationic polymer, and this characteristic distinguishes it from most natural polysaccharides, which are typically neutral or anionic [5, 6]. Its amino groups become protonated under acidic conditions (pK_a ~6.3), giving chitosan its unique solubility profile—soluble in acidic aqueous media and insoluble at neutral or alkaline pH [7, 8]. Structurally, chitosan's linear polymeric chain with functional amino groups offers numerous sites for chemical modifications and conjugation, making it a versatile platform for constructing nanomaterials for biomedical use [9, 10]. Its ability to form gels, nanoparticles, and films further extends its applicability in drug delivery systems [11].

1.2 Biocompatibility, Biodegradability, and Muco-Adhesiveness

One of the most compelling attributes of chitosan is its excellent biocompatibility, which refers to its ability to perform desired functions without eliciting undesirable local or systemic effects in the body [12]. Numerous in vitro and in vivo studies have demonstrated that chitosan is non-toxic to human cells and is generally well tolerated upon administration via various routes, including oral, nasal, ocular, transdermal, and parenteral delivery [13, 14]. Biodegradability is another significant advantage of chitosan. It is enzymatically degraded by lysozyme and other glycosidases in the body into non-toxic, absorbable oligosaccharides and glucosamine, which can be further metabolized or excreted [15]. This natural degradation profile minimizes the risk of accumulation and long-term toxicity, which is a critical requirement for materials used in drug delivery and tissue engineering [16]. Muco-adhesiveness arises from chitosan's cationic nature, which facilitates electrostatic interactions with the negatively charged sialic acid residues and mucins on mucosal surfaces [17]. This interaction prolongs the residence time of chitosan-based formulations at absorption sites such as the gastrointestinal tract, nasal cavity, or ocular tissues, thereby enhancing drug absorption and bioavailability [18].

1.3 Advantages in Drug Delivery Applications

Chitosan offers numerous advantages in drug delivery applications due to its unique physicochemical and biological properties [19]. Its pH-responsiveness, stemming from its cationic nature and solubility in acidic conditions, enables the development of intelligent delivery systems that release drugs in response to acidic microenvironments typical of tumors, inflamed tissues, or intracellular organelles [20]. Chitosan-based nanomaterials can encapsulate a wide variety of therapeutic agents, including hydrophilic and hydrophobic drugs, proteins, peptides, and nucleic acids, allowing for controlled and sustained drug release through various formulation strategies such as ionic crosslinking or polymer blending [21]. Additionally, the presence of reactive amino and hydroxyl groups permits chemical modification and surface functionalization with targeting ligands (e.g., folic acid, peptides, antibodies), enhancing site-specific delivery and minimizing systemic side effects [22]. Chitosan also acts as a permeation enhancer by transiently opening tight junctions between epithelial cells, thereby improving paracellular transport and bioavailability, particularly for orally administered drugs [23]. Furthermore, its ease of processing into diverse forms such as nanoparticles, hydrogels, micelles, and films under mild conditions supports the incorporation of sensitive bioactives without compromising their activity [24]. Collectively, these attributes make chitosan a versatile and powerful platform for designing advanced drug delivery systems tailored to a wide range of therapeutic needs [25].

2 pH-Responsive Mechanisms in Drug Delivery Systems

pH-responsive drug delivery systems have emerged as a promising strategy to achieve site-specific and controlled release of therapeutic agents [26]. This approach capitalizes on the natural pH gradients present in different anatomical and pathological environments of the human body [27]. By designing materials that can respond to these pH shifts, drug carriers can remain stable during circulation and then release their payload precisely at the target site, thereby enhancing therapeutic efficacy while minimizing systemic toxicity [28].

2.1 Physiological pH Variations in the Human Body

The human body exhibits diverse pH microenvironments that can be strategically targeted for drug release [29]. For example, the gastrointestinal (GI) tract features distinct pH zones: the stomach maintains a highly acidic pH ranging from 1.5 to 3.5, while the small intestine has a more neutral to slightly basic pH (6.5–7.5), and the colon has a mildly acidic pH around 6.0–7.0 [30]. Similarly, within cells, acidic organelles such as endosomes (pH ~5.5–6.5) and lysosomes (pH ~4.5–5.0) present microenvironments suitable for intracellular drug release [31]. Pathological conditions also exhibit pH anomalies. For instance, the extracellular pH of tumor tissues typically ranges from 6.0 to 7.0 due to increased glycolysis and lactic acid production, which contrasts with the near-neutral pH (~7.4) of healthy tissue [32]. Likewise, inflamed or infected tissues often exhibit localized acidosis. These physiological and pathological pH gradients provide opportunities for smart, site-specific drug delivery systems that respond to the microenvironment for precise drug unloading [33].

2.2 Chemistry of pH Sensitivity in Chitosan and Derivatives

Chitosan, a linear polysaccharide composed of β-(1→4)-linked D-glucosamine and N-acetyl-D-glucosamine units, is inherently pH-sensitive due to the presence of primary amino groups on the D-glucosamine residues [34]. These amine groups can undergo protonation under acidic conditions, imparting chitosan with a cationic nature and water solubility [35]. At higher pH values (above its pK_a ~6.3), deprotonation of these amino groups occurs, rendering chitosan insoluble and facilitating aggregation, gelation, or phase separation [36]. This reversible pH-sensitive behavior makes chitosan particularly useful for constructing delivery systems that can load drugs at neutral or basic pH and release them under acidic conditions [37].

Moreover, chemical modification of chitosan expands its pH responsiveness and physicochemical stability [38]. For example, the introduction of carboxyl groups (e.g., in carboxymethyl chitosan) can provide amphoteric behavior, allowing responsiveness to both acidic and basic pH levels [39]. Other derivatives, such as N-succinyl chitosan, thiolated chitosan, and acylated or PEGylated chitosan, allow for more sophisticated tuning of solubility, charge density, and degradation profile, enabling tailored responses to specific pH windows found in targeted tissues or intracellular compartments [40]. These modifications are crucial in overcoming the solubility limitations of native chitosan and in enhancing drug encapsulation efficiency, muco-adhesiveness, and biological performance [41].

2.3 *Protonation–Deprotonation Behavior of Amino Groups*

The key mechanism underlying the pH responsiveness of chitosan is the protonation-deprotonation behavior of its free amino groups [42]. Under acidic conditions (pH < pKa ~6.3), these $-NH_2$ groups become protonated to $-NH_3^+$, leading to increased solubility and a positive surface charge [43]. This cationic nature not only improves solubility but also enhances the interaction with negatively charged biological surfaces such as cell membranes or mucosal tissues, promoting cellular uptake and muco-adhesion [44]. In contrast, under neutral or basic conditions, the amino groups revert to their unprotonated $-NH_2$ form, which results in reduced solubility and often leads to nanoparticle aggregation or gelation, triggering the release of the encapsulated drug payload [45].

This protonation–deprotonation dynamic can be harnessed in a variety of nanocarrier designs [46]. For instance, in chitosan nanoparticles, drugs can be loaded and retained at neutral pH but released rapidly in the acidic environments of tumors, endosomes, or the stomach [47]. In hydrogels and films, swelling or contraction behavior in response to pH changes can regulate drug diffusion rates [48]. Moreover, chitosan's ability to act as a polyelectrolyte allows it to form complexes with anionic drugs or nucleic acids, and these complexes can dissociate in acidic environments due to increased repulsion between protonated chitosan chains, facilitating release [49].

3 Design and Synthesis of Chitosan-Based pH-Responsive Nanocarriers

The successful development of chitosan-based pH-responsive drug delivery systems hinges on meticulous design and synthesis of nanocarriers that are capable of encapsulating therapeutic agents, protecting them from premature degradation, targeting specific tissues, and releasing the payload in response to pH changes [50, 51]. This section elaborates on key design principles and synthetic strategies that underpin the fabrication of functional chitosan nanocarriers, focusing on crosslinking methods, surface functionalization, and drug encapsulation-releasing mechanisms [52].

3.1 *Crosslinking Strategies*

Crosslinking is a fundamental step in stabilizing chitosan nanoparticles and tailoring their responsiveness to physiological stimuli [53]. There are two primary types of crosslinking used in the synthesis of chitosan nanocarriers: ionic (physical) crosslinking and covalent (chemical) crosslinking [54]. Ionic crosslinking, often performed using agents like tripolyphosphate (TPP), relies on electrostatic interactions between the positively charged amino groups of chitosan and multivalent anions [55]. This method is widely used due to its simplicity, mild reaction conditions, and preservation of drug activity [55, 56]. It enables the formation of nanoparticles or gels without requiring toxic solvents or harsh reagents, making it suitable for biomedical applications [57]. On the other hand, covalent crosslinking, involves the formation of stable chemical bonds using bifunctional reagents such as glutaraldehyde, genipin, or carbodiimide-based agents (e.g., EDC/NHS) [58]. While it provides improved mechanical strength and long-term stability, covalent crosslinking must be carefully controlled to avoid cytotoxicity from residual crosslinkers [59]. Genipin is particularly favored due to its lower toxicity and ability to form biocompatible networks. The extent of crosslinking directly affects particle size, porosity, degradation rate, and drug release kinetics [60]. Furthermore, pH-sensitive crosslinking systems can be designed by incorporating labile linkers such as Schiff bases, hydrazone bonds, or acid-cleavable acetal and imine bonds, which degrade in acidic conditions. This facilitates smart drug release at targeted sites such as tumors or intracellular organelles [60, 61].

3.2 Functionalization with Targeting Ligands

To enhance the specificity and efficacy of chitosan-based nanocarriers, surface functionalization with targeting ligands is a widely adopted strategy [62]. The reactive amino and hydroxyl groups on chitosan allow for the covalent attachment of a variety of ligands that can recognize and bind to overexpressed receptors on diseased cells or tissues [63]. Common targeting ligands include folic acid, which targets folate receptors frequently overexpressed in many cancers; peptides (e.g., RGD sequences) that bind integrins on tumor vasculature; antibodies or antibody fragments for receptor-specific targeting; aptamers with high binding affinity to cancer biomarkers; and lectins, which can target glycoproteins on mucosal tissues [62–64]. The conjugation of these ligands to chitosan can be achieved via carbodiimide chemistry, click reactions, or other coupling strategies, ensuring that the bioactivity of both the polymer and the ligand is retained [65]. Targeted nanocarriers not only improve therapeutic index by enhancing accumulation at the disease site through receptor-mediated endocytosis but also reduce off-target effects and systemic toxicity. Additionally, multifunctional nanocarriers may combine both targeting moieties and imaging agents to enable theranostic applications [65, 66].

3.3 Encapsulation and Release Profiles

A critical aspect of chitosan-based nanocarrier design is the efficient encapsulation of therapeutic agents and the ability to control their release in response to pH stimuli

[67]. Chitosan can encapsulate a broad spectrum of drugs, including small molecules, proteins, peptides, and nucleic acids. The encapsulation method, whether ionic gelation, emulsion-solvent evaporation, or self-assembly, affects drug loading efficiency, particle morphology, and stability [67, 68].

The drug release profile from chitosan-based carriers is significantly influenced by environmental pH. In neutral or basic conditions (e.g., bloodstream), chitosan remains insoluble and retains the payload within its matrix [69]. Upon exposure to acidic environments (e.g., tumor extracellular matrix, endosomes, or the stomach), the amino groups become protonated, leading to polymer swelling, increased solubility, and subsequent drug release [70]. This pH-triggered release ensures that drugs are preferentially released at the site of action, improving therapeutic outcomes [68–70]. In addition, the release profile can be modulated by adjusting the degree of crosslinking, the nature of drug–polymer interactions, particle size, and the incorporation of co-polymers or responsive linkers [71]. For example, chitosan-hyaluronic acid or chitosan-alginate blends may alter release kinetics, while hydrophobic modification can enhance drug loading and prolong release [72].

4 Chitosan-Based Systems Using Nanosized Porous Materials

To enhance the performance of pH-responsive drug delivery systems, chitosan is increasingly being integrated with nanosized porous materials such as mesoporous silica nanoparticles (MSNs), metal-organic frameworks (MOFs), and zeolites [73]. These hybrid systems combine the structural versatility and responsiveness of chitosan with the high surface area, tunable porosity, and exceptional loading capacities of inorganic or hybrid porous nanomaterials [73, 74]. Such platforms offer significant advantages in terms of drug encapsulation efficiency, protection of labile drugs, and fine control over release kinetics in response to pH and other stimuli [75].

4.1 Incorporation with Mesoporous Silica, MOFs, and Zeolites

The MSNs are among the most widely used inorganic carriers due to their uniform and adjustable pore sizes (typically 2–50 nm), large surface areas (>800 m^2/g), and excellent biocompatibility [76]. Functionalization of MSNs with chitosan creates a hybrid shell–core architecture wherein the chitosan coating serves as a pH-responsive gatekeeper [77]. In acidic environments, protonation of chitosan leads to swelling or partial dissolution of the shell, triggering the release of drug molecules stored in the pores of the silica core [78].

The MOFs are crystalline materials composed of metal ions/clusters coordinated with organic ligands [79]. Their exceptionally high surface areas (often exceeding 1000 m^2/g), controllable pore environments, and tunable surface chemistry make them ideal for drug loading [80]. When combined with chitosan, either by coating the MOF particles or by forming composites, the resulting systems benefit from the pH-sensitive release properties of chitosan and the high payload capacity of MOFs. Some MOFs also exhibit intrinsic biodegradability under acidic or reducing conditions, which further complements the pH-responsiveness of chitosan [80, 81].

Zeolites, which are microporous aluminosilicates, offer high thermal and chemical stability and can accommodate small molecules within their pore systems [82]. Although their pore size (<2 nm) limits the encapsulation of large biomolecules, their integration with chitosan enhances drug retention and modulates release behavior [83]. Surface modification of zeolites with chitosan or chitosan derivatives imparts mucoadhesive and stimuli-responsive characteristics, extending their applicability in biomedical fields [82–84].

4.2 Drug Loading Efficiency and Release Modulation

One of the primary motivations for incorporating chitosan with nanosized porous materials is to maximize drug loading efficiency while achieving precise control over drug release [85]. The porous structures provide abundant internal space for high drug payloads, while the chitosan component acts as a molecular gate that responds to pH stimuli. Drug molecules can be physically adsorbed, electrostatically bound, or covalently tethered within the porous matrix, and their release can be delayed, sustained, or triggered by pH variations [85, 86].

The pH-triggered release is typically facilitated by the swelling or degradation of chitosan under acidic conditions [87]. For instance, at low pH, protonation of chitosan's amino groups leads to increased solubility and permeability, which opens the pores or weakens the polymer barrier, allowing the encapsulated drugs to diffuse out [88]. In more neutral or basic conditions, the deprotonated chitosan remains intact, thereby restricting premature release [89]. The presence of chitosan also improves drug retention by forming hydrogen bonds or ionic interactions with the drug molecules, especially those with acidic or anionic functional groups [90]. Furthermore, the degree of crosslinking, thickness of the chitosan coating, and the physicochemical properties of the incorporated porous materials all contribute to tuning the drug release profile, enabling the design of systems that exhibit zero-order, biphasic, or sustained release behaviors [89–91].

4.3 Synergistic Effects on Responsiveness

The integration of chitosan with nanosized porous materials leads to synergistic enhancements in responsiveness, surpassing the capabilities of either component alone [92]. Chitosan provides environmental sensitivity especially to pH while the porous materials offer high capacity and structural control [93]. Together, they create multifunctional systems capable of responding not only to pH but also to secondary stimuli such as temperature, redox conditions, enzymatic activity, or magnetic fields (when incorporating magnetic MOFs or mesoporous iron oxide) [93, 94].

Additionally, this synergy can improve biocompatibility and stability. Chitosan can protect sensitive porous materials such as silica and MOFs from premature degradation in physiological environments and reduce their potential toxicity [95]. It also enhances cellular uptake and muco-adhesion, facilitating localized delivery and cellular internalization [96]. Some designs exploit this synergy by incorporating targeting ligands on the chitosan shell, while using the porous core for multi-drug loading or combination therapy, making these hybrid systems highly adaptable for complex therapeutic regimens, such as cancer treatment, gene delivery, or vaccine platforms [95–97].

5 Characterization Techniques for Chitosan-Based Systems

The effective development of chitosan-based pH-responsive nanocarriers necessitates comprehensive physicochemical and functional characterization [98]. These analytical methods are critical for confirming the successful synthesis, structural integrity, and functional performance of the nanomaterials [99]. This section outlines key techniques used to evaluate particle size, surface charge, morphology, chemical composition, thermal behavior, and pH-sensitive drug release profiles [98–100].

5.1 Particle Size, Zeta Potential, and Surface Morphology

Particle size plays a pivotal role in determining the biological fate of nanocarriers, including their cellular uptake, circulation time, and tissue penetration [101]. Dynamic light scattering (DLS) is commonly used to assess the hydrodynamic diameter and polydispersity index (PDI) of chitosan nanoparticles in aqueous media [102]. An optimal size range of 100–300 nm is typically preferred for systemic drug delivery, allowing for enhanced permeability and retention (EPR) in tumor tissues while avoiding rapid renal clearance [103].

In addition, the zeta potential measurements, derived from electrophoretic light scattering, provide insights into the surface charge of the nanoparticles, which influences colloidal stability, cellular interaction, and muco-adhesiveness

[104]. Chitosan-based systems usually exhibit a positive surface charge in acidic conditions due to the protonation of amino groups, enhancing their interaction with negatively charged cell membranes or mucosal surfaces [105]. A zeta potential above ±30 mV generally indicates good stability against aggregation [106].

Similarly, the surface morphology and shape are typically evaluated using electron microscopy techniques such as scanning electron microscopy (SEM) and transmission electron microscopy (TEM) [107]. The SEM provides topographical information and surface texture, while TEM offers high-resolution imaging of internal structures and particle dispersion. These methods help distinguish among spherical, rod-like or aggregated particles and reveal structural changes before and after drug loading or pH exposure [108].

5.2 FTIR, XRD, TEM/SEM, and Thermal Analysis

In order to identify functional groups and confirm the successful chemical modification or conjugation of chitosan with targeting ligands, crosslinkers or drugs, the Fourier-transform infrared spectroscopy (FTIR) is employed [109]. Characteristic peaks associated with amine (–NH_2, hydroxyl (–OH), and glycosidic bonds provide evidence of structural integrity, while new peaks or shifts can indicate interactions with loaded drugs or grafted moieties [110]. Similarly, the X-ray diffraction (XRD) is used to determine the crystalline or amorphous nature of the chitosan matrix and its composites [111]. Native chitosan typically exhibits semi-crystalline behavior, while interactions with drugs or incorporation with porous materials such as mesoporous silica or MOFs can alter the diffraction patterns. A loss or broadening of crystalline peaks may suggest successful drug encapsulation or molecular dispersion [111, 112].

To provide valuable information on the thermal stability and decomposition behavior of the nanoparticles, the thermogravimetric analysis (TGA) and differential scanning calorimetry (DSC) are used [113].The TGA measures weight loss as a function of temperature and can help quantify moisture content, organic loading, and degradation profiles. DSC assesses thermal transitions such as melting points, glass transitions, or drug–polymer interactions, confirming the physical state of the encapsulated drug (e.g., crystalline vs. amorphous) [111–113].

5.3 pH-Triggered Release Behavior and In Vitro Testing

The hallmark of chitosan-based pH-responsive systems is their pH-triggered drug release behavior, which is typically studied using in vitro release experiments under simulated physiological and pathological conditions [114]. Drug-loaded nanocarriers are suspended in buffers of varying pH such as pH 7.4 for blood and pH 5.5–6.5 for tumor or endosomal conditions, pH 1.2 for gastric fluids) and incubated at 37 °C. Samples are periodically withdrawn and analyzed typically by UV-Vis spectroscopy or high-performance liquid chromatography (HPLC)to determine the cumulative drug release over time [114, 115]. These studies confirm whether the system remains stable under physiological pH and releases its payload in acidic conditions. pH-dependent release kinetics can be modeled using zero-order, first-order, Higuchi or Korsmeyer–Peppas models to predict release mechanisms such as diffusion, erosion, or swelling [114–116].

In vitro cytotoxicity and biocompatibility assessments are also essential. Common assays include MTT, WST-1 or Alamar Blue, which measure cell viability upon exposure to nanocarriers [117]. These tests are often performed on relevant cell lines such as HeLa, Caco-2, and MCF-7 to validate safety and drug efficacy. Additional functional assays such as cellular uptake such as using fluorescence microscopy or flow cytometry, hemolysis testing, and trans-epithelial transport studies help determine the biological behavior and delivery efficiency of the system [117, 118]. Together, these characterization techniques provide a robust framework for evaluating the physicochemical properties, functional performance, and therapeutic potential of chitosan-based pH-responsive nanocarriers, supporting their further development for clinical translation [117–119].

6 Chitosan-Based Nanomaterials for Liver Cancer Therapy

Carbohydrate-based biodegradable polymers have gained significant attention in the development of nanocarrier systems, primarily due to their inherent biocompatibility, low toxicity, satisfactory physicochemical stability, and the versatile potential for chemical modification and functionalization. Compared to synthetic polymers, these natural materials offer enhanced biological safety and tunability, making them particularly suitable for biomedical applications, including targeted drug delivery and cancer therapy [120]. To address these challenges, the development of advanced drug delivery systems using biocompatible materials has garnered significant interest. Among these, chitosan, a natural, biodegradable, and non-toxic polysaccharide derived from chitin has emerged as a promising nanomaterial for cancer therapies as shown in Fig. 1. Due to its favorable properties, including muco-adhesion, ease of chemical modification, and ability to encapsulate a wide range of therapeutic agents, chitosan-based nanocarriers can improve drug solubility, enhance bioavailability, enable controlled release, and facilitate targeted delivery to tumor tissues [121, 122].

Chitosan is a linear cationic polysaccharide composed of β (1→4) linked D-glucosamine and N-acetyl-D-glucosamine units, as illustrated in Fig. 2. It is obtained through the partial deacetylation of chitin, a naturally abundant biopolymer found in the exoskeletons of crustaceans, insects, and the cell walls of fungi [123]. Chitosan exhibits several advantageous physicochemical and biological properties, including solubility in mildly acidic aqueous environments, the presence of reactive amino and hydroxyl groups amenable to chemical modification and crosslinking, and resistance to oxidative degradation [124]. One of the most

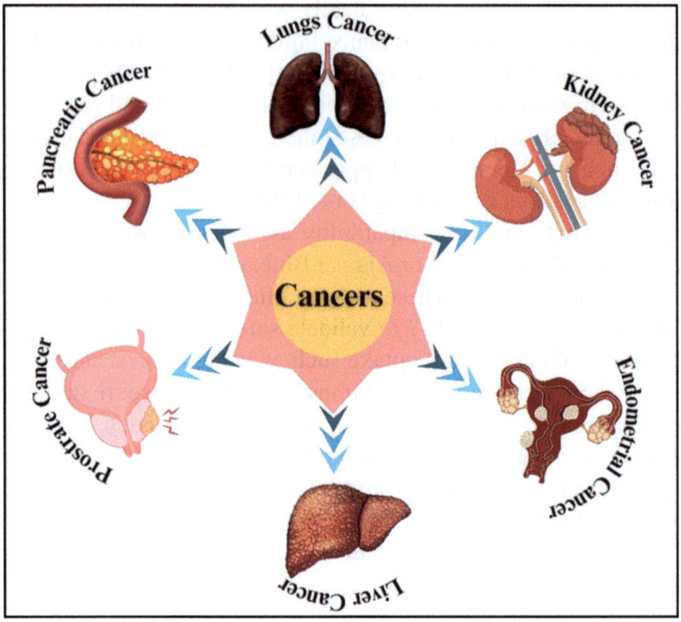

Fig. 1 Types of chitosan used for a variety of cancer therapy

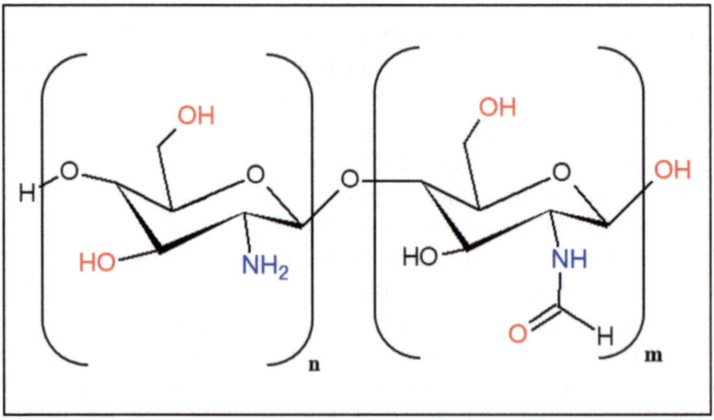

Fig. 2 Structure of chitosan

promising areas for chitosan-based materials is drug delivery, where it serves as a versatile carrier for both small molecules and biotherapeutics such as proteins, peptides, and nucleic acids. Chitosan's cationic nature enables it to form polyelectrolyte complexes with anionic drugs or biological membranes, thereby enhancing cellular uptake, controlled release, and targeted delivery [125]. Recently, researchers have evaluated extensive research on targeted drug delivery of chitosan-based nanomaterials to trigger liver cancer. Owing to this, liver cancer remains one of

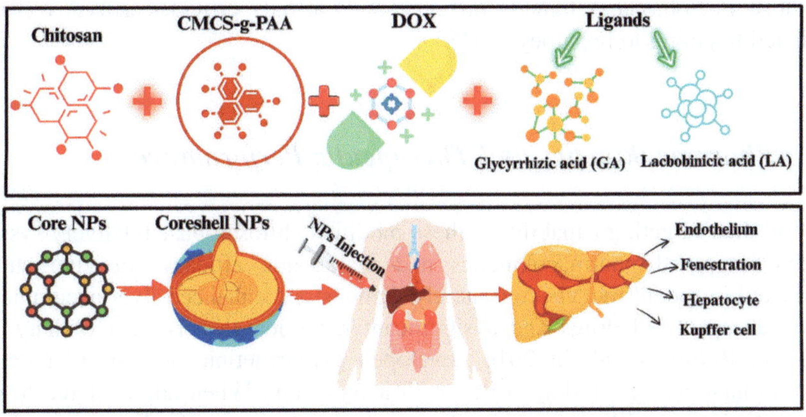

Fig. 3 Glycyrrhetinic acid and LA grafted carboxymethyl chitosan-polyacrylate nano shell delivers DOX, targeting liver cancer cells for anti-tumor effects

the most aggressive malignancies worldwide, with limited treatment options and poor prognosis in advanced stages [126]. Conventional chemotherapeutic approaches often suffer from systemic toxicity, low bioavailability, and a lack of tumor specificity. In this context, nanotechnology has emerged as a promising strategy to improve therapeutic outcomes by enabling targeted drug delivery and controlled release [127].

Chitosan nanoparticles can be functionalized with liver-targeting ligands to improve selective delivery to hepatocellular carcinoma (HCC). Lactobionic (LA) has hepatocellular carcinoma (HCC) cell targeting by binding to Asiatica protein (ASGP) receptors that are overexpressed on the HCC cell surface [128]. The grafting of LA and chitosan can enhance the HCC cell targeting ability of chitosan. In 2020, Hefnawy et al. reported dual-ligand designs using glycyrrhetinic acid (GA) and lactobionic acid (LA, a galactose derivative) as illustrated in Fig. 3. Chitosan polyacrylate core–shell NPs grafted with GA and LA efficiently bind hepatocyte receptors (GA receptor and ASGP receptors) to make a two-ligand nano shell structure for the delivery of DOX [126]. The GA and LA double ligands enhance the HCC cell targeting ability of core-shell nanoparticles and precisely release DOX into tumor cells to achieve their anti-tumor effects. In vitro, these dual-targeted NPs loaded with doxorubicin induced marked apoptosis in HCC cells (upregulating caspases and p53) [129]. Likewise, Lin et al. showed that glycyrrhizin-conjugated chitosan dramatically enhanced uptake in HCC cells: glycyrrhizin-N-caproyl chitosan NPs had a ~10-fold higher internalization in SMMC-7721 cells than unmodified NPs [130]. These NPs also preferentially accumulated in HCC tumors in vivo. Dual-ligand GA/galactose-modified chitosan (e.g., GA + a galactose-terminated polymer) further improves selectivity such carriers specifically internalizing hepatoma cells in vitro and leading to enriched distribution in liver tumors in animal models [129]. Inclusive, leveraging GA/glycyrrhizin and galactose (lactobionate) ligands on chitosan scaffolds provides active, receptor-mediated targeting to hepatocytes [131].

6.1 Pharmacokinetics and Therapeutic Performance

Beyond their targeting capabilities, these modified chitosan-based delivery systems exhibit improved pharmacokinetics and therapeutic efficacy. In vivo studies conducted in rodent models consistently demonstrate enhanced liver accumulation of the encapsulated drugs and a significant reduction in off-target toxicity. For example, Rohilla et al. in 2016 developed glycyrrhetinic acid (GA) modified chitosan nanoparticles (NPs) loaded with atorvastatin. When administered to rats, these GA-chitosan NPs showed significantly higher liver-specific drug accumulation and substantially reduced hepatotoxicity compared to unmodified (plain) chitosan nanoparticles, as shown in Fig. 4a [132]. In a similar approach, glycyrrhizin-functionalized chitosan NPs were used to deliver lamivudine. This formulation

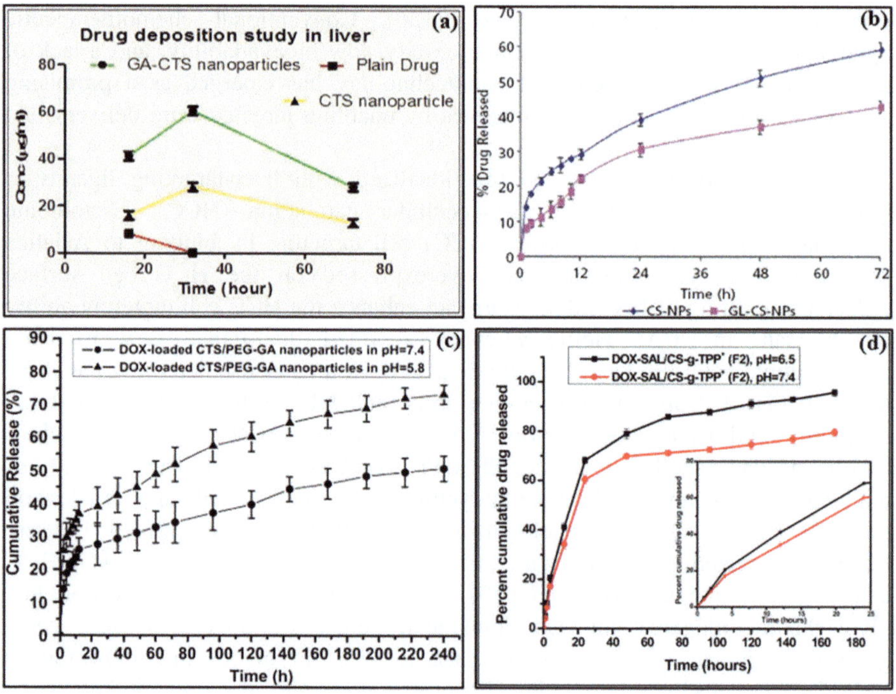

Fig. 4 Drug release and liver targeting performance of chitosan-based nanoparticles. (**a**) GA-modified chitosan NPs show superior liver deposition over time [132]; (**b**) Glycyrrhizin-modified CS-NPs exhibit sustained drug release [133]; (**c**) CTS/PEG-GA NPs display pH-responsive release favoring acidic tumor environments [131]; (**d**) DOX-SAL/CS-g-TPP$^+$ NPs demonstrate higher release at pH 6.5 vs. 7.4 [134]. Results presented as mean ± SD ($n = 3$)

notably prolonged the drug's retention time in hepatic tissue, enhancing its therapeutic potential against liver-targeted conditions. In sum, glycyrrhizin-chitosan NPs loaded with lamivudine extended hepatic drug residence time in rats as revealed in Fig. 4b [133]. A striking example is Tian et al. (2010) who injected GA-chitosan/PEG NPs loaded with doxorubicin into HCC-bearing mice: 51.3% of the dose localized in the liver within 3 h, about 2.6 fold higher than controls, as shown in Fig. 4c [131]. At the cellular level, further refinement of targeting strategies has been achieved by Arafa et al. in 2020, engineered chitosan nanoparticles conjugated with triphenylphosphine (TPP), a moiety known for mitochondrial targeting. These DOX-loaded TPP–chitosan NPs, with a particle size ranging from 70 to 110 nm, demonstrated a sustained drug release profile over 7 days and induced apoptosis in HCC cells at a rate 7.9 times higher than that of free DOX as depicted in Fig. 4d [134].

6.2 *Case Studies and Experimental Evidence*

Recent studies provide concrete data on chitosan carriers in HCC models. Table 1 summarizes representative formulations, targeting motifs, and outcomes. Glycyrrhizin-modified chitosan (e.g., N-caproyl or O-carboxymethyl CS) consistently shows enhanced HCC uptake and tumor suppression [135]. Dual-ligand systems (GA+LA) have demonstrated synergistic targeting and potent in vitro tumor cell killing [126]. Importantly, in vivo data show improved pharmacokinetics and safety: GA chitosan NPs preferentially delivered the drug to the liver with reduced systemic exposure [131, 132]. These findings underscore chitosan's utility as a liver-targeted nanoplatform.

7 Chitosan-Based Nanomaterials for Bone Cancer Therapy

Bone cancers, including primary tumors such as osteosarcoma and secondary bone metastases, present significant therapeutic challenges due to the complex physiology of bone tissue and the limited efficacy of conventional treatments [140]. Systemic chemotherapy, often used as a standard intervention, lacks specificity to bone tissue, leading to off-target toxicity and suboptimal drug concentrations at tumor sites [141]. To address these limitations, nanotechnology-based drug delivery systems have gained prominence, offering the potential for targeted, sustained, and efficient delivery of therapeutics. Among the various nanomaterial platforms explored, chitosan-based nanocarriers stand out due to their intrinsic biocompatibility, biodegradability, mucoadhesive properties, and ease of chemical modification [142]. A key strategy in bone-targeted therapy is the functionalization of nanocarriers with moieties that exhibit high affinity for hydroxyapatite (HAp), the primary mineral component of bone [143]. Ligands such as bisphosphonates (BPs), notably

Table 1 Examples of chitosan-based nanocarriers for liver cancer

Ligand/modification	Chitosan type	Payload	Model (in vitro/in vivo)	Key findings	References
Glycyrrhizin (licorice-derived)	N-caproyl CS	Model drug	SMMC-7721 cells; H22 mouse tumors	~10× higher uptake in HCC cells; ↑ tumor targeting in mice	[136]
Glycyrrhizin (GA)	O-carboxymethyl CS	Lamivudine	Rat model	Prolonged hepatic residence; higher AUC in liver	[133]
Glycyrrhetinic acid (GA)	GA-CS/PEG	Doxorubicin	H22 murine HCC in BALB/c mice	51.3% liver accumulation at 3 h (~2.6× control)	[131]
GA + Lactobionic acid (dual ligand)	CMCS/polyacrylate core-shell	Doxorubicin	HCC cell culture	Enhanced cell uptake; strong induction of apoptosis	[126]
Triphenylphosphine (mitochondria-targeted)	TPP-grafted CS	Doxorubicin	HCC cells in vitro	7.9× higher apoptosis in HCC cells vs. free DOX	[134]
Folate (for HCC gene therapy)	Folate-CS (degree deacetylated)	IP-10 plasmid	H22 tumor-bearing mice	Targeted IP-10 delivery inhibited angiogenesis in tumors	[137]
Folic acid	Chitosan nanoparticles	Sorafenib	HepG2 cells	Enhanced targeting and anticancer efficacy	[138]
Biotin	Chitosan nanoparticles	Plasmid DNA	SMMC-7721 cells	Improved gene delivery and expression	[139]
None	Chitosan nanoparticles	Doxorubicin	HepG2 cells	Increased cellular uptake and cytotoxicity	[135]

Ligands are grafted to chitosan (CS) derivatives for hepatocyte targeting. Therapeutic payloads and key results from recent studies are listed (DOX = doxorubicin)

zoledronate and alendronate-poly-aspartate/glutamate peptides (Asp)_n), and phytic acid have demonstrated strong HAp-binding properties, improving drug accumulation in bone tissue [144, 145]. Studies have shown that BP-modified carriers significantly enhance retention in bone tumors and suppress tumor growth more effectively than their unmodified counterparts [146, 147].

7.1 Bone Targeting Ligands and Surface Modifications

A major challenge for bone cancer (e.g., osteosarcoma, bone metastases) is directing drugs to bone tissue. Bone-targeted nanocarriers exploit moieties that strongly bind hydroxyapatite (HAp), the mineral matrix of bone. Chitosan (CS) offers primary amine groups that readily form amide bonds with the carboxylate of bisphosphonates (e.g., alendronate (ALN), zoledronate (ZOL), and others) chemical structures are shown in Fig. 5. Kang et al. (2022) demonstrated that ALN is grafted to CS, the resulting ALN-CS NPs exhibit striking hydroxyapatite (HAp) binding (KD ≈ 0.15 μM vs. 0.75 μM for plain CS NPs), leading to ~5-fold greater bone affinity in vitro. In an orthotopic osteosarcoma mouse model (HOS/MNNG, intratibial), these targeted nanoclusters showed enhanced tumor accumulation and improved delivery efficiency [147]. Similarly, an intratibial HOS/MNNG murine osteosarcoma model, ALN-CS/PLGA NPs loaded with doxorubicin (DOX) accumulated in tumor-adjacent bone at 3 h post-injection (tissue distribution: 62% ID/g in bone vs. 24% ID/g for control PLGA NPs; $p < 0.01$), reducing tumor volume by 68% over 21 days versus 32% in the control group. In orthotopic K7M2 osteosarcoma and bone metastasis models, these nanoparticles significantly suppressed

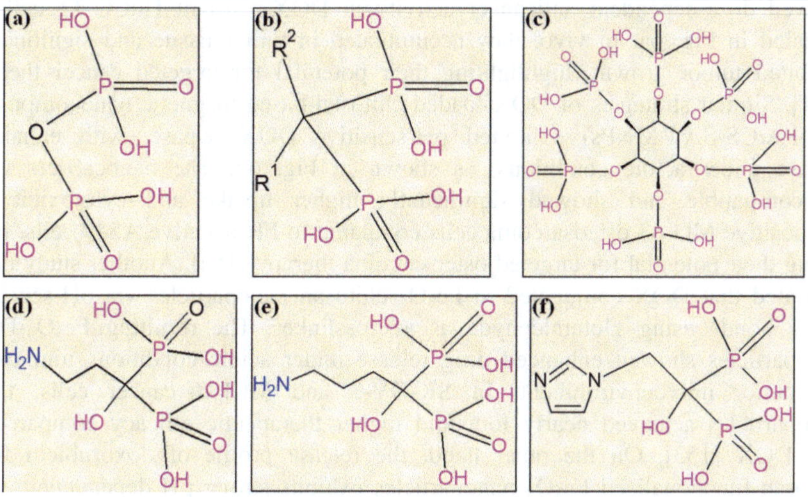

Fig. 5 Molecular structures of targeting ligands of bone tissue engineering. (**a**) Pyrophosphate; (**b**) BPs; (**c**) phytic acid; (**d**) pamidronate; (**e**) ALN; (**f**) ZOL

tumor growth and inhibited metastasis [148]. A study developed, alendronate-modified chitosan NPs prepared via ionic gelation (size: 80–180 nm, DOX EE = 72%) demonstrated pH-responsive release (65% at pH 5.5 vs. 18% at pH 7.4), leading to a 3.2-fold decrease in MG-63 osteosarcoma cell viability compared to free DOX (IC_{50} = 1.8 μg/mL vs. 5.9 μg/mL). While primarily aimed at osteoporosis treatment, the strong affinity of alendronate for bone tissue suggests potential applicability in bone-targeted cancer therapies [149]. Researchers developed a nanoparticle system combining docetaxel and alendronate within chitosan-coated PLGA nanoparticles. In vitro studies on MG-63 osteosarcoma cells showed enhanced cytotoxicity and cellular uptake, indicating the potential of this dual-drug-loaded system for effective osteosarcoma treatment [150]. In chitosan systems, bisphosphonates or acidic polymers can be grafted to CS or its scaffolds to enhance bone affinity. Although specific studies on chitosan along with BP for osteosarcoma are still emerging, these concepts guide ongoing designs of CS-based bone-targeted nanomedicines.

7.2 *Dual pH- and Enzyme-Responsive Systems*

The bone tumor microenvironment (and tumor niches in general) often features acidic pH and proteases (e.g., matrix metalloproteinases). Chitosan's amino groups can be modified into acid-labile bonds (e.g., Schiff bases) or linked via peptide sequences cleavable by tumor enzymes. For instance, Kang et al. developed chitosan-oligosaccharide nanoparticles (ChitoHISss) incorporating histidine and disulfide linkages for pH- and redox-responsive DOX release. These nanoparticles enhanced intracellular DOX delivery under acidic and reductive conditions and showed dose-dependent anticancer activity in DOX-resistant HuCC-T1 cells as revealed in Fig. 6a. In vivo, they accumulated in tumor tissue and significantly inhibited tumor growth, highlighting their potential for targeted cancer therapy [151]. Similar strategies of DOX-loaded chitosan-based magnetic nanocomposites (CS-FA/CS-SA@MNPs) exhibited pH-sensitive DOX release, with enhanced release under acidic conditions, as shown in Fig. 6b. The nanocarriers were cytocompatible and showed significantly higher uptake and cytotoxicity in FR-positive MG-63 osteosarcoma cells compared to FR-negative A549 cells, indicating their potential for targeted osteosarcoma therapy [152]. Another study demonstrated that DOX conjugated to Fe_3O_4 chitosan nanoparticles via pH-sensitive imine bonds using glutaraldehyde as a crosslinker. The resulting Fe_3O_4-DOX nanoparticles showed enhanced drug release under acidic conditions mimicking the tumor microenvironment. In SK-OV-3 and MCF-7 cancer cells, these nanoparticles achieved nearly four-fold higher therapeutic efficacy compared to free DOX [153]. On the other hand, the release profile of doxorubicin from chitosan-functionalized Fe_3O_4 nanoparticles exhibits strong pH dependency, with ~90% release at pH 4.6 and minimal release (~15%) at pH 7.4 over 8 hours as depicted in Fig. 6c. This pH-triggered behavior highlights the nanocarrier's potential

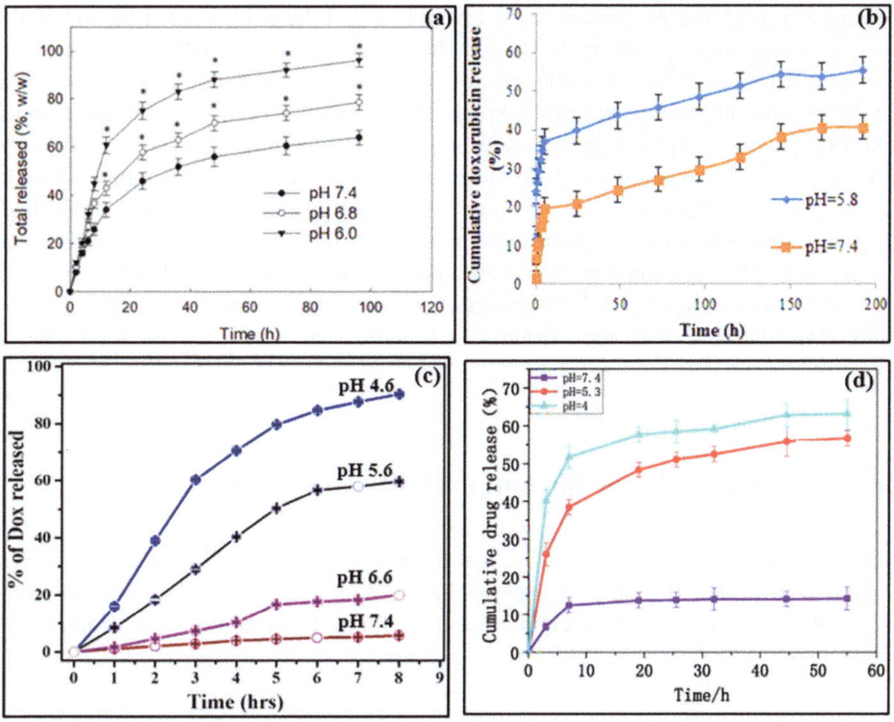

Fig. 6 Drug release pattern of chitosan-based nanoparticles at different pH values. (**a**) Effect of pH on DOX release from ChitoHISss nanoparticles [151]; (**b**) DOX release pattern of CS-FA/CS-SA@MNPs at two different level of pH 5.8 and pH 7.4 [152]; (**c**) The pH effect on drug delivery of Fe_3O_4-Dox nanoparticles at different values of pH (7.4, 6.6, 5.6 and 4.6) [153]; (**d**) Illustration of pH-responsive drug release from Fe_3O_4/CMCS/ALS-DOX nanocomposite at varying pH levels [154]

for targeted drug delivery in acidic tumor microenvironments, enhancing therapeutic specificity and reducing systemic toxicity [154]. Another work revealed that the pH-dependent release of doxorubicin (Dox) from Fe_3O_4/carboxymethyl chitosan/aminated lignosulfonate (Fe_3O_4/CMC/ALS) nanoparticles over 192 hours. At pH 5.8, mimicking the acidic tumor microenvironment, cumulative release exceeded 55%, whereas only ~40% was released at physiological pH 7.4 as shown in Fig. 6d. The enhanced release under mildly acidic conditions underscores the pH-responsive nature of the nanocarrier system, indicating its promise for selective and controlled anticancer drug delivery [155].

7.3 In Vivo and In Vitro Findings

Experimental data on chitosan nanocarriers for bone tumors are still limited, but initial studies are encouraging. Amiryaghoubi et al. in 2022 prepared pH-sensitive chitosan nanoparticles (succinylated CS linked to folic acid) coated on magnetic Fe_3O_4 cores and loaded with DOX. These CS-FA-Fe_3O_4@DOX NPs showed enhanced DOX release at acidic pH and selectively killed folate-receptor-positive MG-63 osteosarcoma cells: cellular uptake and cytotoxicity were markedly higher in MG-63 than in receptor-negative controls [152]. This demonstrates targeted in vitro activity. In vivo, analogous bone-targeting systems (though not chitosan-based) have shown efficacy: e.g., alendronate-PEG-liposomes or ALN modified nanoclusters delivering DOX strongly suppressed tumor growth and metastasis in orthotopic osteosarcoma and bone-metastasis mouse models [147, 148]. By analogy, CS carriers decorated with BPs or acidic peptides are expected to achieve similar targeting. Table 2 lists representative CS nanocarriers relevant to bone cancer therapy.

8 Toxicological and Biocompatibility Considerations

Chitosan is widely recognized for its biocompatibility, making it an attractive material for biomedical applications, including drug delivery. However, a thorough toxicological evaluation is essential, especially for systemic administration and long-term use.

8.1 Hemocompatibility

Most studies report negligible hemolysis or blood toxicity at relevant doses. Chitosan's cationic nature can interact with cells, but in optimized nano formulations (often PEGylated or protonated at physiological pH), no hemolytic activity was observed in standard assays. Similarly, chitosan NPs tend to show selective cytotoxicity toward tumor cells while sparing normal cells [166]. In vivo, chitosan's breakdown into glucosamine is non-acidic and readily metabolized, reducing organ toxicity. For example, long-term rodent feeding studies found very high NOAEL (no-observed-adverse-effect level) (chitosan up to grams/kg). The acute toxicity profiles of chitosan and its oligomers have been well-studied, demonstrating favorable safety margins. Hirano reported an oral LD_{50} of approximately 16 g/kg body weight for chitosan in mice, indicating low acute toxicity [167]. Similarly, chitosan oligomers produced via enzymatic depolymerization were shown to be non-toxic, with no observable clinical signs of toxicity following single oral doses up to 10 g/kg in both male and female Kunming strain mice [168].

Table 2 Examples of chitosan-based nanocarriers for bone-related tumors

Ligand/ modification	Chitosan type	Payload	Model	Key findings	References
Folate (FR targeting)	Succinyl-chitosan + FA	Doxorubicin	MG-63 OS cells (in vitro)	pH-triggered DOX release; high uptake by MG-63 cells; significant cytotoxicity	[152]
Alendronate (bisphosphonate)	Chitosan or liposome + ALN	Doxorubicin	Orthotopic OS mouse model	Targeted nanoparticles inhibited tumor growth and metastases in vivo	[148]
Anti-EGFR antibody	Carboxymethyl chitosan nanoparticles labeled with $Na^{131}I$ (ICED-N)	Doxorubicin (DOX) and Curcumin	MG-63 osteosarcoma cells (3D model)	Enhanced targeting efficacy (513-fold) to EGFR+ cells; increased cytotoxicity and radiosensitivity	[156]
None	Chitosan-loaded hydroxyapatite scaffolds	Chitosan alone	MG-63 osteosarcoma and hFOB osteoblast cells	Decreased osteosarcoma viability by up to 96%; no toxicity to osteoblasts	[157]
Copper ions	Copper-loaded chitosan nanoparticles	Copper	MG-63 osteosarcoma cells	Enhanced anticancer effect; increased ROS generation and caspase-3 expression	[158]
Folic acid	pH-sensitive chitosan-folate magnetic nanoparticles	Doxorubicin (DOX)	Osteosarcoma cells	Targeted delivery; controlled release under acidic conditions	[152]
None	Trimethyl chitosan nanoparticles modified with Poloxamer	Methotrexate	Osteosarcoma cells	Improved drug delivery efficiency; enhanced cytotoxicity	[159]
None	Chitosan-based hydrogel	Pigment epithelium-derived factor (PEDF) gene	Osteosarcoma cells	Sustained gene release; potential for gene therapy applications	[160]
None	Bovine serum albumin-chitosan nanoparticles	EZH2 inhibitor (EPZ6438)	MG-63 osteosarcoma cells; tumor-bearing mice	Induced apoptosis; inhibited tumor growth in vivo	[161]
None	Chitosan nanoparticles	siRNA targeting osteosarcoma cells	In vitro study	Potential siRNA carrier; evaluated inflammatory response	[162]
None	Chitosan-coated iron oxide nanoparticles	None	In vitro study	Potential for magnetic hyperthermia; biocompatible and stable	[163]

(continued)

Table 2 (continued)

Ligand/ modification	Chitosan type	Payload	Model	Key findings	References
None	Chitosan nanoparticles	Doxorubicin (DOX)	Osteosarcoma cells	Improved drug encapsulation efficiency; controlled release	[164]
None	Chitosan-based nanoparticles	Paclitaxel	Osteosarcoma cells	Enhanced solubility and bioavailability of paclitaxel	[165]

Ligands shown are used to enhance bone or tumor cell targeting

8.2 Immunogenicity and Inflammation

Chitosan possesses immunostimulatory properties, which can be both beneficial and limiting. It has been used as a vaccine adjuvant by activating immune cells such as macrophages, partly via NLRP3 inflammasome pathways. This can lead to low-level cytokine induction, which is advantageous in immunotherapy but may require careful control in systemic applications to avoid unwanted inflammation [169]. Chitosan can modulate dendritic cell function. It has been shown to upregulate surface markers like MHC II, CD80, and CD86 via TLR4-dependent pathways, enhancing antigen presentation. However, chitosan alone may not induce cytokine production unless combined with other microbial stimuli [170]. Chitosan and its derivatives have been explored as vaccine adjuvants. They can enhance both humoral and cellular immune responses, including increased antibody titers and T-cell proliferation. Mechanistically, chitosan can induce mitochondrial stress, leading to the activation of the cGAS–STING pathway and subsequent type I interferon production [171]. In wound healing, chitosan-based dressings have demonstrated immunomodulatory properties. They can decrease pro-inflammatory cytokine levels, shorten the inflammatory phase, and accelerate wound closure, making them beneficial in managing chronic wounds [172]. Moreover, Chitosan has been shown to enhance antiviral immune responses by increasing the production of IgG and IgA antibodies against influenza viruses. Additionally, it can stimulate interferon synthesis by macrophages, contributing to its antiviral effects [173].

8.3 Dosage and Regulatory Perspective

Currently, no chitosan-based nanomedicines are FDA-approved for cancer therapy, so all applications are experimental. Chitosan is FDA-recognized (GRAS) as a supplement, but injectable chitosan systems will require full safety testing. Chronic administration of chitosan at high doses may affect fat-soluble vitamin absorption. Researcher eluted that long-term intake of high-dose chitosan (3–9%) alters lipid metabolism and fat-soluble vitamin absorption. The observed reductions in serum vitamin A and E, along with increased $1,25(OH)_2$ vitamin D levels, suggest interference with nutrient bioavailability. Additionally, decreased fat digestion and lowered serum lipids align with chitosan's known lipid-binding properties. These findings emphasize the need for dose optimization in systemic applications to balance therapeutic efficacy with nutritional safety [174]. In practice, chitosan carriers for cancer are used at a much lower percentage of body weight than in feeding studies. Regulatory hurdles include demonstrating batch consistency (molecular weight, degree of deacetylation) and the lack of contaminants.

8.4 Comparison with Other Polymeric Carriers

Compared to synthetic polymers (e.g., PLGA, PEI), chitosan has distinct advantages: its degradation products (glucosamine) are endogenous and non-toxic, whereas polyesters yield acidic monomers. Chitosan is inherently mucoadhesive and can facilitate mucosal absorption (useful for the liver via oral routes). However, unlike inert PEG or PLGA, chitosan's positive charge can cause nonspecific interactions. For example, PLGA–PEG particles are typically neutral and "stealth," while chitosan NPs often require coating to reduce protein adsorption. Immunologically, chitosan may be more stimulatory (as mentioned), whereas synthetic polymers are mostly inert. In terms of cytotoxicity, chitosan NPs often show similar or lower toxicity to normal cells than many cationic polymers (e.g., PEI) [175, 176]. No major systemic toxicity has been reported for chitosan carriers at effective doses, but formal toxicity studies (beyond small animal models) are needed before clinical translation.

9 Implications and Future Directions

The field of stimuli-responsive nanomedicine is rapidly evolving, and chitosan-based platforms are part of this trend. Emerging work focuses on multi-responsive systems that integrate several triggers. For example, the Kang ChitoHISss NPs discussed above simultaneously respond to low pH and high glutathione (a redox cue) in tumors [147]. Such dual (or triple) stimuli designs can achieve more precise spatiotemporal control of drug release. Other groups are combining internal triggers with external ones: e.g., incorporating magnetic cores or photosensitizers into CS carriers so that external magnetic fields or NIR light can heat the nanoparticles and further trigger payload release. Although specific examples of chitosan hyperthermia carriers are still emerging, the strategy is well established in liposomes and polymeric micelles.

9.1 Personalized Medicine and "Smart" Nanoplatforms

Chitosan chemistry allows facile coupling of targeting moieties tailored to an individual's tumor profile. Recent studies outside this review demonstrate multi-ligand targeting, for instance, Feng et al. in 2019 used aptamers for two osteosarcoma targets (CD133 and EGFR) on a nano-carrier, achieving superior OS cell killing [177]. In practice, a future personalized approach might involve assaying a patient's tumor for receptor expression and then customizing a chitosan NP with the optimal ligand (e.g. ASGP-binding glycoconjugate for some HCC, or an OS stem-cell marker aptamer for osteosarcoma).

9.2 *Theranostics and Imaging*

Another trend is combining therapy with diagnostic imaging. Chitosan NPs can be loaded with fluorescent dyes, MRI contrast agents, or radionuclides. Yang et al. in 2022 demonstrated this by loading their dual-responsive NPs with a fluorescent probe (chlorin e6); after intravenous injection, in vivo imaging showed strong nanoparticle accumulation in the tumor site [151]. Such "theranostic" CS platforms would allow clinicians to track biodistribution and adjust dosing in real time, a key feature of smart nanomedicine.

Overall, future chitosan nanocarriers will likely be multi-modal: containing multiple targeting ligands, responding to multiple endogenous stimuli (pH, redox, enzymes), and optionally to external triggers (light, heat). Advances in polymer engineering (e.g., grafting stimuli-cleavable linkers), as well as in biomarker-guided design, will drive these innovations. The integration of machine learning and high-throughput screening can further optimize CS-based nano formulations for individual patient needs. As understanding of the tumor microenvironment deepens, the design of "intelligent" chitosan platforms for personalized liver and bone cancer therapy will become more feasible.

10 Conclusions

Chitosan-based nanoparticles offer a versatile and biocompatible platform for targeted liver and bone cancer therapy. In HCC models, glycyrrhizin- and galactose-modified chitosan's have repeatedly demonstrated enhanced hepatocyte uptake, tumor targeting, and therapeutic efficacy. In bone cancer models, bisphosphonate or acidic peptide modifications similarly enhance bone targeting, and folate- or ligand-conjugated CS NPs have shown selective cytotoxicity in osteosarcoma cells. Chitosan's intrinsic biodegradability and low toxicity (e.g., very high LD_{50}) give it safety advantages over many synthetic carriers. Current findings (both in vitro and in vivo) strongly support the practical relevance of CS nanocarriers: they can deliver chemotherapy preferentially to tumors, spare healthy tissues, and potentially reduce side effects. Looking ahead, the development of multi-stimuli-responsive and personalized chitosan platforms (combining targeted delivery, controlled release, and imaging) promises to advance nanomedicine for liver and bone cancers. Continued research and rigorous safety evaluation will be key to realizing the clinical potential of these "smart" chitosan nano therapies.

References

1. Hisham, F., Akmal, M. M., Ahmad, F., Ahmad, K., & Samat, N. (2024). Biopolymer chitosan: Potential sources, extraction methods, and emerging applications. *Ain Shams Engineering Journal, 15*(2), 102424.

2. Amelia, R., Saptarini, N. M., Halimah, E., Andriani, Y., Nurhasanah, A., Levita, J., & Sumiwi, S. A. (2020). Pharmacology activities and extraction of α-chitin prepared from crustaceans: A review. *Journal of Applied Pharmaceutical Science, 10*(12), 140–149.
3. Muxika, A., Etxabide, A., Uranga, J., Guerrero, P., & De La Caba, K. (2017). Chitosan as a bioactive polymer: Processing, properties, and applications. *International Journal of Biological Macromolecules, 105*, 1358–1368.
4. Khajavian, M., Vatanpour, V., Castro-Muñoz, R., & Boczkaj, G. (2022). Chitin and derivative chitosan-based structures preparation strategies aided by deep eutectic solvents: A review. *Carbohydrate Polymers, 275*, 118702.
5. Wang, H., & Roman, M. (2023). Effects of chitosan molecular weight and degree of deacetylation on chitosan-cellulose nanocrystal complexes and their formation. *Molecules, 28*(3), 1361.
6. Iqbal, Y., Ahmed, I., Irfan, M. F., Chatha, S. A. S., Zubair, M., & Ullah, A. (2023). Recent advances in chitosan-based materials; the synthesis, modifications and biomedical applications. *Carbohydrate Polymers, 321*, 121318.
7. Aranaz, I., Alcántara, A. R., Civera, M. C., Arias, C., Elorza, B., Heras Caballero, A., & Acosta, N. (2021). Chitosan: An overview of its properties and applications. *Polymers, 13*(19), 3256.
8. Ke, C. L., Deng, F. S., Chuang, C. Y., & Lin, C. H. (2021). Antimicrobial actions and applications of chitosan. *Polymers, 13*(6), 904.
9. Yadav, H., Malviya, R., & Kaushik, N. (2024). Chitosan in biomedicine: A comprehensive review of recent developments. *Carbohydrate Polymer Technologies and Applications, 8*, 100551.
10. Bashir, S. M., Ahmed Rather, G., Patrício, A., Haq, Z., Sheikh, A. A., Shah, M. Z. U. H., Singh, H., Khan, A. A., Imtiyaz, S., Ahmad, S. B., & Nabi, S. (2022). Chitosan nanoparticles: A versatile platform for biomedical applications. *Materials, 15*(19), 6521.
11. Ewii, U. E., Attama, A. A., Olorunsola, E. O., Onugwu, A. L., Nwakpa, F. U., Anyiam, C., Chijioke, C., & Ogbulie, T. (2025). Nanoparticles for drug delivery: Insight into *in vitro* and *in vivo* drug release from nanomedicines. *Nano TransMed, 4*, 100083.
12. Rodrigues, S., Dionísio, M., Remunan Lopez, C., & Grenha, A. (2012). Biocompatibility of chitosan carriers with application in drug delivery. *Journal of Functional Biomaterials, 3*(3), 615–641.
13. Desai, N., Rana, D., Salave, S., Gupta, R., Patel, P., Karunakaran, B., Sharma, A., Giri, J., Benival, D., & Kommineni, N. (2023). Chitosan: A potential biopolymer in drug delivery and biomedical applications. *Pharmaceutics, 15*(4), 1313.
14. Vosoughi, P., Naghib, S. M., Jafari, T., & Kangarshahi, B. M. (2025). Chitosan-encapsulated lipid-based nanovesicles for therapeutic applications and tissue engineering: A review. *Carbohydrate Polymer Technologies and Applications, 10*, 100805.
15. Hodde, J. (2002). Naturally occurring scaffolds for soft tissue repair and regeneration. *Tissue Engineering, 8*(2), 295–308.
16. Pamu, S., Prajapati, V. D., Roy, S., & Kondapuram, P. (2025). Bioavailability and toxicity aspects of tissue engineering. In *Electrospun biomaterials for tissue engineering and drug delivery applications* (pp. 521–547). Woodhead Publishing.
17. Mukherjee, B., Satapathy, B. S., Bhattacharya, S., Chakraborty, R., & Mishra, V. P. (2017). Pharmacokinetic and pharmacodynamic modulations of therapeutically active constituents from orally administered nanocarriers along with a glimpse of their advantages and limitations. In *Nano-and microscale drug delivery systems* (pp. 357–375). Elsevier.
18. Lang, X., Wang, T., Sun, M., Chen, X., & Liu, Y. (2020). Advances and applications of chitosan-based nanomaterials as oral delivery carriers: A review. *International Journal of Biological Macromolecules, 154*, 433–445.
19. Azeez, S., Anusha, N., Sathiyaseelan, A., & Nagarajan, S. (2025). Chitosan: A multifaceted biomaterial exploring physicochemical insights and diverse drug delivery applications. *Journal of Drug Delivery Science and Technology, 111*, 107140.

20. Chu, S., Shi, X., Tian, Y., & Gao, F. (2022). pH-responsive polymer nanomaterials for tumor therapy. *Frontiers in Oncology, 12*, 855019.
21. Naskar, S., Sharma, S., & Kuotsu, K. (2019). Chitosan-based nanoparticles: An overview of biomedical applications and its preparation. *Journal of Drug Delivery Science and Technology, 49*, 66–81.
22. Yan, S., Na, J., Liu, X., & Wu, P. (2024). Different targeting ligands-mediated drug delivery systems for tumor therapy. *Pharmaceutics, 16*(2), 248.
23. Rosenthal, R., Günzel, D., Finger, C., Krug, S. M., Richter, J. F., Schulzke, J. D., Fromm, M., & Amasheh, S. (2012). The effect of chitosan on transcellular and paracellular mechanisms in the intestinal epithelial barrier. *Biomaterials, 33*(9), 2791–2800.
24. Rezagholizade-Shirvan, A., Soltani, M., Shokri, S., Radfar, R., Arab, M., & Shamloo, E. (2024). Bioactive compound encapsulation: Characteristics, applications in food systems, and implications for human health. *Food Chemistry: X, 24*, 101953.
25. Pérez-Pacheco, Y., Tylkowski, B., & García-Valls, R. (2025). Chitosan micro/nanocapsules in action: Linking design, production, and therapeutic application. *Molecules, 30*(2), 252.
26. Le, Q. H., Neila, F., Smida, K., Li, Z., Abdelmalek, Z., & Tlili, I. (2023). pH-responsive anticancer drug delivery systems: Insights into the enhanced adsorption and release of DOX drugs using graphene oxide as a nanocarrier. *Engineering Analysis with Boundary Elements, 157*, 157–165.
27. Khalifat, N., Puff, N., Bonneau, S., Fournier, J. B., & Angelova, M. I. (2008). Membrane deformation under local pH gradient: Mimicking mitochondrial cristae dynamics. *Biophysical Journal, 95*(10), 4924–4933.
28. Kurul, F., Turkmen, H., Cetin, A. E., & Topkaya, S. N. (2025). Nanomedicine: How nanomaterials are transforming drug delivery, bio-imaging, and diagnosis. *Next Nanotechnology, 7*, 100129.
29. Gao, W., Chan, J. M., & Farokhzad, O. C. (2010). pH-responsive nanoparticles for drug delivery. *Molecular Pharmaceutics, 7*(6), 1913–1920.
30. San Gabriel, A., & Uneyama, H. (2013). Amino acid sensing in the gastrointestinal tract. *Amino Acids, 45*(3), 451–461.
31. Shinn, J., Kwon, N., Lee, S. A., & Lee, Y. (2022). Smart pH-responsive nanomedicines for disease therapy. *Journal of Pharmaceutical Investigation, 52*(4), 427–441.
32. Lee, D. Y., Kim, J. Y., Ahn, E., Hyeon, J. S., Kim, G. H., Park, K. J., Jung, Y., Lee, Y. J., Son, M. K., Kim, S. W., & Han, S. Y. (2022). Associations between local acidosis induced by renal LDHA and renal fibrosis and mitochondrial abnormalities in patients with diabetic kidney disease. *Translational Research, 249*, 88–109.
33. Zarur, M., Seijo-Rabina, A., Goyanes, A., Concheiro, A., & Alvarez-Lorenzo, C. (2023). pH-responsive scaffolds for tissue regeneration: *In vivo* performance. *Acta Biomaterialia, 168*, 22–41.
34. Kadokawa, J. I., Shimohigoshi, R., Yamashita, K., & Yamamoto, K. (2015). Synthesis of chitin and chitosan stereoisomers by thermostable α-glucan phosphorylase-catalyzed enzymatic polymerization of α-D-glucosamine 1-phosphate. *Organic & Biomolecular Chemistry, 13*(14), 4336–4343.
35. Shikuku, R., Hasnat, M. A., Mashrur, S. B. A., Haque, P., Rahman, M. M., & Khan, M. N. (2024). Chitosan-based pH-sensitive semi-interpenetrating network nanoparticles as a sustained release matrix for anticancer drug delivery. *Carbohydrate Polymer Technologies and Applications, 7*, 100515.
36. Guo, Y., Qiao, D., Zhao, S., Liu, P., Xie, F., & Zhang, B. (2024). Biofunctional chitosan–biopolymer composites for biomedical applications. *Materials Science and Engineering: R: Reports, 159*, 100775.
37. Iglesias, N., Galbis, E., Valencia, C., De-Paz, M. V., & Galbis, J. A. (2018). Reversible pH-sensitive chitosan-based hydrogels, influence of dispersion composition on rheological properties and sustained drug delivery. *Polymers, 10*(4), 392.

38. Shete, A., Chavan, A., Potekar, P., Yadav, G., & Shah, N. (2024). Modification of physicochemical properties of chitosan to improve its pharmaceutical and agrochemical potential applications. *International Journal of Biological Macromolecules, 267*, 131404.
39. Wei, C., Yang, X., Li, Y., Wang, L., Xing, S., Qiao, C., Li, Y., Wang, S., Zheng, J., & Dong, Q. (2024). N-lauric-O-carboxymethyl chitosan: Synthesis, characterization and application as a pH-responsive carrier for curcumin particles. *International Journal of Biological Macromolecules, 256*, 128421.
40. Wang, W., Meng, Q., Li, Q., Liu, J., Zhou, M., Jin, Z., & Zhao, K. (2020). Chitosan derivatives and their application in biomedicine. *International Journal of Molecular Sciences, 21*(2), 487.
41. Kumar, A., Vimal, A., & Kumar, A. (2016). Why Chitosan? From properties to perspective of mucosal drug delivery. *International Journal of Biological Macromolecules, 91*, 615–622.
42. Liu, Y., Lang, C., Ding, Y., Sun, S., & Sun, G. (2023). Chitosan with enhanced deprotonation for accelerated thermosensitive gelation with β-glycerophosphate. *European Polymer Journal, 196*, 112229.
43. Xiao, X., Chen, Z., & Chen, B. (2020). Proton uptake behaviors of organic and inorganic matters in biochars prepared under different pyrolytic temperatures. *Science of the Total Environment, 746*, 140853.
44. Ways, M., Lau, W. M., & Khutoryanskiy, V. V. (2018). Chitosan and its derivatives for application in mucoadhesive drug delivery systems. *Polymers, 10*(3), 267.
45. Okyem, S., Awotunde, O., Ogunlusi, T., Riley, M. B., & Driskell, J. D. (2021). Probing the mechanism of antibody-triggered aggregation of gold nanoparticles. *Langmuir, 37*(9), 2993–3000.
46. Xiao, R., Zhou, G., Wen, Y., Ye, J., Li, X., & Wang, X. (2023). Recent advances on stimuli-responsive biopolymer-based nanocomposites for drug delivery. *Composites Part B: Engineering, 266*, 111018.
47. Biswas, R., Mondal, S., Ansari, M. A., Sarkar, T., Condiuc, I. P., Trifas, G., & Atanase, L. I. (2025). Chitosan and its derivatives as nanocarriers for drug delivery. *Molecules, 30*(6), 1297.
48. Wang, R., Cheng, C., Wang, H., & Wang, D. (2024). Swollen hydrogel nanotechnology: Advanced applications of the rudimentary swelling properties of hydrogels. *ChemPhysMater, 3*(4), 357–375.
49. Hamman, J. H. (2010). Chitosan based polyelectrolyte complexes as potential carrier materials in drug delivery systems. *Marine Drugs, 8*(4), 1305–1322.
50. Stefanache, A., Lungu, I. I., Anton, N., Damir, D., Gutu, C., Olaru, I., PleseaCondratovici, A., Duceac, M., Constantin, M., Calin, G., & Duceac, L. D. (2025). Chitosan nanoparticle-based drug delivery systems: Advances, challenges, and future perspectives. *Polymers, 17*(11), 1453.
51. Du, H., Liu, M., Yang, X., & Zhai, G. (2015). The design of pH-sensitive chitosan-based formulations for gastrointestinal delivery. *Drug Discovery Today, 20*(8), 1004–1011.
52. Lamarra, J., Rivero, S., & Pinotti, A. (2016). Design of chitosan-based nanoparticles functionalized with gallic acid. *Materials Science and Engineering: C, 67*, 717–726.
53. Naghib, S. M., Amiri, S., & Mozafari, M. R. (2024). Stimuli-responsive chitosan-based nanocarriers for drug delivery in wound dressing applications: A review. *Carbohydrate Polymer Technologies and Applications, 7*, 100497.
54. Gierszewska, M., & Ostrowska-Czubenko, J. (2016). Chitosan-based membranes with different ionic crosslinking density for pharmaceutical and industrial applications. *Carbohydrate Polymers, 153*, 501–511.
55. Hong, F., Qiu, P., Wang, Y., Ren, P., Liu, J., Zhao, J., & Gou, D. (2024). Chitosan-based hydrogels: From preparation to applications, a review. *Food Chemistry: X, 21*, 101095.
56. Yusefi, M., Kia, P., Sukri, S. N. A. M., Ali, R. R., & Shameli, K. (2021). Synthesis and properties of chitosan nanoparticles crosslinked with tripolyphosphate. *Journal of Research in Nanoscience and Nanotechnology, 3*(1), 46–52.
57. Alavarse, A. C., Frachini, E. C. G., da Silva, R. L. C. G., Lima, V. H., Shavandi, A., & Petri, D. F. S. (2022). Crosslinkers for polysaccharides and proteins: Synthesis conditions,

mechanisms, and crosslinking efficiency, a review. *International Journal of Biological Macromolecules, 202*, 558–596.
58. Hutomo, D. I., Deandra, F. A., Ketherin, K., García-Gareta, E., Bachtiar, E. W., Amir, L., Tadjoedin, F. M., Widaryono, A., Haerani, N., Lessang, R., & Soeroso, Y. (2024). The effect of carbodiimide crosslinkers on gelatin hydrogel as a potential biomaterial for gingival tissue regeneration. *Gels, 10*(11), 674.
59. Lee, J. M., Edwards, H. H. L., Pereira, C. A., & Samii, S. I. (1996). Crosslinking of tissue-derived biomaterials in 1-ethyl-3-(3-dimethylaminopropyl)-carbodiimide (EDC). *Journal of Materials Science: Materials in Medicine, 7*(9), 531–541.
60. Wahba, M. I. (2024). A comprehensive review on genipin: An efficient natural cross-linker for biopolymers. *Polymer Bulletin, 81*(16), 14251–14305.
61. Wu, W., Luo, L., Wang, Y., Wu, Q., Dai, H. B., Li, J. S., Durkan, C., Wang, N., & Wang, G. X. (2018). Endogenous pH-responsive nanoparticles with programmable size changes for targeted tumor therapy and imaging applications. *Theranostics, 8*(11), 3038.
62. Grewal, A. K., & Salar, R. K. (2024). Chitosan nanoparticle delivery systems: An effective approach to enhancing efficacy and safety of anticancer drugs. *Nano TransMed, 3*, 100040.
63. de Sousa Victor, R., da Cunha, M., Santos, A., Viana de Sousa, B., de Araújo Neves, G., de Lima, N., Santana, L., & Rodrigues Menezes, R. (2020). A review on chitosan's uses as biomaterial: Tissue engineering, drug delivery systems and cancer treatment. *Materials, 13*(21), 4995.
64. Bajracharya, R., Song, J. G., Patil, B. R., Lee, S. H., Noh, H. M., Kim, D. H., Kim, G. L., Seo, S. H., Park, J. W., Jeong, S. H., & Lee, C. H. (2022). Functional ligands for improving anticancer drug therapy: Current status and applications to drug delivery systems. *Drug Delivery, 29*(1), 1959–1970.
65. Kumari, N. U., Chigurupati, S. P. D., Rajana, N., Madan, J., & Mehra, N. K. (2025). Ligand conjugation strategies. In *Ligands for targeted drug delivery* (pp. 159–189). Academic Press.
66. Yu, B. O., Tai, H. C., Xue, W., Lee, L. J., & Lee, R. J. (2010). Receptor-targeted nanocarriers for therapeutic delivery to cancer. *Molecular Membrane Biology, 27*(7), 286–298.
67. Du, H., Yang, X., & Zhai, G. (2014). Design of chitosan-based nanoformulations for efficient intracellular release of active compounds. *Nanomedicine, 9*(5), 723–740.
68. Naskar, S., Kuotsu, K., & Sharma, S. (2019). Chitosan-based nanoparticles as drug delivery systems: A review on two decades of research. *Journal of Drug Targeting, 27*(4), 379–393.
69. Herdiana, Y., Wathoni, N., Shamsuddin, S., & Muchtaridi, M. (2022). Drug release study of the chitosan-based nanoparticles. *Heliyon, 8*(1), 08674.
70. Yoshida, T., Lai, T. C., Kwon, G. S., & Sako, K. (2013). pH-and ion-sensitive polymers for drug delivery. *Expert Opinion on Drug Delivery, 10*(11), 1497–1513.
71. Borandeh, S., van Bochove, B., Teotia, A., & Seppälä, J. (2021). Polymeric drug delivery systems by additive manufacturing. *Advanced Drug Delivery Reviews, 173*, 349–373.
72. Samiraninezhad, N., Asadi, K., Rezazadeh, H., & Gholami, A. (2023). Using chitosan, hyaluronic acid, alginate, and gelatin-based smart biological hydrogels for drug delivery in oral mucosal lesions: A review. *International Journal of Biological Macromolecules, 252*, 126573.
73. Dwitya, S. S., Lin, K. S., Lai, L. J., Mdlovu, N. V., Lin, Z. J., Hsieh, C. C., & Hua, M. D. S. (2025). Intracellular trafficking of nanostructured MOF-303 for targeted liver cancer therapy: A study using soft X-ray tomography. *Materials Today Chemistry, 47*, 102793.
74. Dwitya, S. S., Lin, K. S., Weng, M. T., Mdlovu, N. V., Lai, L. J., & Wu, C. M. (2025). Thermo-and pH-responsive MOF-303 mediated P127 and gelatin coating for combination drug release and liver cancer therapy. *Materials Today Chemistry, 44*, 102548.
75. Wang, Y., Tu, S., Pinchuk, A. N., & Xiong, M. P. (2013). Active drug encapsulation and release kinetics from hydrogel-in-liposome nanoparticles. *Journal of Colloid and Interface Science, 406*, 247–255.

76. Al Tahan, M. A., Marwah, M., El-Zein, H., Al Tahan, S., & Snchez-Aranguren, L. (2025). Exploring mesoporous silica microparticles in pharmaceutical sciences: Drug delivery and therapeutic insights. *International Journal of Pharmaceutics, 678*, 125656.
77. Popat, A., Liu, J., Lu, G. Q. M., & Qiao, S. Z. (2012). A pH-responsive drug delivery system based on chitosan coated mesoporous silica nanoparticles. *Journal of Materials Chemistry, 22*(22), 11173–11178.
78. Wang, Q. Z., Chen, X. G., Liu, N., Wang, S. X., Liu, C. S., Meng, X. H., & Liu, C. G. (2006). Protonation constants of chitosan with different molecular weight and degree of deacetylation. *Carbohydrate Polymers, 65*(2), 194–201.
79. Zohra Zeggai, F., Ait-Touchente, Z., Bachari, K., & Elaissari, A. (2025). Investigation of metal-organic frameworks (MOFs): Synthesis, properties, and applications-an in-depth review. *Chemical Physics Impact, 10*, 100864.
80. Kausar, S., Yousaf, M., Mir, S., Awwad, N. S., Alturaifi, H. A., & Riaz, F. (2024). Mesoporous materials: Synthesis and electrochemical applications amini review. *Electrochemistry Communications, 169*, 107836.
81. Damian-Buda, A. I., Alipanah, N., Bider, F., Sisman, O., Neščáková, Z., & Boccaccini, A. R. (2024). Metal-organic-framework (MOF)-bioactive glass (BG) systems for biomedical applications: A review. *Materials Today Bio, 30*, 101413.
82. Zhu, K., & Zhou, X. (2015). Manipulating the architecture of zeolite catalysts for enhanced mass transfer. *Current Opinion in Chemical Engineering, 9*, 42–48.
83. Akdaşçi, E., Duman, H., Eker, F., Bechelany, M., & Karav, S. (2025). Chitosan and its nanoparticles: A multifaceted approach to antibacterial applications. *Nanomaterials, 15*(2), 126.
84. Amin, M. K., & Boateng, J. S. (2022). Enhancing stability and mucoadhesive properties of chitosan nanoparticles by surface modification with sodium alginate and polyethylene glycol for potential oral mucosa vaccine delivery. *Marine Drugs, 20*(3), 156.
85. Herdiana, Y., Febrina, E., Nurhasanah, S., Gozali, D., Elamin, K. M., & Wathoni, N. (2024). Drug loading in chitosan-based nanoparticles. *Pharmaceutics, 16*(8), 1043.
86. Thananukul, K., Kaewsaneha, C., Opaprakasit, P., Lebaz, N., Errachid, A., & Elaissari, A. (2021). Smart gating porous particles as new carriers for drug delivery. *Advanced Drug Delivery Reviews, 174*, 425–446.
87. Yıldırım, M., Poyraz, S., Acet, Ö., Acet, B. Ö., Karakoç, V., & Odabaşı, M. (2025). Chitosan hydrogels: Versatile platforms for drug delivery in cancer treatment, wound dressing, and 3D bioprinting applications. *International Journal of Biological Macromolecules, 314*, 144367.
88. Herrera-Alonso, A. E., Rodríguez-Chávez, D. F., Toxqui-Terán, A., Rubio-Valle, J. F., Martín-Alfonso, J. E., Longoria-García, S., Gallardo-Blanco, H. L., Sánchez-Domínguez, C. N., & Sánchez-Domínguez, M. (2025). pH-Sensitive chitosan-based hydrogels trap poloxamer micelles as a dual-encapsulating responsive system for the loading and delivery of curcumin. *Polymers, 17*(10), 1335.
89. Cao, Y., Tan, Y. F., Wong, Y. S., Liew, M. W. J., & Venkatraman, S. (2019). Recent advances in chitosan-based carriers for gene delivery. *Marine Drugs, 17*(6), 381.
90. Shu, X. Z., & Zhu, K. J. (2002). Controlled drug release properties of ionically cross-linked chitosan beads: The influence of anion structure. *International Journal of Pharmaceutics, 233*(1–2), 217–225.
91. Luraghi, A., Peri, F., & Moroni, L. (2021). Electrospinning for drug delivery applications: A review. *Journal of Controlled Release, 334*, 463–484.
92. Dorazilová, J., Muchová, J., Šmerková, K., Kočiová, S., Diviš, P., Kopel, P., Veselý, R., Pavliňáková, V., Adam, V., & Vojtová, L. (2020). Synergistic effect of chitosan and selenium nanoparticles on biodegradation and antibacterial properties of collagenous scaffolds designed for infected burn wounds. *Nanomaterials (Basel), 10*(10), 1971.
93. Witoon, T., Tepsarn, S., Kittipokin, P., Embley, B., & Chareonpanich, M. (2011). Effect of pH and chitosan concentration on precipitation and morphology of hierarchical porous silica. *Journal of Non-Crystalline Solids, 357*(19–20), 3513–3519.

94. Armenia, I., Ayllón, C. C., Herrero, B. T., Bussolari, F., Alfranca, G., Grazú, V., & de la Fuente, J. M. (2022). Photonic and magnetic materials for on-demand local drug delivery. *Advanced Drug Delivery Reviews, 191*, 114584.
95. Benny, A., Pai, S. D. K. R., Pinheiro, D., & Chundattu, S. J. (2024). Metal organic frameworks in biomedicine: Innovations in drug delivery. *Results in Chemistry, 7*, 101414.
96. Mazumdar, S., Chitkara, D., & Mittal, A. (2021). Exploration and insights into the cellular internalization and intracellular fate of amphiphilic polymeric nanocarriers. *Acta Pharmaceutica Sinica B, 11*(4), 903–924.
97. Li, M., Wang, Y., Jiang, S., Gao, Y., Zhang, W., Hu, S., Cheng, X., Zhang, C., Sun, P., Ke, W., & Wang, G. (2020). Biodistribution and biocompatibility of glycyrrhetinicacid and galactose-modified chitosan nanoparticles as a novel targeting vehicle for hepatocellular carcinoma. *Nanomedicine, 15*(2), 145–161.
98. Zeng, J., Ren, X., Li, X., Chen, P., & Zhu, S. (2025). Development of pH-responsive active intelligent chitosan film incorporated with pomegranate cellulose nanocrystals and curcumin nanoparticles. *International Journal of Biological Macromolecules, 309*, 142051.
99. Baig, N., Kammakakam, I., & Falath, W. (2021). Nanomaterials: A review of synthesis methods, properties, recent progress, and challenges. *Materials Advances, 2*(6), 1821–1871.
100. Al-Shaeli, M., Benkhaya, S., Al-Juboori, R. A., Koyuncu, I., & Vatanpour, V. (2024). pH-responsive membranes: Mechanisms, fabrications, and applications. *Science of the Total Environment, 946*, 173865.
101. Hoshyar, N., Gray, S., Han, H., & Bao, G. (2016). The effect of nanoparticle size on *in vivo* pharmacokinetics and cellular interaction. *Nanomedicine, 11*(6), 673–692.
102. Tosi, M. M., Ramos, A. P., Esposto, B. S., & Jafari, S. M. (2020). Dynamic light scattering (DLS) of nanoencapsulated food ingredients. In *Characterization of Nanoencapsulated food ingredients* (pp. 191–211). Academic Press.
103. Wu, J. (2021). The enhanced permeability and retention (EPR) effect: The significance of the concept and methods to enhance its application. *Journal of Personalized Medicine, 11*(8), 771.
104. Varenne, F., Coty, J. B., Botton, J., Legrand, F. X., Hillaireau, H., Barratt, G., & Vauthier, C. (2019). Evaluation of zeta potential of nanomaterials by electrophoretic light scattering: Fast field reversal versus slow field reversal modes. *Talanta, 205*, 120062.
105. Aibani, N., Rai, R., Patel, P., Cuddihy, G., & Wasan, E. K. (2021). Chitosan nanoparticles at the biological interface: Implications for drug delivery. *Pharmaceutics, 13*(10), 1686.
106. Souza, I. D., Saez, V., & Mansur, C. R. (2023). Lipid nanoparticles containing coenzyme Q10 for topical applications: An overview of their characterization. *Colloids and Surfaces B: Biointerfaces, 230*, 113491.
107. Aravind, D., Senthilkumar, K., Diwahar, P., Chandrasekar, M., Kumar, T. S. M., Parameswaranpillai, J., Rajini, N., & Siengchin, S. (2022). Review on coir fiber surface modification by various techniques. In *Coir Fiber and its Composites* (pp. 55–78).
108. Samuel, H. S., & Ekpan, F. D. M. (2024). The use of scanning electron microscopy (SEM) for medical application: A mini review. *Eurasian Journal of Science and Technology, 4*(4), 289–294.
109. Mi, Y., Tan, W., Zhang, J., Wei, L., Chen, Y., Li, Q., Dong, F., & Guo, Z. (2018). Synthesis, characterization, and antifungal property of hydroxypropyltrimethyl ammonium chitosan halogenated acetates. *Marine Drugs, 16*(9), 315.
110. Nikonenko, N. A., Buslov, D. K., Sushko, N. I., & Zhbankov, R. G. (2005). Spectroscopic manifestation of stretching vibrations of glycosidic linkage in polysaccharides. *Journal of Molecular Structure, 752*(1–3), 20–24.
111. Wildan, M. W., & Lubis, F. I. (2021). Fabrication and characterization of chitosan/cellulose nanocrystal/glycerol bio-composite films. *Polymers, 13*(7), 1096.
112. Zhu, M., Zhu, Y., Zhang, L., & Shi, J. (2013). Preparation of chitosan/mesoporous silica nanoparticle composite hydrogels for sustained co-delivery of biomacromolecules and small chemical drugs. *Science and Technology of Advanced Materials, 14*(4), 045005.

113. Nurazzi, N. M., Abdullah, N., Norrahim, M. N. F., Kamarudin, S. H., Ahmad, S., Shazleen, S. S., Rayung, M., Asyraf, M. R. M., Ilyas, R. A., & Kuzmin, M. (2022). Thermogravimetric analysis (TGA) and differential scanning calorimetry (DSC) of PLA/cellulose composites. In *Polylactic acid-based nanocellulose and cellulose composites* (pp. 145–164). CRC Press.
114. Chen, H., Tan, W., Tong, T., Shi, X., Ma, S., & Zhu, G. (2023). A pH-responsive asymmetric microfluidic/chitosan device for drug release in infective bone defect treatment. *International Journal of Molecular Sciences, 24*(5), 4616.
115. Domiński, A., Krawczyk, M., Konieczny, T., Kasprow, M., Foryś, A., Pastuch-Gawołek, G., & Kurcok, P. (2020). Biodegradable pH-responsive micelles loaded with 8-hydroxyquinoline glycoconjugates for warburgeffect-basedtumor targeting. *European Journal of Pharmaceutics and Biopharmaceutics, 154*, 317–329.
116. Shafiei, F., Ghavami-Lahiji, M., Kashi, T. S. J., & Najafi, F. (2021). Drug release kinetics and biological properties of a novel local drug carrier system. *Dental Research Journal, 18*(1), 94.
117. Komeri, R., Kasoju, N., & Kumar, P. A. (2022). *In vitro* cytotoxicity and cytocompatibility assays for biomaterial testing under regulatory platform. In *Biomedical product and materials evaluation* (pp. 329–353). Woodhead Publishing.
118. Mohammadi, T., Gheybalizadeh, H., Rahimpour, E., Soleymani, J., & Shafiei-Irannejad, V. (2024). Advanced photoluminescent nanomaterials for targeted bioimaging of cancer cells. *Heliyon, 11*(1), 41566.
119. Ferreira, L. M., & Zucolotto, V. (2024). Chitosan-based nanomedicines: A review of the main challenges for translating the science of polyelectrolyte complexation into innovative pharmaceutical products. *Carbohydrate Polymer Technologies and Applications, 7*, 100441.
120. Liu, Z., Jiao, Y., Wang, Y., Zhou, C., & Zhang, Z. (2008). Polysaccharides-based nanoparticles as drug delivery systems. *Advanced Drug Delivery Reviews, 60*(15), 1650–1662.
121. Khan, A., & Alamry, K. A. (2021). Recent advances of emerging green chitosan-based biomaterials with potential biomedical applications: A review. *Carbohydrate Research, 506*, 108368.
122. GuarÍn-González, Y. A., & Cárdenas-Triviño, G. (2022). New chitosan-based chemo pharmaceutical delivery systems for tumor cancer treatment: Short-review. *Journal of the Chilean Chemical Society, 67*(1), 5425–5432.
123. Quiñones, J. P., Peniche, H., & Peniche, C. (2018). Chitosan based self-assembled nanoparticles in drug delivery. *Polymers, 10*(3), 235.
124. Taghizadeh, M. T., Ashassi-Sorkhabi, H., Afkari, R., & Kazempour, A. (2019). Cross-linked chitosan in nano and bead scales as drug carriers for betamethasone and tetracycline. *International Journal of Biological Macromolecules, 131*, 581–588.
125. Haider, A., Khan, S., Iqbal, D. N., Shrahili, M., Haider, S., Mohammad, K., & Mustafa, G. (2024). Advances in chitosan-based drug delivery systems: A comprehensive review for therapeutic applications. *European Polymer Journal, 210*, 112983.
126. Hefnawy, A., Khalil, I. H., Arafa, K., Emara, M., & El-Sherbiny, I. M. (2020). Dual-ligand functionalized core-shell chitosan-based nanocarrier for hepatocellular carcinoma-targeted drug delivery. *International Journal of Nanomedicine, 15*, 821–837.
127. Anwanwan, D., Singh, S. K., Singh, S., Saikam, V., & Singh, R. (2020). Challenges in liver cancer and possible treatment approaches. *Biochimica et Biophysica Acta (BBA)-Reviews on Cancer, 1873*(1), 188314.
128. Zhang, J., Li, C., Xue, Z. Y., Cheng, H. W., Huang, F. W., Zhuo, R. X., & Zhang, X. Z. (2011). Fabrication of lactobionic-loaded chitosan microcapsules as potential drug carriers targeting the liver. *Acta Biomaterialia, 7*(4), 1665–1673.
129. Li, M., Wang, Y., Jiang, S., Gao, Y., Zhang, W., Hu, S., & Zheng, Q. C. (2020). Biodistribution and biocompatibility of glycyrrhetinic acid and galactose-modified chitosan nanoparticles as a novel targeting vehicle for hepatocellular carcinoma. *Nanomedicine, 15*(2), 145–161.

130. Lin, A., Liu, Y., Huang, Y., Sun, J., Wu, Z., Zhang, X., & Ping, Q. (2008). Glycyrrhizin surface-modified chitosan nanoparticles for hepatocyte-targeted delivery. *International Journal of Pharmaceutics, 359*(1–2), 247–253.
131. Tian, Q., Zhang, C. N., Wang, X. H., Wang, W., Huang, W., Cha, R. T., & Tang, H. (2010). Glycyrrhetinic acid-modified chitosan/poly (ethylene glycol) nanoparticles for liver-targeted delivery. *Biomaterials, 31*(17), 4748–4756.
132. Rohilla, R., Garg, T., Bariwal, J., Goyal, A. K., & Rath, G. (2016). Development, optimization and characterization of glycyrrhetinic acid–chitosan nanoparticles of atorvastatin for liver targeting. *Drug Delivery, 23*(7), 2290–2297.
133. Mishra, D., Jain, N., Rajoriya, V., & Jain, A. K. (2014). Glycyrrhizin conjugated chitosan nanoparticles for hepatocyte-targeted delivery of lamivudine. *Journal of Pharmacy and Pharmacology, 66*(8), 1082–1093.
134. Arafa, K. K., Hamzawy, M. A., Mousa, S. A., & El-Sherbiny, I. M. (2022). Mitochondria-targeted alginate/triphenylphosphonium-grafted-chitosan for treatment of hepatocellular carcinoma. *RSC Advances, 12*(34), 21690–21703.
135. Bonferoni, M. C., Gavini, E., Rassu, G., Maestri, M., & Giunchedi, P. (2020). Chitosan nanoparticles for therapy and theranostics of hepatocellular carcinoma (HCC) and liver-targeting. *Nanomaterials, 10*(5), 870.
136. Lin, A., Chen, J., Liu, Y., Deng, S., Wu, Z., Huang, Y., & Ping, Q. (2009). Preparation and evaluation of N-caproyl chitosan nanoparticles surface modified with glycyrrhizin for hepatocyte targeting. *Drug Development and Industrial Pharmacy, 35*(11), 1348–1355.
137. Hu, Z., Chen, J., Zhou, S., Yang, N., Duan, S., Zhang, Z., & Zhao, Y. (2017). Mouse IP-10 gene delivered by folate-modified chitosan nanoparticles and dendritic/tumor cells fusion vaccine effectively inhibit the growth of hepatocellular carcinoma in mice. *Theranostics, 7*(7), 1942.
138. Ruman, U., Buskaran, K., Pastorin, G., Masarudin, M. J., Fakurazi, S., & Hussein, M. Z. (2021). Synthesis and characterization of chitosan-based nanodelivery systems to enhance the anticancer effect of sorafenib drug in hepatocellular carcinoma and colorectal adenocarcinoma cells. *Nanomaterials (Basel), 11*(2), 497.
139. Cheng, M., Zhu, W., Li, Q., Dai, D., & Hou, Y. (2017). Anti-cancer efficacy of biotinylated chitosan nanoparticles in liver cancer. *Oncotarget, 8*(35), 59068.
140. Gao, X., Li, L., Cai, X., Huang, Q., Xiao, J., & Cheng, Y. (2021). Targeting nanoparticles for diagnosis and therapy of bone tumors: Opportunities and challenges. *Biomaterials, 265*, 120404.
141. Mao, X., Wu, S., Huang, D., & Li, C. (2024). Complications and comorbidities associated with antineoplastic chemotherapy: Rethinking drug design and delivery for anticancer therapy. *Acta Pharmaceutica Sinica B, 14*(7), 2901–2926.
142. Elella, M. H. A., & Kolawole, O. M. (2024). Recent advances in modified chitosan-based drug delivery systems for transmucosal applications: A comprehensive review. *International Journal of Biological Macromolecules, 277*(4), 134531.
143. Hu, B., Zhang, Y., Zhang, G., Li, Z., Jing, Y., Yao, J., & Sun, S. (2022). Research progress of bone-targeted drug delivery system on metastatic bone tumors. *Journal of Controlled Release, 350*, 377–388.
144. Cui, Y., Zhu, T., Li, D., Li, Z., Leng, Y., Ji, X., & Ding, J. (2019). Bisphosphonate-functionalized scaffolds for enhanced bone regeneration. *Advanced Healthcare Materials, 8*(23), 1901073.
145. Ebetino, F. H., Sun, S., Cherian, P., Roshandel, S., Neighbors, J. D., Hu, E., & Russell, R. G. G. (2022). Bisphosphonates: The role of chemistry in understanding their biological actions and structure-activity relationships, and new directions for their therapeutic use. *Bone, 156*, 116289.
146. Shi, S., Duan, H., & Ou, X. (2024). Targeted delivery of anti-osteoporosis therapy: Bisphosphonate-modified nanosystems and composites. *Biomedicine & Pharmacotherapy, 175*, 116699.

147. Kang, N. W., Lee, J. Y., & Kim, D. D. (2022). Hydroxyapatite-binding albumin nanoclusters for enhancing bone tumor chemotherapy. *Journal of Controlled Release, 342*, 111–121.
148. Wu, H., Luo, Y., Xu, D., Ke, X., & Ci, T. (2020). Low molecular weight heparin modified bone targeting liposomes for orthotopic osteosarcoma and breast cancer bone metastatic tumors. *International Journal of Biological Macromolecules, 164*, 2583–2597.
149. Pathak, S., Tripathi, S., Pachouri, C., & Pandey, A. (2022). Encapsulation of alendronate in chitosan based polymeric nanoparticles for effective management of osteoporosis–development to release kinetic study. *International Journal of Medical Nano Research, 9*, 1–8.
150. Liu, Y. F., Liu, R., Li, X. Y., Song, Z., & Zhao, X. H. (2016). Development of docetaxel and alendronate-loaded chitosan-conjugated polylactide-co-glycolide nanoparticles: *In vitro* characterization in osteosarcoma cells. *Tropical Journal of Pharmaceutical Research, 15*, 1353–1360.
151. Yang, J. I., Lee, H. L., Yun, J. J., Kim, J., So, K. H., Jeong, Y. I., & Kang, D. H. (2022). pH and redox-dual sensitive chitosan nanoparticles having methyl ester and disulfide linkages for drug targeting against cholangiocarcinoma cells. *Materials, 15*, 3795.
152. Amiryaghoubi, N., Abdolahinia, E. D., Nakhlband, A., Aslzad, S., Fathi, M., Barar, J., & Omidi, Y. (2022). Smart chitosan–folate hybrid magnetic nanoparticles for targeted delivery of doxorubicin to osteosarcoma cells. *Colloids and Surfaces B: Biointerfaces, 220*, 112911.
153. Adimoolam, M. G., Amreddy, N., Nalam, M. R., & Sunkara, M. V. (2018). A simple approach to design chitosan functionalized Fe_3O_4 nanoparticles for pH responsive delivery of doxorubicin for cancer therapy. *Journal of Magnetism and Magnetic Materials, 448*, 199–207.
154. Liu, Q., Tan, Z., Zheng, D., & Qiu, X. (2023). pH-responsive magnetic Fe_3O_4/carboxymethyl chitosan/aminated lignosulfonate nanoparticles with uniform size for targeted drug loading. *International Journal of Biological Macromolecules, 225*, 1182–1192.
155. Ding, Y., Shen, S. Z., Sun, H., Sun, K., Liu, F., Qi, Y., & Yan, J. (2015). Design and construction of polymerized-chitosan coated Fe_3O_4 magnetic nanoparticles and its application for hydrophobic drug delivery. *Materials Science and Engineering: C, 48*, 487–498.
156. Marshall, S. K., Taweesap, M., Saelim, B., Pachana, V., Benlateh, N., Sangangam, S., Bumrungsin, A., Kholo-Asae, H., & Wongtechanon, I. (2024). Cytotoxicity enhancement in osteosarcoma with multifunctional I-131 radiotherapeutic nanoparticles: *In vitro* three-dimensional spheroid model and release kinetics modelling. *Molecules, 29*, 630.
157. Koski, C., Vu, A. A., & Bose, S. (2020). Effects of chitosan-loaded hydroxyapatite on osteoblasts and osteosarcoma for chemopreventative applications. *Materials Science and Engineering: C, 115*, 111041.
158. Ai, J. W., Liao, W., & Ren, Z. L. (2017). Enhanced anticancer effect of copper-loaded chitosan nanoparticles against osteosarcoma. *RSC Advances, 7*, 15971–15977.
159. Kortam, S., Lu, Z., & Zreiqat, H. (2024). Recent advances in drug delivery systems for osteosarcoma therapy and bone regeneration. *Communications Materials, 5*, 168.
160. Ta, H. T., Dass, C. R., Larson, I., Choong, P. F. M., & Dunstan, D. E. (2009). A chitosan hydrogel delivery system for osteosarcoma gene therapy with pigment epithelium-derived factor combined with chemotherapy. *Biomaterials, 30*, 4815–4823.
161. Liu, C., Zhang, W., Zhu, C., Sun, J., Ding, Y., & Shi, Q. (2024). Inhibitory effect of bovine serum albumin-chitosan nanoparticles loaded with EPZ6438 on osteosarcoma. *Chinese Journal of Tissue Engineering Research, 28*, 1512.
162. Shanmuga, S. S., Natarajan, K., Zsolt, S., Priya, L. S., Ayyar, M., Anboo, S., Nissapatorn, V., Jonna, N., Vasanthapalaniappan, K., & Packirisamy, R. (2024). *In vitro* study on the inflammatory response of chitosan nanoparticles as a potential siRNA carrier targeting towards osteosarcoma cells. *Zeitschrift Für Physikalische Chemie, 238*, 1345–1360.
163. Catalano, E., & Di Benedetto, A. (2017). Characterization of physicochemical and colloidal properties of hydrogel chitosan-coated iron-oxide nanoparticles for cancer therapy. *Journal of Physics: Conference Series, [online] IOP Publishing, 841*, 012010.

164. Imran, H., Tang, Y., Wang, S., Yan, X., Liu, C., Guo, L., Wang, E., & Xu, C. (2023). Optimized dox drug deliveries via chitosan-mediated nanoparticles and stimuli responses in cancer chemotherapy: A review. *Molecules, 29*, 31.
165. Sachdeva, B., Sachdeva, P., Negi, A., Ghosh, S., Han, S., Dewanjee, S., Jha, S. K., Bhaskar, R., Sinha, J. K., Paiva-Santos, A. C., & Jha, N. K. (2023). Chitosan nanoparticles-based cancer drug delivery: Application and challenges. *Marine Drugs, 21*(4), 211.
166. Shanmuganathan, R., Edison, T. N. J. I., LewisOscar, F., Kumar, P., Shanmugam, S., & Pugazhendhi, A. (2019). Chitosan nanopolymers: An overview of drug delivery against cancer. *International Journal of Biological Macromolecules, 130*, 727–736.
167. Hirano, S. (1996). Chitin biotechnology applications. *Biotechnology Annual Review, 2*, 237–258.
168. Qin, C., Gao, J., Wang, L., Zeng, L., & Liu, Y. (2006). Safety evaluation of short-term exposure to chitooligomers from enzymic preparation. *Food and Chemical Toxicology, 44*(6), 855–861.
169. Zhao, Z., Peng, Y., Shi, X., & Zhao, K. (2023). Chitosan derivative composite nanoparticles as adjuvants enhance the cellular immune response via activation of the cGAS-STING pathway. *International Journal of Pharmaceutics, 636*, 122847.
170. Villiers, C., Chevallet, M., Diemer, H., Couderc, R., Freitas, H., Van Dorsselaer, A., & Rabilloud, T. (2009). From secretome analysis to immunology: Chitosan induces major alterations in the activation of dendritic cells via a TLR4-dependent mechanism. *Molecular & Cellular Proteomics, 8*(6), 1252–1264.
171. Masimov, R., & Wasan, E. K. (2024). Chitosan non-particulate vaccine delivery systems. *Journal of Pharmacy & Pharmaceutical Sciences, 27*, 12921.
172. Maita, K. C., Avila, F. R., Torres-Guzman, R. A., Garcia, J. P., Eldaly, A. S., Palmieri, L., & Forte, A. J. (2022). Local anti-inflammatory effect and immunomodulatory activity of chitosan-based dressing in skin wound healing: A systematic review. *Journal of Clinical and Translational Research, 8*(6), 488.
173. Boroumand, H., Badie, F., Mazaheri, S., Seyedi, Z. S., Nahand, J. S., Nejati, M., & Mirzaei, H. (2021). Chitosan-based nanoparticles against viral infections. *Frontiers in Cellular and Infection Microbiology, 11*, 643953.
174. National Toxicology Program. (2017). NTP technical report on the toxicity study of chitosan (CASRN 9012-76-4) Administered in Feed to Sprague Dawley [Crl: Cd (Sd)] Rats.
175. Li, J., Cai, C., Li, J., Li, J., Li, J., Sun, T., & Yu, G. (2018). Chitosan-based nanomaterials for drug delivery. *Molecules, 23*(10), 2661.
176. Wu, K., Yan, Z., Wu, Z., Li, J., Zhong, W., Ding, L., & Jiang, T. (2024). Recent advances in the preparation, antibacterial mechanisms, and applications of chitosan. *Journal of Functional Biomaterials, 15*(11), 318.
177. Feng, S., Wu, Z. X., Zhao, Z., Liu, J., Sun, K., Guo, C., & Wu, Z. (2019). Engineering of bone- and CD44-dual-targeting redox-sensitive liposomes for the treatment of orthotopic osteosarcoma. *ACS Applied Materials & Interfaces, 11*(7), 7357–7368.

[illegible]

Chitosan-Based Hydrogels for Stimuli-Responsive Drug Delivery

Sarita Kumari Yadav and Brahmeshwar Mishra

Abstract The biocompatibility, biodegradability, and stimuli-responsiveness of hydrogels, which possess three-dimensional networks of polymers, make them extremely promising materials for drug delivery applications. Extensive research and commercial interest have been sparked by their capacity to encapsulate medicinal substances and enable controlled, prolonged, or targeted release. Notably, chitosan-based hydrogels have drawn interest due to their superior antibacterial, biodegradable, pH-responsive, and biocompatible qualities, which make them ideal for targeted and oral drug administration. Continuous advancements in hydrogel design are tackling major obstacles in gastrointestinal distribution, even in the face of difficulties with oral formulations like insulin. Drug release kinetics and degradation can be precisely controlled by chemically modifying chitosan hydrogels, which react to various stimuli such as pH, temperature, light, and electric and magnetic fields. Their porous structure and mechanical strength further enhance their utility in advanced delivery systems. In line with sustainability objectives, recent advancements highlight environmentally friendly methods by creating chitosan hydrogels from biomass-derived materials. By permitting site-specific combination treatments and reducing side effects, these hydrogels show promise in oncology for improving the effectiveness of immunotherapy, phototherapy, and chemotherapy. Chitosan-based hydrogels are multipurpose platforms that mark a major breakthrough in the creation of intelligent, stimuli-responsive drug delivery systems designed to satisfy intricate therapeutic requirements in contemporary medicine.

Keywords Light · Magnetic · pH responsive · Targeted drug delivery · Insulin

S. K. Yadav (✉)
Department of Pharmacy, Moti Lal Nehru (MLN) Medical College, Prayagraj, UP, India

B. Mishra (✉)
Department of Pharmaceutical Engineering & Technology, Indian Institute of Technology (BHU), Varanasi, India
e-mail: bmishra.phe@itbhu.ac.in

S. Jana (ed.), *Chitosan-Based Drug Delivery Systems*, AAPS Introductions in the Pharmaceutical Sciences 16, https://doi.org/10.1007/978-3-032-08347-0_8

Abbreviations

AAm-g-HEC	Acrylamide-grafted hydroxyethylcellulose
ALP	Alkaline phosphatase
BSA	Bovine serum albumin
CAGR	Compound annual growth rate
CS	Chitosan
CS–ss–SA	Glycolipid-like nanocarrier
DIP-PCLNPs	Dipyridamole-loaded polycaprolactone nanoparticles
DSC	Differential Scanning Calorimetry
EPR	Enhanced permeability and retention
FA	Folic acid
FITC	Fluorescein isothiocyanate
FTIR	Fourier Transform Infrared Spectroscopy
GIT	Gastrointestinal tract
GSH	Glutathione
HMSNs	Hollow mesoporous silica nanoparticles
IPN	Interpenetrating Network
MMT	montmorillonite
MSRS-CS-MCs	Multi-stimuli responsive smart chitosan-based microcapsules
NMR	Nuclear Magnetic Resonance
OEDes	Ortho ester compounds
PAA	Polyacrylic acid
PDMAEMA	poly(N,N-dimethylaminoethyl methacrylate
PEG	polyethylene glycol
PEGDA	Poly(ethylene glycol) diacrylate
PEO	Freeze-dried chitosan-poly(ethylene oxide)
PHEMA	poly(hydroxyethyl methacrylate)
PLA	Polylactic acid
PNIPAAm	Poly(N-isopropylacrylamide
PVP	Poly(N-vinyl-2-pyrrolidone)
RITC	red fluorescent dye
SEM	Scanning Electron Microscopy
SPIONs,	Short for superparamagnetic iron oxide nanoparticles
TEM	Transmission Electron Microscopy
USD	United States Dollar
XRD	X-ray Diffraction

1 Introduction

Hydrogels, characterized by their three-dimensional polymeric networks, have gained considerable attention in biomedical applications, particularly drug delivery, due to their biocompatibility, biodegradability, adjustable mechanical strength, and responsiveness to diverse stimuli [1]. These distinctive features allow hydrogels to encapsulate and safeguard therapeutic agents, enabling controlled, sustained, or externally triggered release. As a result, they have become highly promising platforms for delivering both small-molecule drugs and biologics [2].

The utilization of hydrogels in drug delivery systems has experienced significant growth in recent years. According to a report by Allied Market Research, the global hydrogel market was valued at USD 22.1 billion in 2019 and is projected to reach USD 31.4 billion by 2027, registering a compound annual growth rate (CAGR) of 6.7% from 2020 to 2027. This expansion is primarily driven by the increasing adoption of hydrogel-based products across various biomedical and pharmaceutical applications, owing to their unique properties such as high water content, biocompatibility, and tunable mechanical strength [3].

Hydrogels exhibit exceptional versatility, making them valuable across a wide range of disciplines due to their unique structural features and adaptability to various environmental conditions [4]. Their biomimetic behavior stems from key attributes such as flexibility, softness, non-toxicity, biocompatibility, and biodegradability, which collectively support their extensive use in biomedical and pharmaceutical applications [5].

The ability of hydrogels to absorb and retain water is attributed to the presence of hydrophilic functional groups such as sulfates, carboxylic acids, hydroxyl groups, and amides, along with cross-linking within polymer chains [6]. These hydrophilic groups facilitate significant swelling behavior, a critical feature for many drug delivery systems. The inherent elasticity and mechanical strength of biocompatible hydrogels not only enable the encapsulation of therapeutic agents but also ensure controlled release and sustained protection, further highlighting their potential in advanced drug delivery applications [7].

Hydrogels have been extensively investigated for their potential in oral insulin delivery as an alternative to traditional parenteral methods. Despite significant research efforts, progress has been limited, and no oral insulin formulation has yet reached commercial success for human use [8]. Various biopolymers, including alginate, hydroxypropyl cellulose, soybean protein, pectin, and cellulose acetate, have been studied for their effectiveness in gastrointestinal tract (GIT) drug delivery applications [9]. Current research focuses on developing cost-effective and efficient hydrogel-based systems capable of achieving controlled drug release via mechanisms tailored to the oral route of administration. Although oral drug delivery poses several challenges, many hydrogel formulations have been designed to overcome these barriers successfully [10].

Today, smart hydrogels are commonly synthesized from both natural and synthetic materials, such as chitosan, cellulose, sodium alginate, hyaluronic acid, and polyethylene glycol (PEG). Among these, chitosan has emerged as a particularly promising candidate due to its excellent biocompatibility. Chitosan-based hydrogels have attracted considerable interest in the biomedical field because of chitosan's inherent biodegradability, low toxicity, and antimicrobial activity. Derived from the deacetylation of chitin, chitosan displays pH-dependent solubility, making it especially suitable for stimuli-responsive drug delivery systems. These hydrogels can be engineered to respond to a wide range of stimuli, including pH, temperature, ionic strength, electric or magnetic fields, light, and specific biomolecules, allowing precise control over drug release profiles [2]. The adaptability of chitosan-based hydrogels lies in their ease of chemical modification and crosslinking, enabling fine-tuning of their properties to meet the specific demands of targeted drug delivery applications.

Chitosan hydrogels, characterized by their unique three-dimensional porous architecture, possess tunable drug-loading capacity, controlled degradation rates, and responsiveness to various stimuli. These features make them highly advantageous for use in drug delivery systems. In recent years, there has been a growing inclination toward the "back-to-nature" approach, promoting the use of biomass-derived materials in the fabrication of chitosan hydrogels for controlled drug release [11]. This trend is driven by their inherent biodegradability, biocompatibility, and non-toxicity, which underscore their suitability for biomedical applications.

Moreover, chitosan hydrogels demonstrate sensitivity to a broad range of environmental triggers, including temperature, pH, ion concentration, light, magnetic fields, redox conditions, ultrasound, and multi-stimuli environments. This responsiveness enables precise and adaptable control over drug release kinetics, highlighting their potential in advanced therapeutic delivery platforms [12]. Furthermore, stimuli-responsive chitosan hydrogels, particularly those sensitive to temperature, have been developed for targeted and effective tumor inhibition, which are discussed in later sections.

Chitosan hydrogels, due to their biocompatibility and biodegradability, serve as effective carriers for delivering drugs, genes, or their combination in cancer therapy. They offer several advantages, including enhanced tumor ablation, minimized off-target effects, and protection of genetic material from degradation. Moreover, chitosan hydrogels have shown potential in enhancing the efficacy of cancer immunotherapy by preventing immune evasion mechanisms. These hydrogels also support photothermal and photodynamic therapies, contributing to tumor suppression. Their multifunctional nature enables the synergistic integration of chemotherapy, immunotherapy, and phototherapy within a single treatment platform [13].

2 Chitosan for Hydrogels

Chitosan, a natural polysaccharide derived from chitin, has gained considerable attention in hydrogel formulation due to its non-toxic nature and mucoadhesive properties. The incorporation of stimuli-responsive elements enhances the ability of chitosan hydrogels to release drugs in a controlled manner under specific physiological conditions.

2.1 Chemical Structure and Functional Groups

Chitosan can be produced from chitin either through chemical treatment with strong alkaline solutions or via enzymatic methods using chitin deacetylases [14]. Chitin, a natural polysaccharide with the chemical formula poly(β-(1–4)-N-acetyl-D-glucosamine), holds significant importance in various biological and industrial applications [15]. The term "chitin" is derived from the Greek word "chiton," meaning "coat of mail." It was first identified by the French chemist Henri Braconnot in 1811. The structure of chitin $(C_8H_{13}O_5N)_n$ resembles cellulose, but with 2-acetamido-2-deoxy-β-D-glucose (NAG) monomer units linked via β(1→4) bonds. Chitosan, a deacetylated derivative of chitin, is soluble in acidic solutions, though solubility can vary based on the degree of deacetylation [16].

Chitin can be converted to chitosan through enzymatic or chemical hydrolysis processes. The global production rate of chitin is estimated at approximately 10^{11} tons annually [17]. It is also known as poly(β-(1→4)-N-acetyl-D-glucosamine), is the second most abundant polysaccharide in nature. It consists of poly-(1→4)-β-linked N-acetyl-D-glucosamine units [18].

Chitin exists in three distinct crystalline allomorphs α-, β-, and γ-forms which differ in the orientation of the microfibrils [19, 20]. Chitin's extensive hydrogen bonding network contributes to its structural rigidity and insolubility. Partial deacetylation of chitin yields chitosan, a linear polysaccharide composed of a random distribution of β-(1→4)-linked D-glucosamine (lacking the N-acetyl group) and N-acetyl-D-glucosamine (containing the N-acetyl group) units [14].

Structurally, chitosan consists of glucosamine units linked by β-(1→4)-glycosidic bonds, resulting in the presence of multiple functional groups, including amino and hydroxyl groups. The amino groups confer pH-sensitive solubility to chitosan: it is soluble in acidic environments (pH below its pKa of approximately 6.5) and becomes insoluble at higher pH levels. This pH-responsive behavior has been widely exploited in the design of signal-directed, on-demand hydrogel films for various biomedical and pharmaceutical applications [21, 22] .

2.2 Drug Delivery Strategies of Chitosan

Chitosan is a naturally derived biopolymer that offers several key advantages, including low cost, abundance, biocompatibility, and biodegradability. Its molecular structure, rich in amino and hydroxyl groups, enables a wide range of chemical interactions, such as hydrogen bonding and electrostatic interactions with hydrophilic molecules. These properties allow chitosan to form complexes with negatively charged DNA and RNA, as well as chelate metal ions, particularly under acidic conditions where its amine groups become protonated and positively charged. Chitosan's functional versatility is further influenced by parameters such as the degree of deacetylation, crystallinity, molecular weight, and chemical modification, all of which affect its antimicrobial activity and ability to promote cell adhesion [23].

The biocompatibility, biodegradability, and chemical reactivity of chitosan make it particularly valuable for biomedical applications. The polymer can be enzymatically degraded by lysozyme, a natural enzyme in the human body, rendering it ideal for drug delivery and tissue engineering purposes [24]. In addition, its cationic nature allows chitosan to disrupt microbial membranes, granting it intrinsic antimicrobial activity. This makes it highly useful in wound healing and food preservation. Furthermore, chitosan exhibits antioxidant activity, broadening its use in pharmaceutical formulations and biodegradable food packaging [25].

Chitosan-based hydrogels are typically formed through cross-linking techniques that generate a three-dimensional network. Physical cross-linking, based on non-covalent interactions such as hydrogen bonding and ionic forces, preserves biocompatibility but may result in hydrogels with weaker mechanical strength. Chemical cross-linking, on the other hand, creates covalent bonds using agents like glutaraldehyde or genipin, yielding more robust and stable hydrogels, though careful purification is required to mitigate potential cytotoxicity. Hybrid cross-linking strategies that combine both methods are increasingly used to balance mechanical integrity with biological performance [26].

2.3 *Applications*

The versatility of chitosan has led to its adoption across multiple domains. In the medical field, they are widely used in drug delivery systems, offering the ability to encapsulate bioactive compounds and release them in a controlled manner. Their antimicrobial and biocompatible properties also make them suitable for wound dressings and scaffolds in tissue engineering, where they support cell attachment and regeneration [24]. In the food industry, chitosan hydrogels are employed in biodegradable packaging and as carriers for functional food ingredients such as probiotics and nutrients [25]. Additionally, environmental applications include their use in water purification, where their porous structure and reactive groups allow for the adsorption of heavy metals and dyes from wastewater.

3 Fundamentals of Hydrogels

3.1 *Definition*

From scientific research to clinical practice, hydrogels have garnered increasing attention due to their transformative potential in biological and medical fields. The term "hydrogel" was first introduced by Van Bemmelen in 1894 [27]. Decades later, in 1960, Wichterle and Lim pioneered the biomedical application of hydrogels by demonstrating the use of poly(2-hydroxyethyl methacrylate) as an implantable filler after eye enucleation and as a material for fabricating contact lenses, drug carriers, and artificial arteries [28]. As research continues to uncover the exceptional physicochemical and biocompatible properties of these hydrophilic polymer networks, their potential to revolutionize a wide array of biomedical applications is becoming increasingly apparent [29].

3.2 Network Formation and Swelling Behavior

Hydrogels, as intricate polymeric networks, exhibit swelling behavior in response to a range of external stimuli, including changes in pH, temperature, electromagnetic radiation, and ionic strength [1]. This swelling is primarily driven by the hydration of hydrophilic and polar functional groups within the polymer chains. Upon hydration, these polar moieties expand, exposing hydrophobic domains that subsequently interact with surrounding water molecules. This interaction promotes further water uptake through osmotic forces, often resulting in extensive network expansion [30].

The porous structure of hydrogels can be readily tailored by adjusting the cross-linking density and the gel's affinity for the aqueous environment during swelling. This structural tunability, combined with the ability to rapidly entrap and release drug molecules, enables hydrogels to function as versatile drug delivery vehicles. They are particularly effective for multiple administration routes, including oral, vaginal, ocular, buccal, and parenteral delivery [31].

3.3 Mechanical and Rheological Properties of Hydrogels

Hydrogels are engineered using a variety of fabrication strategies, each tailored to the specific requirements of their intended application, for example, 3D printing techniques in tissue engineering. The formulation of hydrogels is largely determined by the number and arrangement of monomers, which in turn influences their structural and functional properties. Typically, the manufacturing process involves polymerization and cross-linking reactions, resulting in a diverse array of hydrogel architectures suitable for various biomedical and technological applications [32, 33].

3.4 Classifications of Hydrogels

The classification of hydrogels can be approached from several perspectives, depending on their composition, origin, crosslinking method, physical structure, and responsiveness to stimuli. Each classification provides insight into how hydrogels are synthesized, their structural characteristics, and their potential

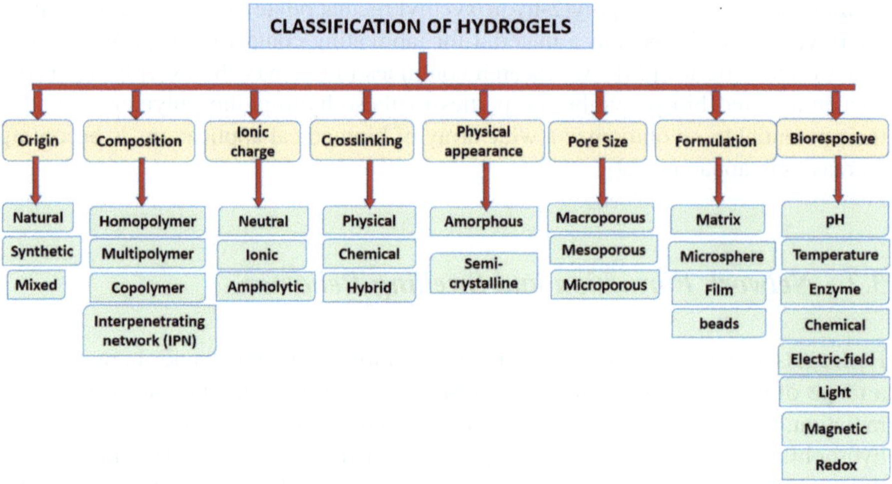

Fig. 1 Classification of hydrogels based on various parameters

applications. Here is an elaborated view of the major classifications, as illustrated in Fig. 1. This chapter focuses on some of the most significant classifications.

1. *Based on Origin*

Hydrogels can be broadly classified by origin into natural, synthetic, and hybrid types. Each class offers distinct advantages depending on the intended application.

(a) *Natural Polymer-Based Hydrogels*

Derived from biological sources such as plants, animals, or microorganisms, natural polymers (e.g., chitosan, collagen, alginate, hyaluronic acid, gelatin) exhibit biocompatibility, biodegradability, and inherent bioactivity. These properties make them ideal for regenerative medicine, tissue engineering, and drug delivery, as they mimic the extracellular matrix and support cell adhesion and proliferation [34, 35]. Chitosan, in particular, has been widely explored for its swelling behavior, 3D structural integrity, and drug encapsulation capabilities [36].

(b) *Synthetic Polymer-Based Hydrogels*

Synthetic hydrogels are created through controlled polymerization processes using monomers like poly(acrylic acid) (PAA), poly(N-isopropylacrylamide) (PNIPAAm), poly(ethylene glycol) (PEG), poly(vinyl alcohol) (PVA), and poly (HEMA) [37–39]. These materials offer tunable mechanical strength, predictable degradation, and functionalization flexibility for stimuli-responsive drug release, biosensing, and wound healing applications.

(c) *Hybrid Polymer-Based Hydrogels*

Hybrid or semi-synthetic hydrogels integrate natural and synthetic polymers to combine the bioactivity of natural materials with the mechanical strength and chemical tunability of synthetics [40, 41]. Two main fabrication strategies are used: physical blending, which relies on non-covalent interactions (e.g., hydrogen bonding), and covalent incorporation, which creates a more stable and homogenous network. Examples include silk fibroin/PVA and chitosan/PEG hydrogels (Ju et al. 2022). These hybrids are customizable in terms of degradation rate, drug release, and mechanical properties, and are particularly useful in tissue engineering, regenerative medicine, and drug delivery [42].

2. *Based on Composition*

Polymer-based hydrogels can also be categorized based on their composition, which influences their structural and functional performance. The main classes include homopolymer, copolymer, multipolymer, and interpenetrating network (IPN) hydrogels.

(a) *Homopolymer Hydrogels*

Formed from a single type of monomer, these hydrogels offer uniform structure and predictable properties. Polymers like polyacrylamide, PEG, Poly (N-isopropylacrylamide (PNIPAAm), agarose, gelatin form homopolymers enabling applications in drug delivery and tissue engineering. However, their limited functionality can restrict advanced applications.

(b) *Copolymer Hydrogels*

Synthesized from two or more monomers, often combining hydrophilic and functional monomers, these hydrogels exhibit enhanced tunability. Examples like Poly (ethylene glycol) diacrylate (PEGDA) and PAA-co-PEG enable biocompatibility, pH responsiveness, and controlled drug release, making them ideal for biomedical applications [43].

(c) *Multipolymer Hydrogels*

These incorporate three or more polymers to integrate diverse properties (e.g., strength, degradability, bioactivity). Systems like chitosan/PVA or PEG/PLA allow for precise control over mechanical, swelling, and degradation behaviors suitable for implants, drug delivery, and regenerative medicine [43].

(d) *Interpenetrating Network (IPN) Hydrogels*

Comprising two or more interlaced polymer networks, IPNs offer superior mechanical strength, stability, and stimuli responsiveness without compromising flexibility. IPNs, either full or semi, can be tailored via sequential or simultaneous polymerization and are used in drug delivery, tissue engineering, wound healing, and biosensing [41].

3. *Based on Crosslinking Mechanisms*

Hydrogels are versatile materials that can be classified into physical hydrogels and chemical hydrogels based on the nature of the crosslinking between polymer chains. This classification significantly influences their mechanical properties, stability, and applications in various fields, including biomedical engineering, drug delivery, and tissue scaffolding.

(a) *Physical Hydrogels*

Physical hydrogels are formed through non-covalent, reversible interactions between polymer chains. These interactions can include ionic, hydrogen bonding, hydrophobic, electrostatic forces, and polymer chain entanglements [44]. Unlike chemical hydrogels, the crosslinking in physical hydrogels is dynamic, and their structure can be altered by external stimuli like pH, temperature, and ionic strength. This reversibility allows for the stimuli-responsive behavior of these hydrogels, such as swelling or deswelling in response to changes in the environment, making them particularly useful in applications like controlled drug release and biosensing [45].

The formation of physical hydrogels typically occurs under mild conditions, meaning they do not require harsh chemicals or extreme temperatures. This makes them ideal for biomedical applications, particularly for injectable drug delivery systems [36].

Despite their advantages in biocompatibility and injectability, physical hydrogels tend to have lower mechanical strength compared to chemical hydrogels, which can limit their use in applications that require robust structural support. Moreover, they can be sensitive to changes in environmental conditions [46]. To enhance their mechanical properties, novel types of dual-crosslinked, nanocomposite, and slide-ring hydrogels are being developed.

(b) *Chemical Hydrogels*

In contrast to physical hydrogels, chemical hydrogels are created through covalent bonds formed between polymer chains, resulting in a permanent, stable three-dimensional network. The crosslinking process in chemical hydrogels is irreversible and typically involves the use of crosslinking agents or functional groups that react to form covalent bonds. Common crosslinking methods for chemical hydrogels include radical polymerization, Schiff base reactions, Michael addition, and epoxy crosslinking [45]. These reactions can be initiated by various stimuli such as heat, light, or chemical initiators.

While chemical hydrogels offer excellent mechanical performance and long-term stability, they often involve harsh synthesis conditions and the use of toxic chemicals. These factors can limit their applicability in sensitive environments, particularly in biomedical applications where biocompatibility is a critical concern [47]. Furthermore, the irreversible nature of chemical crosslinking makes these hydrogels less adaptable to changes in environmental conditions compared to physically crosslinked hydrogels.

The choice between physical and chemical hydrogels depends on the desired properties for a specific application. Physical hydrogels are preferred when stimuli-responsiveness, injectability, or biodegradability are essential. They are particularly useful for applications in drug delivery, tissue engineering, and regenerative medicine, where the ability to respond to specific triggers or undergo controlled degradation is beneficial [46]. On the other hand, chemical hydrogels are ideal for applications that require high mechanical strength, long-term stability, and resistance to degradation, such as in load-bearing implants and scaffolds for tissue engineering [48].

Overall, both physical and chemical hydrogels offer unique advantages and limitations. Their selection and optimization depend on the specific requirements of the application, including mechanical strength, environmental sensitivity, and biocompatibility.

4. *Routes of Administration*

(a) *Injectable Drug Delivery Product*

There has been a growing focus on injectable hydrogels compared to conventional gels in recent years. This is mainly due to their minimally invasive nature during surgery and ability to change shape in real time [49]. Injectable hydrogels can be implanted through a minimally invasive procedure, significantly reducing patient discomfort and pain and decreasing healing time. Furthermore, this approach is cost-effective and applicable to hard-to-reach tissue sites [50]. Injectable hydrogels with multiple functions can be utilized to treat cancer, diabetes, and gene therapy by enabling the effective delivery of numerous pharmaceuticals and other substances, including proteins [51].

(b) *Oral Drug Delivery Product*

The oral route is a popular drug delivery method owing to its numerous benefits, including sustained and controlled drug delivery, ease of administration, feasibility of solid formulation, and high level of patient compliance [52]. Hydrogels are employed in smart oral delivery systems for hydrophobic biological molecules, facilitating site-specific dispersion of medicinal compounds within the intricate GIT. Traditional methods encounter obstacles such as the limited permeability of the epithelial membrane and the denaturation of drugs [53].

Drug delivery systems can be designed to administer therapeutic substances to specific organs in a regulated and predictable manner [7]. There have been studies of controlled release systems used in various pH ranges within the body, including the oral, intestinal, gastric, and periodontal areas [7].

4 Characterizations of Hydrogels

Characterization of hydrogels is crucial because it provides a comprehensive understanding of their physical, chemical, thermal, and mechanical properties, which are essential for their effective use in various applications (Table 1). It helps identify functional groups, crosslinking, and polymer interactions, ensuring that the hydrogel has the correct chemical structure for its intended function. In biomedical applications, such as drug delivery and wound healing, characterization ensures the hydrogel is biocompatible, non-toxic, and exhibits appropriate degradation behavior.

Table 1 Characterization methods of hydrogels [36]

Serial no.	Technique	Method	Outcomes
1	Structural and chemical characterization	FTIR (Fourier Transform Infrared Spectroscopy)	Identifies functional groups and confirms chemical structure Detects crosslinking or chemical modifications in hydrogels
		NMR (Nuclear Magnetic Resonance)	Used for molecular-level structure analysis (if mentioned)
		XRD (X-ray Diffraction)	Evaluates crystallinity Important for mechanical strength and degradation behavior
2	Morphological analysis:	SEM (Scanning Electron Microscopy)	Shows surface and pore morphology Important for determining the internal network, which influences swelling and diffusion
		TEM (Transmission Electron Microscopy)	Nanoscale structural information
3	Thermal stability analysis	DSC (Differential Scanning Calorimetry):	Measures transitions like glass transition temperature (Tg), and melting point Useful for determining thermal stability
		Thermogravimetry analysis (TGA)	Thermal stability of the produced hydrogels Assesses weight loss with temperature To study thermal degradation
4	Mechanical resistance analysis	Texturometer Caliper Tensile testing instrument	To maintain structural integrity Important for applications like tissue engineering scaffolds
5	Viscosity testing	Viscometer	To determine the hydrogel viscosity
6	Swelling and degradation studies:	By imbibing the gels in simulated media and checking after fixed period of time for swelling and degradation	Swelling behavior in different pH/media Water absorption capacity Biodegradation analysis

Mechanical testing evaluates the hydrogel's strength and elasticity, which is particularly important in load-bearing applications like tissue engineering. Swelling studies reveal the hydrogel's capacity to absorb and retain water, a critical factor in applications like hygiene products and agriculture. Thermal analysis ensures the material remains stable under processing or storage conditions. Thus, characterization guides the design, development, and optimization of hydrogels to meet specific performance requirements, ensuring safety, functionality, and reliability in real-world use [54].

5 Stimuli-Responsive Chitosan Hydrogels

Chitosan-based hydrogels can be engineered to respond to various internal and external stimuli, facilitating precise drug delivery (Table 2). The major stimuli-responsive mechanisms include:

5.1 pH-Responsive Hydrogels

Chitosan is inherently pH-sensitive due to the presence of amino groups that become protonated under acidic conditions. This property enables drug release in response to the pH variations in different biological environments, such as the gastrointestinal tract and tumor microenvironment [80].

The most attractive route of controlled release is by pH triggering. The pH gradient in the human GI tract ranges from 1 to 7.5 (saliva, 5–6; stomach, 1–3; small intestine, 6.6–7.5; and colon, 6.4–7.0) [81]. Hydrogels can react to changes in ionic strength, which causes a volume phase transition (e.g., swelling leading to drug release). Here, the chemical properties of the constituent polymer are critical. Neutral or anionic polymers do not experience significant pH-sensitive behavior under acidic conditions, while the cationic chitosan (one of the few positively charged natural polymers) is responsive at low pH. Specifically, chitosan exhibits the pH-sensitive behavior of a weak polybase, dissolving easily at low pH, but remaining insoluble at higher pHs. To take advantage of this phenomenon, hydrogels that dissolve in the stomach have been prepared. Swelling of the hydrogels is due to a change in the charge balance inside of the hydrogel [82, 83].

By selecting polymers or polyelectrolytes based on their isoelectric points, hydrogels swelling and dissolution can be tailored for drug release at varying pHs. In addition, cross-linking the three-dimensional structure can make the polymer network stable in certain pH environments, but unstable under more extreme conditions. For instance, hydrogel swelling is limited in the stomach but triggered in the intestine due to a further change in pH. In one example, ionically cross-linked succinyl chitosan (Suc–Chi)/alginate hydrogel beads showed the release of as little as 11.6% of its nifedipine payload at pH 1.5 (3 h, which is the typical residence time in the stomach), and as much as 76% after 15 h at pH 7.4 [84].

In a study by Berger et al. (2004) on Suc–Chi/alginate hydrogel beads as polymeric carrier, had shown potential for drug delivery in the intestinal area. Such polyelectrolyte complexes were made sensitive to the higher pH due to the large number of carboxylated groups in the networks [85].

Further, research has shown swelling in IPN hydrogels is moderated by the amount of hydrophobic/hydrophilic groups and the cross-linking density [86]. In most cases, the contribution of the primary amines from chitosan is minimal. As a result, hydrogel networks were obtained with lesser swelling at an acidic pH and higher swelling at a pH close to neutral. pH-sensitive chitosan-acrylamide-grafted

Table 2 Elaboration of various stimuli-responsive chitosan-based hydrogels and their research outcomes

Stimuli mechanism	Drug	Formulation	Polymer	Research outcomes	References
Chemical (Glucose)	–	Polyelectrolyte microcapsules/	Alginate/chitosan	Responsive and stable microcapsules with glucose sensitivity	[55]
Glucose	Insulin	Microgels	Chitosan with enzyme nanocapsules	Closed-loop insulin release with glucose control Enzyme-triggered degradation under glucose	[56]
pH and glucose	–	Injectable in situ forming hydrogel	Phenylboronic-chitosan with oxidized dextran	Dual-responsive injectable hydrogel Good biocompatibility	[57]
Glucose	–	Nanoparticles	Glucose-conjugated chitosan	Targeted delivery with enhanced tumor uptake Receptor-mediated uptake by tumor cells	[58]
Glucose	Glucose oxidase	Films	Chitosan loaded with glucose oxidase	Electrocatalytic control of drug release Logic gate-like glucose detection and control	[59]
Glucose (diol binding)	Insulin	Crosslinked nanoparticle	Boronic acid-conjugated chitosan	Sustained insulin release in glucose-responsive manner Controlled release via diol–boronate interaction	[60]
Glucose	Insulin	In situ forming injectable hydrogel	Chitosan and hyluronic acid complex	Promising for adipose tissue regeneration Biodegradation-based release	[49]
Glucose	Insulin	Composite microparticles	Chitosan with concanavalin A and dextran	Effective insulin delivery with glucose responsiveness Competitive binding with glucose	[61]
pH and glucose	–	Nanocomposite hydrogel	Cationic chitosan and graphene oxide	Smart dual-responsive release behavior Swelling and pH-triggered release	[62]

Glucose	–	Hydrogel	Cationic chitosan	Excellent glucose-triggered release behavior Swelling-controlled release	[63]
pH and magnetic	Model compounds)	Injectable hydrogel	Catechol-functionalized chitosan	Swelling/de-swelling triggered by pH/magnetic stimuli Dual responsiveness with tunable mechanical properties for on-demand delivery	[64]
Magnetic and pH-responsive	Methotrexate	Nanoparticles	Magnetite/chitosan (core/shell)	pH-responsive drug release as observed by a sustained drug release over the next 90 h Excellent hemocompatibility and no cytotoxic effects Enhanced the antitumor activity	[65]
pH and oxidation	5-Flurouracil	Nanoparticles	Ferrocene-modified chitosan oligosaccharide	Enhanced drug release and targeted delivery with pH/oxidation responsiveness	[66]
pH/reduction	Doxorubicin	Supramolecular nanogels	Bioreducible crosslinking of β-cyclodextrin-modified chitosan	Enhanced anti-cancer efficacy with leakage-free release in tumor sites	[67]
pH/oxidation-responsive	5-FU	Nanoparticles	Ferrocene-modified chitosan	Applicable for tumor-targeted delivery, imaging, synergistic therapy Enhanced drug release and targeted delivery with pH/oxidation responsiveness	[67]
Temperature- and redox	Doxorubicin	Micelles	Poly(ε-caprolactone)-disulfide-poly(N, N-dimethylaminoethyl methacrylate) (PCL-SS-PDMAEMA), with oleic acid-modified Fe_3O_4 nanoparticles	Dual stimuli responsiveness The PDMAEMA segment imparts temperature sensitivity, while the disulfide linkage provides redox responsiveness, enabling controlled drug release in response to the tumor microenvironment Potential magnetic targeting and hyperthermia applications	[68]
Magnetic	Vancomycin	Microbeads	Chitosan cross-linked with varying lengths of polyethylene glycol dimethacrylate	Stimuli enabled repeatable and controlled drug release	[69]

(continued)

Table 2 (continued)

Stimuli mechanism	Drug	Formulation	Polymer	Research outcomes	References
pH	Model drugs		Chitosan and carboxymethyl cellulose (CMC) electrostatic (polyion complex)	Polysaccharide nanoparticles via. polyion complexation Simple, mild synthesis method suitable for pH-triggered release Swelling and degradation in response to pH and ionic interactions	[70]
Magnetic and passive targeting	Methotrexate & Rhodamine B	Nanocomposites	PEG and chitosan–iron oxide	PEGylated chitosan–iron oxide nanoparticles Magnetic targeting, enhanced cell uptake Enhanced tumor targeting and dual imaging (MRI/fluorescence)	[71]
Magnetic field	Doxorubicin	Nanoparticles	Chitosan–magnetic graphene (cmg) Magnetic	CMG hybrid nanoparticles Co-delivery of drug/genes Tumor-targeted delivery, imaging, synergistic therapy	[72]
Magnetic field	Doxorubicin	nanoparticles	Chitosan-graft-polymer with magnetite core	Sustained release Responsive to external magnetic fields	[73]
Magnetic field	Paclitaxel	micelles	Magnetic superparamagnetic iron oxide nanoparticles (SPIONS embedded in chitosan core)	High loading efficiency, magnetically triggered release Magnetic field controlled release High loading efficiency, magnetically triggered release	[74]
pH and electric field		IPN hydrogel	Chitosan and polyallylamine	Dual responsiveness to pH and electric fields Applications in biomedical devices, such as controlled drug delivery systems, biosensors, and soft actuators	[75]

Redox	Doxorubicin	Micelles	Polycaprolactone-g-chitosan oligosaccharide (PCL-g-COS)	Smart drug delivery system, tumor-specific release Potential to reduce side effects and improve treatment	[76]
Redox	Doxorubicin	Micelles	Glycolipid-like chitosan–stearic acid conjugate (CS–ss–SA)	Tumor-specific release, enhanced selectivity, improved safety	[77]
Redox/pH	Doxorubicin	Nanoparticles	Hollow mesoporous silica (HMSNS) polyethylene glycol and chitosan (CS)	Efficient loading of drug Stable at normal physiological pH and redox conditions Triggered release in acidic and reducing environments, mimicking tumor cells Enhanced cytotoxic effect against cancer cells due to smart release	[78]
Redox/pH/light	Taxol	Polymeric micelle	Photo-responsive fluorescence spiropyran as an imaging agent	Efficient encapsulation and light-triggered release of anticancer drug component also functions as a fluorescent probe for cell imaging	[79]

hydroxyethylcellulose (AAm-g-HEC)-based semi-IPN hydrogel microspheres have been prepared for pH-sensitive drug release, i.e., slower release at pH 1.2 and faster at pH 7.4, illustrating their potential drug release application in the GI tract [87]. Drug release from such an IPN is further affected by temperature if the hydrogel expresses thermoresponsive behavior (e.g., PNiPAM-chitosan) [86].

In addition to PEC and IPN networks, polyblend-based hydrogels also show pH sensitivity. If the blend component is neutral and hydrophilic, such as PEO and poly (vinyl pyrrolidone) (PVP), under acidic conditions, the primary amines of the chitosan are protonated, causing chain–chain repulsion and subsequent swelling or dissolution [88]. Freeze-dried chitosan-poly(ethylene oxide) (PEO) hydrogels were shown to selectively release antibiotics at low pH, indicating their possible use in localized therapeutic release in the stomach. Amoxicillin and metronidazole demonstrated hydrogel swelling in enzyme-free simulated gastric fluid (pH 1.2), allowing the release of most of the amoxicillin (65%) and metronidazole (59%), but showed limited release in simulated intestinal fluid (pH 7.2) after 2 h [88].

Hydrogel system is a promising candidate for on-demand drug release in response to pH stimuli. Bovine serum albumin (BSA) and fluorescein isothiocyanate (FITC)-dextran were used as drug models to assess release profiles in different pH environments. Increased acidity triggered the release of both BSA and FITC-dextran from the hydrogel. Highlighting the advantages of pH-responsive, self-healable, and injectable properties for reducing surgical trauma and improving drug bioavailability [89].

pH changes at the inflammatory site resulted in drug release or pH-responsive hydrogel with inflammatory-responsive characteristics and the capability to passively target macrophages [90]. The inflammatory response is initiated by B and T lymphocytes and sustained by polymorphonuclear leukocytes and macrophages. A naringin-loaded, amphipathic carboxymethyl-hexanoyl chitosan glycerol colloidal hydrogel with pH-responsive, inflammation-targeting properties was shown to be effective in an experimental periodontitis model. pH changes at the inflamed site triggered drug release, leading to reduced periodontal bone loss, decreased inflammatory cell infiltration, and downregulation of toll-like receptors, receptors for advanced glycation end products, and tumor necrosis factor expression [91].

Gao et al. crosslinked carboxymethyl [92] chitosan with an acid-labile ortho ester compound (OEDes) to adjust the pH-sensitivity and mechanical qualities. The hydrogels were wrapped around doxorubicin-loaded gelatin nanoparticles with an average diameter of 50 nm to create a hybrid system that could be inserted into a tumor location in any shape for local chemotherapy. They were administered intravenously and then tested for their ability to prevent tumor growth in mice harboring murine hepatoma tumors as compared to free doxorubicin. The outcomes demonstrated that the OEDes cross-linked chitosan hydrogels could greatly extend the doxorubicin release period as well as improve medication absorption into the tumor site.

The hydrogels, made from chitosan, poly(N-vinyl-2-pyrrolidone) (PVP), and 74% neutralized polyacrylic acid (PAA), showed enhanced swelling behavior depending on the amount of PVP. Increasing PVP content decreased water swelling.

The hydrogels responded most significantly to pH changes, swelling less in acidic conditions and more at neutral and basic pH (with maximum swelling at pH 8). Their biocompatibility and pH-responsive swelling made them suitable for wound healing and controlled drug delivery. Silver sulfadiazine-loaded hydrogels released 91.2% of the drug in a controlled manner within 80 minutes in phosphate buffer saline [93].

5.2 *Temperature-Responsive Hydrogels*

Temperature-responsive hydrogels are as attractive as physical gels because they can be customized for specific purposes. These materials possess a biologically compatible in situ setting process, allowing for the incorporation of cells, bioactive compounds, and even intricate 3-D structures [94]. Hydrogel has been extensively investigated for localized therapy, but few have been commercialized because of its complicated delivery system, difficulties in large-scale production, failure to pass clinical trials, and related side effects [95]. Thermo-responsive chitosan-based hydrogels are designed using temperature-sensitive polymers like poly (N-isopropylacrylamide) (PNIPAAm). These hydrogels undergo sol-gel transition at specific temperatures, facilitating on-demand drug release.

Taymouri et al. developed a thermosensitive chitosan-silk fibroin (CS2/SF0.5) hydrogel for localized, non-invasive drug delivery in bone tissue engineering. They embedded dipyridamole-loaded polycaprolactone nanoparticles (DIP-PCLNPs) into the hydrogel, optimizing nanoparticle formulation for stability and uniformity. The addition of 0.5% silk fibroin improved the hydrogel's gelation time and compressive strength. In vitro studies with MG-63 cells showed that the composite hydrogel enhanced cell viability, osteogenic activity (ALP), and calcium deposition, indicating strong potential for bone regeneration applications [96].

These research findings highlight the potential of the DIP-PCLNPs-CS2/SF0.5 thermosensitive hydrogel as a promising and effective local drug delivery platform for enhancing bone healing and regeneration.

5.3 *Enzyme-Responsive Hydrogels*

Enzyme-sensitive linkers enable targeted drug release by responding to enzymatic activity, particularly useful in tissues with enzyme overexpression. Enzyme-responsive polymers can serve both as metabolic indicators and targeted drug delivery systems. For self-regulated insulin delivery, systems with glucose-sensing and automatic shut-off capabilities are ideal. Incorporating glucose oxidase into pH-responsive, DNA-based hydrogels creates a glucose-triggered matrix for controlled insulin release, mimicking an artificial pancreas. Additionally, hydrogel microcarriers formed by cross-linking oligochitosan, Ca^{2+}, and pectin allow for slow drug release in the upper gastrointestinal tract and rapid release under colon-mimicking conditions [97]. The in situ triggered gel formation and disintegration of polymeric hydrogel provides them with applicability in different areas such as tissue engineering, injectable hydrogels, and controlled release delivery systems.

Targeting localized enzymes has enabled better control of drug release within the body. Chitosan-based and other polysaccharide hydrogels have been utilized for colon-specific drug delivery, exploiting the enzymatic activity in the gastrointestinal tract. The diverse bacterial population in the intestine secretes various polysaccharide-degrading enzymes (e.g., glucosidase, galactosidase, amylase, pectinase, xylanase, xylosidase, dextranase), making these polymers ideal for site-specific degradation and drug release [98, 99].

As such, chitosan polymers are readily cleaved when exposed to these local enzymes, leading to the release of the entrapped drug molecules. These chitosan-based hydrogel capsules have demonstrated success in locally releasing insulin [100]. In a research, chitosan capsules containing insulin (with or without additives like protease inhibitors or absorption enhancers) were coated with hydroxypropyl methylcellulose phthalate for colon-specific drug delivery. These capsules remained stable in the stomach and small intestine of Male Wistar rats but were specifically broken down by microbial activity in the colon (cecal contents), leading to targeted insulin absorption in the colon [101].

Beyond direct polymer degradation, enzymatic activity can induce local pH changes within hydrogel microenvironments, influencing drug release. For example, insulin-loaded hydrogels containing immobilized glucose oxidase generate a pH drop by converting glucose to gluconic acid. The resulting acidic environment enhances insulin solubility, promoting greater drug permeation from the porous hydrogel matrix [80].

5.4 Chemical (Glucose) Responsive Hydrogels

Chitosan-based chemical-responsive hydrogels represent a promising platform for smart drug delivery systems, particularly due to their biocompatibility, biodegradability, and responsive behavior to stimuli such as pH, glucose, and doxorubicin conditions. These hydrogels are particularly effective for diseases like diabetes and cancer, where specific physiological triggers such as glucose levels or pH changes can be harnessed to control drug release.

One of the most significant applications of chitosan-based hydrogels is in glucose-responsive insulin delivery. For instance, Gu et al. (2013) created microgels integrated with enzyme nanocapsules containing glucose oxidase. In this system, the presence of glucose triggered gluconic acid production, lowering the local pH and triggering insulin release in a closed-loop fashion that closely mimics pancreatic function [56]. Similarly, Belbekhouche et al. (2018) developed glucose-sensitive microcapsules using alginate and chitosan polyelectrolyte complexes [55]. These capsules exhibited a selective response to glucose, indicating potential for smart insulin release in diabetic therapy. Other studies, such as Yin et al. (2010), utilized concanavalin A in combination with chitosan and dextran to fabricate microparticles capable of reversible glucose binding, further contributing to glucose-sensitive drug delivery mechanisms [61]. Siddiqui et al. (2016) enhanced this approach by using boronic acid-conjugated chitosan nanoparticles [60]. The boronic acid groups interacted selectively with diol-containing glucose molecules, enabling sustained and glucose-dependent insulin release.

Furthermore, Liu et al. (2014) demonstrated the adaptability of chitosan for more complex bioelectronic systems. They engineered a logic gate system using glucose oxidase entrapped in chitosan films to achieve multi-input/output sensing for future diagnostic or therapeutic devices. Li et al. (2014) took advantage of glucose-conjugated chitosan nanoparticles to target tumor cells through glucose transporter-mediated uptake, highlighting another application beyond glucose sensing alone.

In addition to glucose sensitivity, some chitosan hydrogels are designed to be dual-responsive to both glucose and pH. This is particularly relevant in environments such as tumors or infected tissues, where both glucose concentration and pH are altered. Li et al. (2015) developed an injectable hydrogel system by in situ crosslinking phenylboronic acid-modified chitosan with oxidized dextran. This hydrogel responded to both low pH and high glucose concentrations, enabling controlled and site-specific drug delivery [57]. Zhao et al. (2017) further advanced this concept by incorporating graphene oxide into cationic chitosan hydrogels [62]. This combination improved drug loading and allowed for more precise, dual-stimuli responsive release behavior.

Beyond stimuli responsiveness, injectable and biodegradable hydrogels have shown promise for tissue regeneration and localized therapy. Tan et al. (2010), for instance, created injectable chitosan-hyaluronic acid hydrogels for adipose tissue regeneration, which could also be adapted for responsive drug delivery applications by integrating suitable stimuli-sensitive components [49].

Therefore, chitosan-based chemical-responsive hydrogels offer a flexible, tunable, and highly effective means of achieving targeted, on-demand drug delivery. Their responsiveness to glucose, pH, or both enables precise control over therapeutic release, minimizing side effects and improving patient outcomes. With ongoing research into multifunctional and nanocomposite hydrogels, chitosan continues to be a cornerstone in the development of next-generation drug delivery systems.

5.5 *Electric-Responsive Hydrogels*

Electric-responsive chitosan hydrogels are smart biomaterials that change their physical properties—such as shape, volume, or drug release behavior when exposed to an electric field. These hydrogels are typically made by incorporating chitosan (a biocompatible, biodegradable polymer) with ionic components or forming composite networks (e.g., with PHEMA or montmorillonite) [102, 103].

Polymer hydrogels are elastic, cross-linked networks with fluid-filled spaces that can reversibly change shape and volume in response to environmental stimuli like temperature, pH, or electric fields. Electric field-responsive hydrogels are especially promising for integration into advanced functional systems, including modern robotic control technologies [75, 103].

Electro-sensitive hydrogels, composed of polyelectrolytes and ionic polymer networks, respond to electrical stimuli by swelling, shrinking, or bending. These behaviors have been intensively studied over the past two decades to better understand their underlying mechanisms [103].

Kim et al. (2004) developed chitosan/PHEMA hydrogels, which exhibited reversible electromechanical bending when exposed to an electric field, driven by ion migration and osmotic pressure differences. Their response is influenced by NaCl concentration—low salt levels enhance bending, while high concentrations reduce it due to ionic shielding. These properties make the hydrogels promising smart materials for applications in soft robotics, sensors, and biomedical actuators, especially in low-salt environments [104].

The research by Kaewpirom et al. (2006) explored the electromechanical properties of hydrogels made from chitosan and poly(hydroxyethyl methacrylate) (PHEMA) in sodium chloride (NaCl) solutions under applied electric fields [103]. The goal was to assess their suitability as electrically responsive smart materials for applications such as actuators and sensors.

Such chitosan–PHEMA hydrogels, created through an interpenetrating polymer network, showed improved mechanical strength and swelling capacity. When exposed to an electric field in NaCl solutions, they bent due to ion migration and osmotic pressure differences. The bending response decreased with higher NaCl concentrations, likely due to ionic shielding reducing charge mobility. These findings suggest strong potential for the hydrogels in soft actuators, biosensors, and smart biomedical devices [103].

Both studies explore chitosan-based hydrogels responsive to electric fields in ionic environments, showing potential for smart biomedical devices. Kaewpiromet al. (2006) focused on deformation behavior modulated by PEG content. Kim et al. (2004) demonstrated reversible bending, highlighting soft actuator potential.

The electro-sensitive drug release method uses an external electric field to trigger drug delivery from implanted hydrogels, originally developed for non-chitosan systems. A biodegradable hydrogel containing the drug is implanted subcutaneously, and an electro-conducting patch is applied on the skin to activate release. Once the desired dose is delivered, the patch is removed. Drug release occurs through mechanisms such as deswelling-driven ejection, diffusion, electrophoresis, and electro-induced gel erosion [105].

A chitosan–montmorillonite (MMT) nanocomposite hydrogel was developed to study electrostimulated drug release. At specific MMT concentrations, vitamin B12 showed pseudo-zero-order release, shifting from diffusion-controlled to swelling-controlled under electrical stimulation. Increasing MMT content reduced electroresponsiveness but improved anti-fatigue properties. Notably, the hydrogel with 2 wt.% MMT demonstrated reliable, pulsatile drug release and superior mechanical durability compared to pure chitosan hydrogels [106].

Although electro-sensitive chitosan hydrogels can be used to controlled drug release by modulating an electric field, achieving the desired release rate under physiological conditions remains challenging.

5.6 *Magnetic and Light-Responsive Hydrogels*

The incorporation of magnetic nanoparticles or photothermal agents into chitosan-based hydrogels allows external stimuli, such as magnetic fields or laser irradiation, to trigger drug release with spatial and temporal precision.

A study, developed multi-stimuli responsive chitosan-based microcapsules (MSRS-CS-MCs) for targeted delivery and controlled release of hydrophobic drugs. Fabricated using a sonochemical method from folic acid (FA)-functionalized thiolated chitosan, these microcapsules incorporate: wherein FA for cancer cell targeting (via folate receptors), Rhodamine B for fluorescence tracking, Magnetic nanoparticles (OA-Fe_3O_4) for magnetic responsiveness, Coumarin 6 as a model hydrophobic drug. These 500 nm spherical microcapsules exhibit magnetic and reduction-responsive drug release, targeted delivery to HeLa cells, and fluorescence imaging capability. Their biocompatibility, non-immunogenicity, and multi-functionality make them promising nanocarriers for future biomedical applications [107].

MSRS-CS-MCs were engineered with folic acid (FA) and red fluorescent dye (RITC) on their surface, while magnetic nanoparticles (OA-Fe_3O_4) and Coumarin 6, a model hydrophobic drug, were encapsulated inside. Hydrophobic drugs were easily loaded by dissolving them in soybean oil before sonication. These microcapsules demonstrated strong magnetic responsiveness, targeted delivery to HeLa cells via folate receptors, and reduction-triggered drug release. This makes MSRS-CS-MCs promising multi-stimuli responsive nanocarriers for targeted and controlled delivery of hydrophobic drugs in future biomedical applications [107].

5.7 *Redox-Responsive Hydrogels*

Redox-sensitive chitosan-based hydrogels utilize disulfide bonds that undergo cleavage in the presence of high levels of glutathione (GSH), commonly found in tumor cells. This selective drug release mechanism enhances the therapeutic efficacy of anticancer drugs. Among these stimuli, redox potential is frequently investigated due to the tremendous difference in the concentration of glutathione (GSH) between the extracellular fluids (10 μM) and intracellular fluids (1–10 mM). Moreover, the cytosolic GSH concentration in many tumor cells was at least threefold higher than that in normal cells [108].

The researchers synthesized redox-responsive micelles using a polycaprolactone-grafted chitosan oligosaccharide (PCL-g-COS) copolymer. The design exploits the redox conditions inside cancer cells to trigger controlled release of the drug. Synthesis and characterization of the PCL-g-COS copolymer confirmed successful grafting and self-assembly into micelles. The micelles displayed high encapsulation efficiency for DOX and had an average size suitable for tumor targeting via the enhanced permeability and retention (EPR) effect. In normal conditions: micelles stay stable. Under reducing conditions (simulating the intracellular environment of tumor cells), the micelles disassembled (breakage of disulfide bonds), releasing doxorubicin efficiently. The micelles showed low cytotoxicity in the absence of doxorubicin and enhanced cytotoxicity when loaded with doxorubicin in vitro against cancer cells, indicating effective drug delivery [76].

Previous work developed a chitosan-based glycolipid-like nanocarrier (CS–ss–SA) for redox-responsive drug delivery targeting tumor cells with high glutathione (GSH) levels. The nanocarrier self-assembles into micelles, efficiently loading doxorubicin and remaining stable under normal conditions. In tumor-like environments, the disulfide bonds break due to high GSH, causing micelle destabilization and selective drug release inside cancer cells. The carrier alone shows low toxicity, while doxorubicin-loaded micelles exhibit enhanced cancer cell killing with preferential targeting, reducing side effects on healthy tissue. This makes CS–ss–SA micelles a promising platform for precise, tumor-specific chemotherapy [77]. Guerry et al. focus on a structural innovation (PCL-g-COS) for passive targeting. Hu et al. emphasize intracellular selectivity with a glycolipid-mimetic design for higher specificity in tumor cells.

Multifunctional polymeric micelle systems of Taxol, with photo-responsive fluorescence polymer as an imaging agent, responded to three distinct stimuli: pH, redox conditions, and light (photo-responsiveness) showing controlled anticancer drug delivery and cell imaging. The micelles were synthesized using a spiropyran-based photochromic polymer and linked via boronate ester and disulfide bonds. Disulfide bonds cleave in reductive (high GSH) environments. Boronate ester bonds dissociate in acidic pH (tumor environment). Spiros undergo reversible structure change under UV/visible light, enabling photo-responsiveness. Rapid disassembly and drug release in tumor-mimicking environments (acidic pH, high GSH, or light exposure). Demonstrated both cytotoxicity against cancer cells and imaging capability in vitro [79].

Although redox-responsive DDS based on the disulfide bonds have achieved "on demand" drug release, the dual-stimuli responsive DDS have been designed to further realize optimal drug delivery and release performances in which the gatekeepers could be removed by responding to two stimuli [109]. It is well known that the pH in the microenvironment inside tumors and inflammatory tissues is more acidic than in normal tissues and blood, and endosomes and lysosomes show even lower pH values. The study presented a novel dual-stimuli-responsive nanocarrier system based on hollow mesoporous silica nanoparticles (HMSNs) modified with PEG and chitosan via. disulfide linkers, enabling controlled drug release in response to both pH and redox conditions commonly found in tumor microenvironments [78].

The dual-responsive HMSN-based system provided a promising platform for precise and controlled drug delivery. By responding to both redox and pH stimuli, it enables targeted release within tumor cells, increasing therapeutic efficiency while reducing harm to healthy tissues. HMSNs were modified with PEG and chitosan via disulfide linkages. This created a gatekeeping shell responsive to intracellular GSH (redox) and acidic pH conditions. Redox-triggered release: Disulfide bonds are cleaved in high-Glutathione environments (e.g., tumor cells). pH-responsive release: Chitosan becomes protonated in acidic environments (endosomes), facilitating drug release [78].

6 Merits and Demerits

Hydrogels possess numerous merits that make them highly valuable in biomedical applications, particularly in drug delivery, tissue engineering, and regenerative medicine. One of their key advantages is their high water content, which closely resembles the natural extracellular matrix of biological tissues, thereby promoting excellent biocompatibility and minimizing immune reactions. Hydrogels can be engineered with tunable physical and chemical properties, such as porosity, degradation rate, and mechanical strength, enabling them to be customized for specific therapeutic needs [110]. Their ability to respond to various biological or environmental stimuli, including pH, temperature, glucose, and redox conditions, enables controlled, site-specific drug delivery, which enhances therapeutic efficacy while reducing systemic side effects.

Furthermore, many hydrogels are injectable and can form gels in situ, allowing for minimally invasive administration, which is especially advantageous for delivering drugs to difficult-to-reach areas or for tissue regeneration. They are also effective carriers for encapsulating and protecting sensitive biomolecules such as proteins, peptides, or nucleic acids from degradation, thereby preserving their bioactivity. Hydrogels allow for versatile drug release profiles, including sustained and pulsatile delivery, depending on their structural configuration [29]. In tissue engineering, they provide a supportive environment for cell adhesion, growth, and differentiation, making them suitable as scaffolds for repairing damaged tissues. Moreover, the presence of modifiable functional groups in natural polymers like chitosan or alginate allows easy functionalization of hydrogels with targeting ligands, imaging agents, or other therapeutic components. Collectively, these attributes make hydrogels a versatile and powerful platform in modern biomedical science.

Chitosan-based bioresponsive hydrogels show great promise but face key limitations for clinical use, notably their low mechanical strength. This weakness is due to their high water content and low friction between polymer chains, making them fragile and prone to damage under stress [111]. To overcome this drawback, researchers have developed composite hydrogel structures by incorporating reinforcing elements such as silica nanoparticles to enhance mechanical robustness and structural stability [112].

Another challenge is their sensitivity to environmental conditions, such as pH and ionic strength, which can affect gel stability and drug release or leakage unpredictably. Dehydration is another concern, particularly in thermos-responsive hydrogels, where exposure to temperatures above the lower critical solution temperature (LCST) can lead to shrinkage and stiffening, locking the hydrogel in a less dynamic state and compromising its functionality [113, 114].

Additionally, chitosan's limited solubility at physiological pH reduces its bioavailability and restricts its processing and formulation into stable hydrogel systems without chemical modification. Potential batch-to-batch variability in natural chitosan sources and the need for extensive purification can also impact reproducibility and regulatory approval.

Chitosan is generally biocompatible, but its degradation products can sometimes cause mild inflammation or disrupt the local tissue environment, particularly in sensitive treatments. Additionally, large-scale manufacturing and sterilization of chitosan hydrogels pose technical challenges, limiting their scalability and commercial use. These issues call for further research to optimize their therapeutic potential (Table 3).

Interestingly, some hydrogels can induce localized hypoxia (low oxygen levels), which can be a drawback or benefit depending on the application. For example, hypoxia can promote angiogenesis and tissue regeneration by activating hypoxia-related pathways, aiding healing and neovascularization [115].

Despite these limitations, the overall advantages of hydrogels, such as their biocompatibility, tunable responsiveness, and controlled drug release capabilities, far outweigh their disadvantages, making them highly promising for commercial application in drug delivery systems.

Table 3 Merits and demerits associated with hydrogel drug delivery systems

S. No.	Merits	Demerits
1.	High water content and biocompatibility	Relatively low mechanical strength
2.	Tunable physical and chemical properties	Gel stability and drug release behavior unpredictably
3.	Stimuli-responsive behavior	Unintended drug leakage or loss of structural integrity
4.	Minimal invasiveness and injectable forms	Dehydration
5.	Encapsulation and protection of therapeutics	Chitosan's limited solubility at physiological pH
6.	Versatile drug loading and release profiles	Potential batch-to-batch variability
7.	Support for cell adhesion and growth	Chitosan is generally biocompatible; its degradation products can sometimes induce mild inflammation
8.	Ease of functionalization	Large-scale manufacturing and sterilization of chitosan-based hydrogels remain technically challenging

7 Biomedical Applications

Biomedical bio-responsive chitosan hydrogels are a class of smart materials that respond to specific physiological stimuli such as pH, temperature, redox conditions, enzymes, or glucose levels to enable targeted and controlled drug delivery, tissue regeneration, and diagnostic applications. As discussed earlier, chitosan, a natural polysaccharide derived from chitin, is widely used in biomedical hydrogel systems due to its excellent biocompatibility, biodegradability, non-toxicity, and intrinsic antimicrobial properties. Its reactive functional groups (amine and hydroxyl) allow for chemical modification, enabling the development of tailored, stimuli-sensitive hydrogel systems.

These hydrogels are particularly valuable in drug delivery, where they can remain inert during circulation and release therapeutic agents in response to specific internal triggers like acidic tumor microenvironments, elevated glucose levels (as in diabetes), or oxidative stress. This selective responsiveness enhances drug efficacy, reduces systemic toxicity, and improves patient compliance. In tissue engineering, bio-responsive chitosan hydrogels can adapt to the local environment to support cell proliferation and tissue remodeling, sometimes releasing growth factors or other bioactive agents in response to changes in the surrounding tissue.

Moreover, their tunable swelling, gelation, and degradation behaviors allow for precise control over mechanical properties and functional performance, making them suitable for applications ranging from wound healing and cancer therapy to ocular and cardiovascular treatments. With advances in nanotechnology and polymer chemistry, bio-responsive chitosan hydrogels continue to evolve as multifunctional platforms with potential for personalized and regenerative medicine.

Chitosan, a cationic polysaccharide derived from the deacetylation of chitin, has garnered significant attention in biomedical engineering due to its exceptional biocompatibility, biodegradability, inherent antimicrobial properties, and ease of chemical modification. Chitosan-based hydrogels, in particular, have emerged as versatile platforms for a wide array of biomedical applications, including drug delivery, tissue engineering, wound healing, and biosensing. These hydrogels can be designed to respond to specific stimuli, such as pH, temperature, ionic strength, or the presence of specific biomolecules, enabling controlled and targeted therapeutic interventions. The ability of chitosan to interact with negatively charged cell membranes due to its positive charges enhances its cytocompatibility compared to chitin, making it suitable for biomedical and pharmaceutical applications like sutures, dental and bone implants, and artificial skin [116].

Chitosan hydrogels can be tailored to exhibit responsiveness to a range of biologically relevant stimuli, making them ideal for controlled drug delivery systems. By strategically incorporating stimuli-responsive elements, such as pH-sensitive functional groups (e.g., carboxylic acids or amines) or temperature-sensitive polymers (e.g., Pluronics or poly), the hydrogels can be engineered to release encapsulated drugs in response to specific physiological cues [117]. For instance, in the context of cancer therapy, pH-sensitive chitosan hydrogels can be designed to release chemotherapeutic agents selectively in the acidic microenvironment of tumors, minimizing off-target effects and maximizing therapeutic efficacy [13, 118]. Similarly, temperature-sensitive chitosan hydrogels can be utilized for localized delivery of drugs to inflamed tissues or organs, where the elevated temperature triggers drug release. Chitosan's chemical structure and active chemical groups also allow for modification for tailoring to specific biomedical applications. Moreover, the inherent bioadhesive properties of chitosan facilitate prolonged retention at the site of administration, enhancing drug bioavailability and therapeutic outcomes. The incorporation of nanoparticles and liposomes into chitosan hydrogels allows for the adsorption, absorption, and encapsulation of both drugs of interest and sensitive therapeutics [119].

In the realm of tissue engineering, chitosan hydrogels provide a conducive three-dimensional scaffold for cell adhesion, proliferation, and differentiation, promoting tissue regeneration and repair [120]. The tunable mechanical properties, biocompatibility, and biodegradability of chitosan hydrogels make them an attractive material for creating tissue-engineered constructs for various applications, including cartilage repair, bone regeneration, and skin regeneration. Chitosan's excellent film-forming capabilities, permeability, mechanical strength, and low cost also enhance its value in tissue engineering applications [117, 121]. By incorporating growth factors or other bioactive molecules into the hydrogel matrix, the cellular microenvironment can be further tailored to enhance tissue regeneration. Furthermore, the inherent antimicrobial properties of chitosan can prevent infection at the implantation site, promoting successful tissue integration.

8 Challenges and Future Perspectives

Despite significant progress, challenges remain in the large-scale production, mechanical stability, and long-term biocompatibility of chitosan-based hydrogels (Fig. 2). Future research should focus on optimizing crosslinking techniques, exploring novel hybrid materials, and advancing smart drug delivery systems integrated with biosensors. Chitosan-based hydrogels are particularly attractive due to their excellent biocompatibility, but their clinical application is hampered by complex components and preparation processes.

Chitosan-based bioresponsive hydrogels have shown great promise for applications in controlled drug delivery, tissue engineering, and biosensing. However, several challenges hinder their clinical translation and large-scale application. One primary issue is the batch-to-batch variability of chitosan, stemming from differences in its source and degree of deacetylation. This inconsistency can lead to unpredictable physicochemical and biological properties, affecting reproducibility in research and manufacturing processes. Furthermore, limited mechanical strength and stability under physiological conditions restrict their application, especially in load-bearing or long-term in vivo environments. One major limitation is their mechanical strength, which must be enhanced without sacrificing biocompatibility.

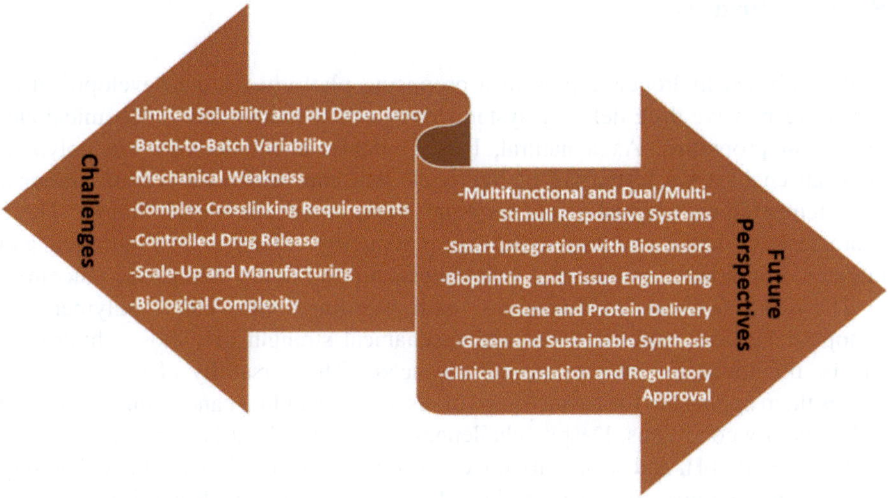

Fig. 2 Diagrammatic representation of various challenges and future perspectives associated with stimuli-responsive chitosan drug delivery systems

Moreover, variations in chitosan source, degree of deacetylation, and molecular weight can result in inconsistent hydrogel properties, underscoring the need for standardized preparation protocols [26]. Scalability also remains a concern for industrial applications. Future research is directed toward addressing these issues by developing stimuli-responsive hydrogels and exploring novel functionalizations for advanced applications in biosensing and regenerative medicine.

Another significant challenge is the precise control over the responsiveness and release kinetics of the hydrogels. Many stimuli-responsive systems show delayed or incomplete responses to environmental triggers such as pH, temperature, or redox changes. In addition, cytotoxicity concerns associated with certain chemical modifications, crosslinking agents, or degradation by-products must be addressed to ensure biocompatibility. Scaling up the production of these hydrogels while maintaining uniformity and functionality is also a key hurdle, especially when integrating multifunctional components like nanoparticles or enzymes.

Looking ahead, the future of chitosan-based bioresponsive hydrogels is promising with several research directions offering potential breakthroughs. Advanced chemical modifications and hybridization with nanomaterials could enhance their responsiveness, mechanical strength, and targeted delivery capabilities. The integration of smart electronics, 3D printing, and biosensing technologies is expected to lead to the development of next-generation intelligent hydrogel systems [118]. Moreover, personalized medicine applications could benefit from hydrogels tailored to individual patient needs through tunable degradation rates and drug release profiles. Finally, addressing regulatory and standardization issues will be crucial for facilitating clinical translation. With continued interdisciplinary collaboration, chitosan-based bioresponsive hydrogels are likely to become a cornerstone in precision biomedical applications.

9 Conclusions

Chitosan-based hydrogels represent a promising platform for the development of stimuli-responsive drug delivery systems due to their unique physicochemical and biological properties. As a natural, biodegradable, and biocompatible polymer, chitosan enables the formation of hydrogels that can respond to various internal (pH, temperature, enzymes) and external (light, magnetic field) stimuli. These characteristics make chitosan hydrogels particularly effective for controlled and targeted drug release, minimizing side effects and improving therapeutic outcomes.

Through chemical modifications and blending with other polymers or nanoparticles, the responsiveness and mechanical strength of chitosan hydrogels can be fine-tuned to suit specific clinical needs. The versatility of these systems makes them applicable for a wide range of diseases, including cancer, infections, and inflammatory conditions. Despite challenges such as batch variability, limited solubility at neutral pH, and scalability concerns, ongoing research and nanotechnology integration continue to advance their functionality and real-world application potential.

Conclusively, chitosan-based stimuli-responsive hydrogels hold significant promise as intelligent drug carriers in next-generation biomedical therapies, offering precise delivery, enhanced patient compliance, and reduced systemic toxicity.

References

1. Farjadian, F., Mohseni, F., Raeisi, A., Sabet, R., & Daneshamouz, S. (2024). Smart stimuli responsive hydrogels; fascinating platform for therapeutic applications. *Journal of Mazandaran University of Medical Sciences., 33*(228), 161–190.
2. Iqbal, Y., Ahmed, I., Irfan, M. F., Chatha, S. A. S., Zubair, M., & Ullah, A. (2023). Recent advances in chitosan-based materials; the synthesis, modifications and biomedical applications. *Carbohydrate Polymers, 321*, 121318.
3. Cascone, S., & Lamberti, G. (2020). Hydrogel-based commercial products for biomedical applications: A review. *International Journal of Pharmaceutics, 573*, 118803.
4. Bahram, M., Mohseni, N., & Moghtader, M. (2016). An introduction to hydrogels and some recent applications. In *Emerging concepts in analysis and applications of hydrogels*. IntechOpen.
5. Ahmad, S., Ahmad, M., Manzoor, K., Purwar, R., & Ikram, S. (2019). A review on latest innovations in natural gums based hydrogels: Preparations & applications. *International Journal of Biological Macromolecules, 136*, 870–890.
6. Kesharwani, P., Bisht, A., Alexander, A., Dave, V., & Sharma, S. (2021). Biomedical applications of hydrogels in drug delivery system: An update. *Journal of Drug Delivery Science and Technology., 66*, 102914.
7. Gull, N., Khan, S. M., Butt, M. T. Z., Zia, S., Khalid, S., Islam, A., et al. (2019). Hybrid cross-linked hydrogels as a technology platform for in vitro release of cephradine. *Polymers for Advanced Technologies, 30*(9), 2414–2424.
8. Chaturvedi, K., Ganguly, K., Nadagouda, M. N., & Aminabhavi, T. M. (2013). Polymeric hydrogels for oral insulin delivery. *Journal of Controlled Release, 165*(2), 129–138.

9. Vashist, A., Vashist, A., Gupta, Y., & Ahmad, S. (2014). Recent advances in hydrogel based drug delivery systems for the human body. *Journal of Materials Chemistry B, 2*(2), 147–166.
10. Farrukh, F., Zaman, M., & Kashif Rehman, M. U. (2023). Pharmaceutical and biomedical applications of hydrogels. *World Journal of Pharmacy and Pharmaceutical Sciences, 12*, 115.
11. Yu, S., Zhang, X., Tan, G., Tian, L., Liu, D., Liu, Y., et al. (2017). A novel pH-induced thermosensitive hydrogel composed of carboxymethyl chitosan and poloxamer cross-linked by glutaraldehyde for ophthalmic drug delivery. *Carbohydrate Polymers, 155*, 208–217.
12. Salahuddin, A., Ashraf, A., Ahmad, K., & Hou, H. (2024). Recent advances in chitosan-based smart hydrogel for drug delivery systems. *International Journal of Biological Macromolecules., 280*, 135803.
13. Zhu, H., Sun, H., Dai, J., Hao, J., & Zhou, B. (2024). Chitosan-based hydrogels in cancer therapy: Drug and gene delivery, stimuli-responsive carriers, phototherapy and immunotherapy. *International Journal of Biological Macromolecules, 282*, 137047.
14. Kaczmarek, M. B., Struszczyk-Swita, K., Li, X., Szczęsna-Antczak, M., & Daroch, M. (2019). Enzymatic modifications of chitin, chitosan, and chitooligosaccharides. *Frontiers in Bioengineering and Biotechnology, 7*, 243.
15. Rinaudo, M. (2006). Chitin and chitosan: Properties and applications. *Progress in Polymer Science, 31*(7), 603–632.
16. Park, B. K., & Kim, M.-M. (2010). Applications of chitin and its derivatives in biological medicine. *International Journal of Molecular Sciences, 11*(12), 5152–5164.
17. Merzendorfer, H., & Zimoch, L. (2003). Chitin metabolism in insects: Structure, function and regulation of chitin synthases and chitinases. *Journal of Experimental Biology, 206*(24), 4393–4412.
18. Kumirska, J., Weinhold, M. X., Thöming, J., & Stepnowski, P. (2011). Biomedical activity of chitin/chitosan based materials—Influence of physicochemical properties apart from molecular weight and degree of N-acetylation. *Polymers, 3*(4), 1875–1901.
19. Pusateri, A. E., McCarthy, S. J., Gregory, K. W., Harris, R. A., Cardenas, L., McManus, A. T., et al. (2003). Effect of a chitosan-based hemostatic dressing on blood loss and survival in a model of severe venous hemorrhage and hepatic injury in swine. *The Journal of Trauma: Injury, Infection, and Critical Care, 54*(1), 177–182.
20. Friedman, A. J., Phan, J., Schairer, D. O., Champer, J., Qin, M., Pirouz, A., et al. (2013). Antimicrobial and anti-inflammatory activity of chitosan–alginate nanoparticles: A targeted therapy for cutaneous pathogens. *Journal of Investigative Dermatology, 133*(5), 1231–1239.
21. Wei, M., Gao, Y., Li, X., & Serpe, M. J. (2017). Stimuli-responsive polymers and their applications. *Polymer Chemistry, 8*(1), 127–143.
22. Wu, L.-Q., Gadre, A. P., Yi, H., Kastantin, M. J., Rubloff, G. W., Bentley, W. E., et al. (2002). Voltage-dependent assembly of the polysaccharide chitosan onto an electrode surface. *Langmuir, 18*(22), 8620–8625.
23. Tian, B., & Liu, J. (2023). Smart stimuli-responsive chitosan hydrogel for drug delivery: A review. *International Journal of Biological Macromolecules, 235*, 123902.
24. Dash, M., Chiellini, F., Ottenbrite, R. M., & Chiellini, E. (2011). Chitosan—A versatile semi-synthetic polymer in biomedical applications. *Progress in Polymer Science, 36*(8), 981–1014.
25. Hong, F., Qiu, P., Wang, Y., Ren, P., Liu, J., Zhao, J., et al. (2024). Chitosan-based hydrogels: From preparation to applications, a review. *Food Chemistry: X., 21*, 101095.
26. Ahmed, E. M. (2015). Hydrogel: Preparation, characterization, and applications: A review. *Journal of Advanced Research, 6*(2), 105–121.
27. Van Bemmelen, J. (1894). Das hydrogel und das krystallinische hydrat des kupferoxyds. *Zeitschrift für anorganische Chemie, 5*(1), 466–483.
28. Wichterle, O., & Lim, D. (1960). Hydrophilic gels for biological use. *Nature, 185*(4706), 117–118.
29. Raeisi, A., & Farjadian, F. (2024). Commercial hydrogel product for drug delivery based on route of administration. *Frontiers in Chemistry, 12*, 1336717.

30. Holback, H., Yeo, Y., & Park, K. (2011). Hydrogel swelling behavior and its biomedical applications. In *Biomedical hydrogels* (pp. 3–24). Elsevier.
31. Erol, O., Pantula, A., Liu, W., & Gracias, D. H. (2019). Transformer hydrogels: a review. *Advanced Materials Technologies, 4*(4), 1900043.
32. Aminabhavi, T., & Dharupaneedi, S. (2017). Production of chitosan-based hydrogels for biomedical applications. In *Chitosan based biomaterials volume 1* (pp. 295–319). Elsevier.
33. Zheng, N., Xu, Y., Zhao, Q., & Xie, T. (2021). Dynamic covalent polymer networks: A molecular platform for designing functions beyond chemical recycling and self-healing. *Chemical Reviews, 121*(3), 1716–1745.
34. Heidari, B. S., Ruan, R., Vahabli, E., Chen, P., De-Juan-Pardo, E. M., Zheng, M., et al. (2023). Natural, synthetic and commercially-available biopolymers used to regenerate tendons and ligaments. *Bioactive Materials, 19*, 179–197.
35. Islam, M. R., Rahman, M. M., Dhar, P. S., Nowrin, F. T., Sultana, N., Akter, M., et al. (2023). The role of natural and semi-synthetic compounds in ovarian cancer: Updates on mechanisms of action, current trends and perspectives. *Molecules, 28*(5), 2070.
36. Thirupathi, K., Raorane, C. J., Ramkumar, V., Ulagesan, S., Santhamoorthy, M., Raj, V., et al. (2022). Update on chitosan-based hydrogels: Preparation, characterization, and its antimicrobial and antibiofilm applications. *Gels, 9*(1), 35.
37. Aimetti, A. A., Tibbitt, M. W., & Anseth, K. S. (2009). Human neutrophil elastase responsive delivery from poly (ethylene glycol) hydrogels. *Biomacromolecules, 10*(6), 1484–1489.
38. Zhong, G., Qiu, M., Zhang, J., Jiang, F., Yue, X., Huang, C., et al. (2023). Fabrication and characterization of PVA@ PLA electrospinning nanofibers embedded with Bletilla striata polysaccharide and Rosmarinic acid to promote wound healing. *International Journal of Biological Macromolecules, 234*, 123693.
39. Yi, Y., Chiao, M., Mahmoud, K. A., Wu, L., & Wang, B. (2022). Preparation and characterization of PVA/PVP conductive hydrogels formed by freeze–thaw processes as a promising material for sensor applications. *Journal of Materials Science, 57*(16), 8029–8038.
40. Borelli, A. N., Young, M. W., Kirkpatrick, B. E., Jaeschke, M. W., Mellett, S., Porter, S., et al. (2022). Stress relaxation and composition of hydrazone-crosslinked hybrid biopolymer-synthetic hydrogels determine spreading and secretory properties of MSCs. *Advanced Healthcare Materials, 11*(14), 2200393.
41. Hu, N., Shi, L., Ji, S., Wang, W., Lei, L., Fan, H., et al. (2023). Adhesive hybrid interpenetrating network hydrogel-based detector to monitor solar radiation dose required for plant growth. *Advanced Engineering Materials, 25*(2), 2201118.
42. Hipwood, L., Clegg, J., Weekes, A., Davern, J. W., Dargaville, T. R., Meinert, C., et al. (2022). Semi-synthetic click-gelatin hydrogels as tunable platforms for 3d cancer cell culture. *Gels., 8*(12), 821.
43. Thang, N. H., Chien, T. B., & Cuong, D. X. (2023). Polymer-based hydrogels applied in drug delivery: An overview. *Gels, 9*(7), 523.
44. Nangia, S., Warkar, S., & Katyal, D. (2018). A review on environmental applications of chitosan biopolymeric hydrogel based composites. *Journal of Macromolecular Science, Part A, 55*(11–12), 747–763.
45. Yang, T., Wang, L., Wu, W.-H., Wei, S., & Zhang, W.-B. (2023). Orchestrating chemical and physical cross-linking in protein hydrogels to regulate embryonic stem cell growth. *ACS Macro Letters, 12*(2), 269–273.
46. Apartsin, E., & Caminade, A.-M. (2023). Single-component physical hydrogels of dendritic molecules. *Journal of Composites Science, 7*(1), 26.
47. Xue, X., Hu, Y., Wang, S., Chen, X., Jiang, Y., & Su, J. (2022). Fabrication of physical and chemical crosslinked hydrogels for bone tissue engineering. *Bioactive Materials, 12*, 327–339.
48. Ghosal, A., & Kaushik, A. (2020). *Intelligent hydrogels in diagnostics and therapeutics*. CRC Press.

49. Tan, H., Rubin, J. P., & Marra, K. G. (2010). Injectable in situ forming biodegradable chitosan-hyaluronic acid based hydrogels for adipose tissue regeneration. *Organogenesis, 6*(3), 173–180.
50. Rizzo, F., & Kehr, N. S. (2021). Recent advances in injectable hydrogels for controlled and local drug delivery. *Advanced Healthcare Materials, 10*(1), 2001341.
51. Huang, X., & Brazel, C. S. (2001). On the importance and mechanisms of burst release in matrix-controlled drug delivery systems. *Journal of Controlled Release, 73*(2–3), 121–136.
52. Lou, J., Duan, H., Qin, Q., Teng, Z., Gan, F., Zhou, X., et al. (2023). Advances in oral drug delivery systems: Challenges and opportunities. *Pharmaceutics, 15*(2), 484.
53. Carr, D. A., Gómez-Burgaz, M., Boudes, M. C., & Peppas, N. A. (2010). Complexation hydrogels for the oral delivery of growth hormone and salmon calcitonin. *Industrial & Engineering Chemistry Research, 49*(23), 11991–11995.
54. Azeera, M., Vaidevi, S., & Ruckmani, K. (2019). Characterization techniques of hydrogel and its applications. In *Cellulose-based superabsorbent hydrogels* (pp. 737–761). Springer.
55. Belbekhouche, S., Charaabi, S., Picton, L., Le Cerf, D., & Carbonnier, B. (2018). Glucose-sensitive polyelectrolyte microcapsules based on (alginate/chitosan) pair. *Carbohydrate Polymers, 184*, 144–153. PubMed PMID: 29352905. Epub 2018/01/22. eng.
56. Gu, Z., Dang, T. T., Ma, M., Tang, B. C., Cheng, H., Jiang, S., et al. (2013). Glucose-responsive microgels integrated with enzyme nanocapsules for closed-loop insulin delivery. *ACS Nano, 7*(8), 6758–6766. PubMed PMID: 23834678. Epub 2013/07/10. eng.
57. Li, J., Hu, W., Zhang, Y., Tan, H., Yan, X., Zhao, L., et al. (2015). pH and glucose dually responsive injectable hydrogel prepared by in situ crosslinking of phenylboronic modified chitosan and oxidized dextran. *Journal of Polymer Science Part A: Polymer Chemistry., 53*(10), 1235–1244.
58. Li, J., Ma, F.-K., Dang, Q.-F., Liang, X.-G., & Chen, X.-G. (2014). Glucose-conjugated chitosan nanoparticles for targeted drug delivery and their specific interaction with tumor cells. *Frontiers of Materials Science, 8*, 363–372.
59. Liu, S., Wang, L., Lian, W., Liu, H., & Li, C. Z. (2015). Logic gate system with three outputs and three inputs based on switchable electrocatalysis of glucose by glucose oxidase entrapped in chitosan films. *Chemistry–An Asian Journal, 10*(1), 225–230.
60. Siddiqui, N. A., Billa, N., Roberts, C. J., & Asantewaa, O. Y. (2016). Cross-linked dependency of boronic acid-conjugated chitosan nanoparticles by diols for sustained insulin release. *Pharmaceutics, 8*(4), 30.
61. Yin, R., Han, J., Zhang, J., & Nie, J. (2010). Glucose-responsive composite microparticles based on chitosan, concanavalin A and dextran for insulin delivery. *Colloids and Surfaces B: Biointerfaces, 76*(2), 483–488.
62. Zhao, X., Zou, X., & Ye, L. (2017). Controlled pH-and glucose-responsive drug release behavior of cationic chitosan based nano-composite hydrogels by using graphene oxide as drug nanocarrier. *Journal of Industrial and Engineering Chemistry., 49*, 36–45.
63. Zou, X., Zhao, X., & Ye, L. (2015). Synthesis of cationic chitosan hydrogel and its controlled glucose-responsive drug release behavior. *Chemical Engineering Journal, 273*, 92–100.
64. Ghadban, A., Ahmed, A. S., Ping, Y., Ramos, R., Arfin, N., Cantaert, B., et al. (2016). Bioinspired pH and magnetic responsive catechol-functionalized chitosan hydrogels with tunable elastic properties. *Chemical Communications, 52*(4), 697–700.
65. Medina-Moreno, A., El-Hammadi, M. M., Martínez-Soler, G. I., Ramos, J. G., García-García, G., & Arias, J. L. (2025). Magnetic and pH-responsive magnetite/chitosan (core/shell) nanoparticles for dual-targeted methotrexate delivery in cancer therapy. *Drug Delivery and Translational Research, 15*(5), 1646–1659.
66. Xu, Y., Wang, L., Li, Y.-K., & Wang, C.-Q. (2014). Oxidation and pH responsive nanoparticles based on ferrocene-modified chitosan oligosaccharide for 5-fluorouracil delivery. *Carbohydrate Polymers, 114*, 27–35.

67. Li, J., & Liu, P. (2018). One-pot fabrication of pH/reduction dual-stimuli responsive chitosan-based supramolecular nanogels for leakage-free tumor-specific DOX delivery with enhanced anti-cancer efficacy. *Carbohydrate Polymers, 201*, 583–590.
68. Zou, H., & Yuan, W. (2015). Temperature-and redox-responsive magnetic complex micelles for controlled drug release. *Journal of Materials Chemistry B, 3*(2), 260–269.
69. Harris, M., Ahmed, H., Barr, B., LeVine, D., Pace, L., Mohapatra, A., et al. (2017). Magnetic stimuli-responsive chitosan-based drug delivery biocomposite for multiple triggered release. *International Journal of Biological Macromolecules, 104*, 1407–1414.
70. Kaihara, S., Suzuki, Y., & Fujimoto, K. (2011). In situ synthesis of polysaccharide nanoparticles via polyion complex of carboxymethyl cellulose and chitosan. *Colloids and Surfaces B: Biointerfaces., 85*(2), 343–348.
71. Lin, J., Li, Y., Li, Y., Wu, H., Yu, F., Zhou, S., et al. (2015). Drug/dye-loaded, multifunctional PEG–chitosan–iron oxide nanocomposites for methotraxate synergistically self-targeted cancer therapy and dual model imaging. *ACS Applied Materials & Interfaces, 7*(22), 11908–11920.
72. Wang, C., Ravi, S., Garapati, U. S., Das, M., Howell, M., Mallela, J., et al. (2013). Multifunctional chitosan magnetic-graphene (CMG) nanoparticles: A theranostic platform for tumor-targeted co-delivery of drugs, genes and MRI contrast agents. *Journal of Materials Chemistry B, 1*(35), 4396–4405.
73. Yuan, Q., Venkatasubramanian, R., Hein, S., & Misra, R. (2008). A stimulus-responsive magnetic nanoparticle drug carrier: Magnetite encapsulated by chitosan-grafted-copolymer. *Acta Biomaterialia, 4*(4), 1024–1037.
74. Mansouri, M., Nazarpak, M. H., Solouk, A., Akbari, S., & Hasani-Sadrabadi, M. M. (2017). Magnetic responsive of paclitaxel delivery system based on SPION and palmitoyl chitosan. *Journal of Magnetism and Magnetic Materials, 421*, 316–325.
75. Jeong, S., Jun Park, S., Shin, M. S., & Kim, S. I. (2002). Characteristics of electrical responsive chitosan/polyallylamine interpenetrating polymer network hydrogel. *Journal of Applied Polymer Science, 86*(9), 2290–2295.
76. Guerry, A., Cottaz, S., Fleury, E., Bernard, J., & Halila, S. (2014). Redox-stimuli responsive micelles from DOX-encapsulating polycaprolactone-g-chitosan oligosaccharide. *Carbohydrate Polymers, 112*, 746–752.
77. Hu, Y.-w., Du, Y.-z., Liu, N., Liu, X., Meng, T.-t., Cheng, B.-l., et al. (2015). Selective redox-responsive drug release in tumor cells mediated by chitosan based glycolipid-like nanocarrier. *Journal of Controlled Release, 206*, 91–100.
78. Jiao, J., Li, X., Zhang, S., Liu, J., Di, D., Zhang, Y., et al. (2016). Redox and pH dual-responsive PEG and chitosan-conjugated hollow mesoporous silica for controlled drug release. *Materials Science and Engineering: C., 67*, 26–33.
79. Lee, S. Y., Lee, H., In, I., & Park, S. Y. (2014). pH/redox/photo responsive polymeric micelle via boronate ester and disulfide bonds with spiropyran-based photochromic polymer for cell imaging and anticancer drug delivery. *European Polymer Journal, 57*, 1–10.
80. Bhattarai, N., Gunn, J., & Zhang, M. (2010). Chitosan-based hydrogels for controlled, localized drug delivery. *Advanced Drug Delivery Reviews, 62*(1), 83–99.
81. DeSesso, J. M., Jacobson, C. F., Williams, A. L., & Lyubimov, A. (2012). Anatomical and physiological parameters that influence gastrointestinal absorption. In *Encyclopedia of drug metabolism and interactions* (Vol. 2, pp. 43–79). Wiley.
82. Wang, H., Li, W., Lu, Y., Wang, Z., & Zhong, W. (1996). Studies on chitosan and poly (acrylic acid) interpolymer complex. II. Solution behaviors of the mixture of water-soluble chitosan and poly (acrylic acid). *Journal of Applied Polymer Science, 61*(12), 2221–2224.
83. Chavasit, V., Kienzle-Sterzer, C., & Antonio, T. J. (1988). Formation and characterization of an insoluble polyelectrolyte complex: Chitosan-polyacrylic acid. *Polymer Bulletin, 19*, 223–230.

84. Dai, Y. N., Li, P., Zhang, J. P., Wang, A. Q., & Wei, Q. (2008). A novel pH sensitive N-succinyl chitosan/alginate hydrogel bead for nifedipine delivery. *Biopharmaceutics & Drug Disposition, 29*(3), 173–184.
85. Berger, J., Reist, M., Mayer, J., Felt, O., & Gurny, R. (2004). Structure and interactions in chitosan hydrogels formed by complexation or aggregation for biomedical applications. *European Journal of Pharmaceutics and Biopharmaceutics, 57*(1), 35–52.
86. Guo, B.-L., & Gao, Q.-Y. (2007). Preparation and properties of a pH/temperature-responsive carboxymethyl chitosan/poly (N-isopropylacrylamide) semi-IPN hydrogel for oral delivery of drugs. *Carbohydrate Research, 342*(16), 2416–2422.
87. Ahmed, A. A.-K., Naik, H. B., & Sherigara, B. (2009). Synthesis and characterization of chitosan-based pH-sensitive semi-interpenetrating network microspheres for controlled release of diclofenac sodium. *Carbohydrate Research, 344*(5), 699–706.
88. Patel, V. R., & Amiji, M. M. (1996). Preparation and characterization of freeze-dried chitosan-poly (ethylene oxide) hydrogels for site-specific antibiotic delivery in the stomach. *Pharmaceutical Research, 13*, 588–593.
89. Wang, X., Johnson, M., Zhang, N., Shen, P., Yang, L., Milne, C., et al. (2023). Stimuli-responsive chitosan-based injectable hydrogel for "on-demand" drug release. *Materials Advances, 4*(23), 6439–6448.
90. Koh, R. H., Jin, Y., Kim, J., & Hwang, N. S. (2020). Inflammation-modulating hydrogels for osteoarthritis cartilage tissue engineering. *Cells, 9*(2), 419.
91. Chang, P. C., Chao, Y. C., Hsiao, M. H., Chou, H. S., Jheng, Y. H., Yu, X. H., et al. (2017). Inhibition of periodontitis induction using a stimuli-responsive hydrogel carrying naringin. *Journal of Periodontology, 88*(2), 190–196.
92. Gao, J., Xu, Y., Zheng, Y., Wang, X., Li, S., Yan, G., et al. (2019). pH-sensitive carboxymethyl chitosan hydrogels via acid-labile ortho ester linkage as an implantable drug delivery system. *Carbohydrate Polymers, 225*, 115237.
93. Rasool, A., Ata, S., & Islam, A. (2019). Stimuli responsive biopolymer (chitosan) based blend hydrogels for wound healing application. *Carbohydrate Polymers, 203*, 423–429.
94. Han, D., Lu, Z., Chester, S. A., & Lee, H. (2018). Micro 3D printing of a temperature-responsive hydrogel using projection micro-stereolithography. *Scientific Reports, 8*(1), 1963.
95. Mishra, B., & Singh, J. (2021). Hydrogel-based drug delivery systems for cancer therapy. In *Advanced drug delivery systems in the management of cancer* (pp. 63–74). Elsevier.
96. Taymouri, S., Amirkhani, S., & Mirian, M. (2021). Fabrication and characterization of injectable thermosensitive hydrogel containing dipyridamole loaded polycaprolactone nanoparticles for bone tissue engineering. *Journal of Drug Delivery Science and Technology, 64*, 102659.
97. Stealey, S., Guo, X., Majewski, R., Dyble, A., Lehman, K., Wedemeyer, M., et al. (2020). Calcium-oligochitosan-pectin microcarrier for colonic drug delivery. *Pharmaceutical Development and Technology, 25*(2), 260–265. PubMed PMID: 31709858. Epub 2019/11/12. eng.
98. Jain, A., Gupta, Y., & Jain, S. K. (2007). Perspectives of biodegradable natural polysaccharides for site-specific drug delivery to the colon. *Journal of Pharmacy & Pharmaceutical Sciences, 10*(1), 86–128. PubMed PMID: 17498397. Epub 2007/05/15. eng.
99. Sinha, V. R., & Kumria, R. (2001). Polysaccharides in colon-specific drug delivery. *International Journal of Pharmaceutics, 224*(1–2), 19–38.
100. Tozaki, H., Komoike, J., Tada, C., Maruyama, T., Terabe, A., Suzuki, T., et al. (1997). Chitosan capsules for colon-specific drug delivery: Improvement of insulin absorption from the rat colon. *Journal of Pharmaceutical Sciences, 86*(9), 1016–1021.
101. Shimono, N., Takatori, T., Ueda, M., Mori, M., Higashi, Y., & Nakamura, Y. (2002). Chitosan dispersed system for colon-specific drug delivery. *International Journal of Pharmaceutics, 245*(1–2), 45–54.
102. Kim, S. J., Park, S. J., Kim, I. Y., Shin, M. S., & Kim, S. I. (2002). Electric stimuli responses to poly (vinyl alcohol)/chitosan interpenetrating polymer network hydrogel in NaCl solutions. *Journal of Applied Polymer Science, 86*(9), 2285–2289.

103. Kaewpirom, S., & Boonsang, S. (2006). Electrical response characterisation of poly (ethylene glycol) macromer (PEGM)/chitosan hydrogels in NaCl solution. *European Polymer Journal, 42*(7), 1609–1616.
104. Kim, S. J., Shin, S. R., Lee, S. M., Kim, I. Y., & Kim, S. I. (2004). Electromechanical properties of hydrogels based on chitosan and poly (hydroxyethylmethacrylate) in NaCl solution. *Smart Materials and Structures, 13*(5), 1036.
105. Murdan, S. (2003). Electro-responsive drug delivery from hydrogels. *Journal of Controlled Release, 92*(1–2), 1–17.
106. Liu, K.-H., Liu, T.-Y., Chen, S.-Y., & Liu, D.-M. (2008). Drug release behavior of chitosan–montmorillonite nanocomposite hydrogels following electrostimulation. *Acta Biomaterialia, 4*(4), 1038–1045.
107. Cui, X., Guan, X., Zhong, S., Chen, J., Zhu, H., Li, Z., et al. (2017). Multi-stimuli responsive smart chitosan-based microcapsules for targeted drug delivery and triggered drug release. *Ultrasonics Sonochemistry, 38*, 145–153.
108. Saito, G., Swanson, J. A., & Lee, K.-D. (2003). Drug delivery strategy utilizing conjugation via reversible disulfide linkages: Role and site of cellular reducing activities. *Advanced Drug Delivery Reviews, 55*(2), 199–215.
109. Stoffelen, C., Voskuhl, J., Jonkheijm, P., & Huskens, J. (2014). Dual stimuli-responsive self-assembled supramolecular nanoparticles. *Angewandte Chemie International Edition, 53*(13), 3400–3404.
110. Ho, T.-C., Chang, C.-C., Chan, H.-P., Chung, T.-W., Shu, C.-W., Chuang, K.-P., et al. (2022). Hydrogels: Properties and applications in biomedicine. *Molecules, 27*(9), 2902.
111. Lin, X., Zhao, X., Xu, C., Wang, L., & Xia, Y. (2022). Progress in the mechanical enhancement of hydrogels: Fabrication strategies and underlying mechanisms. *Journal of Polymer Science, 60*(17), 2525–2542.
112. Xu, P., Shang, Z., Yao, M., & Li, X. (2022). Mechanistic insight into improving strength and stability of hydrogels via nano-silica. *Journal of Molecular Liquids, 357*, 119094.
113. Rizwan, M., Yahya, R., Hassan, A., Yar, M., Azzahari, A. D., Selvanathan, V., et al. (2017). pH sensitive hydrogels in drug delivery: Brief history, properties, swelling, and release mechanism, material selection and applications. *Polymers, 9*(4), 137.
114. Zhang, M., Lee, Y., Zheng, Z., Khan, M. T. A., Lyu, X., Byun, J., et al. (2023). Micro-and nanofabrication of dynamic hydrogels with multichannel information. *Nature Communications, 14*(1), 8208.
115. Park, K. M., & Gerecht, S. (2014). Hypoxia-inducible hydrogels. *Nature Communications, 5*(1), 4075.
116. Cheung, R. C. F., Ng, T. B., Wong, J. H., & Chan, W. Y. (2015). Chitosan: An update on potential biomedical and pharmaceutical applications. *Marine Drugs, 13*(8), 5156–5186.
117. Yadav, S. K., Khan, G., Bansal, M., Thokala, S., Bonde, G. V., Upadhyay, M., et al. (2018). Multiparticulate based thermosensitive intra-pocket forming implants for better treatment of bacterial infections in periodontitis. *International Journal of Biological Macromolecules, 116*, 394–408.
118. Li, Z., Qin, R., Xue, J., Lin, C., & Jiang, L. (2025). Chitosan-based hydrogel beads: Developments, applications, and challenges. *Polymers, 17*(7), 920.
119. Kass, L. E., & Nguyen, J. (2022). Nanocarrier-hydrogel composite delivery systems for precision drug release. *Wiley Interdisciplinary Reviews: Nanomedicine and Nanobiotechnology., 14*(2), e1756.
120. Haider, A., Khan, S., Iqbal, D. N., Khan, S. U., Haider, S., Mohammad, K., et al. (2024). Chitosan as a tool for tissue engineering and rehabilitation: Recent developments and future perspectives–a review. *International Journal of Biological Macromolecules, 278*, 134172.
121. McCarthy, P. C., Zhang, Y., & Abebe, F. (2021). Recent applications of dual-stimuli responsive chitosan hydrogel nanocomposites as drug delivery tools. *Molecules, 26*(16), 4735.

Chitosan-Based Light-Activated Drug Delivery System

Hossein Najafi, Sana Rahimian, Mohammad Doroudian, and Fatemeh Farjadian

Abstract The quest for more precise, safer, and effective drug delivery has accelerated the exploration of stimuli-responsive systems, among which light-activated platforms offer a unique advantage due to their controllability and non-invasiveness. Chitosan, a natural polysaccharide known for its biocompatibility, biodegradability, and chemical modifiability, has emerged as a promising material in the design of advanced drug delivery vehicles. This chapter discusses the critical properties of chitosan that support its application in light-responsive systems, highlighting its structural versatility, mucoadhesive characteristics, and capability to encapsulate a broad range of therapeutic agents. It further explores the mechanisms by which light triggers drug release, emphasizing photochemical and photothermal processes, and examines the synthesis and characterization techniques crucial for the development of chitosan-based nanocarriers. Applications of these systems are reviewed across therapeutic areas such as oncology, antimicrobial treatments, and neurological disorders. This chapter also addresses the challenges facing clinical translation, including stability, targeting specificity, and scalability, and outlines recent innovations in chemical modification of chitosan to enhance its performance. Together, these advances position chitosan-based light-activated systems as a compelling strategy for future precision medicine applications.

H. Najafi
Department of Industrial and Environmental Biotechnology, National Institute of Genetic Engineering and Biotechnology (NIGEB), Tehran, Iran

S. Rahimian
Department of Nanobiotechnology, Faculty of Biological Sciences, Tarbiat Modares University, Tehran, Iran

M. Doroudian (✉)
Department of Cell and Molecular Sciences, Faculty of Biological Sciences, Kharazmi University, Tehran, Iran
e-mail: doroudian@khu.ac.ir

F. Farjadian (✉)
Pharmaceutical Sciences Research Center, School of Pharmacy, Shiraz University of Medical Sciences, Shiraz, Iran
e-mail: farjadian_f@sums.ac.ir

S. Jana (ed.), *Chitosan-Based Drug Delivery Systems*, AAPS Introductions in the Pharmaceutical Sciences 16, https://doi.org/10.1007/978-3-032-08347-0_9

Keywords Chitosan-based nanoparticles · Photothermal and photodynamic therapy · Drug delivery systems · Light-activated drug delivery · Stimuli-responsive nanocarriers

1 Introduction

The pursuit of safer and more effective therapeutics has driven major advances in drug delivery systems, emphasizing controlled and targeted approaches to improve efficacy and reduce toxicity [1, 2]. Traditional delivery methods often face challenges such as poor bioavailability and limited specificity [3], which has led to the development of stimulus-responsive systems that enable precise spatial and temporal drug release [4, 5].

Light has emerged as a particularly attractive external trigger due to its non-invasive nature, adjustable intensity, and tissue-penetrating capabilities, especially in the near-infrared (NIR) range [6, 7]. Light-activated systems allow site-specific drug release, making them especially valuable for treating localized diseases like cancer [8–10].

A promising avenue in the development of light-sensitive platforms lies in the use of natural polymers, particularly chitosan, which has demonstrated considerable potential in biomedical applications [11]. Chitosan is a linear polysaccharide composed mainly of β-(1→4)-linked D-glucosamine units, obtained through the partial deacetylation of chitin as an abundant biopolymer found in the exoskeleton of crustaceans and insects [12]. Its biodegradability, biocompatibility, non-immunogenicity, and mucoadhesive nature have made it an ideal candidate for fabricating nanocarriers in drug delivery [13].

Derived from the deacetylation of chitin, chitosan contains amino and hydroxyl groups that enable chemical modifications and conjugation with photosensitive agents [14]. Its positive charge facilitates interaction with cell membranes and supports intracellular drug delivery [15].

Light-responsive chitosan systems typically incorporate agents like gold nanoparticles or upconversion nanoparticles that convert light into heat or reactive oxygen species (ROS), triggering drug release through structural changes or matrix disruption [16, 17]. NIR-responsive designs are particularly promising due to deeper tissue penetration and lower phototoxicity compared to UV-based systems [18].

Combining chitosan's pH sensitivity with photo-responsiveness has enabled dual-triggered platforms that release drugs selectively in acidic tumor environments and further enhance delivery under light exposure [19, 20]. Such dual-responsive behavior not only increases drug accumulation at the disease site but also significantly reduces premature drug leakage during systemic circulation [21, 22].

Beyond cancer therapy, chitosan's mucoadhesive properties support its use in nasal, ocular, pulmonary, and transdermal delivery [23, 24]. For instance, intranasal delivery of chitosan nanoparticles shows promise for targeting the brain, and integrating light-activation could enhance precision in treating neurological disorders [25]. When light-activation is added to this platform, it may offer a new non-invasive

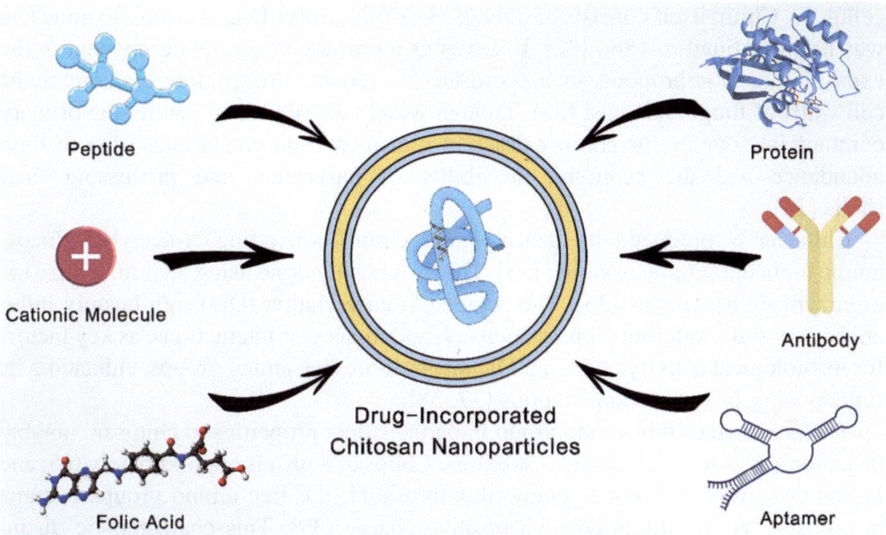

Fig. 1 Schematic illustration of various targeting ligands that can be attached to the surface of chitosan nanoparticles to improve tumor targeting efficiency

strategy for treating neurological disorders with spatial precision and minimal systemic exposure [26].

However, clinical translation faces challenges, including chitosan's limited solubility at physiological pH, modest drug loading capacity, and the need for precise control of light exposure [27]. To address these concerns, researchers have explored strategies such as chemical conjugation with hydrophilic polymers, crosslinking with responsive moieties, and surface engineering techniques that enhance colloidal stability and drug loading capacity [28]. Additionally, the integration of targeting ligands such as folic acid, peptides, or antibodies has further improved the specificity and accumulation of these systems at desired sites (Fig. 1) [29].

An equally important area of advancement lies in diagnostic applications, where chitosan-based nanostructures are being engineered as theranostic agents capable of simultaneous imaging and therapy [30]. Light-activated platforms that combine photothermal or photodynamic effects with drug release offer dual-modality treatment strategies, potentially improving therapeutic outcomes and enabling real-time monitoring of treatment progress [31].

2 Chitosan: Properties and Benefits

2.1 Chemical Structure and Properties

Chitosan is a linear, naturally occurring polysaccharide derived from chitin, a biopolymer that is the second most abundant organic material on Earth after cellulose. Chitin itself consists of poly (β-(1→4)-N-acetyl-D-glucosamine) units and was first identified in 1884 [32]. It serves as a primary structural component in the exoskeletons of arthropods, such as crustaceans (crabs, shrimps, lobsters), and in the cell walls of fungi and yeast [33]. Though widely distributed in nature, the primary commercial sources for chitin extraction remain marine crustaceans, due to their abundance and the economic feasibility of harvesting and processing their shells [34].

Chitosan is produced by deacetylating chitin, converting N-acetyl-D-glucosamine units into D-glucosamine [35]. This is typically done using sodium hydroxide or enzymatic hydrolysis [36]. The degree of deacetylation (DD) significantly influences chitosan's solubility, charge density, and molecular interactions, as key factors for its biological activity. A higher DD means more free amino groups, enhancing its functionality in various applications [37, 38].

The presence of this amino group imparts distinct properties to chitosan, notably its cationic nature under acidic conditions. Chitosan's pKa is approximately 6.5, and in solutions with pH values below this threshold, the free amino groups become protonated, giving the polymer a positive charge [39]. This characteristic distinguishes chitosan from other naturally occurring polysaccharides, which are generally neutral or negatively charged [40].

Its polycationic nature enhances solubility in dilute acids like acetic acid, while it remains insoluble in neutral or alkaline media [41]. This positive charge allows chitosan to interact electrostatically with negatively charged molecules such as nucleic acids, proteins, and phospholipids as a key to its role in drug delivery and biomedical applications [42].

Chitosan's molecular weight (MW) and degree of deacetylation (DD) critically influence its physical and biological properties. Lower MW enhances solubility and permeability, while higher MW offers greater mechanical strength and slower degradation [43]. The DD affects solubility, viscosity, and interactions with biological membranes and macromolecules [44]. Adjusting these parameters allows for fine-tuning chitosan's performance in targeted applications.

A key advantage of chitosan is its amenability to chemical modification, thanks to reactive amino and hydroxyl groups on its backbone, allowing the creation of derivatives with specialized functions [28]. These modifications enable controlled drug release and targeted delivery. For instance, quaternized chitosan retains positive charges over a broad pH range, enhancing membrane interaction and gene transfection efficiency [45], while carboxymethylation improves water solubility and facilitates the formation of hydrogels and nanoparticles for drug delivery [46].

Chitosan's cationic nature contributes to its antimicrobial activity. In acidic environments, its protonated amino groups interact with negatively charged microbial membranes, increasing permeability and causing cell death [44]. This broad-spectrum activity, along with its biocompatibility, supports its use in wound dressings and tissue engineering, where it offers protection and promotes tissue regeneration [47].

Chitosan's capacity to form nanostructures such as nanoparticles, hydrogels, micelles, and microspheres has expanded its utility in drug delivery. These systems can encapsulate various therapeutic agents, including small molecules, proteins, and nucleic acids [48, 49]. Chitosan nanoparticles, typically under 100 nm, enable tissue targeting via enhanced permeability and retention (EPR) effects [50], while hydrogels provide a controlled release and excellent biocompatibility through their water-retaining, 3D polymer networks [51].

2.2 *Advantages of Chitosan in Drug Delivery*

Chitosan possesses a wide array of physicochemical and biological properties that collectively make it one of the most promising biomaterials in the field of drug delivery [52, 53]. Its versatility lies not only in its intrinsic characteristics as a biopolymer but also in its ability to undergo functional modifications that expand its therapeutic potential [54, 55]. Several distinctive advantages position chitosan above many other natural and synthetic polymers commonly used in drug delivery systems [56–58].

2.2.1 Biodegradability and Biocompatibility

One of the most critical advantages of chitosan is its biodegradability and biocompatibility. As a natural polysaccharide derived from chitin, chitosan is readily broken down by enzymes, such as lysozymes, into non-toxic and bioabsorbable oligosaccharides [59]. These degradation products are easily metabolized or excreted from the body without causing adverse effects [60]. This characteristic ensures that chitosan-based drug delivery systems are safe for clinical applications and have minimal risk of long-term toxicity or accumulation in tissues.

2.2.2 Non-Toxic Nature

Chitosan is recognized for its low toxicity profile, making it suitable for both systemic and localized drug delivery [23]. The U.S. Food and Drug Administration (FDA) considers chitosan generally recognized as safe (GRAS), facilitating its use in pharmaceuticals and medical devices [61]. This property is especially valuable in applications where repeated or long-term drug administration is required, such as chronic disease management or vaccine delivery.

2.2.3 Enhanced Drug Permeability and Absorption

A key feature that distinguishes chitosan is its ability to enhance the permeation of drugs across epithelial barriers [62]. Chitosan can transiently open tight junctions between epithelial cells, improving the paracellular transport of macromolecules that would otherwise have limited absorption. This mechanism significantly enhances the bioavailability of drugs administered via mucosal routes, including nasal, ocular, pulmonary, and intestinal pathways [63, 64].

For example, chitosan facilitates the transport of large therapeutic molecules such as peptides, proteins, and even nucleic acids, which are typically restricted by the mucosal barriers of the gastrointestinal tract [48]. By promoting drug permeability, chitosan has enabled the development of non-invasive delivery strategies, such as oral insulin formulations and intranasal peptide therapies [65].

2.2.4 Mucoadhesive Properties for Prolonged Residence Time

Chitosan's polycationic nature allows it to interact electrostatically with negatively charged mucosal surfaces, resulting in strong mucoadhesion. This characteristic prolongs the residence time of drug formulations at the site of administration, increasing drug absorption and reducing dosing frequency [66]. In nasal, ocular, and gastrointestinal drug delivery, this property helps maintain high local drug concentrations and improves therapeutic outcomes [67].

2.2.5 Controlled and Sustained Drug Release

Chitosan-based drug delivery systems offer controlled and sustained release profiles for a wide variety of therapeutic agents. By manipulating factors such as molecular weight, degree of deacetylation, and crosslinking density, chitosan matrices can be tailored to release drugs over extended periods, minimizing fluctuations in drug plasma levels and reducing side effects associated with peak doses (Fig. 2) [23].

The ability of chitosan to form hydrogels, microspheres, nanoparticles, and films enables the design of delivery platforms with specific release kinetics [68, 69]. This capability is particularly advantageous in applications requiring prolonged therapeutic effects, such as chronic pain management, hormone replacement therapy, and anticancer chemotherapy [70, 71].

2.2.6 Capability to Encapsulate a Wide Range of Drugs

Chitosan exhibits excellent encapsulation efficiency for both hydrophilic and hydrophobic drugs. Its molecular structure allows for the entrapment of water-soluble molecules through electrostatic interactions, as well as the incorporation of hydrophobic compounds via chemical modifications or co-formulation with amphiphilic molecules. This versatility makes chitosan a universal carrier for diverse therapeutic agents, including small molecules, biologics, and gene-based therapies [72, 73].

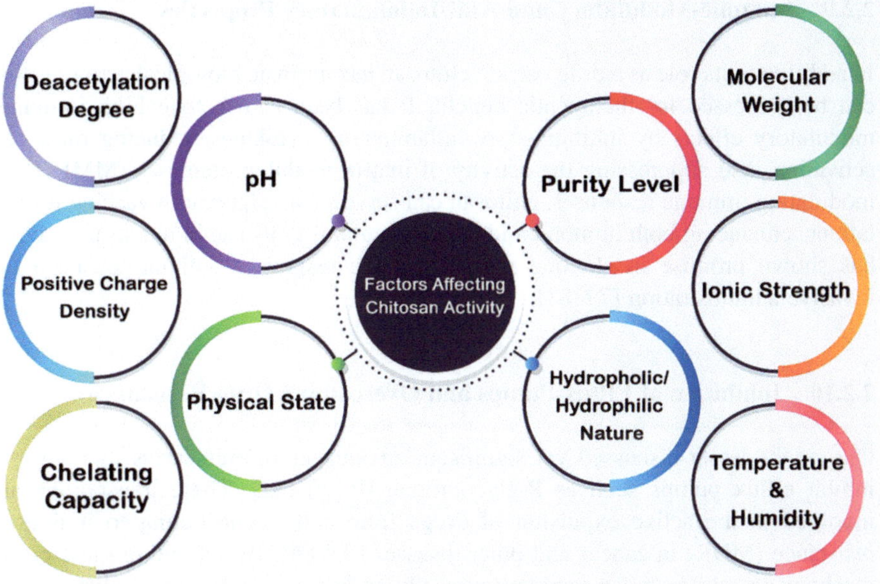

Fig. 2 Overview of key factors influencing the biological activity of chitosan

2.2.7 Responsive to Environmental Stimuli

Chitosan's responsiveness to various environmental stimuli, such as pH, temperature, and enzymatic activity, is another major advantage. Its pH-sensitive solubility enables site-specific drug release in response to changes in the physiological environment, such as the acidic conditions found in tumors, inflamed tissues, or the gastrointestinal tract. Additionally, functional modifications have been introduced to create chitosan-based carriers that respond to external stimuli like light and magnetic fields, enabling precision-controlled drug release strategies [74–76, 348, 352].

2.2.8 Compatibility with Nanotechnology and Advanced Delivery Systems

Chitosan's versatility makes it ideal for nanotechnology-based drug delivery. Its modifiability supports the development of multifunctional nanocarriers for targeted delivery, drug co-delivery, and theranostics [77, 78]. Well-established methods enable the fabrication of nanoparticles, micelles, nanogels, and hybrid nanocomposites with enhanced therapeutic efficacy. Its inherent positive charge promotes cellular uptake via sis, aiding intracellular delivery, especially in gene therapy, where chitosan nanoparticles protect and efficiently transfect nucleic acids [5].

2.2.9 Immune-Modulatory and Anti-Inflammatory Properties

In addition to its role as a drug carrier, chitosan has intrinsic biological activities that can be harnessed for therapeutic benefit. It has been shown to exhibit immune-modulatory effects by inhibiting pro-inflammatory cytokines, reducing mast cell activation, and suppressing the activity of matrix metalloproteinases (MMPs). By modulating immune responses, chitosan can serve as an adjuvant in vaccine formulations, enhancing both humoral and cellular immunity. Its use in mucosal vaccines has shown promise in eliciting robust immune responses without the need for invasive administration [79–81].

2.2.10 Inhibition of Efflux Pumps and Overcoming Drug Resistance

One of the lesser-discussed yet significant advantages of chitosan is its ability to inhibit efflux pumps such as P-glycoprotein (P-gp) [82]. These transporters are involved in the active expulsion of drugs from cells, contributing to multidrug resistance (MDR) in cancer and other diseases [83, 84]. By inhibiting these efflux mechanisms, chitosan can increase intracellular drug concentrations, enhance therapeutic efficacy, and overcome resistance to chemotherapeutic agents [85].

2.2.11 Non-Invasive Routes and Brain Delivery Potential

Chitosan's mucoadhesive properties, along with its ability to enhance permeation across epithelial barriers, make it suitable for non-invasive delivery routes such as intranasal, transdermal, and pulmonary delivery [86]. Intranasal delivery systems based on chitosan have been developed for targeting the central nervous system (CNS), bypassing the blood-brain barrier and enabling direct nose-to-brain drug transport [66]. This approach has significant implications for the treatment of neurological disorders, offering a non-invasive alternative to systemic administration [58].

2.2.12 Ease of Processing and Manufacturing

Chitosan is relatively easy to process and formulate into various drug delivery systems. It can be dissolved in mild acidic solutions, eliminating the need for toxic organic solvents during nanoparticle synthesis [87]. This characteristic makes chitosan processing more environmentally friendly and safer for large-scale manufacturing. Additionally, its abundant availability and low production cost enhance its appeal for widespread use in pharmaceutical applications [88].

3 Light-Activated Drug Delivery Systems

3.1 Stimuli-Responsive Nanocarriers: Toward Precision Drug Delivery

In recent years, stimuli-responsive drug delivery systems have emerged as a strategic platform for enhancing therapeutic efficacy and minimizing systemic side effects. These systems are engineered to release or activate therapeutic agents in response to specific biological or physical cues [89–93]. Internally triggered systems, responsive to factors such as pH, redox gradients, enzymatic activity, or hypoxia, take advantage of the pathological features of diseased tissues [94–97, 350]. However, the intrinsic variability and limited selectivity of these endogenous stimuli can result in premature or non-specific drug release.

To address these limitations, externally controlled stimuli have gained attention for offering better precision and control. Among them, light has proven particularly promising due to its non-invasive nature and high spatiotemporal resolution [98]. Light-responsive nanocarriers enable clinicians to selectively trigger drug release at a specific time and site, significantly improving targeting and reducing harm to surrounding healthy tissues. This is especially beneficial in the treatment of localized diseases, such as solid tumors, where conventional therapies often result in undesirable side effects [99].

3.2 The Clinical Potential of Light as an External Stimulus

Light offers several practical advantages as an external stimulus in drug delivery, including controllability over wavelength, intensity, and exposure duration [100]. It is already used clinically in various therapies such as photodynamic therapy (PDT) and laser ablation, reinforcing its safety and efficacy when appropriately managed [101].

The effectiveness of light-triggered drug delivery depends heavily on the penetration depth of the chosen wavelength [102]. Ultraviolet (UV) and visible light are commonly used to activate many photoresponsive compounds, but both have significant limitations: UV light poses phototoxic risks and both UV and visible light exhibit poor tissue penetration [103, 104]. In contrast, near-infrared (NIR) light in the range of 650–900 nm is more suitable for biomedical applications [105]. This spectral window often referred to as the "NIR biological window" is characterized by minimal absorption by endogenous chromophores such as hemoglobin and melanin, enabling deeper light penetration into tissue, sometimes reaching depths of several millimeters [106, 107].

3.3 *Light-Activated Delivery: Enhancing Specificity and Reducing Side Effects*

Light-activated drug delivery systems enable precise control over the timing and location of therapeutic release. Nanocarriers remain inactive during systemic circulation and are activated only upon exposure to a specific wavelength of light at the target site, thereby minimizing systemic toxicity and enhancing the therapeutic index [108–110]. In cancer therapy, this approach is particularly valuable for targeting tumor tissues while sparing surrounding healthy cells. Additionally, light-triggered systems can induce the generation of reactive oxygen species (as in photodynamic therapy) or localized heating (as in photothermal therapy), which can synergize with drug payloads to improve therapeutic outcomes [111, 112].

4 Mechanisms of Light-Activated Drug Delivery

Light-activated drug delivery systems (LADDS) employ light as a non-invasive external stimulus to achieve highly localized and temporally controlled drug release [113]. This precision addresses the limitations of conventional drug delivery methods, such as off-target toxicity and uncontrolled release kinetics [113, 114]. LADDS are increasingly investigated in fields like oncology, neuromodulation, and immunotherapy due to their ability to integrate therapeutic and diagnostic capabilities in a single platform [6, 115]. These systems typically convert light into chemical or thermal energy, triggering drug release through photochemical reactions or photothermal effects, depending on the properties of the light source and the photoresponsive materials involved (Fig. 3) [116].

4.1 *Types of Light Sources Used*

The performance and suitability of LADDS are largely determined by the choice of light source. Different wavelengths of light possess varying penetration depths, energy intensities, and interactions with biological tissues. Each light source presents distinct advantages and limitations in terms of therapeutic depth, precision, and safety [117].

4.1.1 Ultraviolet (UV) Light

Ultraviolet (UV) light (100–400 nm) has been extensively utilized in early LADDS due to its capacity to induce specific photochemical reactions, including photocleavage of chemical bonds in photosensitive carriers [115, 118]. Many

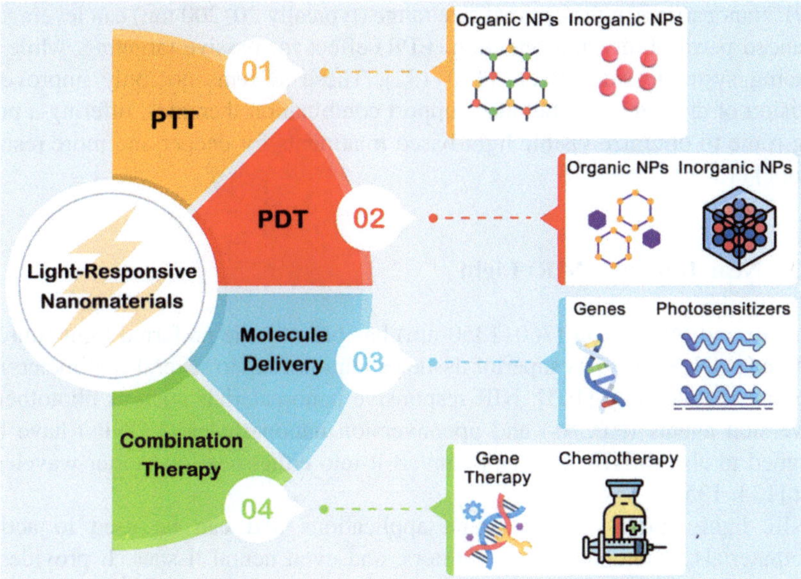

Fig. 3 Representation of light-responsive nanomaterials

UV-responsive nanoplatforms incorporate cleavable linkers such as o-nitrobenzyl and coumarin derivatives, which dissociate under UV irradiation, resulting in controlled drug release [119].

Despite its high photoreactivity, UV light exhibits limited tissue penetration and can cause DNA damage, phototoxicity, and other adverse effects. These factors restrict its use to superficial tissues or applications where localized light exposure can be tightly controlled [120].

4.1.2 Visible Light

Visible light (wavelength range 400–700 nm) is commonly employed in photodynamic therapy (PDT), a clinically validated treatment for certain cancers [121, 122]. Upon light exposure, photosensitizers are activated and initiate the generation of reactive oxygen species (ROS), such as singlet oxygen and superoxide radicals, which damage malignant cells and tissues [123, 124]. Despite its localized action and minimal invasiveness, PDT is hindered by the shallow penetration of visible light into biological tissues and its dependency on oxygen, which limits effectiveness in deep or hypoxic tumors like those found in the pancreas [125, 126].

To improve outcomes, PDT is increasingly being integrated with chemotherapeutic agents, aiming to exploit synergistic effects that can overcome the limitations of light depth and oxygen depletion [127, 128]. Nanotechnology plays a key role in enhancing the selectivity and accumulation of photosensitizers at tumor sites

[129]. Nanocarriers in the optimal size range (typically 20–200 nm) can leverage the enhanced permeability and retention (EPR) effect for passive targeting, while also reducing systemic side effects [130–132]. These systems not only improve the precision of drug delivery but also support combination therapies, offering a promising route to optimize visible light-based treatments for deeper and more resistant tumors [115, 133].

4.1.3 Near-Infrared (NIR) Light

Near-infrared (NIR) light (760–1350 nm) has become the preferred light source in modern LADDS due to its superior tissue penetration (up to several centimeters) and minimal phototoxicity [105]. NIR-responsive nanomaterials such as photothermal conversion agents (PTCAs) and upconversion nanoparticles (UCNPs) have been designed to absorb NIR light and convert it into either heat or shorter-wavelength light [134, 135].

NIR light enables deeper tissue applications and can be used to activate nanomaterials in internal organs, tumors, and even neural tissues. It provides the necessary balance between safety, penetration depth, and precision required for non-invasive therapeutic interventions [136].

4.2 Photochemical and Photothermal Mechanisms

The activation of LADDS relies on the conversion of light energy into chemical or thermal stimuli, which facilitate controlled drug release. These mechanisms are categorized as photochemical and photothermal processes [137].

4.2.1 Photochemical Mechanisms

Photochemical processes involve light-induced chemical reactions that lead to structural changes in nanocarriers, triggering drug release [138].

Photocleavage

Photocleavage involves breaking specific bonds within drug carriers upon exposure to light. Common photocleavable groups include o-nitrobenzyl and coumarin derivatives. Upon irradiation, these linkers dissociate, causing the nanocarrier to release its drug payload [139–141].

Photoisomerization

Photoisomerization describes reversible structural changes in molecules like azobenzene. Light-induced isomerization changes the molecule's physical properties, such as polarity and conformation, which can disrupt nanocarrier integrity and enable drug release [142].

Photocrosslinking and Photodecrosslinking

Light can also initiate crosslinking or decrosslinking reactions in polymer networks. Photocrosslinking stabilizes nanocarriers, while photodecrosslinking can destabilize the network, allowing for precise control over drug release kinetics [143, 144].

Photodynamic Therapy (PDT)

In PDT, photosensitizers are activated by light to produce reactive oxygen species (ROS). These ROS induce oxidative stress, leading to cancer cell death or microbial eradication. Although PDT is clinically approved and effective in certain cancers, its limitations include shallow penetration of light and dependence on oxygen levels within tissues [126, 145].

4.2.2 Photothermal Mechanisms

Photothermal mechanisms exploit nanomaterials' ability to convert light energy into heat. Upon NIR irradiation, photothermal agents generate localized heat, which can disrupt nanocarriers or directly damage pathological tissues [146, 147].

Photothermal Conversion Agents (PTCAs)

PTCAs absorb NIR light and produce heat, which can be used to thermally trigger drug release or induce hyperthermia-mediated cell death. Examples include gold nanoparticles, graphene oxide, copper sulfide, and indocyanine green (ICG) [134, 148–150].

Thermoresponsive Drug Delivery

Thermoresponsive systems integrate PTCAs with temperature-sensitive polymers or linkers that release drugs upon heating. This controlled approach ensures that drug release occurs only under localized thermal conditions, reducing systemic toxicity [151].

4.2.3 Reactive Oxygen Species (ROS) Generation from Light-Activated Polycations

Polycations, including poly(lysine), polyethylenimine, polyamidoamine, and chitosan, have been shown to exhibit photoactivity under certain wavelengths of light, resulting in the generation of ROS. The ability of these materials to carry a positive charge is thought to facilitate their photoactivation and subsequent ROS production. ROS generated from polycation-based systems have demonstrated antimicrobial effects, including disruption of bacterial membranes and interference with essential enzymes and nucleic acids [152].

Chitosan, in particular, has garnered significant interest as a photoactive polycation with inherent antimicrobial properties [153]. Under acidic conditions, chitosan becomes positively charged, enhancing its interaction with negatively charged microbial cell walls and resulting in membrane destabilization and cell death [154, 155]. When combined with photosensitizers or light exposure, chitosan-based materials can further enhance ROS generation, adding an additional layer of antimicrobial or anticancer efficacy [156].

4.2.4 Upconversion Nanoparticles (UCNPs) and Deep Tissue Activation

UCNPs offer a solution to the limited penetration depth of UV and visible light. By absorbing NIR light and emitting higher-energy UV or visible light, they enable activation of conventional photosensitizers and photocleavable linkers in deep tissues. This approach has been successfully used in light-activated systems to trigger photochemical reactions and enhance ROS production for deeper therapeutic intervention [157, 158].

4.2.5 Immunomodulatory Effects

Beyond direct cytotoxic effects, light-activated systems can modulate the immune response. ROS generation and photothermal heating can enhance antigen presentation, promote dendritic cell activation, and create immunogenic tumor microenvironments. Combining LADDS with immune checkpoint inhibitors or other immunotherapies can further amplify antitumor responses and improve therapeutic outcomes [99, 159, 160].

5 Synthesis and Characterization of Chitosan-Based Systems

The methods employed for synthesizing chitosan nanoparticles, as well as their subsequent characterization, critically influence their physicochemical properties and potential biomedical applications [161]. This section outlines the primary synthesis techniques and characterization approaches used for chitosan nanoparticles, with an emphasis on how these strategies impact their functional properties.

5.1 Methods of Synthesis

The synthesis of chitosan nanoparticles involves a variety of physical, chemical, and biological techniques [162]. The choice of method significantly affects the particle size, surface charge, stability, and drug-loading capacity, thereby dictating their suitability for specific applications such as drug delivery, tissue engineering, and wastewater treatment [49, 163]. Broadly, synthesis strategies can be classified into bottom-up and top-down approaches [164].

One of the most widely adopted approaches is ionic gelation, which is favored for its simplicity, non-toxic nature, and absence of organic solvents [165]. This technique utilizes the electrostatic interaction between the protonated amino groups of chitosan and polyanions such as sodium tripolyphosphate (TPP). By varying the chitosan-to-TPP ratio, stirring speed, and pH, it is possible to fine-tune the particle size and surface charge [166]. Nanoparticles formed by ionic gelation exhibit good encapsulation efficiency, biocompatibility, and controlled drug release profiles [167]. This method has been successfully applied for the delivery of proteins, peptides, and nucleic acids due to the mild preparation conditions that preserve biological activity [168].

In contrast to solvent-free ionic gelation, the reverse micellar method (microemulsion) involves the formation of nanosized water-in-oil emulsions stabilized by surfactants [169]. Chitosan is dissolved in an aqueous phase and crosslinked within the micelles using agents such as glutaraldehyde. Although this approach allows the production of ultrafine nanoparticles with narrow size distribution (often below 100 nm), the use of organic solvents and crosslinkers can raise concerns regarding toxicity and environmental impact [170, 171]. Similarly reliant on organic solvents, phase inversion precipitation combines emulsification with solvent displacement to produce nanoparticles. Chitosan dissolved in an aqueous phase is emulsified into an organic solvent, followed by solvent removal through evaporation, leading to nanoparticle formation. Although this method provides high encapsulation efficiency for hydrophobic drugs, the requirement for organic solvents and high-energy input limits its scalability [172, 173].

Diverging from solvent-dependent techniques, self-assembly exploits the spontaneous organization of chitosan molecules into nanoparticles through non-covalent interactions, including electrostatic, hydrophobic, and hydrogen bonding [174, 175]. Chemical modification of chitosan with hydrophobic moieties, such as stearic acid or polyethylene glycol (PEG), enhances the self-assembly process [176]. This method is particularly advantageous for encapsulating both hydrophilic and hydrophobic drugs, providing stability and sustained release [177].

Shifting focus to mechanical processing, top-down approaches include methods such as acid hydrolysis and high-pressure homogenization. Chitin is initially hydrolyzed to chitin nanocrystals, which are subsequently deacetylated to obtain chitosan nanoparticles. While effective, these methods often involve harsh conditions and longer processing times [178]. In contrast to mechanical top-down methods, spray drying involves atomizing a chitosan solution into hot air, resulting in rapid solvent evaporation and particle formation. This method is scalable and efficient for producing chitosan microparticles and nanoparticles. However, it is generally less suitable for heat-sensitive bioactives due to the elevated temperatures involved [179, 180].

To address the limitations of thermal and solvent-based approaches, supercritical carbon dioxide (SC-CO_2) technology offers an environmentally friendly alternative to conventional synthesis methods [181, 182]. Chitosan is dissolved in water under CO_2 pressure, leading to nanoparticle formation upon depressurization and drying [183]. This green approach minimizes the use of harmful solvents and is suitable for applications requiring high purity [182].

5.2 Influence of Synthesis and Characterization on Applications

The relationship between the synthesis strategies and characterization of chitosan-based systems is fundamental to determining their ultimate functionality and effectiveness in various applications [184]. Variations in preparation techniques can dramatically alter key nanoparticle attributes, such as size, surface chemistry, crystallinity, and internal structure. These factors, in turn, govern how chitosan nanoparticles behave in complex biological and environmental systems, affecting their suitability for specific uses [185].

An important consideration is the degree of reproducibility and consistency achieved through different synthesis methods [186]. Small variations in formulation parameters, such as pH, concentration, or crosslinker ratio, can result in substantial differences in the physicochemical profile of chitosan nanoparticles [187, 188]. For applications requiring high precision, such as targeted drug delivery or gene therapy, maintaining consistency in particle characteristics is critical for ensuring predictable biological responses and therapeutic outcomes [189, 190].

The surface chemistry imparted during synthesis significantly influences how these systems interact with biological environments. For example, surface charge not only impacts the stability of nanoparticles in suspension but also plays a role in cellular uptake, biodistribution, and potential immunogenicity [15]. In tissue engineering applications, for instance, the balance between surface charge and hydrophilicity can determine cell adhesion, proliferation, and differentiation on chitosan-based scaffolds [191]. Meanwhile, in drug delivery systems, a finely tuned surface charge can enhance interaction with mucosal tissues or specific cellular receptors, improving bioavailability and therapeutic efficiency [192].

Additionally, the morphological and structural properties revealed through characterization techniques such as electron microscopy and X-ray diffraction have direct implications for how nanoparticles perform in practical applications [193, 194]. The degree of crystallinity, for instance, influences mechanical strength and degradation rate—two parameters that are particularly relevant in scaffold design for regenerative medicine [195]. A more crystalline structure may offer enhanced mechanical integrity, whereas an amorphous form may degrade faster, which could be advantageous for drug release kinetics [17, 196].

Thermal and chemical stability, assessed through thermogravimetric and spectroscopic methods, also determines application potential [197]. Chitosan systems intended for environmental remediation, particularly those used as photocatalysts for degrading pollutants, require high stability under variable temperature and pH conditions [198]. Characterization data confirming robust thermal stability and chemical resistance directly support their use in such demanding applications [199, 200].

Moreover, the integration of characterization findings into synthesis refinement forms a feedback loop that enables the optimization of chitosan-based systems [201, 202]. For example, identifying irregularities in size distribution or surface roughness through microscopy can inform adjustments in synthesis protocols, such as altering crosslinker concentration or modifying reaction times. This iterative approach is vital for producing materials tailored to specific performance requirements [184, 202].

One emerging trend is the development of multifunctional chitosan composites, combining the polymer's inherent properties with other functional agents to broaden application potential [203]. These hybrid systems often require advanced synthesis strategies and comprehensive characterization to validate their structure-function relationships [204]. For example, the inclusion of metal oxides or magnetic nanoparticles within a chitosan matrix has expanded their use into photocatalysis and magnetic targeting, necessitating more sophisticated characterization tools to evaluate these complex systems [205, 206, 349].

6 Applications in Drug Delivery

Chitosan and its derivatives have become increasingly prominent in drug delivery applications due to their versatile chemical structure, biodegradability, biocompatibility, and ability to be modified for specific medical needs [207]. As a naturally derived polysaccharide, chitosan offers unique advantages such as mucoadhesiveness, pH responsiveness, and the ability to enhance drug solubility and permeability [207, 208]. Its polycationic nature, coupled with abundant functional groups, allows for the attachment of targeting ligands and the formation of stable complexes with therapeutic agents [209, 210]. These properties make chitosan a valuable platform for the development of advanced drug delivery systems across a variety of therapeutic areas [210]. This section explores the major applications of chitosan in drug delivery, specifically focusing on cancer therapy, antimicrobial treatments, and other therapeutic applications.

6.1 Cancer Therapy

Cancer remains one of the leading causes of mortality worldwide, driving the need for more effective, targeted, and less toxic therapeutic strategies [211–214, 347, 351]. Chitosan-based drug delivery systems have been extensively explored in oncology due to their inherent biological compatibility and adaptability for functional modifications that enable site-specific drug delivery and controlled release [215, 216].

6.1.1 Targeted and Controlled Drug Delivery

Chitosan nanoparticles have shown significant promise as carriers for anticancer drugs [217, 218]. Their ability to accumulate preferentially in tumor tissues through the enhanced permeability and retention (EPR) effect allows for localized drug concentration at the tumor site [219, 220].

Further enhancement of tumor targeting has been achieved by functionalizing chitosan nanoparticles with ligands such as antibodies, peptides, and small molecules that recognize tumor-specific receptors [50]. This ligand-directed active targeting improves the selective uptake of chitosan nanocarriers by cancer cells while minimizing off-target effects and systemic toxicity [221–223].

6.1.2 Chemotherapy and Combination Therapy

Chitosan nanoparticles can encapsulate a wide variety of chemotherapeutic agents, including hydrophilic and hydrophobic drugs, thereby overcoming the solubility limitations of many anticancer drugs [224]. Chitosan's mucoadhesive properties further enhance drug retention at tumor sites, reducing the frequency of drug administration and improving patient compliance [225, 226].

Furthermore, chitosan derivatives have been explored in combination therapy, where they act as carriers for co-delivery of chemotherapeutic agents and other therapeutic molecules such as small interfering RNA (siRNA) or immune modulators [70, 227]. This strategy can achieve synergistic effects, enhance antitumor efficacy, and overcome multidrug resistance (MDR) [228, 229].

6.2 *Antimicrobial Treatments*

The growing challenge of antimicrobial resistance has led to the exploration of alternative antimicrobial strategies, with chitosan demonstrating significant promise due to its inherent antimicrobial properties [230]. As a polycation, chitosan interacts electrostatically with negatively charged microbial cell membranes, disrupting membrane integrity and leading to cell death [231].

6.2.1 Antibacterial Applications

Chitosan-based delivery systems have been developed to enhance the efficacy of antibiotics, improving their stability, bioavailability, and reducing required dosages [57, 232]. Nanoparticles and hydrogels formulated with chitosan increase the residence time of antibiotics at infection sites through their mucoadhesive properties and can be engineered for sustained drug release, resulting in prolonged antimicrobial activity [192, 233].

Chitosan's chelation of essential metal ions plays a critical role in disrupting bacterial metabolism and inhibiting biofilm formation [234, 235]. By displacing Ca^{2+} ions that stabilize bacterial cell membranes, chitosan induces a loss of membrane fluidity and function, further contributing to its antibacterial effects [231, 235].

6.2.2 Antifungal and Antiviral Applications

Beyond its antibacterial properties, chitosan exhibits antifungal and antiviral activity [236, 237]. Its use in drug delivery systems has been explored for delivering antifungal agents in treating infections such as candidiasis [238, 239]. Chitosan nanoparticles have shown potential for mucosal applications, such as in the vaginal and oral cavities [192].

In antiviral applications, chitosan nanoparticles have been used to deliver antiviral drugs directly to infected tissues [240–242]. For instance, chitosan-coated liposomes have been developed for pulmonary drug delivery, improving the retention and bioavailability of antiviral agents such as oxymatrine in the lungs and showing promise in the treatment of respiratory viral infections [243–245].

6.2.3 Wound Healing and Antimicrobial Dressings

Chitosan's dual role as an antimicrobial and wound healing agent makes it a valuable component in wound care products [233, 246]. Hydrogels, films, and sponges incorporating chitosan and antimicrobial agents provide protective barriers that prevent infection and promote tissue regeneration [247–249]. These dressings are particularly effective in managing chronic wounds and burns, where both infection control and accelerated healing are critical [250, 251].

Moreover, the ability of chitosan and its composites to generate ROS under light activation enhances their antimicrobial efficacy [252, 253]. Light-activated systems using chitosan modified with photosensitizers such as rose bengal or toluidine blue O have demonstrated significant reductions in microbial populations when exposed to specific light wavelengths, making them suitable for sterilizing medical devices like silicone, stainless steel, and glass surfaces [254, 255].

6.3 Other Applications

Chitosan's versatility extends beyond cancer and antimicrobial therapies into a broad spectrum of other therapeutic areas [256, 257]. Chitosan-modified silica has proven to be a stabilizer of insulin that can inhibit insulin fibrillation [258, 259].

6.3.1 Neurological Disorders and Brain Targeting

One of the major challenges in treating neurological diseases is overcoming the blood-brain barrier (BBB) [260–263]. Chitosan nanoparticles have emerged as promising carriers for delivering therapeutic agents to the central nervous system (CNS) [264–266]. Their mucoadhesive properties and positive surface charge enhance their ability to traverse biological barriers, including the nasal mucosa and the BBB [267].

Chitosan-based nanocarriers have been explored for delivering drugs targeting neurodegenerative disorders such as Parkinson's disease, Alzheimer's disease, and gliomas [268–270]. Intranasal delivery using chitosan nanoparticles bypasses the BBB and allows direct drug transport from the nasal cavity to the brain [25, 271]. This non-invasive route offers several advantages, including improved patient compliance and the potential for rapid therapeutic effects [25, 272].

6.3.2 Anti-Inflammatory Therapies

Chitosan exhibits inherent anti-inflammatory activity and can be used to deliver anti-inflammatory drugs for the treatment of inflammatory conditions [273, 274]. Low molecular weight hydrolyzed chitosans, especially those in the 3–5 kDa range, have demonstrated notable inhibitory effects on MMP-2 activity in human dermal fibroblasts [274, 275]. This inhibition is believed to stem from chitosan's strong affinity for zinc ions, which are essential cofactors for MMP-2 enzymatic function [276]. Through its metal-chelating capacity and structural compatibility, chitosan offers a multifaceted strategy for dampening inflammatory responses by regulating MMP activity [55, 277].

6.3.3 Gene Delivery

Chitosan-based nanoparticles are increasingly being used as non-viral vectors for gene therapy [278, 279]. Their ability to complex with negatively charged nucleic acids, such as DNA and siRNA, protects these molecules from enzymatic degradation and enhances cellular uptake [279, 280]. Chemical modifications of chitosan, including quaternization and thiolation, have been shown to improve transfection efficiency and targeting specificity [281].

Gene delivery systems based on chitosan nanoparticles are under investigation for the treatment of various genetic disorders, cancer, and for vaccine delivery [282, 283]. Chitosan's low immunogenicity and biodegradability make it a safe alternative to viral vectors, addressing some of the major limitations associated with viral gene delivery [243, 284].

6.3.4 Gastrointestinal Drug Delivery

Chitosan's ability to open tight junctions between intestinal epithelial cells facilitates the paracellular transport of macromolecules, including peptides and proteins [285, 286], which are typically challenging to deliver orally due to enzymatic degradation and poor permeability [287, 288]. Chitosan-based oral drug delivery systems have been developed to protect drugs from acidic gastric environments and release them in the intestines [285, 289].

Moreover, chitosan's inhibitory effect on efflux transporters, such as P-glycoprotein, helps improve the bioavailability of drugs that are typically pumped out of cells by these proteins [290, 291]. This mechanism makes chitosan-based systems promising for overcoming drug resistance and enhancing the efficacy of oral therapeutics [292].

6.3.5 Vaccines and Therapy

Chitosan nanoparticles have shown great potential in vaccine delivery and immunotherapy due to their biocompatibility, mucoadhesive nature, and immune-enhancing effects [293, 294]. When combined with polyanionic agents like sodium tripolyphosphate or γ-poly-L-glutamic acid, they form stable nanoparticles capable of encapsulating tumor-associated antigens such as MUC1 glycopeptides without the need for complex synthesis. These nanovaccines have demonstrated stronger antigen-specific IgG responses compared to traditional adjuvants [295].

In immunotherapy, chitosan also improves the performance of cytokines like interleukin-12 by prolonging their retention at tumor sites and minimizing systemic toxicity [296, 297]. Intratumoral delivery of chitosan/IL-12 has led to complete tumor regression and durable immune memory in preclinical models, underscoring its ability to trigger potent, long-lasting antitumor responses through activation of CD8+ T cells and NK cells [298, 299].

7 Challenges and Limitations

7.1 Stability and Degradation

Chitosan, a naturally biodegradable biopolymer, has garnered significant attention due to its vast potential in biomedical, pharmaceutical, and industrial applications [54, 300]. However, its susceptibility to environmental and processing conditions presents a critical challenge that affects its long-term stability and functionality [184, 301]. The polymer's structural integrity is highly sensitive to external factors such as temperature fluctuations, moisture content, pH variations, and oxidative conditions, all of which can lead to polymer degradation [301, 302].

Chitosan undergoes enzymatic degradation in vivo, predominantly through the action of lysozyme [303], a non-specific protease present in all mammalian tissues [304]. Lysozyme hydrolyzes the β-1,4-glycosidic bonds of chitosan, producing non-toxic oligosaccharides that are either excreted or incorporated into glycosaminoglycans and glycoproteins [304, 305]. The enzymatic degradation of chitosan contributes to its biocompatibility and biodegradability, making it a suitable candidate for biomedical applications, including drug delivery, wound healing, and tissue engineering [23, 306]. However, its rapid degradation under certain conditions poses a limitation that necessitates strategies to enhance its stability [307].

In vitro degradation of chitosan can be achieved through oxidation, chemical hydrolysis, or enzymatic hydrolysis. These controlled methods are widely used for the preparation of low molecular weight chitosan, which is desirable for various applications [308, 309]. Regardless of the degradation mode, the process typically begins with the random cleavage of β-1,4-glycosidic bonds (depolymerization), followed by deacetylation of N-acetyl linkages. As a result, a decrease in the average molecular weight and an increase in the degree of deacetylation are observed [310–312]. Additionally, the breakdown of chitosan chains often leads to the formation of free radicals, which can trigger oxidative degradation, further altering the polymer's properties [184, 313, 314]. The degradation process may also involve the cleavage of functional groups such as amino, carbonyl, amido, and hydroxyl groups, which significantly influence the polymer's chemical reactivity and physicochemical characteristics [315, 316].

Chitosan's stability is highly dependent on molecular weight, polydispersity, degree of deacetylation, purity, and moisture content [35, 307]. Generally, higher degrees of deacetylation correlate with increased susceptibility to degradation, while a higher degree of crystallinity slows down the degradation process [17, 317, 318]. Furthermore, environmental conditions play a crucial role in determining the degradation rate. For instance, exposure to elevated temperatures, such as 40 °C, results in the loss of mechanical strength and hardness due to dehydration of the polymer [309, 319]. Studies have shown that chitosan solutions stored at ambient or elevated temperatures undergo hydrolysis that follows first-order kinetics, leading to faster degradation. Conversely, when stored at lower temperatures (2–8 °C), chitosan remains stable, with no significant hydrolytic chain scission observed in solutions held at 5 °C. High humidity levels (above 60% RH) also accelerate the degradation process by intensifying water molecule penetration into chitosan chains, leading to swelling, loss of mechanical properties, and increased hydrolytic degradation [307].

The susceptibility of chitosan to external influences and processing conditions limits its shelf life and practical applications in biomedical fields. Therefore, various strategies have been explored to enhance its stability [301]. These strategies include controlling storage conditions, modifying manufacturing parameters (such as temperature and relative humidity), incorporating stabilizing agents, blending chitosan with other polymers, or chemically modifying its structure using ionic or covalent crosslinking methods [307, 320]. Chemical derivatives of chitosan, such as acylated, carboxymethylated, quaternized, and thiolated forms, have demonstrated improved solubility and mechanical properties, thereby broadening its applications [321, 322].

7.2 *Targeting and Specificity Issues in Unmodified Chitosan-Based Nanoparticles*

Chitosan-based nanoparticles have been extensively researched for drug delivery applications due to their biocompatibility, low toxicity, and ability to encapsulate and release therapeutic agents in a controlled manner [323]. However, one of the significant challenges associated with chitosan-based unmodified drug carriers is their limited targeting and specificity [324]. Efficient drug delivery relies on precise targeting mechanisms to ensure that therapeutic agents reach the intended site of action without affecting non-target tissues [325]. However, chitosan-based unmodified carriers often exhibit non-specific interactions, which can result in off-target effects and reduced therapeutic efficacy [324].

The effectiveness of chitosan-based drug delivery systems is influenced by factors such as molecular weight, charge distribution, surface modifications, and conjugation with targeting moieties [326]. To enhance targeting capabilities, researchers have explored various modification strategies, including functionalization with ligands such as antibodies, peptides, or aptamers. These ligands facilitate receptor-mediated endocytosis, improving the selectivity of chitosan-based carriers toward specific cell types (Fig. 3) [327–329].

Chemical modifications of chitosan, such as the introduction of hydrophilic or hydrophobic groups, have also been investigated to optimize drug release profiles and improve bioavailability [330]. Surface functionalization techniques, including PEGylation, have been employed to enhance circulation time and reduce rapid clearance by the immune system [324].

Despite extensive research on chitosan and its derivatives, further studies are needed to develop a systematic approach for optimizing drug targeting and specificity. A comprehensive understanding of pharmacokinetics, in vitro and in vivo toxicity, and long-term safety of chitosan-based biomaterials is essential for their

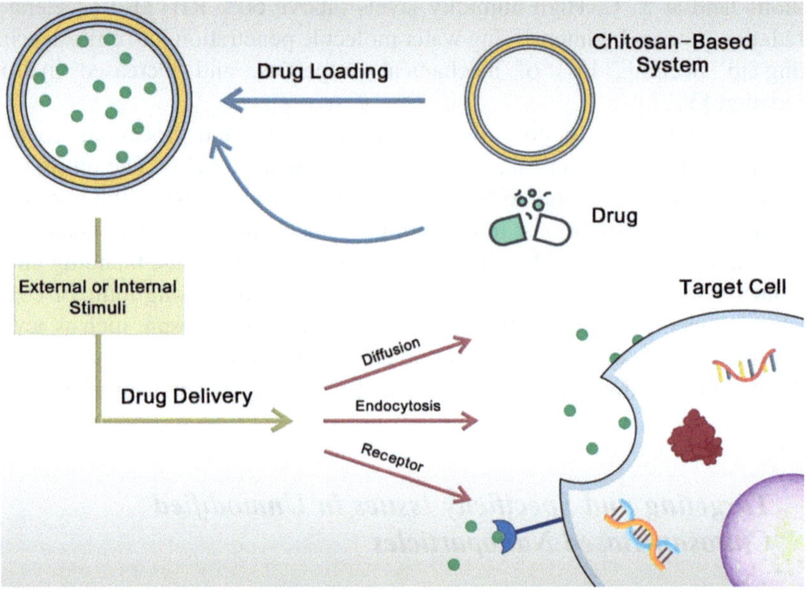

Fig. 4 Schematic representation of drug loading and release processes in chitosan-based systems, highlighting different drug release mechanisms including diffusion, receptor-mediated release, and endocytosis as well as internal and external stimuli-responsive drug delivery pathways

translation into clinical applications. Addressing these challenges will play a pivotal role in expanding the scope of chitosan-based drug delivery systems, paving the way for personalized medicine and advanced therapeutic interventions. The development of novel chitosan-based nanocomposites with improved mechanical strength, thermal stability, and antimicrobial properties holds great promise for overcoming current limitations and enhancing the clinical applicability of chitosan in drug transport and targeted therapy (Fig. 4).

8 Recent Advances and Innovations

8.1 Novel Chitosan Derivatives

The application of Chitosan is often limited by its poor solubility in neutral and alkaline media. To overcome these limitations, chemical modifications have been employed to enhance its properties while retaining the fundamental chitosan skeleton [331]. Various derivatives have been synthesized through modifications at the amino (-NH2) and hydroxyl (-OH) functional groups, leading to improved solubility, bioavailability, and tailored physicochemical properties [332].

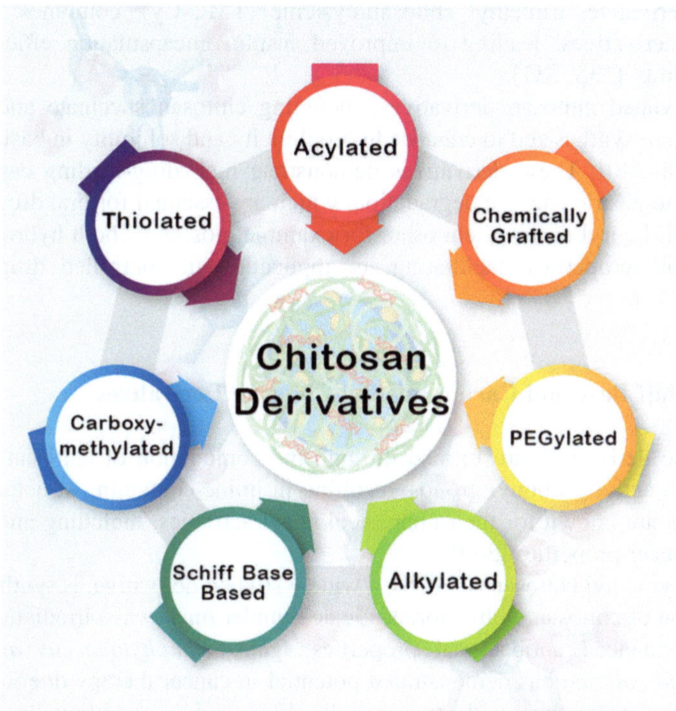

Fig. 5 Summary of different chitosan derivatives used in biological applications

Chemical modifications of chitosan mainly target its amino (NH_2) and hydroxyl (OH) groups, enabling the synthesis of diverse derivatives while preserving its core bioactivity [333]. Key strategies include alkylation, acylation, carboxymethylation, grafting, Carboxymethylation, PEGylation, and thiolation (Fig. 5) [334]. Among these, alkylation yields derivatives like N-trimethyl chitosan (TMC) through methylation, significantly enhancing water solubility and mucoadhesive properties, which are valuable for drug delivery [28]. Acylation introduces hydrophobic chains such as caprylic, lauric, and oleoyl groups, improving chitosan's permeability, enzymatic resistance, and mucoadhesiveness—making these derivatives especially promising for oral drug delivery systems [335, 336].

8.1.1 Thiolated and Carboxylated Chitosan Derivatives

Thiolatedchitosans are synthesized by modifying amine groups with thiol-containing compounds such as cysteine, 2-iminothiolane, thiobutylamidine, or thioglycolic acid. These derivatives display improved mucoadhesive and permeation properties, enabling enhanced drug absorption across biological membranes. One notable derivative, trimethyl chitosan-cysteine (TMC-Cy), combines TMC and thiolated derivatives, leading to improved insulin encapsulation efficiency and bioavailability [335, 337].

Carboxylated chitosan derivatives, including chitosan succinate and chitosan phthalate, are synthesized to enhance hydrophilicity and solubility in basic environments [338–340]. These derivatives demonstrate high drug-loading capacity and resistance to gastric enzyme degradation, which are essential for oral drug formulations [340]. Lauryl succinyl chitosan, for example, possesses both hydrophilic and hydrophobic properties, facilitating its application in controlled drug delivery systems [341].

8.1.2 Schiff Base and Carboxymethyl Chitosan Derivatives

Schiff base derivatives are formed through the conjugation of chitosan's primary amine with active carbonyl groups, resulting in imine-containing structures. These derivatives are known for their pharmacological activities, including antimicrobial and anticancer properties [342].

Carboxymethyl chitosan (CMCS), a widely studied derivative, is synthesized via the reaction of chitosan with chloroacetic acid under microwave irradiation. CMCS exhibits enhanced antibacterial properties against *Staphylococcus aureus* and *Escherichia coli* and has demonstrated potential in cancer therapy due to its strong electrostatic interaction with tumor cells [343, 344]. Additionally, mono-N-carboxymethyl chitosan (MCC) has been developed for improved biocompatibility and has shown promise in facilitating the transport of low molecular weight heparin (LMWH) across epithelial barriers [340].

8.1.3 Grafted and PEGylated Chitosan Derivatives

Chemical grafting techniques enable the attachment of functional groups to the chitosan backbone, producing derivatives with unique biological and physicochemical properties. Grafted chitosan derivatives include chitosan-poly(ethylene glycol) diacrylate and galactose-functionalized chitosan, which have been explored for biomedical applications such as targeted drug delivery and tissue engineering [345, 346].

PEGylation of chitosan, which involves the conjugation of polyethylene glycol (PEG) chains, has been widely employed to enhance chitosan's solubility, biocompatibility, and stability in physiological fluids. PEGylated chitosan derivatives have been extensively investigated for oral peptide and protein delivery, offering improved protection against enzymatic degradation and enhanced intestinal absorption.

9 Conclusion

Chitosan's unique combination of biological compatibility, chemical versatility, and responsiveness to environmental stimuli has solidified its place as a cornerstone material in the evolving field of smart drug delivery systems. In particular, its integration into light-activated platforms represents a significant leap toward achieving precision-controlled therapy, offering the ability to spatially and temporally direct drug release with minimal invasiveness. Despite current challenges, such as limited solubility under physiological conditions and the need for improved targeting specificity, recent innovations in chemical modification and nanotechnology have paved the way for addressing these hurdles. Through careful design of chitosan-based systems, leveraging properties like pH sensitivity, mucoadhesion, and photothermal or photochemical responsiveness, researchers are increasingly able to tailor therapies for a wide range of diseases, from cancer to neurological disorders. Continued advancements in synthesis techniques, surface engineering, and stimuli-responsiveness are likely to expand the clinical potential of these systems. As understanding deepens and interdisciplinary collaborations grow, chitosan-based light-activated drug delivery holds great promise for shaping the future of precision and personalized medicine.

References

1. Rahimian, S., Mirkazemi, K., Kamalinejad, A., & Doroudian, M. (2025). Exosome-based advances in pancreatic cancer: The potential of mesenchymal stem cells. *Critical Reviews in Oncology/Hematology, 207*, 104594.
2. Rahimian, S., Najafi, H., Webber, C. A., & Jalali, H. (2024). Advances in exosome-based therapies for the repair of peripheral nerve injuries. *Neurochemical Research, 49*(8), 1905–1925.
3. Ezike, T. C., Okpala, U. S., Onoja, U. L., Nwike, C. P., Ezeako, E. C., Okpara, O. J., et al. (2023). Advances in drug delivery systems, challenges and future directions. *Heliyon, 9*(6), e17488.
4. Asrorov, A. M., Fang, Y., Wang, H., Mirzaakhmedov, S. Y., & Huang, Y. (2023). Chapter 14 - Stimuli-responsive systems for smart drug delivery systems. In A. K. Nayak, M. S. Hasnain, B. Laha, & S. Maiti (Eds.), *Advanced and modern approaches for drug delivery* (pp. 389–420). Academic Press.
5. Farjadian, F., Ghasemi, S., Akbarian, M., Hoseini-Ghahfarokhi, M., Moghoofei, M., & Doroudian, M. (2022). Physically stimulus-responsive nanoparticles for therapy and diagnosis. *Frontiers in Chemistry, 10*, 952675.
6. Tao, Y., Chan, H. F., Shi, B., Li, M., & Leong, K. W. (2020). Light: A magical tool for controlled drug delivery. *Advanced Functional Materials, 30*(49), 2005029.
7. Pandey, T., & Pandey, V. (2025). A mechanistic understanding to photophysical phenomenon in development of near-infrared (NIR) responsive hydrogels: Advancements in precision drug delivery. *Journal of Drug Delivery Science and Technology, 106*, 106682.
8. Liu, C., Zhou, H., Yu, S., Chen, H., Wang, N., Yao, W., et al. (2022). Sensitive and precise visually guided drug delivery nanoplatform with dual activation of pH and light. *Acta Biomaterialia, 141*, 374–387.

9. El-Tanani, M., Satyam, S. M., Rabbani, S. A., El-Tanani, Y., Aljabali, A. A. A., Al Faouri, I., et al. (2025). Revolutionizing drug delivery: The impact of advanced materials science and technology on precision medicine. *Pharmaceutics [Internet], 17*(3), 375.
10. Al-Jamal, A. N., Al-Hussainy, A. F., Mohammed, B. A., Abbas, H. H., Kadhim, I. M., Ward, Z. H., et al. (2025). Photodynamic Therapy (PDT) in drug delivery: Nano-innovations enhancing treatment outcomes. *Health Sciences Review, 14*, 100218.
11. Ashrafizadeh, M., Hushmandi, K., Mirzaei, S., Bokaie, S., Bigham, A., Makvandi, P., et al. (2023). Chitosan-based nanoscale systems for doxorubicin delivery: Exploring biomedical application in cancer therapy. *Bioengineering & Translational Medicine, 8*(1), e10325.
12. Hisham, F., Maziati Akmal, M. H., Ahmad, F., Ahmad, K., & Samat, N. (2024). Biopolymer chitosan: Potential sources, extraction methods, and emerging applications. *Ain Shams Engineering Journal, 15*(2), 102424.
13. Hamedi, H., Moradi, S., Hudson, S. M., Tonelli, A. E., & King, M. W. (2022). Chitosan based bioadhesives for biomedical applications: A review. *Carbohydrate Polymers, 282*, 119100.
14. Aranaz, I., Alcántara, A. R., Civera, M. C., Arias, C., Elorza, B., Heras Caballero, A., et al. (2021). Chitosan: An overview of its properties and applications. *Polymers [Internet], 13*(19), 3256.
15. Aibani, N., Rai, R., Patel, P., Cuddihy, G., & Wasan, E. K. (2021). Chitosan nanoparticles at the biological interface: Implications for drug delivery. *Pharmaceutics, 13*(10), 1686.
16. Zhao, D., Huang, R., Gan, J.-M., & Shen, Q.-D. (2022). Photoactive nanomaterials for wireless neural biomimetics, stimulation, and regeneration. *ACS Nano, 16*(12), 19892–19912.
17. Herdiana, Y., Wathoni, N., Shamsuddin, S., & Muchtaridi, M. (2022). Drug release study of the chitosan-based nanoparticles. *Heliyon, 8*(1), e08674.
18. Zhang, Y., Qiu, Y., Karimi, A. B., & Smith, B. R. (2025). Systematic review: Mechanisms of photoactive nanocarriers for imaging and therapy including controlled drug delivery. *European Journal of Nuclear Medicine and Molecular Imaging, 52*(4), 1576–1595.
19. Shao, D., Gao, Q., Sheng, Y., Li, S., & Kong, Y. (2022). Construction of a dual-responsive dual-drug delivery platform based on the hybrids of mesoporous silica, sodium hyaluronate, chitosan and oxidized sodium carboxymethyl cellulose. *International Journal of Biological Macromolecules, 202*, 37–45.
20. Narayan, R., Gadag, S., Cheruku, S. P., Raichur, A. M., Day, C. M., Garg, S., et al. (2021). Chitosan-glucuronic acid conjugate coated mesoporous silica nanoparticles: A smart pH-responsive and receptor-targeted system for colorectal cancer therapy. *Carbohydrate Polymers, 261*, 117893.
21. Duan, Q., Shi, J., Zhou, L., Zhang, B., Wang, X., & Sang, S. (2022). pH-responsive and sustained release drug delivery system of BSA coated CDs-DOX. *Journal of Molecular Structure, 1248*, 131358.
22. Lee, Y., & Thompson, D. H. (2017). Stimuli-responsive liposomes for drug delivery. *Wiley Interdisciplinary Reviews. Nanomedicine and Nanobiotechnology, 9*(5), 10.1002/wnan.1450.
23. Desai, N., Rana, D., Salave, S., Gupta, R., Patel, P., Karunakaran, B., et al. (2023). Chitosan: A potential biopolymer in drug delivery and biomedical applications. *Pharmaceutics, 15*(4), 1313.
24. Dahri, M., Beheshtizadeh, N., Seyedpour, N., Nakhostin-Ansari, A., Aghajani, F., Seyedpour, S., et al. (2023). Biomaterial-based delivery platforms for transdermal immunotherapy. *Biomedicine & Pharmacotherapy, 165*, 115048.
25. Omidian, H., Gill, E. J., Dey Chowdhury, S., & Cubeddu, L. X. (2024). Chitosan nanoparticles for intranasal drug delivery. *Pharmaceutics, 16*(6), 746.
26. Bernkop-Schnürch, A., & Dünnhaupt, S. (2012). Chitosan-based drug delivery systems. *European Journal of Pharmaceutics and Biopharmaceutics, 81*(3), 463–469.
27. Onugwu, A. L., Nwagwu, C. S., Onugwu, O. S., Echezona, A. C., Agbo, C. P., Ihim, S. A., et al. (2023). Nanotechnology based drug delivery systems for the treatment of anterior segment eye diseases. *Journal of Controlled Release, 354*, 465–488.

28. Žigrayová, D., Mikušová, V., & Mikuš, P. (2024). Advances in Chitosan derivatives: Preparation, properties and applications in pharmacy and medicine. *Gels [Internet], 10*(11), 701.
29. Kesharwani, P., Halwai, K., Jha, S. K., Al Mughram, M. H., Almujri, S. S., Almalki, W. H., et al. (2024). Folate-engineered chitosan nanoparticles: Next-generation anticancer nanocarriers. *Molecular Cancer, 23*(1), 244.
30. Bonferoni, M. C., Gavini, E., Rassu, G., Maestri, M., & Giunchedi, P. (2020). Chitosan nanoparticles for therapy and theranostics of hepatocellular carcinoma (HCC) and liver-targeting. *Nanomaterials (Basel), 10*(5), 870.
31. Liang, S., Qin, Z., Lu, H., Wen, Z., Raziq, F., Qin, Q., et al. (2025). Combined photothermal-photodynamic therapy by Ce6 loaded on MoO2 functional nanoplatform. *Next Nanotechnology, 8*, 100152.
32. Dohendou, M., Pakzad, K., Nezafat, Z., Nasrollahzadeh, M., & Dekamin, M. G. (2021). Progresses in chitin, chitosan, starch, cellulose, pectin, alginate, gelatin and gum based (nano)-catalysts for the Heck coupling reactions: A review. *International Journal of Biological Macromolecules, 192*, 771–819.
33. Román-Doval, R., Torres-Arellanes, S. P., Tenorio-Barajas, A. Y., Gómez-Sánchez, A., & Valencia-Lazcano, A. A. (2023). Chitosan: Properties and its application in agriculture in context of molecular weight. *Polymers [Internet], 15*(13), 2867.
34. Iber, B. T., Kasan, N. A., Torsabo, D., & Omuwa, J. W. (2021). A review of various sources of Chitin and Chitosan in nature. *Journal of Renewable Materials, 10*(4), 1097–1123.
35. Pellis, A., Guebitz, G. M., & Nyanhongo, G. S. (2022). Chitosan: Sources, processing and modification techniques. *Gels, 8*(7), 393.
36. Novikov, V. Y., Derkach, S. R., Konovalova, I. N., Dolgopyatova, N. V., & Kuchina, Y. A. (2023). Mechanism of heterogeneous alkaline Deacetylation of Chitin: A review. *Polymers [Internet], 15*(7), 1729.
37. Weißpflog, J., Vehlow, D., Müller, M., Kohn, B., Scheler, U., Boye, S., et al. (2021). Characterization of chitosan with different degree of deacetylation and equal viscosity in dissolved and solid state—Insights by various complimentary methods. *International Journal of Biological Macromolecules, 171*, 242–261.
38. Mania, S., Banach-Kopeć, A., Staszczyk, K., Kulesza, J., Augustin, E., & Tylingo, R. (2023). An influence of molecular weight, deacetylation degree of chitosan xerogels on their antimicrobial activity and cytotoxicity. Comparison of chitosan materials obtained using lactic acid and CO2 saturation. *Carbohydrate Research, 534*, 108973.
39. Jhaveri, J., Raichura, Z., Khan, T., Momin, M., & Omri, A. (2021). Chitosan nanoparticles-insight into properties, functionalization and applications in drug delivery and Theranostics. *Molecules [Internet], 26*(2), 272.
40. Benalaya, I., Alves, G., Lopes, J., & Silva, L. R. (2024). A review of natural polysaccharides: Sources, characteristics, properties, food, and pharmaceutical applications. *International Journal of Molecular Sciences [Internet], 25*(2), 1322.
41. Zhang, L., Li, Q., Liang, Y., Zhang, G., Zou, J., Fei, P., et al. (2024). Hydrogels comprising oxidized carboxymethyl cellulose and water-soluble chitosan at varied oxidation levels: Synthesis, characterization, and adsorptive toward methylene blue. *International Journal of Biological Macromolecules, 277*, 134351.
42. Ahmad, K., Zhang, Y., Chen, P., Yang, X., & Hou, H. (2024). Chitosan interaction with stomach mucin layer to enhances gastric retention and mucoadhesive properties. *Carbohydrate Polymers, 333*, 121926.
43. Biswas, R., Mondal, S., Ansari, M. A., Sarkar, T., Condiuc, I. P., Trifas, G., et al. (2025). Chitosan and its derivatives as nanocarriers for drug delivery. *Molecules [Internet], 30*(6), 1297.
44. Carrera, C., Bengoechea, C., Carrillo, F., & Calero, N. (2023). Effect of deacetylation degree and molecular weight on surface properties of chitosan obtained from biowastes. *Food Hydrocolloids, 137*, 108383.

45. Freitas, E. D., Moura, J. C. F., Kerwald, J., & Beppu, M. M. (2020). An overview of current knowledge on the properties, synthesis and applications of quaternary Chitosan derivatives. *Polymers [Internet], 12*(12), 2878.
46. Thanakkasaranee, S., Jantanasakulwong, K., Phimolsiripol, Y., Leksawasdi, N., Seesuriyachan, P., Chaiyaso, T., et al. (2021). High substitution synthesis of Carboxymethyl chitosan for properties improvement of Carboxymethyl Chitosan films depending on particle sizes. *Molecules [Internet], 26*(19), 6013.
47. Yan, D., Li, Y., Liu, Y., Li, N., Zhang, X., & Yan, C. (2021). Antimicrobial properties of Chitosan and Chitosan derivatives in the treatment of enteric infections. *Molecules, 26*(23), 7136.
48. Grewal, A. K., & Salar, R. K. (2024). Chitosan nanoparticle delivery systems: An effective approach to enhancing efficacy and safety of anticancer drugs. *Nano TransMed, 3*, 100040.
49. Mikušová, V., & Mikuš, P. (2021). Advances in chitosan-based nanoparticles for drug delivery. *International Journal of Molecular Sciences [Internet], 22*(17), 9652.
50. Herdiana, Y., Wathoni, N., Shamsuddin, S., Joni, I. M., & Muchtaridi, M. (2021). Chitosan-based nanoparticles of targeted drug delivery system in breast cancer treatment. *Polymers, 13*(11), 1717.
51. Solanki, R., Dhanka, M., Thareja, P., & Bhatia, D. (2024). Self-healing, injectable chitosan-based hydrogels: Structure, properties and biological applications. *Materials Advances, 5*(13), 5365–5393.
52. Li, J., Cai, C., Li, J., Li, J., Li, J., Sun, T., et al. (2018). Chitosan-based nanomaterials for drug delivery. *Molecules, 23*(10), 2661.
53. Tian, B., Hua, S., Tian, Y., & Liu, J. (2020). Chemical and physical chitosan hydrogels as prospective carriers for drug delivery: A review. *Journal of Materials Chemistry B, 8*(44), 10050–10064.
54. Jiménez-Gómez, C. P., & Cecilia, J. A. (2020). Chitosan: A natural biopolymer with a wide and varied range of applications. *Molecules, 25*(17), 3981.
55. Haider, A., Khan, S., Iqbal, D. N., Khan, S. U., Haider, S., Mohammad, K., et al. (2024). Chitosan as a tool for tissue engineering and rehabilitation: Recent developments and future perspectives–a review. *International Journal of Biological Macromolecules, 278*, 134172.
56. Luo, Y., & Wang, Q. (2014). Recent development of chitosan-based polyelectrolyte complexes with natural polysaccharides for drug delivery. *International Journal of Biological Macromolecules, 64*, 353–367.
57. Jafernik, K., Ładniak, A., Blicharska, E., Czarnek, K., Ekiert, H., Wiącek, A. E., et al. (2023). Chitosan-based nanoparticles as effective drug delivery systems—A review. *Molecules, 28*(4), 1963.
58. Mozafari, N., Farjadian, F., Samani, S. M., Azadi, S., & Azadi, A. (2020). Simvastatin-chitosan-citicoline conjugates nanoparticles as the co-delivery system in Alzheimer susceptible patients. *International Journal of Biological Macromolecules, 156*, 1396–1407.
59. Rajinikanth, B. S., Rajkumar, D. S. R., & Vijayaragavan, V. (2024). Chitosan-based biomaterial in wound healing: A review. *Cureus., 16*(2), e55193.
60. Kravanja, G., Primožič, M., Knez, Ž., & Leitgeb, M. (2019). Chitosan-based (Nano)materials for novel biomedical applications. *Molecules [Internet], 24*(10), 1960.
61. Akbarian, M., Chen, S.-H., Kianpour, M., Farjadian, F., Tayebi, L., & Uversky, V. N. (2022). A review on biofilms and the currently available antibiofilm approaches: Matrix-destabilizing hydrolases and anti-bacterial peptides as promising candidates for the food industries. *International Journal of Biological Macromolecules, 219*, 1163–1179.
62. Virmani, T., Kumar, G., Sharma, A., Pathak, K., Akhtar, M. S., Afzal, O., et al. (2023). Amelioration of cancer employing Chitosan, its derivatives, and Chitosan-based nanoparticles: Recent updates. *Polymers [Internet], 15*(13), 2928.
63. Yeh, T. H., Hsu, L. W., Tseng, M. T., Lee, P. L., Sonjae, K., Ho, Y. C., et al. (2011). Mechanism and consequence of chitosan-mediated reversible epithelial tight junction opening. *Biomaterials, 32*(26), 6164–6173.

64. Popa, L., Ghica, M. V., & Dinu-Pîrvu, C. E. (2023). Chitosan biomaterials: Advances and challenges. *International Journal of Molecular Sciences, 24*(22), 16150.
65. Mukhopadhyay, P., Mishra, R., Rana, D., & Kundu, P. P. (2012). Strategies for effective oral insulin delivery with modified chitosan nanoparticles: A review. *Progress in Polymer Science, 37*(11), 1457–1475.
66. Alghareeb, S., Ekenna, I., Asare-Addo, K., Conway, B. R., & Adebisi, A. O. (2025). Chitosan nanoparticles for nasal drug delivery. *Journal of Drug Delivery Science and Technology, 105*, 106623.
67. Hua, S. (2020). Advances in oral drug delivery for regional targeting in the gastrointestinal tract - influence of physiological, pathophysiological and pharmaceutical factors. *Frontiers in Pharmacology, 11*, 524.
68. Bhattarai, N., Gunn, J., & Zhang, M. (2010). Chitosan-based hydrogels for controlled, localized drug delivery. *Advanced Drug Delivery Reviews, 62*(1), 83–99.
69. Tian, B., & Liu, J. (2023). Smart stimuli-responsive chitosan hydrogel for drug delivery: A review. *International Journal of Biological Macromolecules, 235*, 123902.
70. Shakil, M. S., Mahmud, K. M., Sayem, M., Niloy, M. S., Halder, S. K., Hossen, M. S., et al. (2021). Using chitosan or chitosan derivatives in cancer therapy. *Polysaccharides, 2*(4), 795–816.
71. Ding, J., & Guo, Y. (2022). Recent advances in chitosan and its derivatives in cancer treatment. *Frontiers in Pharmacology, 13*, 888740.
72. Kumara, B. N., Gurudatt, N. G., & Prasad, K. S. (2023). Carboxymethyl-hexanoyl chitosan: A promising candidate for hydrophobic and hydrophilic drug delivery. *Carbohydrate Polymer Technologies and Applications, 6*, 100401.
73. Nandgude, T., & Pagar, R. (2021). Plausible role of chitosan in drug and gene delivery against resistant breast cancer cells. *Carbohydrate Research, 506*, 108357.
74. Sabourian, P., Tavakolian, M., Yazdani, H., Frounchi, M., van de Ven, T. G. M., Maysinger, D., et al. (2020). Stimuli-responsive chitosan as an advantageous platform for efficient delivery of bioactive agents. *Journal of Controlled Release, 317*, 216–231.
75. Carreira, A. S., Gonçalves, F. A. M. M., Mendonça, P. V., Gil, M. H., & Coelho, J. F. J. (2010). Temperature and pH responsive polymers based on chitosan: Applications and new graft copolymerization strategies based on living radical polymerization. *Carbohydrate Polymers, 80*(3), 618–630.
76. Fathi, M., Sahandi Zangabad, P., Majidi, S., Barar, J., Erfan-Niya, H., & Omidi, Y. (2017). Stimuli-responsive chitosan-based nanocarriers for cancer therapy. *BioImpacts: BI, 7*(4), 269–277.
77. Hosseini, M., Farjadian, F., & Makhlouf, A. S. H. (2016). Smart stimuli-responsive nano-sized hosts for drug delivery. In: Hosseini, M., & Makhlouf, A. (Eds.), *Industrial applications for intelligent polymers and coatings*. Springer, Cham. https://doi.org/10.1007/978-3-319-26893-4_1
78. Farjadian, F., Mirkiani, S., Ghasemiyeh, P., Kafshboran, H. R., Mehdi-Alamdarlou, S., Raeisi, A., et al. (2024). Smart nanogels as promising platform for delivery of drug, gene, and vaccine; therapeutic applications and active targeting mechanism. *European Polymer Journal, 219*, 113400.
79. Maita, K. C., Avila, F. R., Torres-Guzman, R. A., Garcia, J. P., Eldaly, A. S., Palmieri, L., et al. (2022). Local anti-inflammatory effect and immunomodulatory activity of chitosan-based dressing in skin wound healing: A systematic review. *Journal of Clinical and Translational Research, 8*(6), 488–498.
80. Fong, D., & Hoemann, C. D. (2018). Chitosan immunomodulatory properties: Perspectives on the impact of structural properties and dosage. *Future Science OA, 4*(1), FSO225.
81. Friedman, A. J., Phan, J., Schairer, D. O., Champer, J., Qin, M., Pirouz, A., et al. (2013). Antimicrobial and Anti-Inflammatory Activity of Chitosan–Alginate Nanoparticles: A Targeted Therapy for Cutaneous Pathogens. *Journal of Investigative Dermatology., 133*(5), 1231–1239.

82. Mandracchia, D., Trapani, A., Tripodo, G., Perrone, M. G., Giammona, G., Trapani, G., et al. (2017). In vitro evaluation of glycol chitosan based formulations as oral delivery systems for efflux pump inhibition. *Carbohydrate Polymers, 166*, 73–82.
83. Tian, Y., Lei, Y., Wang, Y., Lai, J., Wang, J., & Xia, F. (2023). Mechanism of multidrug resistance to chemotherapy mediated by P-glycoprotein (review). *International Journal of Oncology, 63*(5), 119.
84. Pote, M. S., & Gacche, R. N. (2023). ATP-binding cassette efflux transporters and MDR in cancer. *Drug Discovery Today, 28*(5), 103537.
85. Dong, Y., Liao, H., Yu, J., Fu, H., Zhao, D., Gong, K., et al. (2019). Incorporation of drug efflux inhibitor and chemotherapeutic agent into an inorganic/organic platform for the effective treatment of multidrug resistant breast cancer. *Journal of Nanobiotechnology, 17*(1), 125.
86. Mura, P., Maestrelli, F., Cirri, M., & Mennini, N. (2022). Multiple roles of Chitosan in mucosal drug delivery: An updated review. *Marine Drugs [Internet], 20*(5), 335.
87. Yanat, M., & Schroën, K. (2021). Preparation methods and applications of chitosan nanoparticles; with an outlook toward reinforcement of biodegradable packaging. *Reactive and Functional Polymers, 161*, 104849.
88. Cheung, R. C., Ng, T. B., Wong, J. H., & Chan, W. Y. (2015). Chitosan: An update on potential biomedical and pharmaceutical applications. *Marine Drugs, 13*(8), 5156–5186.
89. Rahim, M. A., Jan, N., Khan, S., Shah, H., Madni, A., Khan, A., et al. (2021). Recent advancements in stimuli responsive drug delivery platforms for active and passive cancer targeting. *Cancers (Basel), 13*(4), 670.
90. Ghasemi, S., Najafi, M., Doroudian, M., Rastegari, B., Behzad-Behbahani, A., Soltanimehr, H., et al. (2025). Temperature- and pH-responsive Poly(NIPAM-co-HEMA-co-AAm) Nanogel as a smart vehicle for Doxorubicin delivery; combating colorectal cancer. *Gels., 11*(4), 227.
91. Maboudi, A. H., Lotfipour, M. H., Rasouli, M., Azhdari, M. H., MacLoughlin, R., Bekeschus, S., et al. (2024). Micelle-based nanoparticles with stimuli-responsive properties for drug delivery. *Nanotechnology Reviews, 13*(1), 20230218.
92. Doroudian, M., Azhdari, M. H., Goodarzi, N., O'Sullivan, D., & Donnelly, S. C. (2021). Smart Nanotherapeutics and lung cancer. *Pharmaceutics., 13*(11), 1972.
93. Soleymani, S., Doroudian, M., Soezi, M., Beladi, A., Asgari, K., Mobarakshahi, A., et al. (2025). Engendered nanoparticles for treatment of brain tumors. *Oncology Research, 33*(1), 15–26.
94. Qu, M., Jiang, X., Zhou, X., Wang, C., Wu, Q., Ren, L., et al. (2020). Stimuli-responsive delivery of growth factors for tissue engineering. *Advanced Healthcare Materials, 9*(7), e1901714.
95. Ahmadi, S., Rabiee, N., Bagherzadeh, M., Elmi, F., Fatahi, Y., Farjadian, F., et al. (2020). Stimulus-responsive sequential release systems for drug and gene delivery. *Nano Today, 34*, 100914.
96. Shahrivar, R. Y., Fakhr, Z. A., Abbasgholinejad, E., & Doroudian, M. (2023). Smart lipid-based nanoparticles in lung cancer treatment: Current status and future directions. *Advanced Therapeutics, 6*(12), 2300275.
97. Hejabi, F., Abbaszadeh, M. S., Taji, S., O'Neill, A., Farjadian, F., & Doroudian, M. (2022). Nanocarriers: A novel strategy for the delivery of CRISPR/Cas systems. *Frontiers in Chemistry, 10*, 957572.
98. Lin, X., Wu, X., Chen, X., Wang, B., & Xu, W. (2021). Intellective and stimuli-responsive drug delivery systems in eyes. *International Journal of Pharmaceutics, 602*, 120591.
99. Wang, F., Duan, H., Xu, W., Sheng, G., Sun, Z., & Chu, H. (2022). Light-activated nanomaterials for tumor immunotherapy. *Frontiers in Chemistry, 10*, 1031811.
100. Xing, Y., Zeng, B., & Yang, W. (2022). Light responsive hydrogels for controlled drug delivery. *Frontiers in Bioengineering and Biotechnology, 10*, 1075670.
101. Zhao, W., Wang, L., Zhang, M., Liu, Z., Wu, C., Pan, X., et al. (2024). Photodynamic therapy for cancer: Mechanisms, photosensitizers, nanocarriers, and clinical studies. *MedComm (2020)., 5*(7), e603.

102. Guidi, L., Cascone, M. G., & Rosellini, E. (2024). Light-responsive polymeric nanoparticles for retinal drug delivery: Design cues, challenges and future perspectives. *Heliyon, 10*(5), e26616.
103. Weng, W., Xue, G., & Pan, Z. (2024). Development of visible-light-activatable photocaged PROTACs. *European Journal of Medicinal Chemistry, 265*, 116062.
104. Li, G., Wang, C., Jin, B., Sun, T., Sun, K., Wang, S., et al. (2024). Advances in smart nanotechnology-supported photodynamic therapy for cancer. *Cell Death Discovery, 10*(1), 466.
105. Chen, G., Cao, Y., Tang, Y., Yang, X., Liu, Y., Huang, D., et al. (2020). Advanced near-infrared light for monitoring and modulating the spatiotemporal dynamics of cell functions in living systems. *Advanced Science, 7*(8), 1903783.
106. Chihara, T., Umezawa, M., Miyata, K., Sekiyama, S., Hosokawa, N., Okubo, K., et al. (2019). Biological deep temperature imaging with fluorescence lifetime of rare-earth-doped ceramics particles in the second NIR biological window. *Scientific Reports, 9*(1), 12806.
107. Bylinskaya, K. A., Perekatova, V. V., & Turchin, I. V. (2025). Study of minor chromophores in biological tissues by diffuse optical spectroscopy (review). *Sovremennye Tehnologii v Medicine, 17*(1), 146–162.
108. Thakur, N. S., Saleh, N., Khan, A. F., Chakrabarty, B., & Agrahari, V. (2024). Progress and promise of photoresponsive nanocarriers for precision drug delivery in cancer. *Journal of Photochemistry and Photobiology C: Photochemistry Reviews, 59*, 100665.
109. Lagopati, N., Evangelou, K., Falaras, P., Tsilibary, E.-P. C., Vasileiou, P. V. S., Havaki, S., et al. (2021). Nanomedicine: Photo-activated nanostructured titanium dioxide, as a promising anticancer agent. *Pharmacology & Therapeutics, 222*, 107795.
110. Banstola, A., Lin, Z.-T., Li, Y., & Wu, M. X. (2025). PhotoChem interplays: Lighting the way for drug delivery and diagnosis. *Advanced Drug Delivery Reviews, 219*, 115549.
111. Hu, D., Li, Y., Li, R., Wang, M., Zhou, K., He, C., et al. (2024). Recent advances in reactive oxygen species (ROS)-responsive drug delivery systems for photodynamic therapy of cancer. *Acta Pharmaceutica Sinica B, 14*(12), 5106–5131.
112. Shen, S., Qiu, J., Huo, D., & Xia, Y. (2024). Nanomaterial-enabled photothermal heating and its use for cancer therapy via localized hyperthermia. *Small, 20*(7), e2305426.
113. Abdelkader, H., Fathalla, Z., Seyfoddin, A., Farahani, M., Thrimawithana, T., Allahham, A., et al. (2021). Polymeric long-acting drug delivery systems (LADDS) for treatment of chronic diseases: Inserts, patches, wafers, and implants. *Advanced Drug Delivery Reviews, 177*, 113957.
114. Singh, D., Sharma, Y., Dheer, D., & Shankar, R. (2024). Stimuli responsiveness of recent biomacromolecular systems (concept to market): A review. *International Journal of Biological Macromolecules, 261*, 129901.
115. Yuan, Z., Gottsacker, C., He, X., Waterkotte, T., & Park, Y. C. (2022). Repetitive drug delivery using light-activated liposomes for potential antimicrobial therapies. *Advanced Drug Delivery Reviews, 187*, 114395.
116. Liu, B., Sood, R., Wang, F., Zhang, F., Sun, L., Qiu, X., et al. (2024). Principles and clinical applications of transcutaneous laser-assisted drug delivery: A narrative review. *Scars Burns & Healing, 10*, 20595131241234715.
117. Ash, C., Dubec, M., Donne, K., & Bashford, T. (2017). Effect of wavelength and beam width on penetration in light-tissue interaction using computational methods. *Lasers in Medical Science, 32*(8), 1909–1918.
118. Barhoumi, A., Liu, Q., & Kohane, D. S. (2015). Ultraviolet light-mediated drug delivery: Principles, applications, and challenges. *Journal of Controlled Release, 219*, 31–42.
119. Sun, Y., & Davis, E. (2021). Nanoplatforms for targeted stimuli-responsive drug delivery: A review of platform materials and stimuli-responsive release and targeting mechanisms. *Nanomaterials (Basel), 11*(3), 746.

120. Fraikin, G. Y., Belenikina, N. S., & Rubin, A. B. (2024). Photochemical processes of cell DNA damage by UV radiation of various wavelengths: Biological consequences. *Molekuliarnaia Biologiia (Mosk), 58*(1), 3–21.
121. Kirillin, M., Khilov, A., Kurakina, D., Orlova, A., Perekatova, V., Shishkova, V., et al. (2021). Dual-wavelength fluorescence monitoring of photodynamic therapy: From analytical models to clinical studies. *Cancers, 13*(22), 5807.
122. Algorri, J. F., Ochoa, M., Roldan-Varona, P., Rodriguez-Cobo, L., & López-Higuera, J. M. (2021). Light technology for efficient and effective photodynamic therapy: A critical review. *Cancers, 13*(14), 3484.
123. Rajendran, M. (2016). Quinones as photosensitizer for photodynamic therapy: ROS generation, mechanism and detection methods. *Photodiagnosis and Photodynamic Therapy, 13*, 175–187.
124. Zou, Z., Chang, H., Li, H., & Wang, S. (2017). Induction of reactive oxygen species: An emerging approach for cancer therapy. *Apoptosis, 22*, 1321–1335.
125. Mallidi, S., Anbil, S., Bulin, A.-L., Obaid, G., Ichikawa, M., & Hasan, T. (2016). Beyond the barriers of light penetration: Strategies, perspectives and possibilities for photodynamic therapy. *Theranostics, 6*(13), 2458.
126. Gunaydin, G., Gedik, M. E., & Ayan, S. (2021). Photodynamic therapy—Current limitations and novel approaches. *Frontiers in Chemistry, 9*, 691697.
127. Menilli, L., Milani, C., Reddi, E., & Moret, F. (2022). Overview of nanoparticle-based approaches for the combination of Photodynamic Therapy (PDT) and chemotherapy at the preclinical stage. *Cancers, 14*(18), 4462.
128. Yi, J., Liu, L., Gao, W., Zeng, J., Chen, Y., Lan, M., et al. (2024). Advances and perspectives in phototherapy-based combination therapy for cancer treatment. *Journal of Materials Chemistry B., 12*, 6285.
129. Doroudian, M., MacLoughlin, R., Poynton, F., Prina-Mello, A., & Donnelly, S. C. (2019). Nanotechnology based therapeutics for lung disease. *Thorax, 74*(10), 965–976.
130. Shinde, V. R., Revi, N., Murugappan, S., Singh, S. P., & Rengan, A. K. (2022). Enhanced permeability and retention effect: A key facilitator for solid tumor targeting by nanoparticles. *Photodiagnosis and Photodynamic Therapy, 39*, 102915.
131. Onzi, G., Guterres, S. S., Pohlmann, A. R., & Frank, L. A. (2022). Passive targeting and the enhanced permeability and retention (EPR) effect. In *The ADME encyclopedia: A comprehensive guide on biopharmacy and pharmacokinetics* (pp. 753–766). Cham: Springer International Publishing.
132. Attia, M. F., Anton, N., Wallyn, J., Omran, Z., & Vandamme, T. F. (2019). An overview of active and passive targeting strategies to improve the nanocarriers efficiency to tumour sites. *Journal of Pharmacy and Pharmacology, 71*(8), 1185–1198.
133. Muniyandi, K., George, B., Parimelazhagan, T., & Abrahamse, H. (2020). Role of photoactive Phytocompounds in photodynamic therapy of cancer. *Molecules [Internet], 25*(18), 4102.
134. Zhang, X., Wang, S., Cheng, G., Yu, P., & Chang, J. (2022). Light-responsive nanomaterials for cancer therapy. *Engineering, 13*, 18–30.
135. Cui, S., Chen, H., Zhu, H., Tian, J., Chi, X., Qian, Z., et al. (2012). Amphiphilic chitosan modified upconversion nanoparticles for in vivo photodynamic therapy induced by near-infrared light. *Journal of Materials Chemistry, 22*(11), 4861–4873.
136. Yu, N., Huang, L., Zhou, Y., Xue, T., Chen, Z., & Han, G. (2019). Near-infrared-light Activatable nanoparticles for deep-tissue-penetrating wireless Optogenetics. *Advanced Healthcare Materials, 8*(6), e1801132.
137. Almadani, I. F., Almadani, M. F., AlSawaftah, N., Abuwatfa, W. H., & Husseini, G. A. (2024). Nanocarriers responsive to light—A review. *Micro [Internet], 4*(4), 827–844.
138. Fomina, N., Sankaranarayanan, J., & Almutairi, A. (2012). Photochemical mechanisms of light-triggered release from nanocarriers. *Advanced Drug Delivery Reviews, 64*(11), 1005–1020.

139. Liu, J., Kang, W., & Wang, W. (2022). Photocleavage-based photoresponsive drug delivery. *Photochemistry and Photobiology, 98*(2), 288–302.
140. Liu, T., Bao, B., Li, Y., Lin, Q., & Zhu, L. (2023). Photo-responsive polymers based on o-Nitrobenzyl derivatives: From structural design to applications. *Progress in Polymer Science, 146*, 101741.
141. Muralidhar, J. R., Kodama, K., Hirose, T., Ito, Y., & Kawamoto, M. (2022). Photocleavage behavior of a polythiophene derivative containing a coumarin unit. *Polymer Journal, 54*(2), 191–198.
142. Rickhoff, J., Arndt, N. B., Böckmann, M., Doltsinis, N. L., Ravoo, B. J., & Kortekaas, L. (2022). Reversible, Red-Shifted photoisomerization in protonated Azobenzenes. *The Journal of Organic Chemistry, 87*(16), 10605–10612.
143. Karlsson, J., Tzeng, S. Y., Hemmati, S., Luly, K. M., Choi, O., Rui, Y., et al. (2021). Photocrosslinked bioreducible polymeric nanoparticles for enhanced systemic siRNA delivery as cancer therapy. *Advanced Functional Materials., 31*(17), 2009768.
144. Salkho, N. M., Awad, N. S., Pitt, W. G., & Husseini, G. A. (2022). Photo-induced drug release from polymeric micelles and liposomes: Phototriggering mechanisms in drug delivery systems. *Polymers [Internet], 14*(7), 1286.
145. Alvarez, N., & Sevilla, A. (2024). Current advances in photodynamic therapy (PDT) and the future potential of PDT-combinatorial cancer therapies. *International Journal of Molecular Sciences [Internet], 25*(2), 1023.
146. Cui, X., Ruan, Q., Zhuo, X., Xia, X., Hu, J., Fu, R., et al. (2023). Photothermal nanomaterials: A powerful light-to-heat converter. *Chemical Reviews, 123*(11), 6891–6952.
147. Oudjedi, F., & Kirk, A. G. (2025). Near-infrared nanoparticle-mediated photothermal cancer therapy: A comprehensive review of advances in monitoring and controlling thermal effects for effective cancer treatment. *Nano Select, 6*(3), e202400107.
148. Munjal, T., & Dutta, S. (2021). Biocompatible nanoreactors of catalase and nanozymes for anticancer therapeutics. *Nano Select, 2*(10), 1849–1873.
149. Xu, Q., Shen, Y., Zhang, Y., & Shao, X. (2019). PEGylated cyanine dye nanoparticles as photothermal agents for mosquito and cancer cell control. *Bioorganic & Medicinal Chemistry Letters, 29*(16), 2398–2404.
150. Choi, H. W., Lim, J. H., Kim, C. W., Lee, E., Kim, J.-M., Chang, K., et al. (2022). Near-infrared light-triggered generation of reactive oxygen species and induction of local hyperthermia from Indocyanine green encapsulated mesoporous silica-coated graphene oxide for colorectal cancer therapy. *Antioxidants [Internet], 11*(1), 174.
151. Ghosh Dastidar, D., & Chakrabarti, G. (2019). Chapter 6 - Thermoresponsive drug delivery systems, characterization and application. In S. S. Mohapatra, S. Ranjan, N. Dasgupta, R. K. Mishra, & S. Thomas (Eds.), *Applications of targeted Nano drugs and delivery systems* (pp. 133–155). Elsevier.
152. Gagon, A. T., Britt, D. W., & Bastarrachea, L. J. (2020). Antimicrobial light-activated polypropylene modified with Chitosan: Characterization and reusability. *Journal of Agricultural and Food Chemistry, 68*(46), 13076–13082.
153. Facchinatto, W. M., Araujo, L. O., Moraes, T. B., Abelha, T. F., Lima, T. H. N., Santos, D. M., et al. (2022). Antimicrobial and photoantimicrobial activities of Chitosan/CNPPV nanocomposites. *International Journal of Molecular Sciences [Internet], 23*(20), 12519.
154. Matica, M. A., Aachmann, F. L., Tøndervik, A., Sletta, H., & Ostafe, V. (2019). Chitosan as a wound dressing starting material: Antimicrobial properties and mode of action. *International Journal of Molecular Sciences [Internet], 20*(23), 5889.
155. Yilmaz Atay, H. (2019). Antibacterial activity of Chitosan-based systems. In S. Jana & S. Jana (Eds.), *Functional Chitosan: Drug delivery and biomedical applications* (pp. 457–489). Springer Singapore.
156. Wang, M., Wang, Y., Chen, G., Gao, H., & Peng, Q. (2024). Chitosan-based multifunctional biomaterials as active agents or delivery systems for antibacterial therapy. *Bioengineering (Basel)., 11*(12), 1278.

157. Shah, N., Squire, J., Guirguis, M., Saha, D., Hoyt, K., Wang, K. K., et al. (2022). Deep-tissue activation of photonanomedicines: An update and clinical perspectives. *Cancers [Internet], 14*(8), 2004.
158. Liang, G., Wang, H., Shi, H., Wang, H., Zhu, M., Jing, A., et al. (2020). Recent progress in the development of upconversion nanomaterials in bioimaging and disease treatment. *Journal of Nanobiotechnology, 18*(1), 154.
159. Kang, W., Liu, Y., & Wang, W. (2023). Light-responsive nanomedicine for cancer immunotherapy. *Acta Pharmaceutica Sinica B, 13*(6), 2346–2368.
160. Xu, W., Liu, W., Yang, J., Lu, J., Zhang, H., & Ye, D. (2024). Stimuli-responsive nanodelivery systems for amplifying immunogenic cell death in cancer immunotherapy. *Immunological Reviews, 321*(1), 181–198.
161. Jha, R., & Mayanovic, R. A. (2023). A review of the preparation, characterization, and applications of Chitosan nanoparticles in Nanomedicine. *Nanomaterials (Basel) [Internet], 13*(8), 1302.
162. Joudeh, N., & Linke, D. (2022). Nanoparticle classification, physicochemical properties, characterization, and applications: A comprehensive review for biologists. *Journal of Nanobiotechnology, 20*(1), 262.
163. Isibor, P. O. (2024). Nanochitosan synthesis, optimization, and characterization. In P. O. Isibor, A. O. Adeogun, & A. A. Enuneku (Eds.), *Nanochitosan-based enhancement of fisheries and aquaculture: Aligning with sustainable development goal 14—life below water* (pp. 65–112). Springer Nature Switzerland.
164. Thirugnanasambandan, T., & Gopinath, S. C. B. (2023). Laboratory to industrial scale synthesis of chitosan-based nanomaterials: A review. *Process Biochemistry, 130*, 147–155.
165. Hoang, N. H., Le Thanh, T., Sangpueak, R., Treekoon, J., Saengchan, C., Thepbandit, W., et al. (2022). Chitosan nanoparticles-based ionic gelation method: A promising candidate for plant disease management. *Polymers [Internet], 14*(4), 662.
166. Hamidi, F., Mojtaba, A. A., Fatemeh, J., Hadi, M. M., & Baghani, A. N. (2022). Ionic gelation synthesis, characterization and adsorption studies of cross-linked chitosan-tripolyphosphate (CS-TPP) nanoparticles for removal of As (V) ions from aqueous solution: Kinetic and isotherm studies. *Toxin Reviews, 41*(3), 795–805.
167. Silva, N. C., Chevigny, C., Domenek, S., Almeida, G., Assis, O. B. G., & Martelli-Tosi, M. (2025). Nanoencapsulation of active compounds in chitosan by ionic gelation: Physicochemical, active properties and application in packaging. *Food Chemistry, 463*, 141129.
168. Wang, H., Deng, H., Gao, M., & Zhang, W. (2021). Self-assembled Nanogels based on ionic gelation of natural polysaccharides for drug delivery. *Frontiers in Bioengineering and Biotechnology, 9*, 703559.
169. Yang, Y., Gupta, V. K., Amiri, H., Pan, J., Aghbashlo, M., Tabatabaei, M., et al. (2023). Recent developments in improving the emulsifying properties of chitosan. *International Journal of Biological Macromolecules, 239*, 124210.
170. Oliveira, D. A. J., Amaral, J. G., Garcia, L. B., dos Santos, M. S., Silva, L. A. O., Almeida, M. P., et al. (2021). Associating chitosan and microemulsion as a topical vehicle for the administration of herbal medicines. *Carbohydrate Polymers, 255*, 117482.
171. Gosala, R., Raghunandhakumar, S., & Subramanian, B. (2025). Modulating drug delivery with Nano-selenium capped Chitosan reverse micelles for anticancer potential. *Journal of Drug Delivery Science and Technology., 108*, 106860.
172. Jafernik, K., Malinowska, M. A., Bialik-Wąs, K., Sharafan, M., Ekiert, H., Sikora, E., et al. (2023). Chitosan Nanoparticles: A Potential Biomedical Device. In S. K. Swain & A. Biswal (Eds.), *Chitosan nanocomposites: Bionanomechanical applications* (pp. 135–159). Springer Nature Singapore.
173. Mahmoudi, Z. (2023). Chapter 22 - Microencapsulation: Phase inversion precipitation. In F. Sefat, G. Farzi, & M. Mozafari (Eds.), *Principles of biomaterials encapsulation : Volume 1* (pp. 443–457). Woodhead Publishing.

174. Nicolle, L., Journot, C. M., & Gerber-Lemaire, S. (2021). Chitosan functionalization: Covalent and non-covalent interactions and their characterization. *Polymers, 13*(23), 4118.
175. Yu, Y., Tian, R., Zhao, Y., Qin, X., Hu, L., Zou, J. J., et al. (2023). Self-assembled corrole/chitosan photothermal nanoparticles for accelerating infected diabetic wound healing. *Advanced Healthcare Materials., 12*(16), 2201651.
176. Zhang, Y., Zhang, Y., Ding, R., Zhang, K., Guo, H., & Lin, Y. (2024). Self-assembled nanocarrier delivery systems for bioactive compounds. *Small, 20*(26), 2310838.
177. Liu, Y., Sun, M., Wang, T., Chen, X., & Wang, H. (2021). Chitosan-based self-assembled nanomaterials: Their application in drug delivery. *View., 2*(1), 20200069.
178. Miron, A., Sarbu, A., Zaharia, A., Sandu, T., Iovu, H., Fierascu, R. C., et al. (2022). A top-down procedure for synthesizing calcium carbonate-enriched Chitosan from Shrimp Shell wastes. *Gels [Internet], 8*(11), 742.
179. de la Paz, N., Fernández, M., López, O., Garcia, C., Nogueira, A., Torres, L., et al. (2021). Spray drying of chitosan acid salts: Process development, scaling up and physicochemical material characterization. *Marine Drugs., 19*(6), 329.
180. Faghmous, N., Bouzid, D., Ammouchi, H., Alioua, I., Boukellal, B., & Boyron, O. (2025). Optimization of spray drying process conditions for producing insulin-loaded alginate/chitosan polyelectrolyte complexes. *Drying Technology, 43*, 1–15.
181. Costa, C., Grenho, L., Casimiro, T., Gomes, P. S., Fernandes, M. H., Corvo, M. L., et al. (2025). Optimization of quercetin-liposomal dry powders for pulmonary delivery using supercritical CO_2-assisted spray drying. *The Journal of Supercritical Fluids., 222*, 106591.
182. Peng, H.-H., Wang, Z.-D., Guan, Y.-X., & Yao, S.-J. (2021). Supercritical CO2 assisted preparation of chitosan-based nano-in-microparticles with potential for efficient pulmonary drug delivery. *Journal of CO2 Utilization, 46*, 101486.
183. Chaschin, I. S., Perepelkin, E. I., Levin, E. E., Abramchuk, S. S., Anuchina, N. M., Kizas, O. A., et al. (2024). Green synthesis of silver nanoparticles using Chitosan in carbonic acid solutions: Effect of pressure and temperature on the structure and antimicrobial properties. *Journal of Inorganic and Organometallic Polymers and Materials, 35*, 1–13.
184. Wang, J., & Zhuang, S. (2022). Chitosan-based materials: Preparation, modification and application. *Journal of Cleaner Production, 355*, 131825.
185. Iqbal, Y., Ahmed, I., Irfan, M. F., Chatha, S. A. S., Zubair, M., & Ullah, A. (2023). Recent advances in chitosan-based materials; the synthesis, modifications and biomedical applications. *Carbohydrate Polymers, 321*, 121318.
186. Dadashi, H., Vandghanooni, S., Karamnejad-Faragheh, S., Karimian-Shaddel, A., Eskandani, M., & Jahanban-Esfahlan, R. (2024). A rapid protocol for synthesis of chitosan nanoparticles with ideal physicochemical features. *Heliyon, 10*(11), e32228.
187. Alehosseini, E., Shahiri Tabarestani, H., Kharazmi, M. S., & Jafari, S. M. (2022). Physicochemical, thermal, and morphological properties of chitosan nanoparticles produced by ionic gelation. *Food, 11*(23), 3841.
188. Poznanski, P., Hameed, A., & Orczyk, W. (2023). Chitosan and chitosan nanoparticles: Parameters enhancing antifungal activity. *Molecules, 28*(7), 2996.
189. Mitchell, M. J., Billingsley, M. M., Haley, R. M., Wechsler, M. E., Peppas, N. A., & Langer, R. (2021). Engineering precision nanoparticles for drug delivery. *Nature Reviews Drug Discovery, 20*(2), 101–124.
190. Manzari, M. T., Shamay, Y., Kiguchi, H., Rosen, N., Scaltriti, M., & Heller, D. A. (2021). Targeted drug delivery strategies for precision medicines. *Nature Reviews Materials, 6*(4), 351–370.
191. Esmaeili, J., Jalise, S. Z., Pisani, S., Rochefort, G. Y., Ghobadinezhad, F., Mirzaei, Z., et al. (2024). Development and characterization of Polycaprolactone/chitosan-based scaffolds for tissue engineering of various organs: A review. *International Journal of Biological Macromolecules., 272*, 132941.
192. Mura, P., Maestrelli, F., Cirri, M., & Mennini, N. (2022). Multiple roles of chitosan in mucosal drug delivery: An updated review. *Marine Drugs, 20*(5), 335.

193. Wadhwa, P., Sharma, S., Sahu, S., Sharma, A., & Kumar, D. (2022). A review of nanoparticles characterization techniques. *Current Nanomaterials, 7*(3), 202–214.
194. Garavand, F., Cacciotti, I., Vahedikia, N., Rehman, A., Tarhan, Ö., Akbari-Alavijeh, S., et al. (2022). A comprehensive review on the nanocomposites loaded with chitosan nanoparticles for food packaging. *Critical Reviews in Food Science and Nutrition, 62*(5), 1383–1416.
195. Podgorbunskikh, E., Kuskov, T., Rychkov, D., Lomovskii, O., & Bychkov, A. (2022). Mechanical amorphization of chitosan with different molecular weights. *Polymers, 14*(20), 4438.
196. Babaei-Ghazvini, A., Patel, R., Vafakish, B., Yazdi, A. F. A., & Acharya, B. (2024). Nanocellulose in targeted drug delivery: A review of modifications and synergistic applications. *International Journal of Biological Macromolecules, 278*, 135200.
197. Souza, N. L. G. D., Brandão, H. M., & de Oliveira, L. F. C. (2011). Spectroscopic and thermogravimetric study of chitosan after incubation in bovine rumen. *Journal of Molecular Structure, 1005*(1), 186–191.
198. Yan, B., Dai, Y., Xin, L., Li, M., Zhang, H., Long, H., et al. (2024). Research progress in the degradation of printing and dyeing wastewater using chitosan based composite photocatalytic materials. *International journal of biological macromolecules., 263*, 130082.
199. Liu, Z. (2021). Review and prospect of thermal analysis technology applied to study thermal properties of energetic materials. *FirePhysChem, 1*(3), 129–138.
200. Barra, G., Guadagno, L., Raimondo, M., Santonicola, M. G., Toto, E., & Vecchio, C. S. (2023). A comprehensive review on the thermal stability assessment of polymers and composites for aeronautics and space applications. *Polymers, 15*(18), 3786.
201. Isa, M. T., Abdulkarim, A. Y., Bello, A., Bello, T. K., & Adamu, Y. (2024). Synthesis and characterization of chitosan for medical applications: A review. *Journal of Biomaterials Applications, 38*(10), 1036–1057.
202. Isibor, P. O. (2024). Nanochitosan synthesis, optimization, and characterization. In *Nanochitosan-based enhancement of fisheries and aquaculture: Aligning with sustainable development goal 14–life below water* (pp. 65–112). Springer.
203. Guo, Y., Qiao, D., Zhao, S., Zhang, B., & Xie, F. (2024). Advanced functional chitosan-based nanocomposite materials for performance-demanding applications. *Progress in Polymer Science., 157*, 101872.
204. Kudiyarasu, S., Perumal, M. K. K., Renuka, R. R., & Natrajan, P. M. (2024). Chitosan composite with mesenchymal stem cells: Properties, mechanism, and its application in bone regeneration. *International Journal of Biological Macromolecules., 275*, 133502.
205. Ahmad, I., Aftab, M. A., Fatima, A., Mekkey, S. D., Melhi, S., & Ikram, S. (2024). A comprehensive review on the advancement of transition metals incorporated on functional magnetic nanocomposites for the catalytic reduction and photocatalytic degradation of organic pollutants. *Coordination Chemistry Reviews, 514*, 215904.
206. Jiang, M., Althomali, R. H., Ansari, S. A., Saleh, E. A. M., Gupta, J., Kambarov, K. D., et al. (2023). Advances in preparation, biomedical, and pharmaceutical applications of chitosan-based gold, silver, and magnetic nanoparticles: A review. *International Journal of Biological Macromolecules, 251*, 126390.
207. Mikušová, V., & Mikuš, P. (2021). Advances in chitosan-based nanoparticles for drug delivery. *International Journal of Molecular Sciences, 22*(17), 9652.
208. Iacob, A. T., Lupascu, F. G., Apotrosoaei, M., Vasincu, I. M., Tauser, R. G., Lupascu, D., et al. (2021). Recent biomedical approaches for chitosan based materials as drug delivery nanocarriers. *Pharmaceutics, 13*(4), 587.
209. Kumar, R., Santa Chalarca, C. F., Bockman, M. R., Bruggen, C. V., Grimme, C. J., Dalal, R. J., et al. (2021). Polymeric delivery of therapeutic nucleic acids. *Chemical Reviews, 121*(18), 11527–11652.
210. Biswas, R., Mondal, S., Ansari, M. A., Sarkar, T., Condiuc, I. P., Trifas, G., et al. (2025). Chitosan and its derivatives as nanocarriers for drug delivery. *Molecules, 30*(6), 1297.

211. Ajam-Hosseini, M., Heydari, R., Rasouli, M., Akhoondi, F., Asadi Hanjani, N., Bekeschus, S., et al. (2024). Lactic acid in macrophage polarization: A factor in carcinogenesis and a promising target for cancer therapy. *Biochemical Pharmacology, 222*, 116098.
212. Hosseinkazemi, H., Samani, S., O'Neill, A., Soezi, M., Moghoofei, M., Azhdari, M. H., et al. (2022). Applications of iron oxide nanoparticles against breast cancer. *Journal of Nanomaterials, 2022*(1), 6493458.
213. Ajam-Hosseini, M., Akhoondi, F., & Doroudian, M. (2023). Nano based-oncolytic viruses for cancer therapy. *Critical Reviews in Oncology/Hematology, 185*, 103980.
214. Hanjani, N. A., Esmaelizad, N., Zanganeh, S., Gharavi, A. T., Heidarizadeh, P., Radfar, M., et al. (2022). Emerging role of exosomes as biomarkers in cancer treatment and diagnosis. *Critical Reviews in Oncology/Hematology, 169*, 103565.
215. Zhong, L., Li, Y., Xiong, L., Wang, W., Wu, M., Yuan, T., et al. (2021). Small molecules in targeted cancer therapy: Advances, challenges, and future perspectives. *Signal Transduction and Targeted Therapy, 6*(1), 201.
216. Haider, A., Khan, S., Iqbal, D. N., Shrahili, M., Haider, S., Mohammad, K., et al. (2024). Advances in chitosan-based drug delivery systems: A comprehensive review for therapeutic applications. *European Polymer Journal, 210*, 112983.
217. Alhodieb, F. S., Barkat, M. A., Barkat, H. A., Ab Hadi, H., Khan, M. I., Ashfaq, F., et al. (2022). Chitosan-modified nanocarriers as carriers for anticancer drug delivery: Promises and hurdles. *International Journal of Biological Macromolecules., 217*, 457–469.
218. Wang, G., Li, R., Parseh, B., & Du, G. (2021). Prospects and challenges of anticancer agents' delivery via chitosan-based drug carriers to combat breast cancer: A review. *Carbohydrate Polymers, 268*, 118192.
219. Zi, Y., Yang, K., He, J., Wu, Z., Liu, J., & Zhang, W. (2022). Strategies to enhance drug delivery to solid tumors by harnessing the EPR effects and alternative targeting mechanisms. *Advanced Drug Delivery Reviews, 188*, 114449.
220. Huang, D., Sun, L., Huang, L., & Chen, Y. (2021). Nanodrug delivery systems modulate tumor vessels to increase the enhanced permeability and retention effect. *Journal of Personalized Medicine, 11*(2), 124.
221. Niitsoo, L. (2022). *Novel insights into the formation of folate conjugated glycol chitosan nanoparticles for active tumour targeting and as a potential platform for ligand multiplexing*. UCL (University College London).
222. Agwa, M. M., Marzouk, R. E., & Sabra, S. A. (2024). Advances in active targeting of ligand-directed polymeric nanomicelles via exploiting overexpressed cellular receptors for precise nanomedicine. *RSC Advances, 14*(32), 23520–23542.
223. Soltani, A., Faramarzi, M., Farjadian, F., Parsa, S. A. M., & Panahi, H. A. (2022). pH-responsive glycodendrimer as a new active targeting agent for doxorubicin delivery. *International Journal of Biological Macromolecules, 221*, 508–522.
224. Kurczewska, J. (2023). Chitosan-based nanoparticles with optimized parameters for targeted delivery of a specific anticancer drug—A comprehensive review. *Pharmaceutics, 15*(2), 503.
225. Ways, T. M., Lau, W. M., & Khutoryanskiy, V. V. (2018). Chitosan and its derivatives for application in mucoadhesive drug delivery systems. *Polymers, 10*(3), 267.
226. Pandey, M., Choudhury, H., Ying, J. N. S., Ling, J. F. S., Ting, J., Ting, J. S. S., et al. (2022). Mucoadhesive nanocarriers as a promising strategy to enhance intracellular delivery against oral cavity carcinoma. *Pharmaceutics., 14*(4), 795.
227. Teixeira, P. V., Fernandes, E., Soares, T. B., Adega, F., Lopes, C. M., & Lúcio, M. (2023). Natural compounds: Co-delivery strategies with chemotherapeutic agents or nucleic acids using lipid-based nanocarriers. *Pharmaceutics, 15*(4), 1317.
228. Gong, J., Shi, T., Liu, J., Pei, Z., Liu, J., Ren, X., et al. (2023). Dual-drug codelivery nanosystems: An emerging approach for overcoming cancer multidrug resistance. *Biomedicine & Pharmacotherapy, 161*, 114505.
229. Duan, C., Yu, M., Xu, J., Li, B.-Y., Zhao, Y., & Kankala, R. K. (2023). Overcoming Cancer Multi-drug Resistance (MDR): Reasons, mechanisms, nanotherapeutic solutions, and challenges. *Biomedicine & Pharmacotherapy, 162*, 114643.

230. Dhlamini, K. S., Selepe, C. T., Ramalapa, B., Tshweu, L., & Ray, S. S. (2024). Reimagining Chitosan-based antimicrobial biomaterials to mitigate antibiotic resistance and alleviate antibiotic overuse: A review. *Macromolecular Materials and Engineering, 309*(9), 2400018.
231. Ardean, C., Davidescu, C. M., Nemeş, N. S., Negrea, A., Ciopec, M., Duteanu, N., et al. (2021). Factors influencing the antibacterial activity of chitosan and chitosan modified by functionalization. *International Journal of Molecular Sciences., 22*(14), 7449.
232. Saadh, M. J., Hsu, C.-Y., Mustafa, M. A., Mutee, A. F., Kaur, I., Ghildiyal, P., et al. (2024). Advances in chitosan-based blends as potential drug delivery systems: A review. *International Journal of Biological Macromolecules., 273*, 132916.
233. Hemmingsen, L. M., Škalko-Basnet, N., & Jøraholmen, M. W. (2021). The expanded role of chitosan in localized antimicrobial therapy. *Marine Drugs., 19*(12), 697.
234. Khubiev, O. M., Egorov, A. R., Kirichuk, A. A., Khrustalev, V. N., Tskhovrebov, A. G., & Kritchenkov, A. S. (2023). Chitosan-based antibacterial films for biomedical and food applications. *International Journal of Molecular Sciences., 24*(13), 10738.
235. Nasaj, M., Chehelgerdi, M., Asghari, B., Ahmadieh-Yazdi, A., Asgari, M., Kabiri-Samani, S., et al. (2024). Factors influencing the antimicrobial mechanism of chitosan action and its derivatives: A review. *International Journal of Biological Macromolecules, 277*, 134321.
236. Confederat, L. G., Tuchilus, C. G., Dragan, M., Sha'at, M., & Dragostin, O. M. (2021). Preparation and antimicrobial activity of chitosan and its derivatives: A concise review. *Molecules, 26*(12), 3694.
237. Aghasadeghi, R., Heidari, M. H., Sadat, M. S., Irani, S., Amini, S., Siadat, D. S., et al. (2013). Lamivudine-PEGylated chitosan: A novel effective nanosized antiretroviral agent. *Current HIV Research., 11*(4), 309–320.
238. Kraisit, P., Yonemochi, E., Furuishi, T., Mahadlek, J., & Limmatvapirat, S. (2022). Chitosan film containing antifungal agent-loaded SLNs for the treatment of candidiasis using a Box-Behnken design. *Carbohydrate Polymers, 283*, 119178.
239. Muangsawat, S., Chaiyosang, P., Sinkanarak, P., Sukted, J., Thanyasrisung, P., & Matangkasombut, O. (2024). Effects of efflux pumps on antifungal activity of chitosan against Candida albicans. *Journal of Oral Microbiology, 16*(1), 2357976.
240. Boroumand, H., Badie, F., Mazaheri, S., Seyedi, Z. S., Nahand, J. S., Nejati, M., et al. (2021). Chitosan-based nanoparticles against viral infections. *Frontiers in Cellular and Infection Microbiology., 11*, 643953.
241. Safer, A.-M., & Leporatti, S. (2021). Chitosan nanoparticles for antiviral drug delivery: A novel route for COVID-19 treatment. *International Journal of Nanomedicine, 16*, 8141–8158.
242. Jaber, N., Al-Remawi, M., Al-Akayleh, F., Al-Muhtaseb, N., Al-Adham, I. S., & Collier, P. J. (2022). A review of the antiviral activity of Chitosan, including patented applications and its potential use against COVID-19. *Journal of Applied Microbiology, 132*(1), 41–58.
243. Žigrayová, D., Mikušová, V., & Mikuš, P. (2023). Advances in antiviral delivery systems and chitosan-based polymeric and nanoparticulate antivirals and antiviral carriers. *Viruses, 15*(3), 647.
244. Alwahsh, W., Sahudin, S., Alkhatib, H., Bostanudin, M. F., & Alwahsh, M. (2024). Chitosan-based Nanocarriers for pulmonary and intranasal drug delivery systems: A comprehensive overview of their applications. *Current Drug Targets, 25*(7), 492–511.
245. Fernández-García, R., & Fraguas-Sánchez, A. I. (2024). Nanomedicines for pulmonary drug delivery: Overcoming barriers in the treatment of respiratory infections and lung cancer. *Pharmaceutics, 16*(12), 1584.
246. Tiboni, M., Elmowafy, E., El-Derany, M. O., Benedetti, S., Campana, R., Verboni, M., et al. (2022). A combination of sugar esters and chitosan to promote in vivo wound care. *International Journal of Pharmaceutics, 616*, 121508.
247. Elangwe, C. N., Morozkina, S. N., Olekhnovich, R. O., Krasichkov, A., Polyakova, V. O., & Uspenskaya, M. V. (2022). A review on chitosan and cellulose hydrogels for wound dressings. *Polymers, 14*(23), 5163.

248. Che, X., Zhao, T., Hu, J., Yang, K., Ma, N., Li, A., et al. (2024). Application of chitosan-based hydrogel in promoting wound healing: A review. *Polymers, 16*(3), 344.
249. Mohite, P., Rahayu, P., Munde, S., Ade, N., Chidrawar, V. R., Singh, S., et al. (2023). Chitosan-based hydrogel in the management of dermal infections: A review. *Gels., 9*(7), 594.
250. Yao, Y., Zhang, A., Yuan, C., Chen, X., & Liu, Y. (2021). Recent trends on burn wound care: Hydrogel dressings and scaffolds. *Biomaterials Science, 9*(13), 4523–4540.
251. Mishra, A., Kushare, A., Gupta, M. N., & Ambre, P. (2024). Advanced dressings for chronic wound management. *ACS Applied Bio Materials, 7*(5), 2660–2676.
252. Facchinatto, W. M., Araujo, L. O., Moraes, T. B., Abelha, T. F., Lima, T. H., Santos, D. M., et al. (2022). Antimicrobial and photoantimicrobial activities of Chitosan/CNPPV nanocomposites. *International Journal of Molecular Sciences., 23*(20), 12519.
253. Shen, H., Jiang, C., Li, W., Wei, Q., Ghiladi, R. A., & Wang, Q. (2021). Synergistic photodynamic and photothermal antibacterial activity of in situ grown bacterial cellulose/MoS2-chitosan nanocomposite materials with visible light illumination. *ACS Applied Materials & Interfaces, 13*(26), 31193–31205.
254. Krishnaswami, V., Rajan, E., Dutta, G., Muruganantham, S., & Sugumaran, A. (2024). Chitosan mediated smart photodynamic therapy based novel drug delivery systems-a futuristic view. *Carbohydrate Polymer Technologies and Applications., 7*, 100510.
255. Dube, E. (2024). Antimicrobial photodynamic therapy: Self-disinfecting surfaces for controlling microbial infections. *Microorganisms, 12*(8), 1573.
256. Lunawat, A. K., Thakur, S., Kurmi, B. D., Gupta, G. D., Patel, P., & Raikwar, S. (2024). Revolutionizing cancer treatment: The role of chitosan nanoparticles in therapeutic advancements. *Journal of Drug Delivery Science and Technology, 96*, 105661.
257. Wang, Z., Yu, F., & Hu, F. (2024). Functional chitosan and its derivative-related drug delivery systems for nano-therapy: Recent advances. *Pharmaceutics, 16*(3), 337.
258. Akbarian, M., Tayebi, L., Mohammadi-Samani, S., & Farjadian, F. (2020). Mechanistic assessment of functionalized mesoporous silica-mediated insulin fibrillation. *The Journal of Physical Chemistry B, 124*(9), 1637–1652.
259. Akbarian, M., & Farjadian, F. (2024). Chitosan-coated glass beads as stabilizer of insulin; a novel strategy for the storage of pharmaceutical proteins. *ChemistrySelect, 9*(40), e202402813.
260. Sánchez-Dengra, B., González-Álvarez, I., Bermejo, M., & González-Álvarez, M. (2023). Access to the CNS: Strategies to overcome the BBB. *International Journal of Pharmaceutics, 636*, 122759.
261. Wu, D., Chen, Q., Chen, X., Han, F., Chen, Z., & Wang, Y. (2023). The blood–brain barrier: Structure, regulation and drug delivery. *Signal Transduction and Targeted Therapy, 8*(1), 217.
262. Jalali, H., Rahimian, S., Shahsavarian, N., Norouzi, R., Ahmadiyeh, Z., Najafi, H., et al. (2024). The organoid modeling approach to understanding the mechanisms underlying neurodegeneration: A comprehensive review. *Life Sciences., 358*, 123198.
263. Ghasempour, E., Hesami, S., Movahed, E., Keshel, S. H., & Doroudian, M. (2022). Mesenchymal stem cell-derived exosomes as a new therapeutic strategy in the brain tumors. *Stem Cell Research & Therapy, 13*(1), 527.
264. Khodaverdi, K., Bakhshi, A., Mozafari, M., & Naghib, S. M. (2024). A review of chitosan-based nanocarriers as drug delivery systems for brain diseases: Critical challenges, outlooks and promises. *International Journal of Biological Macromolecules., 278*, 134962.
265. Ekhator, C., Qureshi, M. Q., Zuberi, A. W., Hussain, M., Sangroula, N., Yerra, S., et al. (2023). Advances and opportunities in nanoparticle drug delivery for central nervous system disorders: A review of current advances. *Cureus., 15*(8), e44302.
266. Vahab, S. A., & KI A Kumar, V. S. (2024). Exploring chitosan nanoparticles for enhanced therapy in neurological disorders: A comprehensive review. *Naunyn-Schmiedeberg's Archives of Pharmacology., 398*, 1–17.
267. Gabold, B. (2023). *Chitosan nanoparticles for nose-to-brain delivery of biologics: lmu*
268. Shayganfard, M. (2022). A review on chitosan in drug delivery for treatment of neurological and psychiatric disorders. *Current Pharmaceutical Biotechnology, 23*(4), 538–551.

269. Eslahi, M., Dana, P. M., Asemi, Z., Hallajzadeh, J., Mansournia, M. A., & Yousefi, B. (2021). The effects of chitosan-based materials on glioma: Recent advances in its applications for diagnosis and treatment. *International Journal of Biological Macromolecules, 168*, 124–129.
270. Ait Hamdan, Y., Elouali, S., Maloui, A. B., El-Mansoury, B., Oudadesse, H., & Rhazi, M. (2023). Chitosan and its derivatives as potential neuro-protective agents for the treatment of Parkinson's disease. In *Experimental and clinical evidence of the neuropathology of Parkinson's disease* (pp. 253–274). IGI Global.
271. Jeong, S.-H., Jang, J.-H., & Lee, Y.-B. (2023). Drug delivery to the brain via the nasal route of administration: Exploration of key targets and major consideration factors. *Journal of Pharmaceutical Investigation, 53*(1), 119–152.
272. Zhang, Y.-B., Xu, D., Bai, L., Zhou, Y.-M., Zhang, H., & Cui, Y.-L. (2022). A review of non-invasive drug delivery through respiratory routes. *Pharmaceutics, 14*(9), 1974.
273. Nguyen, H. T., Do, N. H., Lac, H. D., Nguyen, P. L., & Le, P. K. (2023). Synthesis, properties, and applications of chitosan hydrogels as anti-inflammatory drug delivery system. *Journal of Porous Materials., 30*(2), 655–670.
274. Paczkowska-Walendowska, M., & Cielecka-Piontek, J. (2022). Chitosan as a functional carrier for the local delivery anti-inflammatory systems containing scutellariae baicalensis radix extract. *Pharmaceutics, 14*(10), 2148.
275. Jin, Z., Zhang, P., Huang, H., Liu, J., Jiang, C., Zhang, H., et al. (2024). Food-derived skincare ingredient as a promising strategy for skin aging: Current knowledge and future perspectives. *Colloids and Surfaces B: Biointerfaces, 244*, 114170.
276. Emam-Djomeh, Z., & Hajikhani, M. (2022). Chitosan/poly (ethylene glycol)/ZnO bionanocomposite for wound healing application. In *Biodegradable and environmental applications of bionanocomposites* (pp. 31–65). Springer.
277. Razdan, K., Garcia-Lara, J., Sinha, V., & Singh, K. K. (2022). Pharmaceutical strategies for the treatment of bacterial biofilms in chronic wounds. *Drug Discovery Today, 27*(8), 2137–2150.
278. Trivedi, M., & Kumar, S. (2022). Potential of chitosan-based nanocomposites for biomedical application in gene therapy. In *Chitosan-based nanocomposite materials: Fabrication, characterization and biomedical applications* (pp. 121–144). Springer.
279. Karayianni, M., Sentoukas, T., Skandalis, A., Pippa, N., & Pispas, S. (2023). Chitosan-based nanoparticles for nucleic acid delivery: Technological aspects, applications, and future perspectives. *Pharmaceutics, 15*(7), 1849.
280. Al-Absi, M. Y., Caprifico, A. E., & Calabrese, G. (2022). Chitosan and its structural modifications for siRNA delivery. *Advanced Pharmaceutical Bulletin, 13*(2), 275.
281. Chen, Q., Qi, Y., Jiang, Y., Quan, W., Luo, H., Wu, K., et al. (2022). Progress in research of chitosan chemical modification technologies and their applications. *Marine Drugs, 20*(8), 536.
282. Antoniou, V., Mourelatou, E. A., Galatou, E., Avgoustakis, K., & Hatziantoniou, S. (2024). Gene therapy with Chitosan nanoparticles: Modern formulation strategies for enhancing cancer cell transfection. *Pharmaceutics, 16*(7), 868.
283. Rodolfo, C., Eusébio, D., Ventura, C., Nunes, R., Florindo, H. F., Costa, D., et al. (2021). Design of experiments to achieve an efficient chitosan-based DNA vaccine delivery system. *Pharmaceutics., 13*(9), 1369.
284. Gong, X., Gao, Y., Shu, J., Zhang, C., & Zhao, K. (2022). Chitosan-based nanomaterial as immune adjuvant and delivery carrier for vaccines. *Vaccine, 10*(11), 1906.
285. Pathomthongtaweechai, N., & Muanprasat, C. (2021). Potential applications of chitosan-based nanomaterials to surpass the gastrointestinal physiological obstacles and enhance the intestinal drug absorption. *Pharmaceutics., 13*(6), 887.
286. Liu, S., Wen, X., Zhang, X., & Mao, S. (2023). Oral delivery of biomacromolecules by overcoming biological barriers in the gastrointestinal tract: An update. *Expert Opinion on Drug Delivery, 20*(10), 1333–1347.
287. Verma, S., Goand, U. K., Husain, A., Katekar, R. A., Garg, R., & Gayen, J. R. (2021). Challenges of peptide and protein drug delivery by oral route: Current strategies to improve the bioavailability. *Drug Development Research, 82*(7), 927–944.

288. Zhu, Q., Chen, Z., Paul, P. K., Lu, Y., Wu, W., & Qi, J. (2021). Oral delivery of proteins and peptides: Challenges, status quo and future perspectives. *Acta Pharmaceutica Sinica B, 11*(8), 2416–2448.
289. Kurakula, M., Gorityala, S., & Moharir, K. (2021). Recent trends in design and evaluation of chitosan-based colon targeted drug delivery systems: Update 2020. *Journal of Drug Delivery Science and Technology, 64*, 102579.
290. Nguyen, T.-T.-L., Duong, V.-A., & Maeng, H.-J. (2021). Pharmaceutical formulations with P-glycoprotein inhibitory effect as promising approaches for enhancing oral drug absorption and bioavailability. *Pharmaceutics, 13*(7), 1103.
291. Gandla, K., Islam, F., Zehravi, M., Karunakaran, A., Sharma, I., Haque, M. A., et al. (2023). Natural polymers as potential P-glycoprotein inhibitors: Pre-ADMET profile and computational analysis as a proof of concept to fight multidrug resistance in cancer. *Heliyon, 9*(9), e19454.
292. Zuhairah Zainuddin, S., & Abdul Hamid, K. (2021). Chitosan-based oral drug delivery system for peptide, protein and vaccine delivery. *IntechOpen*. https://doi.org/10.5772/intechopen.95771
293. Li, X., Xing, R., Xu, C., Liu, S., Qin, Y., Li, K., et al. (2021). Immunostimulatory effect of chitosan and quaternary chitosan: A review of potential vaccine adjuvants. *Carbohydrate Polymers, 264*, 118050.
294. Dmour, I., & Islam, N. (2022). Recent advances on chitosan as an adjuvant for vaccine delivery. *International Journal of Biological Macromolecules, 200*, 498–519.
295. Chen, P.-G., Huang, Z.-H., Sun, Z.-Y., Gao, Y., Liu, Y.-F., Shi, L., et al. (2017). Chitosan nanoparticles based nanovaccines for cancer immunotherapy. *Pure and Applied Chemistry, 89*(7), 931–939.
296. Lian, H., Ma, S., Zhao, D., Zhao, W., Cui, Y., Hua, Y., et al. (2022). Cytokine therapy combined with nanomaterials participates in cancer immunotherapy. *Pharmaceutics., 14*(12), 2606.
297. Dong, C., Tan, D., Sun, H., Li, Z., Zhang, L., Zheng, Y., et al. (2024). Interleukin-12 delivery strategies and advances in tumor immunotherapy. *Current Issues in Molecular Biology, 46*(10), 11548–11579.
298. Habiba, U., Rafiq, M., Khawar, M. B., Nazir, B., Haider, G., & Nazir, N. (2022). The multifaceted role of IL-12 in cancer. *Advances in Cancer Biology-Metastasis., 5*, 100053.
299. Lima, B. V., Oliveira, M. J., Barbosa, M. A., Goncalves, R. M., & Castro, F. (2021). Immunomodulatory potential of chitosan-based materials for cancer therapy: A systematic review of in vitro, in vivo and clinical studies. *Biomaterials Science, 9*(9), 3209–3227.
300. Kedir, W. M., Abdi, G. F., Goro, M. M., & Tolesa, L. D. (2022). Pharmaceutical and drug delivery applications of chitosan biopolymer and its modified nanocomposite: A review. *Heliyon, 8*(8), e10196.
301. Thambiliyagodage, C., Jayanetti, M., Mendis, A., Ekanayake, G., Liyanaarachchi, H., & Vigneswaran, S. (2023). Recent advances in chitosan-based applications—A review. *Materials, 16*(5), 2073.
302. Gonçalves, C., Ferreira, N., & Lourenço, L. (2021). Production of low molecular weight chitosan and chitooligosaccharides (COS): A review. *Polymers, 13*(15), 2466.
303. Reay, S. L., Jackson, E. L., Ferreira, A. M., Hilkens, C. M., & Novakovic, K. (2022). In vitro evaluation of the biodegradability of chitosan–genipin hydrogels. *Materials Advances, 3*(21), 7946–7959.
304. Ferraboschi, P., Ciceri, S., & Grisenti, P. (2021). Applications of lysozyme, an innate immune defense factor, as an alternative antibiotic. *Antibiotics, 10*(12), 1534.
305. Verčimáková, K., Karbowniczek, J., Sedlář, M., Stachewicz, U., & Vojtová, L. (2024). The role of glycerol in manufacturing freeze-dried chitosan and cellulose foams for mechanically stable scaffolds in skin tissue engineering. *International Journal of Biological Macromolecules, 275*, 133602.
306. Kean, T., & Thanou, M. (2010). Biodegradation, biodistribution and toxicity of chitosan. *Advanced Drug Delivery Reviews, 62*(1), 3–11.

307. Szymańska, E., & Winnicka, K. (2015). Stability of chitosan—A challenge for pharmaceutical and biomedical applications. *Marine Drugs, 13*(4), 1819–1846.
308. Niu, Y., & Hu, W. (2024). Preparation, characterization and application in environmental protection of low-molecular-weight chitosan: A review. *Sustainable Environment Research, 34*(1), 29.
309. Pandit, A., Indurkar, A., Deshpande, C., Jain, R., & Dandekar, P. (2021). A systematic review of physical techniques for chitosan degradation. *Carbohydrate Polymer Technologies and Applications, 2*, 100033.
310. Ali, G., Sharma, M., Salama, E.-S., Ling, Z., & Li, X. (2024). Applications of chitin and chitosan as natural biopolymer: Potential sources, pretreatments, and degradation pathways. *Biomass Conversion and Biorefinery, 14*(4), 4567–4581.
311. Qiu, S., Zhou, S., Tan, Y., Feng, J., Bai, Y., He, J., et al. (2022). Biodegradation and prospect of polysaccharide from crustaceans. *Marine Drugs., 20*(5), 310.
312. Al-Jbour, N. D., Beg, M., Gimbun, J., & Alam, A. (2021). Preparation and characterization of low molecular weight chitosan with different degrees of deacetylation by the acid hydrolysis method. *International Journal of Applied Pharmaceutics, 13*(2), 153–164.
313. Hsu, S.-C., Don, T.-M., & Chiu, W.-Y. (2002). Free radical degradation of chitosan with potassium persulfate. *Polymer Degradation and Stability, 75*(1), 73–83.
314. Ma, F., Zhang, S., Li, P., Sun, B., Xu, Y., Tao, D., et al. (2021). Investigation on the role of the free radicals and the controlled degradation of chitosan under solution plasma process based on radical scavengers. *Carbohydrate Polymers, 257*, 117567.
315. Carboué, Q., Fadlallah, S., Lopez, M., & Allais, F. (2022). Progress in degradation behavior of most common types of functionalized polymers: A review. *Macromolecular Rapid Communications., 43*(13), 2200254.
316. Lucas, N., Bienaime, C., Belloy, C., Queneudec, M., Silvestre, F., & Nava-Saucedo, J.-E. (2008). Polymer biodegradation: Mechanisms and estimation techniques–a review. *Chemosphere, 73*(4), 429–442.
317. Islam, N., Dmour, I., & Taha, M. O. (2019). Degradability of chitosan micro/nanoparticles for pulmonary drug delivery. *Heliyon, 5*(5), e01684.
318. Wang, H., & Roman, M. (2023). Effects of Chitosan molecular weight and degree of Deacetylation on chitosan−cellulose nanocrystal complexes and their formation. *Molecules [Internet], 28*(3), 1361.
319. No, H. K., Kim, S. H., Lee, S. H., Park, N. Y., & Prinyawiwatkul, W. (2006). Stability and antibacterial activity of chitosan solutions affected by storage temperature and time. *Carbohydrate Polymers, 65*(2), 174–178.
320. Cavallaro, G., Micciulla, S., Chiappisi, L., & Lazzara, G. (2021). Chitosan-based smart hybrid materials: A physico-chemical perspective. *Journal of Materials Chemistry B, 9*(3), 594–611.
321. Jain, A., Gulbake, A., Shilpi, S., Jain, A., Hurkat, P., & Jain, S. K. (2013). A new horizon in modifications of chitosan: Syntheses and applications. *Critical Reviews™ in Therapeutic Drug Carrier Systems, 30*(2), 91.
322. Wang, W., Meng, Q., Li, Q., Liu, J., Zhou, M., Jin, Z., et al. (2020). Chitosan derivatives and their application in biomedicine. *International Journal of Molecular Sciences, 21*(2), 487.
323. Jafernik, K., Ładniak, A., Blicharska, E., Czarnek, K., Ekiert, H., Wiącek, A. E., et al. (2023). Chitosan-based nanoparticles as effective drug delivery systems—A review. *Molecules [Internet], 28*(4), 1963.
324. Zaiki, Y., Iskandar, A., & Wong, T. W. (2023). Functionalized chitosan for cancer nano drug delivery. *Biotechnology Advances, 67*, 108200.
325. Rahimian, S., Najafi, H., Afzali, B., & Doroudian, M. (2024). Extracellular vesicles and exosomes: Novel insights and perspectives on lung cancer from early detection to targeted treatment. *Biomedicines [Internet], 12*(1), 123.
326. Abu Elella, M. H., & Kolawole, O. M. (2024). Recent advances in modified chitosan-based drug delivery systems for transmucosal applications: A comprehensive review. *International Journal of Biological Macromolecules, 277*, 134531.

327. Iftode, L., Cadinoiu, A. N., Rață, D. M., Atanase, L. I., Vochița, G., Rădulescu, L., et al. (2025). Double peptide-functionalized Carboxymethyl Chitosan-coated liposomes loaded with dexamethasone as a potential strategy for active targeting drug delivery. *International Journal of Molecular Sciences [Internet], 26*(3), 922.
328. Fathi-Karkan, S., Mirinejad, S., Ulucan-Karnak, F., Mukhtar, M., Ghahramani Almanghadim, H., Sargazi, S., et al. (2023). Biomedical applications of aptamer-modified chitosan nanomaterials: An updated review. *International Journal of Biological Macromolecules, 238*, 124103.
329. Herdiana, Y., Wathoni, N., Shamsuddin, S., Joni, I. M., & Muchtaridi, M. (2021). Chitosan-based nanoparticles of targeted drug delivery system in breast cancer treatment. *Polymers (Basel), 13*(11), 1717.
330. El-Saadony, M. T., Saad, A. M., Alkafaas, S. S., Dladla, M., Ghosh, S., Elkafas, S. S., et al. (2025). Chitosan, derivatives, and its nanoparticles: Preparation, physicochemical properties, biological activities, and biomedical applications—A comprehensive review. *International Journal Of Biological Macromolecules, 313*, 142832.
331. Fonseca-Santos, B., & Chorilli, M. (2017). An overview of carboxymethyl derivatives of chitosan: Their use as biomaterials and drug delivery systems. *Materials Science and Engineering: C, 77*, 1349–1362.
332. Shoueir, K. R., El-Desouky, N., Rashad, M. M., Ahmed, M. K., Janowska, I., & El-Kemary, M. (2021). Chitosan based-nanoparticles and nanocapsules: Overview, physicochemical features, applications of a nanofibrous scaffold, and bioprinting. *International Journal of Biological Macromolecules, 167*, 1176–1197.
333. Madera-Santana, T. J., Herrera-Méndez, C. H., & Rodríguez-Núñez, J. R. (2018). An overview of the chemical modifications of chitosan and their advantages. *Green Materials, 6*(4), 131–142.
334. Yadav, M., Kaushik, B., Rao, G. K., Srivastava, C. M., & Vaya, D. (2023). Advances and challenges in the use of chitosan and its derivatives in biomedical fields: A review. *Carbohydrate Polymer Technologies and Applications, 5*, 100323.
335. Sivanesan, I., Hasan, N., Kashif Ali, S., Shin, J., Gopal, J., Muthu, M., et al. (2022). Novel Chitosan derivatives and their multifaceted biological applications. *Applied Sciences [Internet], 12*(7), 3267.
336. Manohar, M., Shetty, A., Lobo, C. L., Jain, P., Hebbar, S., Dhas, N., et al. (2025). Emerging trends in chitosan based colloidal drug delivery systems: A translational journey from research to practice. *Carbohydrate Polymers, 360*, 123604.
337. Han, X., Tian, L., Chen, Q., Wang, L., Mi, Y., Li, Q., et al. (2025). Preparation of novel thiolated chitosan with significant antioxidant activity. *International Journal of Biological Macromolecules, 308*, 142536.
338. Al Bakain, R. Z., Abulateefeh, S. R., & Taha, M. O. (2015). Synthesis and characterization of chitosan-lactate–phthalate and evaluation of the corresponding zinc-and aluminum-crosslinked beads as potential controlled release matrices. *European Polymer Journal., 73*, 402–412.
339. Harutyunyan, L., Harutyunyan, R., Gabrielyan, G., & Lasareva, E. (2019). Modification of chitosan and chitosan succinate by surfactants and investigation of their properties. *Colloids and Surfaces A: Physicochemical and Engineering Aspects., 578*, 123622.
340. Sivanesan, I., Tasneem, S., Hasan, N., Shin, J., Muthu, M., Gopal, J., et al. (2022). Surveying the oral drug delivery avenues of novel Chitosan derivatives. *Polymers (Basel), 14*(11), 2131.
341. Rekha, M. R., & Sharma, C. P. (2009). Synthesis and evaluation of lauryl succinyl chitosan particles towards oral insulin delivery and absorption. *Journal of Controlled Release, 135*(2), 144–151.
342. Antony, R., Arun, T., & Manickam, S. T. D. (2019). A review on applications of chitosan-based Schiff bases. *International Journal of Biological Macromolecules, 129*, 615–633.
343. Shariatinia, Z. (2018). Carboxymethyl chitosan: Properties and biomedical applications. *International Journal of Biological Macromolecules, 120*, 1406–1419.

344. Mawazi, S. M., Kumar, M., Ahmad, N., Ge, Y., & Mahmood, S. (2024). Recent applications of Chitosan and its derivatives in antibacterial, anticancer, wound healing, and tissue engineering fields. *Polymers (Basel), 16*(10), 1351.
345. Savin, C.-L., Popa, M., Delaite, C., Costuleanu, M., Costin, D., & Peptu, C. A. (2019). Chitosan grafted-poly(ethylene glycol) methacrylate nanoparticles as carrier for controlled release of bevacizumab. *Materials Science and Engineering: C, 98*, 843–860.
346. Merzendorfer, H. (2019). Chitosan derivatives and grafted adjuncts with unique properties. In: Cohen, E., & Merzendorfer, H. (Eds.), *Extracellular sugar-based biopolymers matrices. Biologically-inspired systems* (vol 12). Springer, Cham. https://doi.org/10.1007/978-3-030-12919-4_3
347. Rahimian, S., Najafi, H., Doroudian, M. (2025). CXCR4-targeted theranostics in acute leukemia: Disrupting leukemic cell-microenvironment interactions with pentixafor and pentixather medical. *Oncology, 42*(9). https://doi.org/10.1007/s12032-025-02924-w
348. Raeisi, A., Doroudian, M., Rastegari, B., Mohammadi-Samani, S., Behzad-Behbahani, A., & Farjadian, F. (2025). pH-responsive modified dextran nanogel for liver targeted doxorubicin delivery. *Gels, 11*(10), 784. https://doi.org/10.3390/gels11100784
349. Farjadian, F., Moradi, S., & Hosseini, M. (2017). Thin chitosan films containing superparamagnetic nanoparticles with contrasting capability in magnetic resonance imaging. *Journal of Materials Science: Materials in Medicine, 28*, 47. https://doi.org/10.1007/s10856-017-5854-2
350. Hajiaghapour Asr, M., Dayani, M,. Segherloo, F. S., Kamedi, A., Neill, O. N., MacLoughlin, R., & Doroudian, M. (2023). Lipid nanoparticles as promising carriers for mRNA vaccines for viral lung infections. *Pharmaceutics, 15*(4), 1127. https://doi.org/10.3390/pharmaceutics15041127
351. Nahand, J. S., Vandchali, N. R., Darab, H., Doroudian, M., Banafshe, H. R., Moghoofei, M., Babaei, F., Salmaninejad, A., & Mirzaei, H. (2020). Exosomal MicroRNAs: Novel players in cervical cancer. *Epigenomicsm 12*(18), 1651–1660. https://doi.org/10.2217/epi-2020-0026
352. Doroudian, M., Neill, A. O., Loughlin, R. M., Prina-Mello, A., Volkov, Y., & Donnelly, S. C. (2021). Nanotechnology in pulmonary medicine. *Current Opinion in Pharmacology, 56*, 85–92. https://doi.org/10.1016/j.coph.2020.11.002

Electrospun Chitosan-Based Systems in Drug Delivery

Semsi Muratoglu, Fatima Al-Sulaiti, Murat Inal, Tugba Gulsun, and Selma Sahin

Abstract Electrospun chitosan drug delivery systems (DDS) carry great fortune to the next era of therapeutic technologies and biomedical applications. These systems have been successfully applied to a wide range of biomedical and therapeutic applications due to their unique advantages and biocompatible characteristics. Advanced DDS are rapidly revolutionizing in the modern era of therapeutics, to fill the gaps and drawbacks associated with conventional pharmaceutical formulations. Advanced DDS are techniques or devices fabricated to deliver therapeutic materials to target diseased body sites, increasing treatment efficacy and safety by regulating formulation properties (e.g., release rate, timing, targeting to tissue). Advanced DDS aim to enhance pharmacological efficacy, balancing challenges associated with traditional pharmaceuticals such as low/inconsistent solubility, permeability, and bioavailability, while decreasing toxicity and adverse drug reactions (ADRs). Electrospinning allows the fabrication of continuous polymer fibers, within the micrometer or nanometer scale, by exposing a polymer solution or melt to high voltages. Electrospinning is a very effective technique used to fabricate chitosan nanofibers. Electrospinning shows high versatility; it allows incorporating a wide range of therapeutic agents, including low solubility drugs, proteins, and DNA, and allows sustained and controlled-release designs. Chitosan is a non-toxic, naturally occurring biopolymer that exhibits unique characteristics that are useful for numerous applications, such as advanced DDS and biomedical technologies (e.g., tissue engineering and tissue regeneration). Inherently, it possesses antimicrobial, antioxidant, anticancer, non-toxic, biocompatible, and biodegradable properties. Combining electrospinning methods with chitosan results in advanced DDS with synergistic

S. Muratoglu · F. Al-Sulaiti · T. Gulsun · S. Sahin (✉)
Department of Pharmaceutical Technology, Faculty of Pharmacy, Hacettepe University, Ankara, Turkey
e-mail: semsi.muratoglu@hacettepe.edu.tr; fatimaalsulaiti@hacettepe.edu.tr; tgulsun@hacettepe.edu.tr; selmas@hacettepe.edu.tr

M. Inal
Bioengineering Department, Faculty of Engineering and Natural Sciences, Kırıkkale University, Kırıkkale, Turkey
e-mail: minal@kku.edu.tr

S. Jana (ed.), *Chitosan-Based Drug Delivery Systems*, AAPS Introductions in the Pharmaceutical Sciences 16, https://doi.org/10.1007/978-3-032-08347-0_10

performance by utilizing the unique properties of both. Electrospun-based chitosan fibers are used in advanced DDS due to several advantageous characteristics, such as optimized porosity, mechanical characteristics, boosted surface functionality, vast surface area, and pore size distribution across various scales. Functionalized electrospun chitosan-based nanofibers are modernly utilized to produce top-notch DDS and boosted scaffolds for the new era of medicine, such as regenerative medical applications, wound treatment, targeted drug therapy, personalized medicine, and antimicrobial coatings. As a result of their multifunctionality, biocompatibility, and capacity for controlled release, chitosan-based electrospun DDS represent an up-and-coming platform for diverse therapeutic and biomedical applications. This chapter discusses the characteristics, production, revolution, advancements, application, limitations, and challenges of electrospun chitosan advanced DDS. This chapter aims to comprehensively analyze chitosan-based electrospun DDS, focusing on their fabrication techniques, critical process parameters, properties, and biomedical applications in targeted and sustained therapeutics. Associated drug release and delivery mechanisms are explained. Applications of electrospun chitosan in drug delivery, including wound dressings, tissue engineering, bone regeneration, other diseases, and biomedical applications are explained in depth and suggested strategic solutions and future perspectives are proposed.

Keywords Chitosan · Drug delivery systems · Electrospinning · Nanofibers

Abbreviations

ADR(s)	Adverse drug reaction(s)
AFM	Atomic force microscope
API	Active pharmaceutical ingredient
APTT	Activated partial thromboplastin time
BSA	Bovine serum albumin
coPLA	Poly(l-lactide-co-d,l-lactide)
CRES	Chemical reactive electrospinning
DDS	Drug delivery system(s)
DSC	Differential scanning calorimetry
FESEM	Field emission scanning electron microscopy
FTIR	Fourier transform infrared spectroscopy
HPβCD	Hydroxypropyl-β-cyclodextrin
LMWC	Low molecular weight chitosan
PBS	Phosphate-buffered saline
PCL	Polycaprolactone
PEO	Polyethylene oxide
PLA	Polylactic acid
PLGA	Poly(lactide-co-glycolic acid)
PRES	Photoreactive electrospinning

PT	Prothrombin time
PVA	Polyvinyl alcohol
PVP	Polyvinylpyrrolidone
RES	Reactive electrospinning
SEM	Scanning electron microscope
TGA	Thermogravimetric analysis
WVP	Water vapor permeation rate
WVTR	Water vapor transmission rate
XRD	X-ray diffraction analysis
XPS	X-ray photoelectron spectroscopy
↑	Improving, increase(d), more or high(er)
↓	Decrease(d), less(er) or low(er)
1H NMR	Proton nuclear magnetic resonance

1 Introduction

1.1 Drug Delivery Systems (DDS): An Overview

Drug delivery systems (DDS) are techniques or devices fabricated to deliver therapeutic materials to target diseased body sites, increasing treatment efficacy and safety by regulating formulation properties (e.g., release rate, timing, targeting to tissue) [1–4]. DDS aim to enhance pharmacological efficacy, balancing challenges associated with traditional pharmaceuticals such as low/inconsistent solubility, permeability, or bioavailability, while decreasing toxicity and adverse drug reactions (ADRs) [1, 2, 4]. DDS are categorized according to (1) types and (2) routes of administration. DDS can be administered enterally (i.e., oral, sublingual, buccal, rectal routes), parenterally (i.e., intravenous, intramuscular, subcutaneous routes), or topically via direct skin application [5]. Advanced types of DDS include nanotechnology-based systems, microparticles, vesicular systems and personalized medicine applications [4, 6, 7]. DDS are composed of multiple materials like polymers (e.g., dextran, chitosan, cellulose), lipids (liposomes, nanoemulsions), along with inorganic nanocrystals [3, 5]. Polymer science advances in the modern era allow designing specialized DDS with special characteristics that improve drug delivery and overcome challenges with traditional pharmaceutical formulations [2, 3].

Advanced DDS holds an essential place in contemporary medicine by improving the effectiveness and safety profiles of therapeutics and pharmaceuticals. Through advanced drug delivery mechanisms (i.e., controlled and targeted delivery), advanced DDS overcome drawbacks and challenges associated with conventional pharmaceuticals and traditional drug administration techniques, paving the pathway for more positive health, societal, and economic outcomes. Advanced DDS delivers therapeutic agents via controlled-release or targeted delivery, utilizing nanotechnology advancements. Controlled drug release assures a predetermined rate of release, decreasing dosing frequency, and enhancing patient adherence [2, 3, 8, 9]. Targeted delivery utilizes molecular techniques to direct therapeutic agents to specific in vivo receptors at the cellular or molecular level, decreasing unwanted interactions and improving therapeutic efficacy [1, 2, 4, 6, 9]. Nanotechnology involves fabricating nanoparticles to optimize stability, bioavailability, and targeted delivery, especially in complex disease therapeutics (e.g., oncology, immunology), [4, 5, 10]. Advanced DDS is a major and significant part of personalized medicine, fulfilling individualized patient-specific therapy needs [4, 11].

1.2 Electrospinning Technology: An Overview

Electrospinning describes fabricating continuous polymer fibers, possessing diameters within the micrometers to nanometers scale, by exposing polymer solution or melt to high voltages [12–14]. Electrospinning methods utilize electrohydrodynamic forces to form a charged jet of the polymer solution, stretching and solidifying in the form of fibers during solvent evaporation [12, 15]. Key parameters impacting the procedure include: (1) polymer solution attributes (i.e., viscosity, conductivity, surface tension), (2) applied voltage, (3) space between nozzle and the collector and (4) environmental conditions [13, 14]. There are various types of electrospun fiber structures and morphologies produced by electrospinning. Nanofibers are structurally very fine fibers with nanometer-scale diameters, providing maximized surface area and porosity [16–18]. Core-shell structures manifest as fibers, including a core material encapsulated by a shell; these structures are beneficial for dual-drug loaded systems and controlled release DDS [19–21]. Differently, hollow fiber structures possess a hollow interior, which is useful for filtration and other applications [22, 23]. Porous fibers are characterized by superficial or interior pores, providing various functionality across a wide range of applications [22, 23].

Electrospinning methods provide many benefits to modern DDS. First, this technique offers increased surface area, thus improving solubility, drug loading capacity, and maximizing bioavailability [17, 18, 24]. Also, it allows fabricating sustained and controlled drug release, enhancing therapeutic outcomes and reducing ADRs [19, 25, 26]. This method shows high versatility; it allows incorporating numerous therapeutic agents, together with low solubility drugs, proteins, and DNA [17, 18, 24]. This is due to its ability to protect biomolecules. To exemplify, core-shell morphologies protect sensitive biomolecules from degradation and allow controlling the release kinetics [19, 20, 26]. In the modern era of therapeutics, electrospinning technology provides a promising future for many biomedical and therapeutic applications. Electrospinning is highly relevant to biomedical and therapeutic applications since it synthesizes fibers imitating the extracellular matrix; this improves cell adhesiveness and growth [27–29]. In tissue engineering, electrospun scaffolds are employed to provide supportive tissue regeneration and repair, allowing a healthy and enhancing physiologic environment for cell growth

[27, 28, 30]. In drug delivery, electrospun fibers allow designing controlled and sustained release DDS of therapeutic agents [16, 17, 25]. In wound treatment, wound dressings composed of electrospun nanofibers are preferred secondary to the amplified surface area, porosity, and easiness of incorporating antimicrobials [20, 31]. Furthermore, electrospun fibers reduce toxicity in oncology by locally delivering chemotherapeutics, decreasing systemic ADRs while maximizing therapeutic efficacy [19, 32].

1.3 Chitosan: Rationale for Use in Advanced DDS

Chitosan is a polyaminosaccharide made from β-(1→4)-glycosidic bonds linking glucosamine and N-acetylglucosamine units [33–35]. Chitosan possesses various reactive functional units, such as hydroxyl and amino groups, paving the way to multiple chemical modification capabilities [36–38]. Molecular weight and acetylation levels are essential influencers of chitosan's characteristics [36, 39, 40]. Chitosan is taken from chitin, a substance present within exoskeletons of crustaceans (i.e., shrimp, lobster, crab) [33, 41, 42], or fungi and algae [43, 44]. Chemical, enzymatic, or microwave-based schemes are used to extract chitosan from chitin via deacetylation methods [33, 42].

Chitosan possesses unique properties (biocompatibility, biodegradability, mucoadhesiveness, etc.) that overcome challenges and limitations associated with traditional pharmaceutical formulations. Due to being of natural origins, chitosan is biodegradable and biocompatible, confirming compatibility with living tissues and physiologic environments, paving the way to wide biomedical applications [36, 45, 46]. Rapid drug degradation, as seen with conventional pharmaceuticals, requires multiple patient doses. Chitosan-based controlled-release formulations overcome rapid degradation problems by forming a drug-protecting shield, with its biodegradable nature allowing sustained release and decreasing dosing frequency requirements while maintaining therapeutic levels. Moreover, it has a highly safe profile with low allergenicity [45, 47]. Chitosan shows significant antimicrobial effects (i.e., antibacterial and antifungal activity) that are influenced by multiple variables (i.e., pH, temperature, molecular weight) [34, 40, 48]. Solubility properties are associated with acetylation levels as well as molecular weight. Generally, it is soluble in acidic solutions [36, 39, 43]. Many active pharmaceutical ingredients (API) exhibit low aqueous solubility, resulting in poor gastrointestinal absorption and lower bioavailability. Chitosan can improve these solubility issues by forming nanoparticles and improving drug dispersion. Its mucoadhesive characteristics also help better absorption through biological membranes, improving bioavailability. Chitosan's viscosity is relatively high; yet low molecular weight chitosan (LMWC) has better in vivo activity and absorption [34, 35]. Its bio-adhesive properties facilitate advanced DDS [40, 47]. Chitosan is durable as it shows good thermal stability [36, 49]. Chitosan allows encapsulating drugs sensitive to environmental factors (i.e., pH, temperature, and light), preserving efficacy and increasing shelf-life due to prolonged stability.

Conventional pharmaceuticals exhibit non-specific drug distribution, causing ADRs. Chitosan is capable of conjugating with target ligands, making it a part of targeted DDS, which improves the safety of high-risk drugs. These promising properties make chitosan very promising in tissue engineering, drug delivery, wound healing, and dental implant uses due to its biocompatibility and bioactivity [36, 40, 45, 46].

1.4 Synergy of Electrospinning and Chitosan in Advanced DDS

Combining electrospinning methods with chitosan leverages the unique attributes of both elements, resulting in advanced DDS with synergized performance. As a result of their multifunctionality, biocompatibility, and capacity for controlled release, chitosan-based electrospun DDS represents an up-and-coming platform for diverse therapeutic and biomedical applications. This chapter aims to discuss electrospun chitosan-based DDS, focusing on their fabrication techniques, properties, and potential applications in targeted and sustained therapeutics. The obstacles and limitations of electrospun chitosan DDS are discussed in depth, and suggested strategic solutions and future perspectives are provided.

2 Electrospinning: Process and Parameters

Electrospinning technology holds a significant place in drug delivery research [50, 51]. It provides valuable benefits to build a DDS, like providing high surface area, solubility enhancement, and the capability of drug release manipulations [52–54]. This technology is based on electrically driven jets [55, 56]. Liquids tend to spin and flow through an electrical field that is named a "liquid jet." This jet travels from the nozzle to the collector in the electrospinning process. This jet occurs owing to the liquid droplet that formed at the tip of the nozzle, named "Taylor cone" [56]. Electrospinning generates an electric field as the jet starts to spin due to whipping instability provided by the charged system. During this spinning process, the solvent system evaporates while solid fibers accumulate on the collector. The accumulated fibers create a nonwoven fiber mat, used to develop further tailored materials such as orally disintegrating films [57], scaffolds [58], and patches [59].

In drug delivery and material research, electrospinnable liquid systems consist of polymers such as polyvinyl alcohol (PVA), polyvinylpyrrolidone (PVP), polycaprolactone (PCL), and chitosan [60–63]. Choosing the polymer composition is the crucial step in designing the capabilities of a DDS. For instance, soluble polymers such as PVP provide immediate release mechanisms to the DDS [64]. On the contrary, insoluble or slow-degrading polymers such as PCL or chitosan provide a controlled release mechanism and mechanical strength [65, 66]. Additionally,

Eudragit polymers provide pH-responsive behavior [67], alginates provide swelling capacity [68], while cellulose-based polymers enhance biocompatibility [69]. In view of these capabilities of polymers, the first step in the development of a DDS should be the determination of the polymer composition. After determining the polymer solution components, the researcher should build an electrospinning scheme that considers the parameters impacting the critical quality attributes of electrospun DDS.

The main components of the electrospinning set-up are as follows: (1) power supply, (2) collector, (3) syringe pump, and (4) a nozzle [70]. The type and features of these components affect the critical quality attributes of electrospun fibers. Generally, the main process variables are pump rate, applied voltage, working distance, and collector type. Moreover, the needle size, current type, humidity, temperature, and spinning solution attributes (i.e., conductivity, viscosity, and surface tension) are fundamental to obtaining the desired product [70, 71]. These variables should be well established for process integrity and uniform continuous production. Any change or manipulation in these components changes the process and the product. Table 1 shows the critical parameters of electrospinning processes.

Common parameters, such as working distance and applied voltage, are considered more predictable and manageable for the researcher. Yet, it should be highlighted that all components and parameters should be well-established. For instance, the power supply is usually used to create a direct current, but there are further options such as the alternating current [82]. An alternative current scheme can change the fiber formation and the volume of the final product [82, 83]. There are solutions to needle clogging problems, such as the needleless ultrasound-enhanced electrospinning technique [84]. The collector type is the other critical component of the electrospinning set-up. Flat and drum collectors are commonly used for different purposes [70]. There are also wire or mandrel collector options [85]. Conductivity is the main requirement of the collector. Any material possessing sufficient conductivity may be used as a collector to be coated by electrospun fibers, such as implantable stents [86].

There are several techniques based on nozzle type for electrospinning. The commonly used method is the monoaxial technique, which has basic needles. In this method, a polymer solution filled in a syringe is coupled to a monoaxial nozzle to prepare monolayered fibers. The polymer solution can contain one polymer or a mixture. As a design approach, there are more nozzle-related options, such as coaxial and triaxial techniques [87–89]. These nozzles are designed as interlocking single-center nozzles. It is possible to produce core-sheath structured fibers owing to these methods [90]. The other approach is side-by-side electrospinning, creating Janus fiber structures [61, 91]. Such approaches allow manipulation of the release behavior of the advanced DDS, such as biphasic release [92]. Figure 1 provides a schematic presentation of commonly used electrospinning techniques to produce electrospun chitosan-based DDS.

Chitosan is a low-spinnable polymer [93, 94]. Bhattarai et al. demonstrated that pure chitosan (2%) presented insufficient electrospinning, acting like an electrospray technique. Scanning electron microscope (SEM) imaging revealed that only sprayed

Table 1 Critical parameters of the electrospinning process

Parameter		Increase (↑) effect	Decrease (↓) effect	References
Polymer solution-based properties	Viscosity	↓Spinnability Needle clogging problems ↑Fiber diameter	Spray-like manufacturing Formation of droplets ↓Fiber diameter	[72, 73]
	Conductivity	↓Fiber diameter	Loss of spinnability due to the insufficient conductivity Undesired irregularities such as bead formation on fibers ↑Fiber diameter	[72, 74]
	Surface tension	Jet instability Droplet and bead formation on fibers	Requiring ↓ voltage to spin Finer ↓fiber diameter	[75, 76]
Process & environment-based properties	Applied voltage	↓ Fiber diameter ↓Droplet size on the needle tip Bead defects with irregular fiber formation	↑ Fiber diameter Bead and droplet formation on fibers Insufficient spinnability	[77, 78]
	Pump rate	Bead formation on fibers due to the insufficient drying process ↑ Fiber diameter	↓Fiber diameter	[62]
	Needle size	↑Fiber diameter	↓ Fiber diameter	[79]
	Working distance	Unable to obtain uniform fibers above optimum level	Bead formation on fibers due to the insufficient drying process below the optimum level Unable to obtain uniform fibers below the optimum level	[76, 80]
	Humidity	↓Fiber diameter	Allows a faster drying process of the jet Clogging problems of the nozzle ↑ Fiber diameter	[76, 81]
	Temperature	Affects polymer properties (viscosity, conductivity & surface tension)	Affects polymer properties (viscosity, conductivity & surface tension)	[58]

↓: decrease(d), less(er) or low(er)
↑: increase(d), more or high(er)

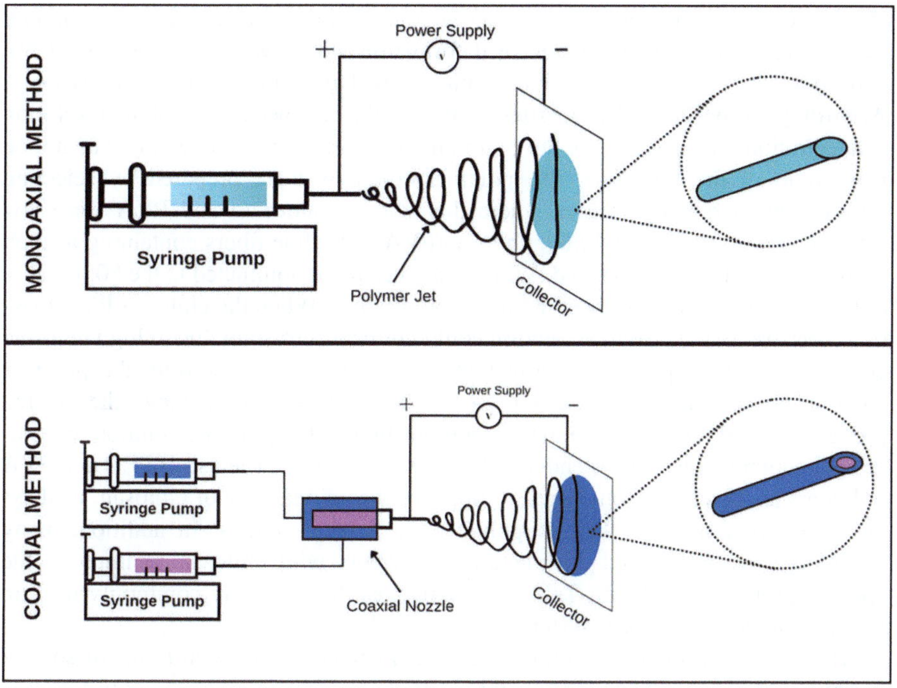

Fig. 1 Schematic presentation of commonly used electrospinning techniques to produce electrospun chitosan-based DDS

droplets accumulated on the collector. This reveals the fact that pure chitosan cannot be directly electrospun. The increased viscosity of the chitosan solution explains this issue. Electrospinning highly viscous polymer solutions is difficult due to several problems, such as clogging of the nozzle and the need for relatively high voltage use [95]. Generally, blending with PVA or polyethylene oxide (PEO) is often used as a strategy to overcome this drawback [96].

Bhattarai et al. used this strategy and blended chitosan with PEO, a co-electrospinning polymer. Increasing the PEO ratio in the blend caused a decrease in viscosity. The -OH and -NH_2units of chitosan polymer chains manifested in high viscosity secondary to strong hydrogen bonds. PEO modifies the molecular interactions of chitosan and water by forming new hydrogen bonds with OH units present in PEO and water. These new hydrogen bonds result in a decrease in polymer blend viscosity while increasing the water solubility of chitosan [95]. Similarly, PVA interacts with amino and hydroxyl units present in chitosan chains. Zhang et al. evaluated PVA as a blend polymer with chitosan. They focused on chitosan/PVA solution concentration (4% to 8%) and chitosan/PVA mass ratio (20:80; 40:60; 50:50; and 60:40). Prepared polymer solutions were evaluated in terms of viscosity, surface tension, and conductivity. Initially, solution concentration studies revealed that increasing the concentrations from 4% to 8% fixed chitosan/PVA mass ratios

(50:50) resulted in higher viscosity. At 4% concentration, the low viscosity polymer solution resulted in predominant bead formation with few fibers. The use of 8% concentration resulted in a few fibers with a thick diameter due to the high viscosity. According to the desired properties of the produced fibers, 7% polymer solution concentration was selected as the optimum concentration. In order to evaluate the mass ratio of the polymers, 7% polymer solutions were prepared by selecting different mass ratios. In cases where the ratios of chitosan and PVA were low (mass ratios of 20:80 and 40:60; chitosan/PVA), ultrafine fibers containing a small number of beads were produced. While bead formation dominated in the 50:50 ratio, only bead formation was observed in the 60:40 ratio. When the chitosan/PVA mass ratio was 80:20, the polymer solution could not be electrospun due to high chitosan content. The main parameters affecting the electrospun fibers were the polymer solution concentration and mass ratio of polymers [97]. To achieve the desired process, it is critical to determine the composition of the polymer solution, considering the interaction of the blended polymer with chitosan. The use of PEO and PVA polymers as a blend is considered to improve the spinnability of chitosan, as these polymers were used as "guest" polymers in several studies. In addition, many polymers have been used as functional polymers with chitosan to improve the specific properties of DDS. Table 2 summarizes the polymers commonly used in electrospun chitosan-based systems.

The chitosan polymer solution should be prepared carefully in terms of solvent attributes, chitosan concentration, and molecular weight. Acetic acid concentration is a crucial parameter when electrospinning pure chitosan alone. Due to the impact of increasing the jet charge and decreasing the surface tension of polymer solution, increasing acetic acid concentration facilitates electrospinning [120]. Generally, chitosan is used in monoaxial electrospinning techniques. There is also coaxial electrospun chitosan-based DDS. For instance, dos Santos et al. produced PVA/chitosan nonwoven mats for periodontal drug delivery. Chitosan is used as a shell layer to manipulate drug delivery. The crosslinked chitosan layer delayed the permeation of tetracycline hydrochloride through the core layer [121]. All these indicate that it is possible to design DDS based on electrospun nanofiber mats using chitosan.

3 Advantages of Electrospun Chitosan in DDS

Electrospinning provides desirable benefits secondary to fiber production, such as high surface area, improved drug loading capacity, and continuous production [122]. However, there are several limitations related to the process, the polymer solution, or the nature of the desired product. In this context, the electrospun chitosan approach may provide a way to overcome these undesirable limitations. Chitosan, a natural-based polymer, consists of polysaccharides that enable the production of derivatives that impart various properties such as biocompatibility, mucoadhesion,

Table 2 Commonly used polymers in chitosan-based electrospun DDS

Polymer name	Polymer type	Electrospinning type	Function in the chitosan-based system	Common use in advanced DDS	References
Starch	Natural	As a monoaxial blend	↑ Water absorption capacity of DDS	Wound dressing	[98]
Gelatin	Natural	As a monoaxial blend	Useful biodegradability/biocompatibility, adhesion capability properties	General DDS & tissue engineering scaffolds	[99, 100]
Poly-cyclodextrin	Semi-synthetic	As a monoaxial blend	↑ Solubility of lipophilic drugs &ensuring prolonged release	Implant coating	[60]
HPβCD	Semi-synthetic	2-layered sandwich system	Solubility enhancer & taste masker in a sandwiched design nanofiber mat 1st layer: PVP/HPβCD, 2nd layer: chitosan-cysteine/PVA	Oral candidiasis patch	[59, 101]
PVA	Synthetic	As a monoaxial blend	↑ Chitosan spinnability	Wound dressing and tissue engineering scaffolds	[102–106]
PCL	Synthetic	Monoaxial; single or as a blend	Supporting base of the system ↑ mechanical strength ↑ chitosan spinnability	Wound dressing, tissue engineering scaffolds	[107–110]
PEO	Synthetic	As a monoaxial blend	↑ Chitosan spinnability	Gastro-retentive delivery	[111, 112]
Eudragit RL100	Synthetic	Monoaxial single electrospinning	Coating material for the DDS	Ocular DDS	[113]
PLA	Synthetic	As a monoaxial blend and As a coaxial core solution	↑ Mechanical strength and ensuring biodegradability & biocompatibility	Wound dressing and Implant Coating	[114–116]
PLGA	Synthetic	As a monoaxial blend	↑ Mechanical strength and have moderate biodegradability	Local drug delivery scaffolds	[117]
PVP	Synthetic	As a monoaxial blend	↑ Chitosan spinnability	General DDS	[118, 119]

↑: improving, increase (d), more or high (er)

DDS drug delivery systems, *HPβCD* Hydroxypropyl-β-cyclodextrin, *PCL* Polycaprolactone, *PEO* Poly (ethylene oxide), *PLA* Polylactic acid, *PLGA* Poly(lactic-co-glycolic acid), *PVA* Polyvinyl alcohol, *PVP* Polyvinylpyrrolidone and biodegradability [123, 124]. The use of these derivatives as a component of electrospun materials may be an interesting strategy in drug development.

3.1 Mucoadhesion

The mucoadhesion ability of chitosan derivatives is the most interesting feature of the polymer for the drug development process. Mucosal membranes are covered by mucus, which contains mucins (glycoproteins) [125]. Chitosan derivatives provide mucoadhesion with mucins through electrostatic interactions and hydrogen bonds [126]. Mucin is negatively charged, and the cationic nature of the chitosan provides strong mucoadhesion ability. Wound dressing materials, mucosal DDS, and tissue regeneration materials should have sufficient mucoadhesion capability [127, 128]. Chitosan is a very useful natural polymer when developing these DDS.

A study by Anisiei and colleagues aimed to produce a wound dressing material based on electrospun chitosan blended with N,N,N-trimethyl chitosan and PEO at various mass ratios. The in vitro bioadhesion and mucoadhesion experiments demonstrated that all chitosan-based formulations presented excellent adhesion capabilities [129].

Bozdag and workers reported the application as a bone tissue regeneration material [130]. In their study, an electrospun nanofiber was composed of four different polymers based on their unique properties. PVA and hydroxyapatite were chosen to provide good mechanical features and a bone-like structure. Chitosan was used in this study as a component of electrospinning solutions. The rationale behind using chitosan in this study was to provide bio-adhesion in addition to its biocompatibility, biodegradation, and gelation behavior [130].

The mucoadhesive behavior of chitosan may be a favorable option to ensure drug delivery to other mucosal sites. For example, Samprasit et al. developed mucoadhesive electrospun nanofibers and applied them to porcine buccal mucosa in in vitro texture analysis studies to determine their mucoadhesion properties. Chitosan/PVA nanofibrous mats showed approximately 11 g detachment force and were considered as a good mucoadhesive system for the oral cavity [131]. Stie et al. also demonstrated the adhesive abilities of chitosan with chitosan/PEO nanofibers. The ex vivo studies on porcine buccal mucosa revealed that chitosan/PEO nanofibers provided strong mucoadhesion performance as a sublingual DDS compared to the negative control (hydrophobic ethyl cellulose film) [112].

3.2 Mechanical Strength

Electrospun drug delivery materials often suffer from poor mechanical strength issues. Chitosan with PVA may provide desirable mechanical properties to DDS [132]. Chitosan lactate is blended with PVA to increase the tensile strength and flexibility of wound dressings [107]. Adeliet al. produced electrospun mats containing different combinations of PVA, chitosan, and starch as wound dressing, and evaluated them for mechanical strength under wet and dry conditions. Although the tensile strength of the mats decreased in the wet state, all mats exhibited good mechanical strength, which is promising for the product's life-cycle management in terms of handling and patient comfort [98].

3.3 Crosslinking Ability

Chitosan derivatives may provide several advantages to the DDS, such as crosslinking capability. Crosslinking is a process in which functional units in electrospun scaffolds are linked together by various strategies [133]. Generally, after electrospinning, the electrospun mat is exposed to crosslinkers [134, 135]. Due to crosslinking, the system gains better mechanical properties, increased stability, and modified drug release [111, 134, 136]. Darbasizadeh et al. studied ranitidine hydrochloride in crosslinked chitosan/PEO nanofibrous mats and demonstrated that crosslinked samples exhibited higher thermal stability and mechanical behavior compared to non-crosslinked samples. Secondary to the low solubility of the crosslinked system, the developed system showed lower swelling capacity with sustained release of ranitidine hydrochloride [111]. Cui et al. prepared crosslinked PVA/chitosan nanofibers for transdermal drug delivery. According to drug release studies, crosslinked nanofibers reduced the burst release of ampicillin sodium by almost two-fold compared to non-crosslinked nanofibers (29% to 63% burst release). Mechanical properties were also improved by crosslinking. Young modulus in cross-linked nanofibers containing 0.5% glutaraldehyde as cross-linker increased more than five-fold compared to non-cross-linked nanofibers [137]. In drug development, crosslinking strategies can be a valuable option to customize the desired product.

3.4 Pharmacological Advantages

When designing an advanced DDS, pharmacologically active polymers attract attention. Due to the antibacterial [106], anti-inflammatory [138], antioxidant [139], anticancer [140], and other activities of chitosan derivatives, researchers tend to use chitosan as an electrospinning polymer. For example, in a study aiming to produce an electrospun membrane for diabetic wound healing, chitosan was selected due to its antioxidant and antibacterial properties [141]. Hadjianfar et al. aimed to produce electrospun DDS for cancer treatment and developed PCL/chitosan-based 5-fluorouracil-loaded nanofibers to achieve this goal. Chitosan was selected for its anticancer potential and its ability to reduce the systemic effects of anticancer drugs due to its pH sensitivity [109].

4 Characterization Approaches of Electrospun Chitosan DDS

The choice of characterization techniques for electrospun DDS depends on (1) the type of the designed DDS and (2) the purpose and usage of the DDS. These techniques generally reveal the overall material properties and may include product-specific characterizations (e.g., swelling capacity for wound dressings). Figure 2 summarizes the main characterization approaches of electrospun chitosan DDS discussed in this section.

4.1 Imaging-Based Morphological Analysis

The electrospinning technique provides continuous production of fibers under specific conditions. Fiber morphology should be investigated in the drug development

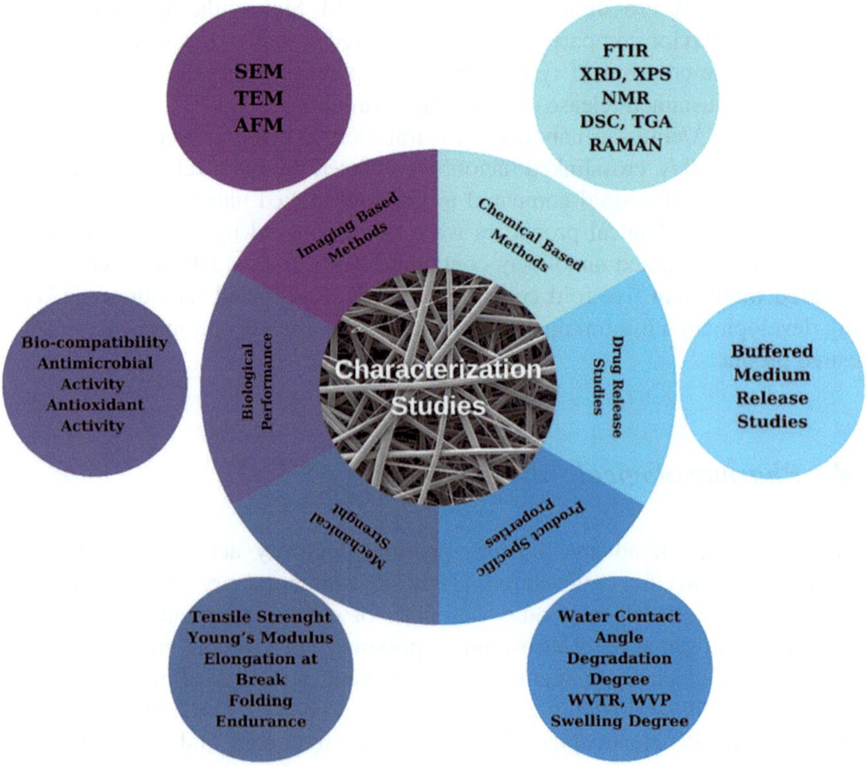

Fig. 2 Schematic diagram summarizing the main characterization methods of electrospun chitosan DDS

process and further studies. Electrospun fibers should be uniform and free from irregularities such as bead formation and droplets [142]. SEM is a widely used imaging strategy to visualize the surface morphology and structure of electrospun chitosan nanofibers. SEM helps determine fiber diameter, uniformity, and the presence of any beads or defects [99, 119, 143, 144]. It is the most preferred technique for elucidating the fiber formation in electrospun materials. In one study, SEM imaging revealed that the fiber diameter decreased in response to increasing chitosan content in the chitosan/silk fibroin blend [145]. Similar to SEM, field emission scanning electron microscopy (FESEM) provides high-resolution images that facilitate in-depth examination of the surface morphology of produced electrospun chitosan nanofibers [143]. Other options for evaluating the morphology of nanofibers are optical microscopy and atomic force microscopy (AFM) technologies. Optical microscopy provides information that aids in process optimization [146]. AFM characterizes the surface morphology of materials, such as determining possible phase separation of fibers [147].

4.2 Chemical and Structural Characterization

Various instrumental analysis methods are used to investigate the chemical and structural attributes of electrospun chitosan-based DDS. Fourier transform infrared spectroscopy (FTIR) determines the integrity of the chemical structure of chitosan, its components, and drugs added to the DDS by detecting functional groups and chemical bonds in evaluated structures. FTIR provides data on material compatibility with manufacturing processes and stability [98]. FTIR verifies the integrity of medicinal products within the nanofibers and ensures that no chemical interactions exist between DDS components [119, 148]. To exemplify, FTIR analysis demonstrated that chitosan and PVP interact via H-bonds, which was supported by Raman analysis in one of the reports [118]. Proton nuclear magnetic resonance (1H NMR) enables effective grafting or adjustment of chitosan as well as analysis of molecular structure [144].

X-ray-based characterizations such as X-ray diffraction analysis (XRD) and X-ray photoelectron spectroscopy (XPS) are widely used for characterization of electrospun chitosan-based DDS. Generally, XRD is performed to determine the crystallinity of a product. XRD explores the crystalline structure of the electrospun fibers and helps ensure successful medicinal product incorporation into nanofibrous DDS [144]. XPS is another option to determine the surface compositions of nanofiber composites in terms of elements. To exemplify, a study aiming to develop DDS containing 5-fluorouracil showed that XPS analysis helped determine the PVP content, but failed to detect the exact location of chitosan on the nanofibrous scaffold [118].

Thermal analysis methods such as differential scanning calorimetry (DSC) and thermogravimetric analysis (TGA) are widely used to determine the thermal behavior of the DDS [149]. TGA quantifies the thermal stability and composition of the electrospun nanofibers by evaluating weight changes secondary to heat exposure [144]. Thermal analysis methods give information about the transition and degradation temperatures of molecules. Using this information, it is possible to detect any incompatibility due to any of the fabrication process parameters or the polymers used. These methods also help determine the optimum formulation approaches. Bahtiyar et al. presented that crosslinking of chitosan/PVA fibers with gallic acid resulted in higher system thermostability according to TGA [150].

4.3 Drug Release Studies

In vitro drug release investigations are fundamental methods for projecting the in vivo performance of a DDS. There are various methods to determine the release behavior. Selecting the release medium is the first step in planning a drug release study. Usually, buffered solutions such as phosphate-buffered saline (PBS) (pH 7.4) are utilized [117]. These studies involve in vitro drug release testing and fitting to kinetic models. In drug release tests, the drug-loaded delivery system is immersed in a relevant medium, allowing quantification of the drug released as a function of time. Validated analytical techniques (e.g., high-performance liquid chromatography and ultraviolet-visible spectrophotometry) are employed to determine the explored drug levels [99, 143, 144]. Fitting to kinetic models is then analyzed to characterize drug release kinetics and dissolution mechanisms. These mechanisms may include polymer relaxation or diffusion, which can be characterized by exploring multiple kinetic models (e.g., Ritger-Peppas model) [119, 151].

4.4 Physico-Mechanical Properties

Mechanical behavior testing evaluates nanofiber-based DDS. To determine the mechanical strength and properties of the nanofiber mat, thickness, tensile strength, elongation at break and elastic modulus properties should be examined [145, 152]. Tensile strength tests explore the mechanical attributes (i.e., strength, elasticity) of the nanofibers, which are vital characteristics impacting performance in drug delivery applications [144]. Water contact angle quantification explores the wettability and hydrophilicity of the nanofibers, which are properties that can affect drug release rates and biocompatibility [144]. For example, in a study conducted to produce Fe_3O_4 nanoparticle-loaded electrospun wound dressings consisting of chitosan/gelatin blend, the optimum Fe_3O_4 ratio was determined by mechanical behavior. Interestingly, exceeding 1% Fe_3O_4 content was associated with a decrease in Young's modulus results. To attain the desired tensile properties and toughness, 1% Fe_3O_4 content was considered optimum [152].

Determining the degree of degradation of a fabricated DDS is beneficial for understanding the system's in vivo behavior. For instance, chitosan/PVA nanofibers produced by Ibrahim et al. were shown to be resistant to degradation due to the crosslinked system, which decreased drug release. The degree of degradation can be computed using Eq. 1 [143].

$$\%\text{Degree of Degradation} = \frac{W_i - W_l}{W_i} \times 100 \tag{1}$$

W_i: Initial weight measurement of the electrospun system
W_l: Final weight measurement of the electrospun system

Wound dressings should have good swelling ability. The water vapor permeation rate (WVP) provides key information needed to evaluate the wound dressing. It can be performed by gravimetric and volumetric methods [153]. Determining the water vapor transmission rate (WVTR) is used to assess water transmission considering the time and material area. WVP reflects the interaction ability of the material with water, calculated while considering the thickness of the material. WVTR and WVP are computed according to Eqs. 2 and 3 [154, 155].

$$\text{WVTR} = \frac{W_i - W_l}{A \times t} \tag{2}$$

$$\text{WVP} = \frac{\text{WVTR}}{\Delta P} \times X \tag{3}$$

W_i: Initial weight measurement of the electrospun DDS
W_l: Final weight measurement of the electrospun DDS
A: Area available for permeation in the nanofiber film
t: Experiment time
X: Thickness of the nanofiber film
ΔP: The difference in partial water vapor pressure across both film sides

Swelling studies are generally carried out in a water-based medium such as PBS solution. After immersing in the PBS solution, the increase in material weight helps to calculate the swelling degree [144]. The swelling degree calculation is applied according to Eq. 4.

$$\%\text{Swelling degree} = \frac{(W_S - W_D)}{W_D} \times 100 \tag{4}$$

W_D: measured weight of nanofiber mat DDS under dry conditions
W_S: measured weight of nanofiber mat DDS under swelling

Water contact angle studies are generally applied to these materials to determine the wettability of wound dressings [108]. The contact angle of water helps determine wettability properties. Decreased contact angle indicates improved wettability. For instance, a study reported that chitosan and β-cyclodextrin have water contact angle measurements of 110° and 74°, respectively. The fabricated electrospun shows superhydrophilicity, indicated by a zero water contact angle [144].

4.5 Biological Evaluation

These tests encompass a wide range of efficacy-focused and DDS-specific methods. Cytocompatibility and cell proliferation assays are crucial to confirm biocompatibility with cell tissues and the lack of cellular growth-inhibition effects. These tests are fundamental in tissue engineering and wound healing uses [144, 148]. Antimicrobial activity is another factor explored as part of characterization. Antibacterial activity tests determine the antibacterial characteristics of the drug-loaded chitosan nanofibers by exposure to a wide range of bacterial strains to determine coverage across microbes and give insights towards effectiveness in infection treatment or prophylaxis [119]. Electrospun chitosan-based DDS generally aim to demonstrate antibacterial properties. To determine antibacterial activity, agar well diffusion assays using *Staphylococcus aureus* (i.e., gram-positive) in addition to *Escherichia coli* (i.e., gram negative) bacteria are utilized. Comparing inhibition zones versus the standard control (ampicillin) gives the activity index of the system [119]. The activity index is computed according to Eq. 5 [119].

$$\%\text{Activity index} = \frac{R_1}{R_2} \times 100 \tag{5}$$

R_1: The diameter of the electrospun material inhibition zone
R_2: The diameter of the control antibiotic (ampicillin) inhibition zone

Specific formulations may require specific characterization studies. For example, the fabricated tranexamic acid-loaded hemostatic patch should be compatible with blood [156]. A study examined the hemocompatibility of these drug-loaded patches; tranexamic acid loading on electrospun chitosan reduced hemolysis [156]. This study also examined product-specific conditions such as blood clotting duration, plasma recalcification duration, and platelet adhesion [156]. In a study evaluating the nanofiber-coated stents, a stent crimping study was conducted [60]. The coated stents were crimped for installation in the catheter. To examine the crimping process compatibility of the prepared coated stents, post-process visualization of the nanofiber layers was performed using SEM imaging and normal eye observation. There should be no visible changes in the integrity of fibers to assure the activity of the DDS. There are other characterization approaches based on the intended use, such as anticancer efficacy (cytotoxicity assay) [157], antioxidant activity [158, 159], and angiogenic activity (related to the healing process) [160].

5 Drug Delivery Mechanisms of Electrospun Chitosan DDS

Chitosan-based electrospun systems may provide drug delivery with diffusion and degradation-controlled mechanisms [161]. It is also possible to exhibit drug release by swelling behavior or a combination of these mechanisms [162]. In a study,

PCL/chitosan electrospun membranes were fabricated for controlled drug delivery. The researchers compared PCL nanofibers to PCL/chitosan nanofibers. During in vitro release examinations, drug release was controlled by desorption from the surface and diffusion mechanism during the initial days. Subsequently, chitosan-related swelling reduced the membrane pore size, decreasing the release rate. As the swelling continues, pores disappear, and the diffusion of azithromycin gets longer. The final release mechanism of chitosan-containing membranes was hydrolytic degradation of the polymer [66].

To determine the drug release mechanism of the system, it is beneficial to calculate the mathematical models of drug release kinetics. Drug release kinetics can be described by numerous models [163]. These include, but are not limited to, zero-order, first-order, and Higuchi models. These drug release kinetic models describe the release mechanisms such as Korsmeyer-Peppas, Fickian diffusion, and non-Fickian diffusion models [163, 164]. A study developed a wound dressing for dual drug delivery using a core-sheath system composed of gelatin nanoparticles dispersed in chitosan/PVA fibers was utilized [103]. The experimental studies showed that the gelatine nanoparticles released more drugs before loading into the fiber mats than after loading. This core-sheath design ensured controlled drug release. To determine the release kinetics of the dual DDS, kinetic release constants of zero order, Korsmeyer-Peppas and Higuchi models were calculated. Results fitted with Korsmeyer-Peppas kinetics for both drugs (lidocaine and erythromycin) [103]. Another report exploring Fenbufen-loaded electrospun poly(D,L-lactide-co-glycolide) (PLGA)/chitosan blended nanofibrous scaffolds reported the association between drug release and elevations in chitosan content. A possible explanation for this observation is that increased chitosan resulted in higher swelling capacity, a phenomenon that resulted in higher drug release [117]. Figure 3 summarizes the main drug release mechanisms and kinetic models associated with electrospun chitosan-based DDS.

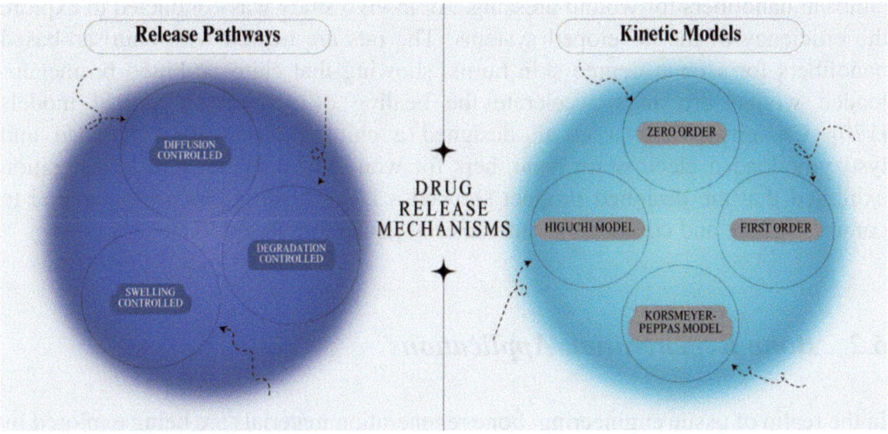

Fig. 3 Main drug release mechanisms and kinetic models related to electrospun chitosan-based DDS

6 Applications of Electrospun Chitosan in Drug Delivery

Conventional DDS and studies face several shortcomings due to the nature of the drug and the dosage form. With the help of polymer technology, it is possible to design a well-structured DDS to overcome these challenges. Chitosan, combined with electrospinning technology, holds a significant position in modern therapeutics as a valuable option for developing well-designed drug delivery materials like wound dressings [165], tissue engineering scaffolds [166], and various local applications [156]. Natural biopolymers demonstrate better biocompatibility than synthetic materials [167], making them preferable for human-related applications. Electrospun chitosan fibers attract considerable interest in tissue regeneration and engineering, surgical technologies, body implants, and advanced DDS [167]. This section details the clinical and biomedical utility of electrospun chitosan-based DDS, including nanofibers in addition to scaffolds.

6.1 *Wound Dressings*

Chitosan presents valuable attributes like being non-toxic, biocompatible, and biodegradable. Also, it shows pharmacological features such as anti-inflammatory, antibacterial, and hemostatic properties [108]. It accelerates the activity of macrophages to support wound healing. Additionally, it is known that chitosan improves the granulation tissue construction and collagen production [98]. It endures a rapid healing process and prevents scar formation [168]. Owing to these features, chitosan plays a vital role in electrospun wound dressing applications [169]. A well-developed wound dressing should prevent water loss and dryness of the wound area. With the help of chitosan-based systems, it is possible to maintain the moisture stability of the wound [146]. Bayat et al. developed electrospun bromelain-loaded chitosan nanofibers for wound dressing. An in vivo study was conducted to explore the efficiency of the developed systems. The rats are treated with chitosan-based nanofibers for second degree skin burns, showing that chitosan-based bromelain-loaded wound dressings accelerate the healing of wounds in animal models [170]. Charernsriwilaiwat et al. designed a chitosan-based PVA blended and lysozyme-loaded electrospun nanofibers for wound healing. In vivo investigation exhibited that the designed product showed a faster healing process compared to controls (gauze and commercial antibacterial gauze dressing) [171].

6.2 *Bone Regeneration Applications*

In the realm of tissue engineering, bone regeneration materials are being explored by researchers [172]. Electrospun materials occupy a notable place. Electrospun scaffolds offer the chance of mimicking the osteogenic processes and environment [173, 174]. In addition to the properties of chitosan, the cationic nature of the polymer accelerates tissue regeneration [105, 130, 175]. Frohbergh et al. conducted a study aiming to develop a crosslinked electrospun scaffold with chitosan for bone regeneration material. In cell culture studies with 7F2 cells (osteoblast-like cells), crosslinked nanofibrous scaffolds behaved as a bone extracellular matrix that enhanced osteo-induction. The system was proposed as a regenerative material for maxillofacial injuries [176]. Simvastatin has been studied for osteogenic applications in electrospun chitosan nanofibers. It is shown that chitosan-based electrospun membranes can provide a valuable platform for hydrophobic drugs such as simvastatin for osteogenic indications such as bone regeneration [177]. Zhu et al. designed an electrospun PCL/chitosan membrane loaded with metformin. The crosslinked membranes showed suitable cell adhesion and proliferation properties. Metformin presented a 23-day release after a burst release. It is shown that metformin-loaded chitosan-based electrospun membranes may be used as a bone regeneration material [110].

6.3 Hemostatic Applications for Hemorrhage Control

Chitosan nanofibers are promising in settings of bleeding and blood coagulation diseases. Chitosan nanofibers are used to produce hemostatic dressings [167]. They possess the potential to form cationic clusters, which interact with anions found on erythrocytes, allowing termination or decrease of hemorrhage [167, 178]. The hemostatic effect accelerates platelet and erythrocyte aggregation [167, 179]. These mechanisms occur due to the external factor (chitosan nanofibers) and are not dependent on the functionality of the physiologic internal clotting system [167]. Electrospun chitosan-based DDS showed enhanced hematologic coagulation speed and optimized tensile attributes [167]. Sasmal and workers fabricated tranexamic acid-loaded chitosan–PVA electrospun nanofibers for hemorrhage management [156]. They noticed that chitosan content was associated with coagulation speed [156].

6.4 Thrombolytic and Anti-Thrombogenic Capabilities

Thrombosis, the process of blood clot formation, can result in obstructions in the vascular system, leading to lethality and morbidity (i.e., stroke, myocardial infarction, venous thromboembolism) [167, 180]. Traditional pharmacotherapy involves thrombolytics (e.g., urokinase, streptokinase). These conventional medications possess several disadvantages, such as unwanted ADRs, short half-life, and non-specificity [167]. Advanced DDS, utilizing nanotechnology, has the potential to bypass these hindrances. Electrospun nanofibers can encapsulate biomolecules, showing antithrombotic and anti-thrombogenic activities [167]. Chitosan derivatives, such as H-chitosan, were associated with anti-thrombogenic capabilities [167, 181].

6.5 Vascular and Neural Tissue Regeneration and Engineering

Conventional vascular implants are widely used in vascular diseases in clinical settings. These conventional implants pose the risks of infection, thrombogenicity, and immunogenicity. Vascular tissue engineering provides a new technology to overcome these risks associated with conventional approaches [167]. Recent research explores the utilization of vascular growth factors, electrospun biodegradable nanostructured scaffolds, and autologous cells in vascular tissue regeneration and repair [182–186]. For this purpose, nanoparticles synthesized from various materials (e.g., PLA, PCL, collagen, chitosan, or silk) have been explored to date [167]. Electrospun chitosan-based nanofibers were utilized as scaffolds in this regard [185]. Researchers reported that PLA/chitosan nanofibers were associated with more activated partial thromboplastin time (APTT) and prothrombin time (PT) versus PLA fibers [185]. SEM imaging revealed almost no erythrocyte attachments to chitosan-based coating, accompanied by heparin adjustment. Therefore, this fiber mesh has the potential for success as a vascular gasket [185]. Conventional therapies failed to provide solutions to permanent damage to the nervous system. Advanced electrospun-based nanotechnologies, including chitosan-based platforms, have potential for neural tissue regeneration and engineering [167, 187]. Jafarisavari et al. developed a nerve guidance conduit based on electrospun PCL and chitosan. Chitosan was chosen for its high biocompatibility and bioadhesion properties, which may positively affect nerve regeneration. To obtain tubular-shaped scaffolds, a cylindrical collector was used. The PCL solution and chitosan-PEO blend solutions were separately filled into two separate syringe pumps for electrospinning. PEO was used to overcome chitosan's low spinnability as it increases viscosity. Vitamin B12 and piracetam are also mixed with chitosan-PEO solution. The obtained tubular scaffold was crosslinked with glutaraldehyde, and then the internal lumen of the nerve guidance conduit was filled with hyaluronic acid (2%) and crosslinked. After all these steps, a well-designed DDS was successfully attained. The developed nanofibrous tubular conduits were evaluated in a Wistar rat model suffering from a sciatic nerve defect. After applying these scaffolds, results demonstrated that the tubular system may act as nerve tissue. It supported myelination, axon repair, and functional healing of sciatic nerve injury. Indeed, these platforms can be designed for specific goals such as modified drug delivery and tissue engineering [188].

6.6 Promising Use in Cosmeceutical Research

Due to its anti-aging properties, chitosan has potential applications in cosmetic products through electrospinning [167]. Chitosan is desirable in oral care, hair care, and skin care [189]. To provide effective cosmetic profits, there are potential combinations with functional components. When the feature of functional benefits of cosmetics is on the table, these products are called cosmeceuticals. A cosmeceutical should combine aesthetic benefit and also therapeutic behavior [190]. As a cosmeceutical material, Charoenying et al. fabricated arbutin-loaded electrospun chitosan/PVA nanofibers for dermal delivery. Arbutin acts as a skin whitening material. Highly loaded electrospun chitosan/PVA nanofibers were fabricated successfully with rapid release behavior [191]. Hu et al. prepared a facial mask containing collagen peptides with chitosan on an electrospun fiber film. The obtained masks presented good moisturizing performance compared to commercial masks [192]. It is a promising strategy to develop chitosan-based electrospun nanofibers as a drug delivery scheme for cosmeceutical products.

6.7 Macromolecule-Loaded Systems

Loading macromolecules (e.g., protein-based molecules) onto electrospun nanofibers is possible. Bozdag et al. and Nicknejad et al. loaded bovine serum albumin (BSA) onto chitosan-based nanofibers [130, 193]. Nicknejad et al. achieved controlled release as a potential DDS for proteins [193]. Roozbahani et al. designed chitosan/PCL-based electrospun nanofibers for BSA delivery. This study revealed the impact of chitosan concentration on protein release, with increased concentrations resulting in decreased initial release behavior [194]. As a macromolecule, insulin may be a good candidate for the electrospinning process. In one study, an electrospun-chitosan scaffold loaded with insulin was fabricated for buccal drug delivery. The manufactured products presented up to 16-fold permeability results compared to free insulin [195]. In another study, monolayered and multilayered chitosan-based nanofiber mats were prepared for oral insulin delivery. The in vitro release studies indicated pH-dependent release behavior due to the chitosan's nature. At pH 6.8 and 7.4, the system released more insulin than at pH 1.2 because of the lower interaction of insulin with chitosan in basic medium [196].

6.8 Other Local Applications

Chitosan-based electrospun systems have other applications intended for topical use. For example, Mendes et al. studied potential platforms for topical drug delivery based on electrospun chitosan blend with phospholipids [156]. This study used curcumin, vitamin B12, and diclofenac as model therapeutics. It has been observed that promising potential platforms have been developed that exhibit the desirable biocompatibility with L929 cells. The in vitro model drugs' release behaviors showed variabilities related to their solubility. The designed platforms provided the ability to manipulate drug release due to phospholipid content, which may be a strategy to build a transdermal DDS [197]. In another investigation, Sasmal et al. aimed to prepare a hemostatic patch containing tranexamic acid, which is an antifibrinolytic agent. The chitosan-based PVA blended electrospun mat loaded with tranexamic acid demonstrated that it is possible to design local patches with different release profiles of tranexamic acid, and as a result, different hemostasis times can be achieved [156]. Toshkova et al. prepared electrospun quaternized chitosan-based implants containing chitosan, poly(l-lactide-co-D,L-lactide) (coPLA), and doxorubicin [198]. These tablet shaped implants are obtained by nonwoven nanofibrous mats via hydraulic press compression. In order to determine the local antitumor activity of the implants, an in vivo study was conducted. The developed system presented the most significant antitumor activity in the local tumor area and after tumor removal surgery, while a reduction in metastases and recurrence events was demonstrated successfully [198]. Elsewhere, simvastatin-loaded electrospun chitosan/poly-cyclodextrin-based nanofiber-coated stents were fabricated [60]. Similarly, PEO in a mixture with chitosan (1:9; w/w) was utilized to overcome the limited spinnability of chitosan. The in vitro drug release investigations showed that coated stents were able to ensure prolonged release of simvastatin. This study showed that the potential superiority of nanofiber-coated stents compared to conventional coated stents is due to the high drug loading capacity of nanofibers [60]. Due to the mucoadhesive property of chitosan, researchers developed anaspirin-loaded, fast-dissolving oral film as a model drug [199]. Chitosan use ensures mucoadhesion to buccal mucosa to enhance the bioavailability of the drug. The chitosan/pullulan blend solution was used as an electrospinning solution. The fabricated films dissolved within 1 min [199]. Mirzaeei et al. prepared monolayered and multilayered electrospun nanofibrous ocular DDS [113]. Ofloxacin is used as an antibacterial fluoroquinolone to treat conjunctivitis. The nanofiber system provided excellent entrapment efficiency of ofloxacin, producing smaller ocular inserts. The fabricated inserts delayed drug release nearly 103 h, a promising finding that may allow for reduced dosing frequency. Local irritation tests on the rabbit eye showed good tolerability (i.e., negligible or no irritation) [113].

7 Challenges and Limitations of Electrospun Chitosan DDS

7.1 Solubility Challenges

Chitosan is insoluble in aqueous physiologic medium, influencing antibacterial efficacy and effectiveness in various drug delivery platforms [200, 201]. Due to its low solubility, chitosan shows low interaction with the wound area. It should have good swelling capability to achieve clinical efficacy. Solubility challenges with chitosan require chemically modified chitosan-based derivatives, which may involve complex processes, prolonging and complicating the production process [201]. To improve chitosan's solubility and antibacterial activity in physiological environments, researchers investigated chemical alterations targeting the amino and hydroxyl functional units found within chitosan [202]. Therefore, the interest in modified chitosan has been increasing since modified chitosan shows more promising properties under physiological conditions compared to unmodified chitosan. Frequent alterations include adding amino, hydroxyl, carboxyalkyl, and quaternary alkyl functional units [203–207]. To exemplify, quaternized chitosan, which permanently carries positive charges due to incorporating quaternary alkyl units, shows improved solubility and antibacterial efficacy under neutral and basic media [200, 208].

7.2 Mechanical Strength Issues

Electrospun chitosan nanofibers show relatively poor mechanical strength and durability, limiting their direct use in treatment techniques requiring load-bearing systems. It has low spinnability, requiring blending with PVA or other polymers. Therefore, extra supporting materials or crosslinking techniques are incorporated to enhance mechanical durability [133, 200]. Crosslinking is a process in which functional units within electrospun scaffolds are interconnected, which can be achieved by various strategies [133]. Crosslinking and reactive electrospinning (RES) are strategies employed to overcome the physicochemical and mechanical drawbacks of fabricated nanofibers, achieving a more biocompatible, stable product that is tailored to advanced and intelligent DDS [133]. The RES technique is a versatile strategy for producing functional fibers with numerous properties tailored to exact biomedical uses. It allows the fabrication of fibers with adjustable mechanical characteristics by applying variable crosslinking techniques and including different chemical or biochemical reactions during the spinning procedures. RES is applied throughout fiber generation (in situ) or after completion, gaining amplified structural integrity and enhanced durability. The concept of RES was established in 2005, by Kim et al. [209], a process that achieved products possessing mechanical and aqueous stability as a function of time [210]. Chemical reactive electrospinning (CRES) and photoreactive electrospinning (PRES) are types of RES [133]. During CRES, chemical reactions result in crosslinking. In PRES, photoinitiated crosslinking is applied [133]. Using these strategies aids in achieving durable products that are stable under bioenvironmental hardships (e.g., pH alterations, enzymes, and biomechanical stress).

7.3 Stability and Structural Integrity Issues

Electrospun chitosan nanofibers are vulnerable to degradation under physiological aqueous media, which may affect the integrity of the DDS and lead to potential loss of functionality [124, 200]. Strategies to overcome this challenge are warranted [124, 200]. According to the intended use, chitosan nanofibers must possess well-defined, predetermined specific morphology and volume [124]. Given the numerous experimental variables (i.e., solution viscosity and concentration, solvent evaporation rate, electrospinning processing factors (i.e., voltage, flow rate)), accomplishing uniform and controllable fiber morphology and structure may be highly challenging [124]. To assure uniformity and structural integrity, these variables must be stringently controlled. Moreover, mechanical and environmental durability should be achieved to assure that no degradation occurs, threatening structural integrity. Chitosan nanofibers may be vulnerable to moisture or other environmental conditions, causing degradation, limiting shelf-life, and functionality. Moisture and environmental resistance of chitosan nanofibers can be amplified through using strategies such as crosslinking methods, encapsulation approaches, or surface modifications [211]. Focusing on the functionalization of chitosan nanofibers is not easy. Functionalizing involves incorporating functional additives of chemical groups to improve activity and biocompatibility [124]. While doing so, morphological and structural stability should be preserved without compromise, since such processes may alter surface aspects of chitosan nanofibers [212].

7.4 Safety Concerns and Potential Toxicities

Despite chitosan being biocompatible, adding other materials during fabrication or drugs can bring potential toxicity concerns. Electrospun chitosan DDS shows altered degrees of potential cytotoxicity, related to composition, type of chemical modifications, procedure variables, etc. Cautious design and optimization are warranted to prevent potential toxicities while achieving positive clinical outcomes with electrospun chitosan DDS. To exemplify, the safety and toxicity profile of loaded drugs as well as by-products of chemical reactions during the fabrication process, needs an in-depth assessment [213, 214]. Chemical alterations can alter the cytotoxicity properties. Increasing the degree of trimethylation on chitosan derivatives generally amplified cytotoxicity. Also, polymeric chitosan derivatives exhibited more toxicity compared to oligomeric derivatives at similar trimethylation levels [215]. Contrastingly, crosslinking chitosan nanofibers with genipin enhanced the mechanical aspects and stability, allowing sustained drug release with no cytotoxicity in fibroblasts [121]. Potential toxicity of nanofibers remains a dark side, causing fears and hurdles towards application [124]. Maximized surface area and minimized sizes can result in ADRs and cellular toxicity [216]. Also, surface charge characteristics and distribution, cross reactivity, and components used during fabrication impact safety profiles [217]. Despite the numerous in vitro and in vivo reports suggesting toxicity concerns linked to chitosan-based nanofibers, the exact mechanisms of toxicity are not fully revealed [124, 218]. Reactive oxygen species may induce oxidative stress, forcing degenerative processes at the cellular level [219]. Other mechanisms may include inhibition or interference with stages of the cell cycle that are mediated by alterations to gene expression [220]. These toxicity concerns lead researchers to propose methods to overcome or minimize these toxicities. One of the strategies to combat these risks is to incorporate biodegradable polymers within the formulations. For example, chitosan/polyhydroxybutyrate was used for ulcers after surgeries [221]. Elsewhere, chitosan-carboxylated cellulose nanofiber sponge was used for wound care [222]. Surface coating with biocompatible elements is another strategy to improve safety. Fine-tuning surface attributes such as charge, density and hydrophobicity can minimize toxicity risks [124]. Indeed, more research is needed to discover more strategies to tackle this issue and understand the underlying mechanisms associated with toxicities.

7.5 *Complex Fabrication Processes: Scaling-up Production and Cost-Effectiveness*

The manufacturing protocol requires stringently specific conditions and alterations to improve chitosan's attributes, which adds complexity. This also encompasses the need for post-modification processing to gain target fiber morphology and functionality [223]. There are multiple parameters and factors (i.e., polymer solution intrinsic attributes and electrospinning properties) involved in electrospun chitosan DDS [124, 133]. Since each parameter setting is specific to the respective DDS and not generalizable to all DDS, scaling up production, ensuring batch-to-batch uniformity becomes very challenging. Usually, production costs are relatively high due to the necessity of adding extra materials during design/fabrication or the requirements of extensive modification procedures [213]. These factors limit commercial survival, increase costs, and limit translation from bench to bedside. Process optimization is required before scale-up production. There is a risk that laboratory results may not translate to the same outcomes upon large-scale production. To minimize this risk, to make sure that the fiber quality is consistent between and within batches, all relevant design parameters (e.g., chitosan composition and concentration, voltage, flow rate, viscosity, and evaporation rate) must be stringently controlled, defined, and optimized before moving to the production stage [224]. Production requires special infrastructure and equipment preparations. A continuous supply chain of chitosan and all needed raw materials needs to be guaranteed to prevent disruptions, by designing efficient renewable sourcing technologies with low costs. The infrastructure of the producing facility needs to be assessed and modified accordingly. Cost-efficiency may be an issue given the increasing material costs, large equipment investments, and amplified energy utilization requiring special power supply set-ups.

Strong clinical rationale and justifications are needed to support the increased costs associated with chitosan DDS to assure market survival and demand.

7.6 Achieving Predefined Drug Release Profiles

Designing, controlling, and sustaining predefined drug release behavior from electrospun chitosan DDS can be problematic. A significant concern is burst release, resulting in suboptimal clinical outcomes while exposing the patient to ADRs [213, 225]. Optimizing drug release kinetics to guarantee extended and controlled release remains a major challenge [225, 226]. Difficulties controlling drug release from chitosan-based inserts related to drug collection on surfaces of the DDS or the presence of uncontrolled bursts seen in the drug release profiles have been reported [226–229]. In one of the fabricated formulations, triamcinolone acetonide electrospun nanofibers exhibited burst release (around 30% within 60 mins), failing to achieve optimum controlled delivery [226]. Therefore, formulation optimization parameters must be maintained cautiously to attain targeted drug release behaviors.

7.7 Engineering Mucoadhesive Properties

Mucoadhesive chitosan-based electrospun nanofibers present a platform for wide therapeutic sublingual uses [112]. Although chitosan shows mucoadhesive properties, deacetylation levels do not significantly help configure this property, limiting the ability to fine-tune adhesion levels and mechanisms [112]. This may impact the effectiveness of drug delivery for mucosal applications. Stie and workers explored the mucoadhesive attributes of chitosan/PEO nanofibers in terms of the degree of deacetylation (53–96%) of chitosan [112]. They reported that increased deacetylation led to enhanced aqueous morphological stability but did not impact mucoadhesiveness characteristics of chitosan/PEO nanofibers. Decreased deacetylation was associated with destroyed fibrous morphology completely or partially in aqueous media. This work reported that mucoadhesiveness is independent of deacetylation levels and is mainly attributed to swelling mechanisms [112].

8 Future Perspectives

Electrospun chitosan-based DDS are attracting increasing interest among clinicians, biomedical engineers, and pharmaceutical researchers. The associated unique properties pave the way to promising therapeutic and biomedical applications of high potential. The relatively high encapsulation efficiency and controlled release capabilities make them a perfect platform for targeted drug delivery. This covers a gap that could not be solved with conventional pharmaceuticals. Indeed, this allows delivering sensitive or high-risk therapeutics (e.g., antibiotics, chemotherapy, macromolecules) while protecting from ADRs and degradation risks within physiologic environments. These DDS have versatility in drug loading features, meaning that they are highly compatible with a wide platform of therapeutics, such as antimicrobials, chemotherapeutic agents, proteins, and anti-inflammatory agents. This versatility maximizes utility in many medical treatments. Under these systems, future research will be geared to develop stimuli-responsive systems. Stimuli-responsive chitosan nanofibers will help control therapeutic agent release according to exposure to predefined triggers (e.g., pH changes, temperature, or light). This will improve the controllability of drug release, maximize efficacy, and minimize toxicity. Technological advancements are expected to focus on advanced electrospinning methods (e.g., coaxial electrospinning, emulsion electrospinning), while incorporating green manufacturing strategies. These advanced electrospinning approaches facilitate the creation of core-shell nanofibers, which allow core encapsulation of drugs, functioning as a protective barrier while controlling release. Future directions will also pay special attention to ensure that all methods are environmentally friendly. Green fabrication methods like green electrospinning are anticipated to emerge. These approaches fulfill the rising need for sustainable and biocompatible DDS.

Electrospun chitosan DDS are expected to excel in various biomedical areas, especially in wound healing, tissue engineering, tissue regeneration, and individualized targeted drug therapy in oncology. Chitosan nanofibers are promising because of their biocompatibility, biodegradability, and antibacterial activity. Their applications include scaffolds that facilitate tissue regeneration as well as drug-eluting dressings that promote wound repair. In oncology, chitosan nanofibers are very promising by paving the pathway to target chemotherapy directly to tumor tissues, bypassing unwanted systemic toxicity. The primary significant challenges with these DDS are hardships regarding translation from bench to bedside, costs, scalability, and commercialization. Future efforts need to focus on improving cost-efficiency by optimizing production procedures to gain more cost-effective manufacturing. Also, more studies exploring ways to enhance functionalization of chitosan nanofibers are needed. Chemical alterations can improve unwanted properties and amplify utility areas.

9 Conclusion

Electrospun chitosan DDS carry great fortune to the next era of therapeutics and biomedical applications. To date, they have been successfully applied to a wide range of biomedical and therapeutic applications due to their unique advantages and biocompatible characteristics. Ongoing studies and technological advancements are anticipated to successfully bypass the present challenges, leading to more efficient, effective, and versatile drug delivery strategies. Integration of stimuli-responsive properties and green fabrication techniques is anticipated to further improve their potential, underscoring their importance in the new-generation therapeutic platforms. Ongoing studies and technological advancements are expected to effectively overcome current challenges and increase efficiency. Integration of stimuli-responsive properties and green manufacturing techniques is expected to further enhance their potential, underscoring their importance in the new generation of therapeutic platforms.

Declarations of Interest None.

References

1. Li, C., Wang, J., Wang, Y., Gao, H., Wei, G., Huang, Y., et al. (2019). Recent progress in drug delivery. *Acta Pharmaceutica Sinica B, 9*(6), 1145–1162.
2. Raj, R. M., & Raj, V. (2023). Polymers in drug delivery and targeting. In *Handbook of polymers in medicine* (pp. 595–634).
3. Bassyouni, F., Elhalwany, N., Abdel Rehim, M., & Neyfeh, M. (2015). Advances and new technologies applied in controlled drug delivery system. *Research on Chemical Intermediates, 41*(4), 2165–2200.
4. Bagade, O., & Sampathi, S. (2024). Restoration and sustenance of nano drug delivery systems: Potential, challenges, and limitations. In *Biosystems, biomedical & drug delivery systems characterization, restoration and optimization* (pp. 105–139).
5. Agnihotri, T. G., Jain, R., Naga Jothi Prasath, V. R., Jadhav, P. V., Gomte, S. S., & Jain, A. (2023). Protein and peptide delivery through chitin, chitosan, and starch. In *Peptide and protein drug delivery using polysaccharides* (pp. 169–195).
6. Yadav, N., & Sangwan, N. (2022). Biotechnology approaches in developing novel drug-delivery systems. In *Biotechnology in healthcare, volume 1: Technologies and innovations* (pp. 125–146).
7. Hassounah, I. A., Shehata, N. A., Kimsawatde, G. C., Hudson, A. G., Sriranganathan, N., Joseph, E. G., et al. (2016). Designing and testing single tablet for tuberculosis treatment through electrospinning. In *Fabrication and self-assembly of nanobiomaterials applications of nanobiomaterials Volume 1* (pp. 335–365).
8. Arafat, M. (2015). Approaches to achieve an oral controlled release drug delivery system using polymers: A recent review. *International Journal of Pharmacy and Pharmaceutical Sciences, 7*(7), 16–21.
9. Abraham, T. R., Jose, A. D., Parveena, A., & Thomas, S. (2024). Cubosomes – excellent pharmaceutical carriers. In *Nanostructured materials for biomedical applications* (pp. 161–177).
10. Verma, A. K. (2023). An insight into collagen-based nano biomaterials for drug delivery applications. In *Engineering materials* (pp. 397–428).
11. Sivakumar, M., Muthu, Y., & Elumalai, K. (2024). Advancements in drug delivery systems: A focus on microsphere-based targeted delivery. *Biomedical Materials & Devices*, 1–28. https://doi.org/10.1007/s44174-024-00245-6
12. Ji, D., Lin, Y., Guo, X., Ramasubramanian, B., Wang, R., Radacsi, N., et al. (2024). Electrospinning of nanofibres. *Nature Reviews Methods Primers., 4*(1), 1.
13. Shepa, I., Mudra, E., & Dusza, J. (2021). Electrospinning through the prism of time. *Materials Today Chemistry, 21*, 100543.
14. Garg, K., Sell, S. A., & Bowlin, G. L. (2009). Electrospinning and its influence on the structure of polymeric nanofibers. In *Handbook of textile fibre structure* (pp. 460–483).

15. Lukáš, D., Sarkar, A., Martinová, L., Vodsed'álková, K., Lubasová, D., Chaloupek, J., et al. (2009). Physical principles of electrospinning (electrospinning as a nano-scale technology of the twenty-first century). *Textile Progress., 41*(2), 59–140.
16. Guarino, V., Altobelli, R., Cirillo, V., Cummaro, A., & Ambrosio, L. (2015). Additive electrospraying: A route to process electrospun scaffolds for controlled molecular release. *Polymers for Advanced Technologies, 26*(12), 1359–1369.
17. Torres-Martínez, E. J., Bravo, J. M. C., Medina, A. S., González, G. L. P., & Gómez, L. J. V. (2018). A summary of electrospun nanofibers as drug delivery system: Drugs loaded and biopolymers used as matrices. *Current Drug Delivery, 15*(10), 1360–1374.
18. Torres-Martínez, E. J., Pérez-González, G. L., Serrano-Medina, A., Grande, D., Vera-Graziano, R., Cornejo-Bravo, J. M., et al. (2019). Drugs loaded into electrospun polymeric nanofibers for delivery. *Journal of Pharmacy and Pharmaceutical Sciences, 22*(1), 313–331.
19. Yang, Y., Zhang, R., Liang, Z., Guo, J., Chen, B., Zhou, S., et al. (2024). Application of electrospun drug-loaded nanofibers in cancer therapy. *Polymers, 16*(4), 504.
20. Suner, S. C., Oral, A., & Yildirim, Y. (2024). Design of Poly(lactic) acid/gelatin core-shell bicomponent systems as a potential wound dressing material. *Journal of the Mechanical Behavior of Biomedical Materials, 150*, 106255.
21. Ma, L., Shi, X., Zhang, X., & Li, L. (2019). Preparation of the controllable core-shell structured electrospun polymer fibers and their application. *Progress in Chemistry, 31*(9), 1213–1220.
22. Li, Z., & Wang, C. (2013). Applications of electrospun nanofibers. In *SpringerBriefs in materials* (pp. 75–139).
23. Srivastava, R. K. (2017). Electrospinning of patterned and 3D nanofibers. In *Electrospun Nanofibers* (pp. 399–447).
24. Zare, M., Dziemidowicz, K., Williams, G. R., & Ramakrishna, S. (2021). Encapsulation of pharmaceutical and nutraceutical active ingredients using electrospinning processes. *Nanomaterials, 11*(8), 1968.
25. Karuppannan, S. K., Mani, S., Bushion, J., Dowlath, M. J. H., & Arunachalam, K. D. (2023). Electrospun nanofibers in drug delivery. In *Advances in novel formulations for drug delivery* (pp. 263–278).
26. Ghosal, K., Augustine, R., Zaszczynska, A., Barman, M., Jain, A., Hasan, A., et al. (2021). Novel drug delivery systems based on triaxial electrospinning based nanofibers. *Reactive and Functional Polymers, 163*, 104895.
27. Malik, S., Sundarrajan, S., Hussain, T., Nazir, A., Berto, F., & Ramakrishna, S. (2021). Electrospun biomimetic polymer nanofibers as vascular grafts. *Material Design and Processing Communications., 3*(6), e203.
28. Rajput, M., Tyeb, S., & Chatterjee, K. (2023). Techniques to fabricate electrospun nanofibers for controlled release of drugs and biomolecules. In *Advances in polymer science* (Vol. 291, pp. 37–68). Springer.
29. Mao, Y., Shen, W., Wu, S., Ge, X., Ao, F., Ning, Y., et al. (2023). Electrospun polymers: Using devices to enhance their potential for biomedical applications. *Reactive and Functional Polymers, 186*, 105568.
30. Uyar, T., & Kny, E. (2017). *Electrospun materials for tissue engineering and biomedical applications: Research, design and commercialization* (pp. 1–428).
31. Cheng, H., Yang, X., Che, X., Yang, M., & Zhai, G. (2018). Biomedical application and controlled drug release of electrospun fibrous materials. *Materials Science and Engineering C, 90*, 750–763.
32. Mohammadinejad, R., Madamsetty, V. S., Kumar, A., Varzandeh, M., Dehshahri, A., Zarrabi, A., et al. (2021). Electrospun nanocarriers for delivering natural products for cancer therapy. *Trends in Food Science and Technology, 118*, 887–904.
33. Chatterjee, R., Maity, M., Hasnain, M. S., & Nayak, A. K. (2021). Chitosan: Source, chemistry, and properties. In *Chitosan in drug delivery* (pp. 1–22).

34. Martirosova, E. I., Grebenkina, N. A., & Plashchina, I. G. (2015). A study on the effect of alkylresorcinols (AR) - methylresorcinol (MR) and hexylresorcinol (HR) on the enzymatic activity of two hydrolases: Glycosidases (lysozyme) and amidohydrolase (papain) in the reaction of chitosan hydrolysis. In *Chemical technology: Key developments in applied chemistry, biochemistry and materials science* (pp. 119–128).
35. Martirosova, E. I., Grebenkina, N. A., & Plashchina, I. G. (2015). Effect of alkylresorcinols on the chitolytic activity of lysozyme and papain. In *Life chemistry research: Biological systems* (pp. 275–283).
36. Desai, R., Pachpore, R., Patil, A., Jain, R., & Dandekar, P. (2021). Review of the structure of chitosan in the context of other sugar-based polymers. *Advances in Polymer Science, 287*, 23–74.
37. Elamri, A., Zdiri, K., Hamdaoui, M., & Harzallah, O. (2023). Chitosan: A biopolymer for textile processes and products. *Textile Research Journal, 93*(5–6), 1456–1484.
38. Gomes, L. P., Paschoalin, V. M. F., & Del Aguila, E. M. (2017). Chitosan nanoparticles: Production, physicochemical characteristics and nutraceutical applications. *Revista Virtual de Quimica, 9*(1), 387–409.
39. Lizardi-Mendoza, J., Argüelles Monal, W. M., & Goycoolea Valencia, F. M. (2016). Chemical characteristics and functional properties of chitosan. In *Chitosan in the preservation of agricultural commodities* (pp. 3–31).
40. Mahmood, A., Maher, N., Amin, F., Alqutaibi, A. Y., Kumar, N., & Zafar, M. S. (2024). Chitosan-based materials for dental implantology: A comprehensive review. *International Journal of Biological Macromolecules, 268*, 131823.
41. Espinosa-Cavazos, K. G., Sáenz-Galindo, A., & Castañeda-Facio, A. O. (2020). Chitosan films properties and aplications. *Afinidad, 77*(591), 203–208.
42. Santos, V. P., Marques, N. S. S., Maia, P. C. S. V., de Lima, M. A. B., Franco, L. O., & de Campos-Takaki, G. M. (2020). Seafood waste as attractive source of chitin and chitosan production and their applications. *International Journal of Molecular Sciences, 21*(12), 1–17.
43. Antonino, R. S. C. M. D. Q., Fook, B. R. P. L., Lima, V. A. D. O., Rached, R. I. D. F., Lima, E. P. N., Lima, R. J. D. S., et al. (2017). Preparation and characterization of chitosan obtained from shells of shrimp (Litopenaeus vannamei Boone). *Marine Drugs., 15*(5), 141.
44. Minakov, D. V., Egorova, E. Y., Markin, V. I., & Bazarnova, N. G. (2024). Prospects for modifying the structure of chitin and chitosan of higher mushroom to expand their application potential (areview). *Russian Journal of Bioorganic Chemistry, 50*(7), 2711–2729.
45. Arfin, T. (2017). Chitosan and its derivatives: Overview of commercial applications in diverse fields. In *Chitosan: Derivatives, composites and applications* (pp. 115–149).
46. Jiang, T., James, R., Kumbar, S. G., & Laurencin, C. T. (2014). Chitosan as a biomaterial: Structure, properties, and applications in tissue engineering and drug delivery. In *Natural and synthetic biomedical polymers* (pp. 91–113).
47. Hafizi, T., Shahriari, M. H., Abdouss, M., & Kahdestani, S. A. (2023). Synthesis and characterization of vancomycin-loaded chitosan nanoparticles for drug delivery. *Polymer Bulletin, 80*(5), 5607–5621.
48. Riaz Rajoka, M. S., Mehwish, H. M., Wu, Y., Zhao, L., Arfat, Y., Majeed, K., et al. (2020). Chitin/chitosan derivatives and their interactions with microorganisms: A comprehensive review and future perspectives. *Critical Reviews in Biotechnology, 40*(3), 365–379.
49. Mohd Affandy, M. A., & Rovina, K. (2024). Characterization of chitosan derived from mushroom sources: Physicochemical, morphological, thermal analysis. *Sustainable Chemistry and Pharmacy, 40*, 101624.
50. Tayebi-Khorrami, V., Rahmanian-Devin, P., Fadaei, M. R., Movaffagh, J., & Askari, V. R. (2024). Advanced applications of smart electrospun nanofibers in cancer therapy: With insight into material capabilities and electrospinning parameters. *International Journal of Pharmaceutics, 8*, 100265.

51. Lian, S., Lamprou, D., & Zhao, M. (2024). Electrospinning technologies for the delivery of biopharmaceuticals: Current status and future trends. *International Journal of Pharmaceutics, 651*, 123641.
52. Xu, L., Li, W., Sadeghi-Soureh, S., Amirsaadat, S., Pourpirali, R., & Alijani, S. (2022). Dual drug release mechanisms through mesoporous silica nanoparticle/electrospun nanofiber for enhanced anticancer efficiency of curcumin. *Journal of Biomedical Materials Research Part A, 110*(2), 316–330.
53. Gulsun, T., Inal, M., Akdag, Y., Izat, N., Oner, L., & Sahin, S. (2022). The development and characterization of electrospun gelatin nanofibers containing indomethacin and curcumin for accelerated wound healing. *Journal of Drug Delivery Science and Technology, 67*, 103000.
54. Kamali, H., Farzadnia, P., Movaffagh, J., & Abbaspour, M. (2022). Optimization of curcumin nanofibers as fast dissolving oral films prepared by emulsion electrospinning via central composite design. *Journal of Drug Delivery Science and Technology, 75*, 103714.
55. Taylor, G. (1969). Electrically driven jets. *Proceedings of the Royal Society of London, 313*(1515), 453–475.
56. Taylor, G. I. (1964). Disintegration of water drops in an electric field. *Proceedings of the Royal Society of London Series A Mathematical and Physical Sciences, 280*(1382), 383–397.
57. Muratoglu, S., Inal, M., Gulsun, T., & Sahin, S. (2024). Development and characterization of orodispersible films containing amlodipine besylate and rosuvastatin calcium based on electrospun fibers. *Journal of Drug Delivery Science and Technology, 101*, 106225.
58. Samraj, S. M. D., & Vadodaria, K. (2022). Development and optimization of gelatin nanofibrous scaffold for tissue engineering applications. *Journal of Drug Delivery Science and Technology., 71*, 103283.
59. Tonglairoum, P., Ngawhirunpat, T., Rojanarata, T., Panomsuk, S., Kaomongkolgit, R., & Opanasopit, P. (2015). Fabrication of mucoadhesive chitosan coated polyvinylpyrrolidone/cyclodextrin/clotrimazole sandwich patches for oral candidiasis. *Carbohydrate Polymers, 132*, 173–179.
60. Kersani, D., Mougin, J., Lopez, M., Degoutin, S., Tabary, N., Cazaux, F., et al. (2020). Stent coating by electrospinning with chitosan/poly-cyclodextrin based nanofibers loaded with simvastatin for restenosis prevention. *European Journal of Pharmaceutics and Biopharmaceutics, 150*, 156–167.
61. Wang, M., Li, D., Li, J., Li, S., Chen, Z., Yu, D.-G., et al. (2020). Electrospun Janus zein–PVP nanofibers provide a two-stage controlled release of poorly water-soluble drugs. *Materials & Design, 196*, 109075.
62. Najim, M. A., Khalil, B. I., & Hameed, A. A. (2022). Characterizing optimum electrospinning conditions for graft urethanized poly(vinyl alcohol)(U-PVA). *Heliyon, 8*(11), e11423.
63. Asci, H., Savran, M., Cengiz Callıoglu, F., Sahin, S., Hasseyid, N., Kaynak, M., et al. (2022). Supralingual administration of paracetamol embedded in polyvinyl alcohol nanofibers: A pharmacokinetic study. *Journal of Drug Delivery Science and Technology, 67*, 102948.
64. Gültekin, Y., Korkmaz, Ç., Öztürk, N., Filazi, A., Deniz, A., Doğan, O., et al. (2024). Preparation and evaluation of fast-dissolving albendazole sulfoxide and praziquantel-loaded polyvinylpyrrolidone nanofiber films by electrospinning. *Journal of Drug Delivery Science and Technology, 92*, 105304.
65. Farzaei, M. H., Derayat, P., Pourmanouchehri, Z., Kahrarian, M., Samimi, Z., Hajialyani, M., et al. (2023). Characterization and evaluation of antibacterial and wound healing activity of naringenin-loaded polyethylene glycol/polycaprolactone electrospun nanofibers. *Journal of Drug Delivery Science and Technology, 81*, 104182.
66. Alimohammadi, M., Fakhraei, O., Moradi, A., Kabiri, M., Moradi, A., Passandideh-Fard, M., et al. (2022). Controlled release of azithromycin from polycaprolactone/chitosan nanofibrous membranes. *Journal of Drug Delivery Science and Technology, 71*, 103246.
67. Karthikeyan, K., Guhathakarta, S., Rajaram, R., & Korrapati, P. S. (2012). Electrospun zein/eudragit nanofibers based dual drug delivery system for the simultaneous delivery of aceclofenac and pantoprazole. *International Journal of Pharmceutics, 438*(1–2), 117–122.

68. Najafiasl, M., Osfouri, S., Azin, R., & Zaeri, S. (2020). Alginate-based electrospun core/shell nanofibers containing dexpanthenol: A good candidate for wound dressing. *Journal of Drug Delivery Science and Technology, 57*, 101708.
69. Kerwald, J., de Moura Junior, C. F., Freitas, E. D., de Moraes Segundo, J. D. D. P., Vieira, R. S., & Beppu, M. M. (2022). Cellulose-based electrospun nanofibers: A review. *Cellulose, 29*(1), 25–54.
70. Muratoglu, S., Inal, M., Akdag, Y., Gulsun, T., & Sahin, S. (2024). Electrospun nanofiber drug delivery systems and recent applications: An overview. *Journal of Drug Delivery Science and Technology, 92*, 105342.
71. Li, Y., Zhu, J., Cheng, H., Li, G., Cho, H., Jiang, M., et al. (2021). Developments of advanced electrospinning techniques: A critical review. *Advanced Materials Technologies., 6*(11), 2100410.
72. Amariei, N., Manea, L., Bertea, A., Bertea, A., & Popa, A. (Eds.). (2017). *The influence of polymer solution on the properties of electrospun 3D nanostructures*. IOP Conference Series: Materials Science and Engineering. IOP Publishing.
73. Huan, S. Q., Liu, G. X., Han, G. P., Cheng, W. L., Fu, Z. Y., Wu, Q. L., et al. (2015). Effect of experimental parameters on morphological, mechanical and hydrophobic properties of electrospun polystyrene fibers. *Materials, 8*(5), 2718–2734.
74. Gudkova, V., Krumme, A., Martson, T., Rikko, M., Tarassova, E., & Viirsalu, M. (2014). The impact of 1-butyl-3-methylimidazolium chloride on electrospinning process of SAN polymer solutions and electrospun fiber morphology. *Journal of Electrostatics, 72*(6), 433–436.
75. Zhang, Y. P., Deng, L. L., Zhong, H., Pan, J. J., Li, Y., & Zhang, H. (2020). Superior water stability and antimicrobial activity of electrospun gluten nanofibrous films incorporated with glycerol monolaurate. *Food Hydrocolloid, 109*, 106116.
76. Bhardwaj, N., & Kundu, S. C. (2010). Electrospinning: A fascinating fiber fabrication technique. *Biotechnology Advances, 28*(3), 325–347.
77. Deitzel, J. M., Kleinmeyer, J., Harris, D., & Tan, N. C. B. (2001). The effect of processing variables on the morphology of electrospun nanofibers and textiles. *Polymer, 42*(1), 261–272.
78. Dorati, R., Chiesa, E., Pisani, S., Genta, I., Modena, T., Bruni, G., et al. (2020). The effect of process parameters on alignment of tubular electrospun nanofibers for tissue regeneration purposes. *Journal of Drug Delivery Science and Technology, 58*, 101781.
79. Pisani, S., Genta, I., Dorati, R., Modena, T., Chiesa, E., Bruni, G., et al. (2022). A design of experiment (DOE) approach to correlate PLA-PCL electrospun fibers diameter and mechanical properties for soft tissue regeneration purposes. *Journal of Drug Delivery Science and Technology, 68*, 103060.
80. Khan, I., Pandit, J., Ahmed, S., Zameer, S., Nikita, Ahmad, S., et al. (2022). Development and evaluation of biodegradable polymeric lomustine nanofibres for the efficient tumor targeting: In vitro characterization, ex vivo permeation and degradation study. *Journal of Drug Delivery Science and Technology., 75*, 103685.
81. Pelipenko, J., Kristl, J., Jankovic, B., Baumgartner, S., & Kocbek, P. (2013). The impact of relative humidity during electrospinning on the morphology and mechanical properties of nanofibers. *International Journal of Pharmaceutics, 456*(1), 125–134.
82. Kalous, T., Holec, P., Jirkovec, R., Batka, O., Zabka, P., Pokorny, P., et al. (2023). The effect of frequency change on the alternating current electrospinning of polyamide 6 and its productivity. *Journal of Environmental Chemical Engineering, 11*(2), 109543.
83. Holec, P., Jirkovec, R., Kalous, T., Baťka, O., Brožek, J., & Chvojka, J. (2022). The potential for the direct and alternating current-driven electrospinning of polyamides. *Nanomaterials, 12*(4), 665.
84. Nieminen, H. J., Laidmae, I., Salmi, A., Rauhala, T., Paulin, T., Heinamaki, J., et al. (2018). Ultrasound-enhanced electrospinning. *Scientific Reports, 8*(1), 4437.
85. Alfaro De Prá, M. A., Ribeiro-do-Valle, R. M., Maraschin, M., & Veleirinho, B. (2017). Effect of collector design on the morphological properties of polycaprolactone electrospun fibers. *Materials Letters, 193*, 154–157.

86. Lee, C. H., Hsieh, M. J., Liu, S. C., Chen, J. K., Liu, S. J., Hsieh, I. C., et al. (2018). Novel bifurcation stents coated with bioabsorbable nanofibers with extended and controlled release of rosuvastatin and paclitaxel. *Materials Science and Engineering: C, 88*, 61–69.
87. Davani, F., Alishahi, M., Sabzi, M., Khorram, M., Arastehfar, A., & Zomorodian, K. (2021). Dual drug delivery of vancomycin and imipenem/cilastatin by coaxial nanofibers for treatment of diabetic foot ulcer infections. *Materials Science and Engineering: C, 123*, 111975.
88. Repanas, A., & Glasmacher, B. (2015). Dipyridamole embedded in Polycaprolactone fibers prepared by coaxial electrospinning as a novel drug delivery system. *Journal of Drug Delivery Science and Technology, 29*, 132–142.
89. Yu, D. G., Li, X. Y., Wang, X., Yang, J. H., Bligh, S. W. A., & Williams, G. R. (2015). Nanofibers fabricated using triaxial electrospinning as zero order drug delivery systems. *ACS Applied Materials and Interfaces, 7*(33), 18891–18897.
90. Viry, L., Moulton, S. E., Romeo, T., Suhr, C., Mawad, D., Cook, M., et al. (2012). Emulsion-coaxial electrospinning: Designing novel architectures for sustained release of highly soluble low molecular weight drugs. *Journal of Materials Chemistry, 22*(22), 11347–11353.
91. Yu, D. G., Yang, C., Jin, M., Williams, G. R., Zou, H., Wang, X., et al. (2016). Medicated Janus fibers fabricated using a teflon-coated side-by-side spinneret. *Colloids and Surfaces B: Biointerfaces, 138*, 110–116.
92. Zheng, X., Kang, S., Wang, K., Yang, Y., Yu, D. G., Wan, F., et al. (2021). Combination of structure-performance and shape-performance relationships for better biphasic release in electrospun Janus fibers. *International Journal of Pharmaceutics, 596*, 120203.
93. Homayoni, H., Ravandi, S. A. H., & Valizadeh, M. (2009). Electrospinning of chitosan nanofibers: Processing optimization. *Carbohydrate Polymers, 77*(3), 656–661.
94. Sedghi, R., Shaabani, A., Mohammadi, Z., Samadi, F. Y., & Isaei, E. (2017). Biocompatible electrospinning chitosan nanofibers: A novel delivery system with superior local cancer therapy. *Carbohydrate Polymers, 159*, 1–10.
95. Bhattarai, N., Edmondson, D., Veiseh, O., Matsen, F. A., & Zhang, M. (2005). Electrospun chitosan-based nanofibers and their cellular compatibility. *Biomaterials, 26*(31), 6176–6184.
96. Ohkawa, K., Cha, D., Kim, H., Nishida, A., & Yamamoto, H. (2004). Electrospinning of chitosan. *Macromolecular Rapid Communications, 25*(18), 1600–1605.
97. Zhang, Y., Huang, X., Duan, B., Wu, L., Li, S., & Yuan, X. (2007). Preparation of electrospun chitosan/poly (vinyl alcohol) membranes. *Colloid and Polymer Science, 285*, 855–863.
98. Adeli, H., Khorasani, M. T., & Parvazinia, M. (2019). Wound dressing based on electrospun PVA/chitosan/starch nanofibrous mats: Fabrication, antibacterial and cytocompatibility evaluation and in vitro healing assay. *International Journal of Biological Macromolecules, 122*, 238–254.
99. Talebian, A., & Mansourian, A. (2017). Release of vancomycin from electrospun gelatin/chitosan nanofibers. *Materials Today: Proceedings, 4*, 7065.
100. Ulker Turan, C., & Guvenilir, Y. (2022). Electrospun poly(ω-pentadecalactone-co-ε-caprolactone)/gelatin/chitosan ternary nanofibers with antibacterial activity for treatment of skin infections. *European Journal of Pharmaceutical Sciences, 170*, 106113.
101. Tonglairoum, P., Ngawhirunpat, T., Rojanarata, T., Kaomongkolgit, R., & Opanasopit, P. (2014). Fast-acting clotrimazole composited PVP/HPβCD nanofibers for oral candidiasis application. *Pharmaceutical Research, 31*, 1893–1906.
102. Thien, D. V. H. (2016). Electrospun chitosan/PVA nanofibers for drug delivery. *Vietnam Journal of Science and Technology, 54*(4B), 185.
103. Fathollahipour, S., Abouei Mehrizi, A., Ghaee, A., & Koosha, M. (2015). Electrospinning of PVA/chitosan nanocomposite nanofibers containing gelatin nanoparticles as a dual drug delivery system. *Journal of Biomedical Materials Research Part A, 103*(12), 3852–3862.
104. Charernsriwilaiwat, N., Rojanarata, T., Ngawhirunpat, T., & Opanasopit, P. (2014). Electrospun chitosan/polyvinyl alcohol nanofibre mats for wound healing. *International Wound Journal, 11*(2), 215–222.

105. Salim, S. A., Taha, A. A., Khozemy, E. E., El-Moslamy, S. H., & Kamoun, E. A. (2022). Electrospun zinc-based metal organic framework loaded-PVA/chitosan/hyaluronic acid interfaces in antimicrobial composite nanofibers scaffold for bone regeneration applications. *Journal of Drug Delivery Science and Technology, 76*, 103823.
106. Qureshi, D., Pattanaik, S., Mohanty, B., Anis, A., Kulikouskaya, V., Hileuskaya, K., et al. (2022). Preparation of novel poly (vinyl alcohol)/chitosan lactate-based phase-separated composite films for UV-shielding and drug delivery applications. *Polymer Bulletin., 79*, 1–38.
107. Fahimirad, S., Satei, P., Ganji, A., & Abtahi, H. (2023). Wound healing capability of the double-layer polycaprolactone/polyvinyl alcohol-chitosan lactate electrospun nanofiber incorporating Echinacea purpurea extract. *Journal of Drug Delivery Science and Technology, 87*, 104734.
108. Patole, V., Bhosale, P., Ingavle, G., Behere, I., Vyawahare, N., Ottoor, D., et al. (2024). In vitro and in vivo assessment of gallic acid-chitosan/polycaprolactone conjugate electrospun nanofibers for wound healing. *Journal of Drug Delivery Scienceand Technology, 95*, 105569.
109. Hadjianfar, M., Semnani, D., & Varshosaz, J. (2018). Polycaprolactone/chitosan blend nanofibers loaded by 5-fluorouracil: An approach to anticancer drug delivery system. *Polymers for Advanced Technologies, 29*(12), 2972–2981.
110. Zhu, J., Ye, H., Deng, D., Li, J., & Wu, Y. (2020). Electrospun metformin-loaded polycaprolactone/chitosan nanofibrous membranes as promoting guided bone regeneration membranes: Preparation and characterization of fibers, drug release, and osteogenic activity in vitro. *Journal of Biomaterials Applications, 34*(9), 1282–1293.
111. Darbasizadeh, B., Motasadizadeh, H., Foroughi-Nia, B., & Farhadnejad, H. (2018). Tripolyphosphate-crosslinked chitosan/poly (ethylene oxide) electrospun nanofibrous mats as a floating gastro-retentive delivery system for ranitidine hydrochloride. *Journal of Pharmaceutical and Biomedical Analysis, 153*, 63–75.
112. Stie, M. B., Gätke, J. R., Wan, F., Chronakis, I. S., Jacobsen, J., & Nielsen, H. M. (2020). Swelling of mucoadhesive electrospun chitosan/polyethylene oxide nanofibers facilitates adhesion to the sublingual mucosa. *Carbohydrate Polymers, 242*, 116428.
113. Mirzaeei, S., Taghe, S., Asare-Addo, K., & Nokhodchi, A. (2021). Polyvinyl alcohol/chitosan single-layered and polyvinyl alcohol/chitosan/eudragit rl100 multi-layered electrospun nanofibers as an ocular matrix for the controlled release of ofloxacin: An in vitro and in vivo evaluation. *AAPS PharmSciTech, 22*(5), 170.
114. Xu, J., Zhang, J., Gao, W., Liang, H., Wang, H., & Li, J. (2009). Preparation of chitosan/PLA blend micro/nanofibers by electrospinning. *Materials Letters, 63*(8), 658–660.
115. Afshar, S., Rashedi, S., Nazockdast, H., & Ghazalian, M. (2019). Preparation and characterization of electrospun poly(lactic acid)-chitosan core-shell nanofibers with a new solvent system. *International Journal of Biological Macromolecules, 138*, 1130–1137.
116. Khashi, M., Hassanajili, S., & Golestaneh, S. I. (2018). Electrospun poly-lactic acid/chitosan nanofibers loaded with paclitaxel for coating of a prototype polymeric stent. *Fibers and Polymers, 19*, 1444–1453.
117. Meng, Z. X., Zheng, W., Li, L., & Zheng, Y. F. (2011). Fabrication, characterization and in vitro drug release behavior of electrospun PLGA/chitosan nanofibrous scaffold. *Materials Chemistry and Physics, 125*(3), 606–611.
118. Grant, J. J., Pillai, S. C., Perova, T. S., Hehir, S., Hinder, S. J., McAfee, M., et al. (2021). Electrospun fibres of chitosan/PVP for the effective chemotherapeutic drug delivery of 5-fluorouracil. *Chemosensors, 9*(4), 70.
119. Gouda, M., Khalaf, M. M., Shaaban, S., & El-Lateef, H. M. A. (2022). Fabrication of chitosan nanofibers containing some steroidal compounds as a drug delivery system. *Polymers, 14*(10), 2094.
120. Geng, X., Kwon, O.-H., & Jang, J. (2005). Electrospinning of chitosan dissolved in concentrated acetic acid solution. *Biomaterials, 26*(27), 5427–5432.
121. dos Santos, D. M., Chagas, P. A. M., Leite, I. S., Inada, N. M., de Annunzio, S. R., Fontana, C. R., et al. (2020). Core-sheath nanostructured chitosan-based nonwovens as a potential drug

delivery system for periodontitis treatment. *International Journal of Biological Macromolecules, 142*, 521–534.
122. AnjiReddy, K., & Karpagam, S. (2017). Chitosan nanofilm and electrospun nanofiber for quick drug release in the treatment of Alzheimer's disease: In vitro and in vivo evaluation. *International Journal of Biological Macromolecules, 105*, 131–142.
123. Ahmed, T. A., & Aljaeid, B. M. (2016). Preparation, characterization, and potential application of chitosan, chitosan derivatives, and chitosan metal nanoparticles in pharmaceutical drug delivery. *Drug Design, Development and Therapy, 10*, 483–507.
124. Kumar, V., Sharma, N., Janghu, P., Pasrija, R., Umesh, M., Chakraborty, P., et al. (2023). Synthesis and characterization of chitosan nanofibers for wound healing and drug delivery application. *Journal of Drug Delivery Science and Technology, 87*, 104858.
125. Sogias, I. A., Williams, A. C., & Khutoryanskiy, V. V. (2008). Why is chitosan mucoadhesive? *Biomacromolecules, 9*(7), 1837–1842.
126. Sudhakar, Y., Kuotsu, K., & Bandyopadhyay, A. (2006). Buccal bioadhesive drug delivery—A promising option for orally less efficient drugs. *Journal of Controlled Release, 114*(1), 15–40.
127. Morales, J. O., & McConville, J. T. (2011). Manufacture and characterization of mucoadhesive buccal films. *European Journal of Pharmaceutics and Biopharmaceutics, 77*(2), 187–199.
128. Ways, T. M., Lau, W. M., & Khutoryanskiy, V. V. (2018). Chitosan and its derivatives for application in mucoadhesive drug delivery systems. *Polymers, 10*(3), 267.
129. Anisiei, A., Andreica, B.-I., Mititelu-Tartau, L., Coman, C. G., Bilyy, R., Bila, G., et al. (2023). Biodegradable trimethyl chitosan nanofiber mats by electrospinning as bioabsorbable dressings for wound closure and healing. *International Journal of Biological Macromolecules, 249*, 126056.
130. Bozdag, M., Urek, F., Cesur, S., Sahin, A., & Gunduz, O. (2025). Bovine serum albumin (BSA)-loaded polyvinyl alcohol (PVA) / chitosan (CH) / hydroxyapatite (HA) electrospun nanofibers for bone tissue regeneration. *Journal of Drug Delivery Science and Technology, 106*, 106712.
131. Samprasit, W., Kaomongkolgit, R., Sukma, M., Rojanarata, T., Ngawhirunpat, T., & Opanasopit, P. (2015). Mucoadhesive electrospun chitosan-based nanofibre mats for dental caries prevention. *Carbohydrate Polymers, 117*, 933–940.
132. Salleh, N. A. M., Afifi, A. M., Zuki, F. M., & SalehHudin, H. S. (2025). Enhancing mechanical properties of chitosan/PVA electrospun nanofibers: A comprehensive review. *Beilstein Journal of Nanotechnology, 16*, 286–307.
133. Younes, H. M., Kadavil, H., Ismail, H. M., Adib, S. A., Zamani, S., Alany, R. G., et al. (2024). Overview of tissue engineering and drug delivery applications of reactive electrospinning and crosslinking techniques of polymeric nanofibers with highlights on their biocompatibility testing and regulatory aspects. *Pharmaceutics, 16*(1), 32.
134. Austero, M. S., Donius, A. E., Wegst, U. G., & Schauer, C. L. (2012). New crosslinkers for electrospun chitosan fibre mats. I. Chemical analysis. *Journal of The Royal Society Interface, 9*(75), 2551–2562.
135. Schiffman, J. D., & Schauer, C. L. (2007). Cross-linking chitosan nanofibers. *Biomacromolecules, 8*(2), 594–601.
136. Fadaie, M., Mirzaei, E., Asvar, Z., & Azarpira, N. (2019). Stabilization of chitosan based electrospun nanofibers through a simple and safe method. *Materials Science and Engineering: C, 98*, 369–380.
137. Cui, Z., Zheng, Z., Lin, L., Si, J., Wang, Q., Peng, X., et al. (2018). Electrospinning and crosslinking of polyvinyl alcohol/chitosan composite nanofiber for transdermal drug delivery. *Advances in Polymer Technology, 37*(6), 1917–1928.
138. Kim, S. (2018). Competitive biological activities of chitosan and its derivatives: Antimicrobial, antioxidant, anticancer, and anti-inflammatory activities. *International Journal of Polymer Science, 2018*(1), 1708172.

139. Aytekin, A. O., Morimura, S., & Kida, K. (2011). Synthesis of chitosan–caffeic acid derivatives and evaluation of their antioxidant activities. *Journal of Bioscience and Bioengineering, 111*(2), 212–216.
140. Azuma, K., Osaki, T., Minami, S., & Okamoto, Y. (2015). Anticancer and anti-inflammatory properties of chitin and chitosan oligosaccharides. *Journal of Functional Biomaterials, 6*(1), 33–49.
141. He, J., Zhou, S., Wang, J., Sun, B., Ni, D., Wu, J., et al. (2024). Anti-inflammatory and anti-oxidative electrospun nanofiber membrane promotes diabetic wound healing via macrophage modulation. *Journal of Nanobiotechnology, 22*(1), 116.
142. Valachová, K., El Meligy, M. A., & Šoltés, L. (2022). Hyaluronic acid and chitosan-based electrospun wound dressings: Problems and solutions. *International Journal of Biological Macromolecules, 206*, 74–91.
143. Ibrahim, S., Sayed, H. M., El-Rafei, A. M., El Amir, A., Ismail, M., & Allam, N. K. (2016). Improved genistein loading and release on electrospun chitosan nanofiber blends. *Journal of Molecular Liquids, 223*, 1056–1061.
144. Norouzi, Z., & Abdouss, M. (2023). Electrospun nanofibers using beta-cyclodextrin grafted chitosan macromolecules loaded with indomethacin as an innovative drug delivery system. *International Journal of Biological Macromolecules., 233*, 123518.
145. Cai, Z.-X., Mo, X.-M., Zhang, K.-H., Fan, L.-P., Yin, A.-L., He, C.-L., et al. (2010). Fabrication of chitosan/silk fibroin composite nanofibers for wound-dressing applications. *International Journal of Molecular Sciences, 11*(9), 3529–3539.
146. Naseri, N., Algan, C., Jacobs, V., John, M., Oksman, K., & Mathew, A. P. (2014). Electrospun chitosan-based nanocomposite mats reinforced with chitin nanocrystals for wound dressing. *Carbohydrate Polymers, 109*, 7–15.
147. Urbanek, O., Sajkiewicz, P., & Pierini, F. (2017). The effect of polarity in the electrospinning process on PCL/chitosan nanofibres' structure, properties and efficiency of surface modification. *Polymer, 124*, 168–175.
148. Dorati, R., Pisani, S., Maffeis, G., Conti, B., Modena, T., Chiesa, E., et al. (2018). Study on hydrophilicity and degradability of chitosan/polylactide-co-polycaprolactone nanofibre blend electrospun membrane. *Carbohydrate Polymers, 199*, 150–160.
149. Zhao, R., Li, X., Sun, B., Zhang, Y., Zhang, D., Tang, Z., et al. (2014). Electrospun chitosan/sericin composite nanofibers with antibacterial property as potential wound dressings. *International Journal of Biological Macromolecules, 68*, 92–97.
150. Bahtiyar, C., Cakir, N. T., Kahveci, M. U., Acik, G., & Altinkok, C. (2024). Fabrication of gallic acid containing poly(vinyl alcohol)/chitosan electrospun nanofibers with antioxidant and drug delivery properties. *International Journal of Biological Macromolecules, 281*, 136055.
151. Herdiana, Y., Wathoni, N., Shamsuddin, S., & Muchtaridi, M. (2022). Drug release study of the chitosan-based nanoparticles. *Heliyon, 8*(1), e08674.
152. Cai, N., Li, C., Han, C., Luo, X., Shen, L., Xue, Y., et al. (2016). Tailoring mechanical and antibacterial properties of chitosan/gelatin nanofiber membranes with Fe3O4 nanoparticles for potential wound dressing application. *Applied Surface Science, 369*, 492–500.
153. Trinca, R. B., Westin, C. B., da Silva, J. A. F., & Moraes, Â. M. (2017). Electrospun multilayer chitosan scaffolds as potential wound dressings for skin lesions. *European Polymer Journal, 88*, 161–170.
154. Ebrahimzadeh, S., Bari, M. R., Hamishehkar, H., Kafil, H. S., & Lim, L.-T. (2021). Essential oils-loaded electrospun chitosan-poly(vinyl alcohol) nonwovens laminated on chitosan film as bilayer bioactive edible films. *LWT, 144*, 111217.
155. Yusof, N. L. B. M., Wee, A., Lim, L. Y., & Khor, E. (2003). Flexible chitin films as potential wound-dressing materials: Wound model studies. *Journal of Biomedical Materials Research Part A, 66*(2), 224–232.
156. Sasmal, P., & Datta, P. (2019). Tranexamic acid-loaded chitosan electrospun nanofibers as drug delivery system for hemorrhage control applications. *Journal of Drug Delivery Science and Technology, 52*, 559–567.

157. Radmansouri, M., Bahmani, E., Sarikhani, E., Rahmani, K., Sharifianjazi, F., & Irani, M. (2018). Doxorubicin hydrochloride - loaded electrospun chitosan/cobalt ferrite/titanium oxide nanofibers for hyperthermic tumor cell treatment and controlled drug release. *International Journal of Biological Macromolecules, 116*, 378–384.
158. Silva Iahnke, A. O., Monte, M. L., Sant'Anna Cadaval, T. R., & Almeida Pinto, L. A. (2025). Electrospinning of gelatin fine fibers loaded with hop waste extract: Development, characterization, and application as antioxidant chitosan bilayer films. *Journal of Molecular Liquids, 426*, 127364.
159. Ignatova, M., Manolova, N., Rashkov, I., & Markova, N. (2016). Quaternized chitosan/κ-carrageenan/caffeic acid–coated poly(3-hydroxybutyrate) fibrous materials: Preparation, antibacterial and antioxidant activity. *International Journal of Pharmaceutics, 513*(1), 528–537.
160. Mirhaj, M., Varshosaz, J., Labbaf, S., Emadi, R., Marcus Seifalian, A., & Sharifianjazi, F. (2023). An antibacterial multi-layered scaffold fabricated by 3D printing and electrospinning methodologies for skin tissue regeneration. *International Journal of Pharmaceutics, 645*, 123357.
161. Zeng, R., Tu, M., Liu, H.-W., Zhao, J.-H., Zha, Z.-G., & Zhou, C.-R. (2009). Preparation, structure and drug release behaviour of chitosan-based nanofibres. *IET Nanobiotechnology, 3*(1), 8–13.
162. Fu, Y., & Kao, W. J. (2010). Drug release kinetics and transport mechanisms of non-degradable and degradable polymeric delivery systems. *Expert Opinion on Drug Delivery, 7*(4), 429–444.
163. Kir, F., Al-Sulaiti, F. K., & Sahin, S. (2024). Evaluation of in vitro dissolution profiles of modified-release metoprolol succinate tablets crushed using mortar and pestle technique. *European Journal of Pharmaceutical Sciences, 194*, 106694.
164. Wu, J., Zhang, Z., Gu, J. G., Zhou, W., Liang, X., Zhou, G., et al. (2020). Mechanism of a long-term controlled drug release system based on simple blended electrospun fibers. *Journal of Controlled Release, 320*, 337–346.
165. Liu, C., Zhu, Y., Lun, X., Sheng, H., & Yan, A. (2022). Effects of wound dressing based on the combination of silver @curcumin nanoparticles and electrospun chitosan nanofibers on wound healing. *Bioengineered, 13*(2), 4328–4339.
166. Sill, T. J., & von Recum, H. A. (2008). Electrospinning: Applications in drug delivery and tissue engineering. *Biomaterials, 29*(13), 1989–2006.
167. Tamilarasi, G. P., Sabarees, G., Manikandan, K., Gouthaman, S., Alagarsamy, V., & Solomon, V. R. (2023). Advances in electrospun chitosan nanofiber biomaterials for biomedical applications. *Materials Advances, 4*(15), 3114–3139.
168. Jayakumar, R., Prabaharan, M., Sudheesh Kumar, P. T., Nair, S. V., & Tamura, H. (2011). Biomaterials based on chitin and chitosan in wound dressing applications. *Biotechnology Advances, 29*(3), 322–337.
169. Shiravandi, A., Ashtiani, M. K., & Daemi, H. (2023). Fabrication of affinity-based drug delivery systems based on electrospun chitosan sulfate/poly(vinyl alcohol) nanofibrous mats. *International Journal of Biological Macromolecules, 252*, 126438.
170. Bayat, S., Amiri, N., Pishavar, E., Kalalinia, F., Movaffagh, J., & Hashemi, M. (2019). Bromelain-loaded chitosan nanofibers prepared by electrospinning method for burn wound healing in animal models. *Life Sciences, 229*, 57–66.
171. Charernsriwilaiwat, N., Opanasopit, P., Rojanarata, T., & Ngawhirunpat, T. (2012). Lysozyme-loaded, electrospun chitosan-based nanofiber mats for wound healing. *International Journal of Pharmaceutics, 427*(2), 379–384.
172. Ranganathan, S., Balagangadharan, K., & Selvamurugan, N. (2019). Chitosan and gelatin-based electrospun fibers for bone tissue engineering. *International Journal of Biological Macromolecules, 133*, 354–364.
173. Xie, J., Lou, X., Wang, X., Yang, L., & Zhang, Y. (2015). Electrospun nanofibers of hydroxyapatite/collagen/chitosan promote osteogenic differentiation of the induced pluripotent stem cell-derived mesenchymal stem cells. *Journal of Controlled Release, 213*, e53.

174. Zhang, Y., Venugopal, J. R., El-Turki, A., Ramakrishna, S., Su, B., & Lim, C. T. (2008). Electrospun biomimetic nanocomposite nanofibers of hydroxyapatite/chitosan for bone tissue engineering. *Biomaterials, 29*(32), 4314–4322.
175. Su, H., Liu, K.-Y., Karydis, A., Abebe, D. G., Wu, C., Anderson, K. M., et al. (2016). In vitro and in vivo evaluations of a novel post-electrospinning treatment to improve the fibrous structure of chitosan membranes for guided bone regeneration. *Biomedical Materials, 12*(1), 015003.
176. Frohbergh, M. E., Katsman, A., Botta, G. P., Lazarovici, P., Schauer, C. L., Wegst, U. G. K., et al. (2012). Electrospun hydroxyapatite-containing chitosan nanofibers crosslinked with genipin for bone tissue engineering. *Biomaterials, 33*(36), 9167–9178.
177. Murali, V. P., Fujiwara, T., Gallop, C., Wang, Y., Wilson, J. A., Atwill, M. T., et al. (2020). Modified electrospun chitosan membranes for controlled release of simvastatin. *International Journal of Pharmaceutics, 584*, 119438.
178. Nakielski, P., & Pierini, F. (2019). Blood interactions with nano- and microfibers: Recent advances, challenges and applications in nano- and microfibrous hemostatic agents. *Acta Biomaterialia, 84*, 63–76.
179. Huang, J., Cheng, Y., Wu, Y., Shi, X., Du, Y., & Deng, H. (2019). Chitosan/tannic acid bilayers layer-by-layer deposited cellulose nanofibrous mats for antibacterial application. *International Journal of Biological Macromolecules, 139*, 191–198.
180. Thomas, H., Diamond, J., Vieco, A., Chaudhuri, S., Shinnar, E., Cromer, S., et al. (2018). Global atlas of cardiovascular disease 2000-2016: The path to prevention and control. *Global Heart, 13*(3), 143–163.
181. Hirano, S., & Noishiki, Y. (1985). The blood compatibility of chitosan and N-acylchitosans. *Journal of Biomedical Materials Research, 19*(4), 413–417.
182. Naito, Y., Shinoka, T., Duncan, D., Hibino, N., Solomon, D., Cleary, M., et al. (2011). Vascular tissue engineering: Towards the next generation vascular grafts. *Advanced Drug Delivery Reviews, 63*(4–5), 312–323.
183. Mitchell, S. L., & Niklason, L. E. (2003). Requirements for growing tissue-engineered vascular grafts. *Cardiovascular Pathology, 12*(2), 59–64.
184. Lee, S. J., Yoo, J. J., Lim, G. J., Atala, A., & Stitzel, J. (2007). In vitro evaluation of electrospun nanofiber scaffolds for vascular graft application. *Journal of Biomedical Materials Research Part A, 83*(4), 999–1008.
185. Wang, T., Ji, X., Jin, L., Feng, Z., Wu, J., Zheng, J., et al. (2013). Fabrication and characterization of heparin-grafted poly-L-lactic acid-chitosan core-shell nanofibers scaffold for vascular gasket. *ACS Applied Materials & Interfaces, 5*(9), 3757–3763.
186. Kurobe, H., Maxfield, M. W., Breuer, C. K., & Shinoka, T. (2012). Concise review: Tissue-engineered vascular grafts for cardiac surgery: Past, present, and future. *Stem Cells Translational Medicine, 1*(7), 566–571.
187. Willerth, S. M., & Sakiyama-Elbert, S. E. (2007). Approaches to neural tissue engineering using scaffolds for drug delivery. *Advanced Drug Delivery Reviews, 59*(4–5), 325–338.
188. Jafarisavari, Z., Ai, J., Abbas Mirzaei, S., Soleimannejad, M., & Asadpour, S. (2024). Development of new nanofibrous nerve conduits by PCL-chitosan-hyaluronic acid containing piracetam-vitamin B12 for sciatic nerve: A rat model. *International Journal of Pharmaceutics, 655*, 123978.
189. Aranaz, I., Acosta, N., Civera, C., Elorza, B., Mingo, J., Castro, C., et al. (2018). Cosmetics and cosmeceutical applications of chitin, chitosan and their derivatives. *Polymers, 10*(2), 213.
190. Morganti, P., & Paglialunga, S. (2008). EU borderline cosmetic products review of current regulatory status. *Clinics in Dermatology, 26*(4), 392–397.
191. Charoenying, T., Ngawhirunpat, T., Rojanarata, T., Tonglairoum, P., & Opanasopit, P. (2018). Chitosan/polyvinyl alcohol electrospun nanofiber mats as carriers for arbutin. *Thai Journal of Pharmaceutical Sciences, 2018*(42), 211–214.
192. Hu, W., Yin, H., Guo, Y., Gao, Y., & Zhao, Y. (2024). Fabrication of multifunctional facial masks from phenolic acid grafted chitosan/collagen peptides via aqueous electrospinning. *International Journal of Biological Macromolecules, 267*, 131443.

193. Nicknejad, E. T., Ghoreishi, S. M., & Habibi, N. (2015). Electrospinning of cross-linked magnetic chitosan nanofibers for protein release. *AAPS PharmSciTech, 16*(6), 1480–1486.
194. Roozbahani, F., Sultana, N., Almasi, D., & Naghizadeh, F. (2015). Effects of chitosan concentration on the protein release behaviour of electrospun poly (ε-caprolactone)/chitosan nanofibers. *Journal of Nanomaterials, 2015*(1), 747420.
195. Lancina, M. G., III, Shankar, R. K., & Yang, H. (2017). Chitosan nanofibers for transbuccal insulin delivery. *Journal of Biomedical Materials Research Part A, 105*(5), 1252–1259.
196. Fouad, S. A., Ismail, A. M., Abdel Rafea, M., Abu Saied, M. A., & El-Dissouky, A. (2024). Preparation and characterization of chitosan nanofiber: Kinetic studies and enhancement of insulin delivery system. *Nanomaterials, 14*(11), 952.
197. Mendes, A. C., Gorzelanny, C., Halter, N., Schneider, S. W., & Chronakis, I. S. (2016). Hybrid electrospun chitosan-phospholipids nanofibers for transdermal drug delivery. *International Journal of Pharmaceutics, 510*(1), 48–56.
198. Toshkova, R., Manolova, N., Gardeva, E., Ignatova, M., Yossifova, L., Rashkov, I., et al. (2010). Antitumor activity of quaternized chitosan-based electrospun implants against Graffi myeloid tumor. *International Journal of Pharmaceutics, 400*(1), 221–233.
199. Qin, Z. Y., Jia, X. W., Liu, Q., Kong, B. H., & Wang, H. (2019). Fast dissolving oral films for drug delivery prepared from chitosan/pullulan electrospinning nanofibers. *International Journal of Biological Macromolecules, 137*, 224–231.
200. Chiu, P. H., Wu, Z. Y., Hsu, C. C., Chang, Y. C., Huang, C. M., Hu, C. T., et al. (2024). Enhancement of antibacterial activity in electrospun fibrous membranes based on quaternized chitosan with caffeic acid and berberine chloride for wound dressing applications. *RSC Advances, 14*(47), 34756–34768.
201. Rostami, E. (2021). Chitosan-based nanoparticles for drug delivery. In *Nanoengineering of biomaterials: Drug delivery & biomedical applications, 2 volumes* (pp. 1–32).
202. Sahariah, P., & Másson, M. (2017). Antimicrobial chitosan and chitosan derivatives: A review of the structure-activity relationship. *Biomacromolecules, 18*(11), 3846–3868.
203. Lahmer, R. A., Jones, D. L., Townsend, S., Baker, S., & Williams, A. P. (2014). Susceptibility of Escherichia coli O157 to chitosan-arginine in beef liquid purge is affected by bacterial cell growth phase. *International Journal of Food Science and Technology., 49*(2), 515–520.
204. Peng, Y., Han, B., Liu, W., & Xu, X. (2005). Preparation and antimicrobial activity of hydroxypropyl chitosan. *Carbohydrate Research, 340*(11), 1846–1851.
205. Badawy, M. E. I., & Rabea, E. I. (2012). Characterization and antimicrobial activity of water-soluble N-(4-carboxybutyroyl) chitosans against some plant pathogenic bacteria and fungi. *Carbohydrate Polymers, 87*(1), 250–256.
206. Sahariah, P., Óskarsson, B. M., Hjálmarsdóttir, M., & Másson, M. (2015). Synthesis of guanidinylated chitosan with the aid of multiple protecting groups and investigation of antibacterial activity. *Carbohydrate Polymers, 127*, 407–417.
207. Fu, X., Shen, Y., Jiang, X., Huang, D., & Yan, Y. (2011). Chitosan derivatives with dual-antibacterial functional groups for antimicrobial finishing of cotton fabrics. *Carbohydrate Polymers, 85*(1), 221–227.
208. Tan, H., Ma, R., Lin, C., Liu, Z., & Tang, T. (2013). Quaternized chitosan as an antimicrobial agent: Antimicrobial activity, mechanism of action and biomedical applications in orthopedics. *International Journal of Molecular Sciences, 14*(1), 1854–1869.
209. Kim, S. H., Kim, S.-H., Nair, S., & Moore, E. (2005). Reactive electrospinning of cross-linked poly (2-hydroxyethyl methacrylate) nanofibers and elastic properties of individual hydrogel nanofibers in aqueous solutions. *Macromolecules, 38*(9), 3719–3723.
210. Ghosh, T., Das, T., & Purwar, R. (2021). Review of electrospun hydrogel nanofiber system: Synthesis, properties and applications. *Polymer Engineering & Science, 61*(7), 1887–1911.
211. Rasouli, R., Barhoum, A., Bechelany, M., & Dufresne, A. (2019). Nanofibers for biomedical and healthcare applications. *Macromolecular Bioscience., 19*(2), 1800256.
212. Alavi, M., & Nokhodchi, A. (2022). Antimicrobial and wound healing activities of electrospun nanofibers based on functionalized carbohydrates and proteins. *Cellulose, 29*(3), 1331–1347.

213. Iqbal, H., Khan, B. A., Khan, Z. U., Razzaq, A., Khan, N. U., Menaa, B., et al. (2020). Fabrication, physical characterizations and in vitro antibacterial activity of cefadroxil-loaded chitosan/poly(vinyl alcohol) nanofibers against Staphylococcus aureus clinical isolates. *International Journal of Biological Macromolecules, 144*, 921–931.
214. Amiri, N., Ajami, S., Shahroodi, A., Jannatabadi, N., Amiri Darban, S., Fazly Bazzaz, B. S., et al. (2020). Teicoplanin-loaded chitosan-PEO nanofibers for local antibiotic delivery and wound healing. *International Journal of Biological Macromolecules, 162*, 645–656.
215. Kean, T., Roth, S., & Thanou, M. (2005). Trimethylated chitosans as non-viral gene delivery vectors: Cytotoxicity and transfection efficiency. *Journal of Controlled Release, 103*(3), 643–653.
216. Najahi-Missaoui, W., Arnold, R. D., & Cummings, B. S. (2020). Safe nanoparticles: Are we there yet? *International Journal of Molecular Sciences, 22*(1), 385.
217. Egbuna, C., Parmar, V. K., Jeevanandam, J., Ezzat, S. M., Patrick-Iwuanyanwu, K. C., Adetunji, C. O., et al. (2021). Toxicity of nanoparticles in biomedical application: Nanotoxicology. *Journal of Toxicology., 2021*(1), 9954443.
218. Sonin, D., Pochkaeva, E., Zhuravskii, S., Postnov, V., Korolev, D., Vasina, L., et al. (2020). Biological safety and biodistribution of chitosan nanoparticles. *Nanomaterials, 10*(4), 810.
219. Horie, M., & Tabei, Y. (2021). Role of oxidative stress in nanoparticles toxicity. *Free Radical Research, 55*(4), 331–342.
220. Tuli, H. S., Joshi, R., Kaur, G., Garg, V. K., Sak, K., Varol, M., et al. (2023). Metal nanoparticles in cancer: From synthesis and metabolism to cellular interactions. *Journal of Nanostructure in Chemistry, 13*(3), 321–348.
221. Amini, F., Semnani, D., Karbasi, S., & Banitaba, S. N. (2019). A novel bilayer drug-loaded wound dressing of PVDF and PHB/Chitosan nanofibers applicable for post-surgical ulcers. *International Journal of Polymeric Materials and Polymeric Biomaterials, 68*(13), 772–777.
222. Chen, J., Zhao, L., Ling, J., Yang, L.-Y., & Ouyang, X.-k. (2023). A quaternized chitosan and carboxylated cellulose nanofiber-based sponge with a microchannel structure for rapid hemostasis and wound healing. *International Journal of Biological Macromolecules., 233*, 123631.
223. Welchoff, M. A., Wittenberg, A. T., Jimenez, J. C., Kamireddi, D., Snelling, E. K., Street, R. M., et al. (2023). Post-modification of electrospun chitosan fibers. *Polymer Engineering and Science, 63*(7), 1921–1931.
224. Singh, Y. P., Dasgupta, S., Nayar, S., & Bhaskar, R. (2020). Optimization of electrospinning process & parameters for producing defect-free chitosan/polyethylene oxide nanofibers for bone tissue engineering. *Journal of Biomaterials Science, Polymer Edition, 31*(6), 781–803.
225. Ahmady, A. R., Solouk, A., Saber-Samandari, S., Akbari, S., Ghanbari, H., & Brycki, B. E. (2023). Capsaicin-loaded alginate nanoparticles embedded polycaprolactone-chitosan nanofibers as a controlled drug delivery nanoplatform for anticancer activity. *Journal of Colloid and Interface Science, 638*, 616–628.
226. Mirzaeei, S., Berenjian, K., & Khazaei, R. (2018). Preparation of the potential ocular inserts by electrospinning method to achieve the prolong release profile of triamcinolone acetonide. *Advanced Pharmaceutical Bulletin, 8*(1), 21–27.
227. Fulgêncio Gde, O., Viana, F. A., Ribeiro, R. R., Yoshida, M. I., Faraco, A. G., & Cunha-Júnior Ada, S. (2012). New mucoadhesive chitosan film for ophthalmic drug delivery of timolol maleate: In vivo evaluation. *Journal of Ocular Pharmacology and Therapeutics, 28*(4), 350–358.
228. Verestiuc, L., Nastasescu, O., Barbu, E., Sarvaiya, I., Green, K. L., & Tsibouklis, J. (2006). Functionalized chitosan/NIPAM (HEMA) hybrid polymer networks as inserts for ocular drug delivery: Synthesis, in vitro assessment, and in vivo evaluation. *Journal of Biomedical Materials Research Part A, 77*(4), 726–735.
229. Foureaux, G., Franca, J. R., Nogueira, J. C., Fulgencio, G. O., Ribeiro, T. G., Castilho, R. O., et al. (2015). Ocular inserts for sustained release of the angiotensin-converting enzyme 2 activator, diminazene aceturate, to treat glaucoma in rats. *PLoS One, 10*(7), e0133149.

Chitosan-Based Nanoparticles for Brain Drug Delivery

Shadi Askari, Mahnaz Eskandari, and Ali Zarrabi

Abstract This chapter reviews chitosan-based nanoparticles for brain drug delivery, focusing on their ability to cross the blood-brain barrier (BBB) and improving treatment for neurological disorders. It covers chitosan nanotechnology, its properties, and fabrication methods. This chapter also discusses transport pathways, such as adsorptive-mediated transcytosis and receptor-mediated transport. It explores therapeutic uses in neurodegenerative diseases, cancer, and psychiatric disorders, emphasizing hybrid nanocarriers and gene delivery systems. Challenges in manufacturing and regulation are highlighted, along with future trends in personalized medicine, neural stem cell-based delivery systems, and CRISPR-Cas9 technology.

Keywords Chitosan-based nanoparticles · Blood-brain barrier (BBB) · Neurological disorders · Gene delivery systems · Nanotechnology · Neurodegenerative diseases

Abbreviations

AI	Artificial Intelligence
BBB	Blood-Brain Barrier
CNS	Central Nervous System
CRISPR	Clustered Regularly Interspaced Short Palindromic Repeats
DDA	Degree of Deacetylation
MMP	Matrix Metalloproteinases

S. Askari · M. Eskandari (✉)
Department of Biomedical Engineering, Amirkabir University of Technology (Tehran Polytechnic University), Tehran, Iran
e-mail: eskandarim@aut.ac.ir

A. Zarrabi
Department of Biomedical Engineering, Faculty of Engineering and Natural Sciences, Istinye University, Istanbul, Turkiye

S. Jana (ed.), *Chitosan-Based Drug Delivery Systems*, AAPS Introductions in the Pharmaceutical Sciences 16, https://doi.org/10.1007/978-3-032-08347-0_11

MRI	Magnetic Resonance Imaging
NSC	Neural Stem Cell
PEG	Polyethylene Glycol
PLGA	Poly(lactic-co-glycolic acid)
QbD	Quality by Design
RNP	Ribonucleoprotein
siRNA	Small Interfering RNA
SLN	Solid Lipid Nanoparticles
TPP	Tripolyphosphate
TfR	Transferrin Receptor

1 Introduction

The blood-brain barrier (BBB) is a major challenge in modern therapeutics. It acts as a highly selective interface that blocks over 98% of small-molecule drugs and nearly all biologic elements from entering the central nervous system (CNS) [1]. The BBB consists of tightly joined endothelial cells, astrocytic foot processes, and efflux transporters like P-glycoprotein. It evolved to protect the brain from toxins, but now limits treatment options for neurological disorders [2]. Traditional methods, such as direct brain injections or temporary BBB disruption with hyperosmotic agents, carry risks like infection, inflammation, and off-target effects [3]. Even drugs given systemically that cross the BBB poorly often fail to reach effective brain concentrations due to fast clearance and poor distribution.

Due to existing challenges, researchers are exploring new ways to deliver drugs to the brain. Chitosan-based nanoparticles are a promising non-invasive option. Chitosan is made by deacetylating chitin, a natural substance in crustacean shells. It offers several benefits for drug delivery, including biocompatibility, biodegradability, and mucoadhesiveness [4, 5]. Chitosan is positively charged, which helps it bind to the negatively charged surfaces of the nasal epithelium. This property allows drugs to bypass the BBB through the olfactory and trigeminal nerve pathways [6]. Chitosan can also temporarily open tight junctions between epithelial cells, improving paracellular transport. Its surface can be modified with targeting ligands to support receptor-mediated transcytosis across the BBB [7]. These features make chitosan-based nanoparticles effective in increasing drug delivery to the brain and targeting specific neural areas.

Recent progress in nanoparticle design has improved the use of chitosan nanoparticles. Researchers can now control particle size to avoid nasal mucociliary clearance, increase drug loading (up to 85% for hydrophobic drugs), and ensure sustained drug release [8]. Preclinical studies show strong results in different brain disorders. For example, rivastigmine-loaded nanoparticles increased brain bioavailability by 4.7 times compared to oral doses in Alzheimer's models [9].In glioblastoma studies, pH-responsive temozolomide carriers reduced tumor growth by 92%

in mice [10]. For neuroinflammation, curcumin nanoformulations lowered interleukin-6 levels by 68% after traumatic brain injury [11]. Despite major progress, key challenges remain, such as batch-to-batch differences in chitosan sources, possible nasal irritation with long-term use, and the need for scalable manufacturing methods [4].

Now, advances in molecular biology, materials science, and pharmacokinetics are shaping the next generation of chitosan systems. New developments include zwitterionic surface coatings to reduce protein buildup, CRISPR-loaded nanoparticles for gene editing in brain diseases, and machine-learning tools for designing advanced carriers [12]. As central nervous system (CNS) disorders are expected to affect 15% of the population by 2040, creating effective chitosan-based treatments is not just a scientific goal, it is a public health priority [13]. This chapter systematically examines the scientific foundations, therapeutic applications, and translational potential of these nanoscale vehicles, providing a roadmap for researchers aiming to bridge the gap between laboratory breakthroughs and clinically viable brain-targeted therapies.

2 Fundamental Concepts of Chitosan Nanotechnology

Chitosan is a polysaccharide derived from the partial deacetylation of chitin. This polymer has a unique cationic nature due to its primary amino groups, distinguishing chitosan from most neutral or anionic natural polymers [14]. The degree of deacetylation (DDA) controls the positive charge density and influences key properties like solubility, mucoadhesion, and drug-loading efficiency [14, 15]. At low pH (<6.5), protonation of glucosamine units enhances solubility. Higher DDA (>85%) results in more compact nanoparticles with slower degradation, ideal for sustained drug release [14]. The molecular weight, ranging from 3 to 300 kDa, also affects nanoparticle characteristics. Low-molecular-weight chitosan (81 kDa) forms smaller, more uniform particles (238 nm), while high-molecular-weight variants (6000 kDa) create larger aggregates (319 nm) with reduced electrophoretic mobility [15].

The interplay between DDA and molecular weight plays a critical role in drug-loading capacity. High-DDA chitosan (88%) has a higher amino group density (6.1 mol/kg), enhancing electrostatic interactions with anionic drugs like nucleic acids or proteins. In contrast, intermediate DDA (75–85%) improves hydrophobic drug encapsulation due to increased chain flexibility. Ionic gelation, the primary fabrication method, uses chitosan's affinity for polyanions like tripolyphosphate (TPP) to form sub-200 nm particles through charge-mediated crosslinking. This process is pH-sensitive; below pH 6.5, excessive protonation can cause particle aggregation, while near-neutral conditions stabilize nanoparticles for biomedical use [15, 16].

Advanced modifications address some limitations. Thiolation adds sulfhydryl groups, boosting mucoadhesion threefold through disulfide bonding with mucosal glycoproteins. PEGylation reduces rapid clearance by decreasing opsonization, while ligand conjugation (e.g., transferrin) facilitates receptor-mediated BBB penetration. These engineered variants maintain chitosan's biocompatibility and biodegradability, which are essential for FDA-unapproved but widely studied drug delivery systems, while overcoming solubility issues at physiological pH. The polymer's semi-crystalline structure, with β-(1-4)-linked glucosamine units, also allows for tunable mechanical properties using ceramic or polymer composites, broadening applications in neural scaffolds and multifunctional nanocarriers [16].

2.1 Chitosan Nanoparticle Fabrication Methods

Nanoparticle fabrication methods are essential for optimizing drug delivery systems, especially for challenging applications like brain-targeted therapies. Among these methods, ionic gelation is widely used due to its simplicity and scalability (Fig. 1).

Ionic gelation is the most common technique for chitosan nanoparticle synthesis. This method relies on electrostatic interactions between protonated chitosan amino groups and polyanionic crosslinkers like sodium tripolyphosphate (TPP), forming nanoparticles through charge-mediated self-assembly. Recent improvements include turbulent mixing protocols, which allow nanoparticle formation at polymer concentrations up to 5 mg/mL while keeping particle sizes under 300 nm [17]. Electrospraying and spinning disc processing have further improved size uniformity, reducing polydispersity indices to below 0.2 with controlled shear forces [17, 18]. A key advancement involves modulating NaCl concentration during synthesis. This adjustment allows for systematic size tuning from 100 to 900 nm by altering the ionic strength's impact on polymer chain entanglement [19]. While traditional dropwise TPP addition risks localized over-crosslinking, the dilution

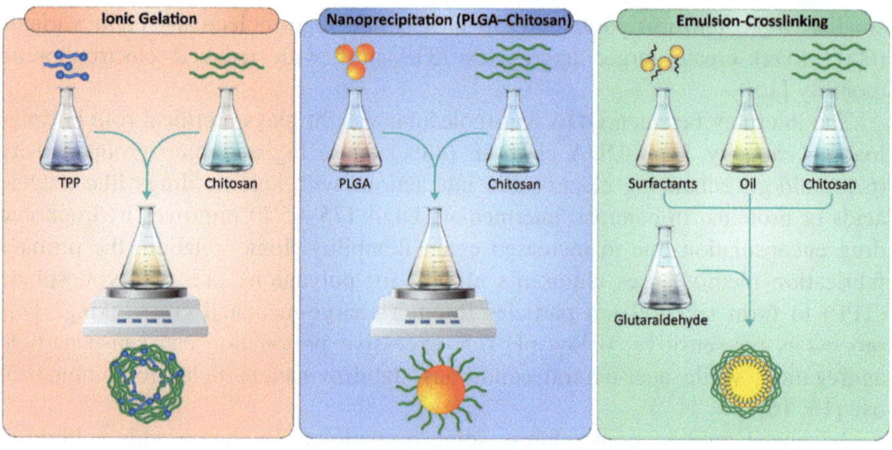

Fig. 1 Chitosan nanoparticles fabrication methods for blood-brain barrier delivery

method, where chitosan and TPP are pre-mixed at gelation-inhibiting ionic strengths before gradual salt reduction, offers better control over nucleation kinetics, achieving less than 10% batch variability [19].

Nanoprecipitation is a key method for combining PLGA's controlled release with chitosan's mucoadhesive properties in hybrid PLGA-chitosan systems. In optimized protocols, PLGA films are activated with carbodiimide chemistry (EDC/NHS) and then react with chitosan solutions, forming covalent amide bonds between PLGA carboxyl groups and chitosan amines [20]. This process produces hybrid polymers with adjustable degradation rates. Chitosan coatings increase surface charge from −25 mV to +32 mV and extend mucosal residence time by four times compared to unmodified PLGA nanoparticles [21]. The molecular weight and deacetylation degree of chitosan significantly affect hybridization efficiency. For instance, 15 kDa chitosan with 92% DDA creates 180 nm particles with 89% encapsulation efficiency for hydrophobic drugs, while 60 kDa chitosan forms 240 nm particles [21, 22].

Emulsion-crosslinking overcomes the limits of ionic gelation for hydrophobic drugs by using water-in-oil emulsions stabilized with surfactants like Span 80. Hydrodynamic cavitation reactors improve this method by cutting energy use by 78% compared to ultrasonic techniques. These reactors also produce particles around 286 nm with a polydispersity of 0.34. A key improvement is the timing of glutaraldehyde addition. Adding it before emulsification forms denser networks and leads to 92% drug retention, compared to 74% when added after. This also reduces leftover surfactants. The type of oil matters too;medium-chain triglycerides yield 310 nm particles, while mineral oil gives 420 nm, due to differences in interfacial tension and droplet breakup during cavitation [23]. These changes support the scalable production of uniform nanoparticles and maintain chitosan's pH-sensitive release.

Fabrication methods play a key role in delivering chitosan nanoparticles across the BBB. Control over particle size, surface charge, and drug encapsulation is essential for overcoming the BBB's physiological barriers. The BBB, with tight endothelial junctions and selective permeability, only allows nanoparticles smaller than 200 nm to pass through and avoid immune detection [24]. Ionic gelation, nanoprecipitation, and emulsion-crosslinking allow precise adjustment of these parameters to support brain targeting. Chitosan's mucoadhesive nature, enhanced through these techniques, increases retention time and improves interaction with BBB endothelial cells. These methods also support the encapsulation of both hydrophilic and hydrophobic drugs, enabling broad therapeutic applications for conditions such as neurodegenerative diseases and brain tumors. Table 1 compares three primary nanoparticle fabrication techniques, ionic gelation, nanoprecipitation, and emulsion-crosslinking, highlighting their critical process parameters, resulting nanoparticle characteristics, unique advantages, and suitability for brain-targeted drug delivery applications.

Table 1 Summary of key process parameters and characteristics of chitosan-based nanoparticle fabrication methods for brain-targeted drug delivery

Fabrication method	Key process parameters	Nanoparticle characteristics	Advantages	Applications
Ionic gelation	Chitosan-TPP electrostatic interaction Chitosan concentration: ≤ 5 mg/mL NaCl concentration tuning Mixing protocol (e.g., turbulent mixing) Dilution vs. dropwise TPP addition	Size: <300 nm Tunable size: 100–900 nm Polydispersity index: <0.2 Batch variability: <10%	Simple Scalable High control over particle size and uniformity	Brain-targeted drug delivery Hydrophilic drug encapsulation
Nanoprecipitation (PLGA–chitosan)	PLGA activation via EDC/NHS chemistry Chitosan molecular weight and degree of deacetylation Covalent bonding conditions	Size: 180–240 nm Zeta potential shift: −25 mV to +32 mV Encapsulation efficiency: up to 89%	Controlled degradation Improved mucoadhesion Hybrid structure	Controlled-release brain therapies Hydrophobic drugs
Emulsion-crosslinking	Water-in-oil emulsion Surfactant: Span 80 Reactor type: Hydrodynamic cavitation Timing of glutaraldehyde addition Type of oil used	Size: 286–420 nm Polydispersity: ~0.34 Drug retention: up to 92%	Suitable for hydrophobic drugs Energy-efficient Tunable network density	Long-retention brain therapies Hydrophobic drug encapsulation

2.2 *Characterization Parameters*

Characterizing chitosan nanoparticles is crucial for improving drug delivery performance, especially for crossing the BBB. Key parameters such as particle size, zeta potential, and drug loading efficiency affect biodistribution, stability, and therapeutic outcomes.

Particle size plays a major role in determining BBB penetration and distribution. Nanoparticles between 50 and 200 nm avoid nasal mucociliary clearance and support effective brain-section targeting. Electrospray fabrication allows precise size control, producing 105 nm particles by adjusting chitosan concentration (0.5% w/v), flow rate (0.05 mL/h), and applied voltage (15 kV). This combination reduces polydispersity to 0.2 and maintains zeta potential at +59.3 mV [25]. In ionic gelation, the chitosan-to-TPP ratio affects size: a 3:1 ratio produces 132 nm particles, while a 5:1 ratio increases size to 208 nm due to lower crosslinking density [26]. Molecular weight also impacts size. Low-molecular-weight chitosan (81 kDa) forms 238 nm particles with a narrow distribution, while 6000 kDa chitosan creates aggregates over 300 nm [27]. Dynamic light scattering and TEM imaging reveal polyhedral shapes, such as pentagonal or hexagonal forms, caused by crystallization during TPP-mediated gelation [26].

Zeta potential analysis reveals important details about mucoadhesion. Values above +30 mV support strong interactions with negatively charged sialic acid in nasal mucus. Acrylated chitosan nanoparticles show 2.3 times more mucosal retention than unmodified versions due to higher surface charge (+49 mV vs. +32 mV) [28]. Insulin-loaded nanoparticles maintain +21.6 mV zeta potential with optimized chitosan/insulin ratios [29]. Environmental stability depends on the isoelectric point at pH 9.0, which helps retain positive surface charge across a physiological pH range of 3.0–7.4 [26]. Flow-through tests show higher adhesion efficiency for acrylated nanoparticles, with 78% retention after 60 min, compared to 34% for standard chitosan particles [28].

Drug loading efficiency depends on chitosan's molecular structure. High-molecular-weight chitosan (300 kDa) achieves 89.9% paclitaxel encapsulation using hydrophobic interactions in modified OCC carriers [27]. Lyophilization with 5% sucrose preserves 92% BSA activity after redispersion [30]. For hydrophobic drugs, saturation incubation of preformed nanoparticles achieves 34.6% loading, while oligonucleotides reach 87.6% encapsulation through electrostatic interactions [27, 31]. FTIR spectroscopy confirms drug–polymer bonding, with shifts in the amide I band (from 1645 to 1628 cm^{-1}) showing stable ionic crosslinking during in vitro release [30]. Release profiles depend on chitosan matrix degradation. High-degree deacetylation (92%) nanoparticles sustain drug levels for 72 h, while those with 75% DDA show release within 24 h [32].

3 Mechanisms of BBB Penetration

Chitosan-based nanoparticles cross the blood–brain barrier (BBB) using three main methods: adsorptive-mediated transcytosis, receptor-mediated transport, and paracellular diffusion by altering tight junctions (Fig. 2). Each method involves different molecular and physical interactions, with specific benefits and limitations for delivering drugs to the brain.

3.1 Adsorptive-Mediated Transcytosis

Chitosan nanoparticles have a cationic surface charge due to protonated amino groups at physiological pH. This charge interacts with the negatively charged glycocalyx on brain endothelial cells, triggering adsorptive transcytosis. In this process, nanoparticles bind to cell membranes and are taken up by endocytosis. Studies show that chitosan nanoparticles with zeta potentials above +30 mV have 3.2 times higher endothelial uptake than neutral particles, measured by fluorescence intensity in bEnd3 cell models [33]. The process includes clathrin-coated pit

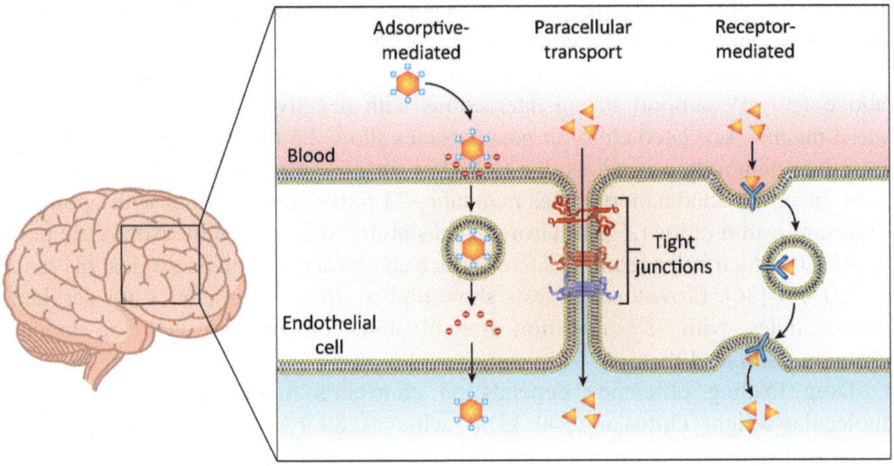

Fig. 2 Schematic of BBB penetration mechanism

formation, vesicular transport through the cytoplasm, and exocytosis on the abluminal side. Particle size affects this pathway, with 50–100 nm particles achieving 89% transcytosis efficiency in mice. Larger particles (>200 nm) are degraded by lysosomes [34].

The degree of deacetylation (DDA) and molecular weight of chitosan influence transcytosis rates. High-DDA chitosan (92%) nanoparticles show 2.7 times greater brain accumulation than those with 75% DDA, due to a higher charge density (6.1 vs. 4.8 mmol/g amino groups) [33]. However, excessive positive charge (> +50 mV) can lead to nonspecific protein adsorption and opsonization, reducing bioavailability by 41% in circulation. PEGylation can help coating chitosan-TPP nanoparticles with 5% PEG 2000 maintaining a +28 mV charge and increasing plasma half-life from 2.1 to 8.7 h [34].

3.2 Receptor-Mediated Transport

Ligand-functionalized chitosan nanoparticles allow for targeted delivery through receptor-mediated transcytosis. A common example is transferrin receptor (TfR) targeting, as brain endothelial cells have 10–100 times more TfR than peripheral tissues. By covalently conjugating transferrin to chitosan nanoparticles using strain-promoted azide-alkyne cycloaddition, 110–150 nm particles are formed with 98% ligand integrity, confirmed by SDS-PAGE [35]. In co-culture models, these nanoparticles show 4.5 times higher uptake by glioblastoma (U87) cells than non-targeted particles, with transcytosis rates of 12.3 ng/mg protein/hour across RPMI 2650 nasal epithelial barriers [35]. Surface plasmon resonance (SPR) reveals a nanomolar binding affinity (KD = 38 nM) to TfR, enabling displacement of endogenous transferrin.

In addition to transferrin, chitosan nanoparticles functionalized with lactoferrin, leptin, or anti-ABCG2 antibodies show potential. Lactoferrin-coated nanoparticles achieve 92% BBB penetration in Alzheimer's rat models and reduce amyloid-β plaques by 67% at one-tenth the free drug dose [34]. Dual-ligand systems enhance uptake by combining TfR-binding peptides and cell-penetrating peptides (e.g., TAT). TAT/RVG29-chitosan nanoparticles show 3.8 times greater brain accumulation than single-ligand particles in ischemic stroke models [34].

3.3 Paracellular Route via Tight Junction Modulation

Chitosan can temporarily open tight junctions, allowing paracellular transport of nanoparticles. At concentrations ≤0.5% (w/v), it disrupts zonula occludens-1 (ZO-1) protein networks, increasing pore size from 8 Å to 34 Å within 30 min in Caco-2 monolayers [36, 37]. This reversible change raises the permeability of 150 kDa dextran (6.2 nm radius) from 0.8×10^{-6} to 5.3×10^{-6} cm/s. In vivo, chitosan nanoparticles lower transepithelial electrical resistance (TEER) by 78% in nasal tissue, boosting insulin absorption 4.1 times compared to solution forms [36].

Atomic force microscopy shows a two-phase response. Low doses (0.1%) cause F-actin depolymerization via Rac1/RhoA signaling. Higher doses (0.5%) activate protein kinase C (PKC), shifting claudin-5 to the lateral membranes [37]. The process depends on pH. At pH < 6.5, protonated chitosan binds sialic acid on occludin, causing calcium influx and protein phosphorylation. Tight junctions recover within 2 h after removal [36, 37].

In primates, transport efficiency ranks as receptor-mediated > adsorptive > paracellular. Transferrin receptor (TfR)-targeted nanoparticles reach 3.8% of the injected dose per gram of brain, compared to 1.2% for cationic chitosan and 0.7% for paracellular carriers [34, 35]. Paracellular transport, though less efficient, offers fast absorption (Tmax 15 min vs. 45 min for transcytosis), useful in urgent neurological treatment. Triple-action nanoparticles, combining charge adhesion, TfR targeting, and junction opening, raise brain levels of antiretrovirals 12-fold over single-mechanism systems [33, 34].

Adsorptive transcytosis is simple but nonspecific, needing advanced pharmacokinetic modeling to reduce off-target exposure. Receptor-mediated systems struggle with batch variability in ligand conjugation (±18% in GMP conditions). Paracellular delivery needs tight control of mucosal contact time. Thermosensitive chitosan-g-PNIPAM keeps junctions open for 90 ± 12 min, matching nasal clearance rates [36, 37]. Phase I trials (NCT04804709) show intranasal chitosan-insulin nanoparticles are safe, with no long-term barrier damage over 6 months.

3.4 Key Enhancing Factors

Modifying surface charge is key to improving chitosan nanoparticles for brain delivery. Chitosan's natural positive charge, from protonated amino groups at physiological pH, helps it stick to negatively charged mucosal surfaces and endothelial membranes [34, 38]. However, a high surface charge (+45 mV) leads to rapid opsonization, reducing circulation time by 58% in mice. PEGylation with 2 kDa methoxy-PEG reduces zeta potential to +28 mV and increases plasma half-life from 1.5 to 7.3 h by adding steric protection [34]. Thiolation adds sulfhydryl groups (200 μmol SH/cm^2), which form disulfide bonds with mucin, extending nasal residence time 3.2 times compared to unmodified chitosan. Ligand attachment boosts targeting, transferrin-labeled nanoparticles cross the blood–brain barrier (BBB) 89% more efficiently while keeping a +21 mV charge for mucoadhesion [38].

Mucoadhesion depends on chitosan's molecular weight and functional groups. High-MW chitosan (300 kDa) sticks 78% more to mucosa than low-MW (50 kDa) due to deeper polymer-mucin interaction, shown by quartz crystal microbalance tests [21]. Thiolated versions perform even better, 92% of insulin-loaded nanoparticles stay in the olfactory epithelium after 60 min, versus 34% for non-thiolated ones [38]. Higher viscosity also slows clearance; 0.5% chitosan solutions reduce mucociliary clearance by 68% compared to saline [8]. Thermosensitive gels, which solidify at nasal temperatures, extend retention up to 90 ± 12 min.

Controlled release depends on cross-linking and material makeup. Stronger TPP cross-linking (5:1 chitosan:TPP) extends donepezil release to 72 h with zero-order kinetics ($R^2 = 0.98$), while weaker (3:1) ratios cause a 12-h burst [39]. pH-responsive systems use chitosan's solubility change at pH 6.5. For example, glioma-targeted carriers release 89% of temozolomide at pH 5.5 (tumor) versus 23% at normal pH [34]. Core-shell designs with PLGA cores and chitosan shells provide a two-phase release: 40% burst from the surface, followed by 60% slow release over 5 days through erosion [21].

Hybrid systems combine these benefits. Dual-functional nanoparticles with PEGylated surfaces (+18 mV) and thiolated chitosan show 92% nasal retention and 48-h rivastigmine release in Alzheimer's models. Machine learning helps optimize charge and mucoadhesion balance, improving bioavailability 3.4 times over standard formulations. These advances tackle three key challenges: precise targeting, long mucosal contact, and steady therapeutic release, vital for bringing chitosan systems into clinical neurology [34, 38].

4 Therapeutic Applications in Neurological Disorders

Chitosan-based nanoparticles are gaining significant attention in the treatment of neurological disorders due to their ability to enhance drug delivery mechanisms and optimize therapeutic outcomes. The unique properties of chitosan enable targeted delivery of drugs to the central nervous system (CNS), bypassing the BBB and minimizing peripheral side effects. This section explores the application of chitosan-based formulations in the treatment of neurodegenerative diseases like Alzheimer's and Parkinson's, highlighting their potential to improve drug efficacy, reduce neuroinflammation, and enhance cognitive functions. Additionally, the role of chitosan in facilitating direct CNS delivery, through both intranasal and injectable systems, is examined, alongside a comparative analysis of different chitosan-based formulations in terms of drug release profiles, encapsulation efficiency, and long-term stability.

4.1 Alzheimer's Disease

Chitosan-based nanoparticles show strong potential for treating neurodegenerative disorders by improving drug delivery. In Alzheimer's disease, galantamine-loaded chitosan nanoparticles improved therapeutic results. In a scopolamine-induced rat model, intranasal galantamine-chitosan nanoparticles (G-NP) enhanced spatial memory and exploratory behavior. They also reduced β-amyloid deposits by 50% compared to controls [40]. These nanoparticles, made using ionic gelation, released the drug over 8 h and stayed stable for over a year at 5 °C. This stability addresses the problem of rapid drug elimination seen in oral formulations [41]. Luteolin-loaded chitosomes (412.8 nm, +37.4 mV) also performed well. They had 86.6% entrapment efficiency and cut hyperphosphorylated tau protein levels by half in a sporadic Alzheimer's mouse model. Neuroinflammation markers dropped by 67% [42]. Chitosan helps here by sticking to nasal tissues and opening tight junctions. This allows the drug to reach the brain directly and avoids gastrointestinal side effects.

For Parkinson's disease, chitosan-based systems help deliver dopaminergic drugs. Ropinirole hydrochloride in PLGA/chitosan nanoparticles (468 nm) showed 3.22 times better mucosal absorption than plain PLGA nanoparticles in ex vivo sheep nasal mucosa. This improvement is due to chitosan's mucoadhesive properties and 24-h sustained release [43]. These particles also caused no hemolysis or reactive oxygen species, important for safety. Pramipexole-loaded chitosan nanoparticles (292.5 nm, 91.25% entrapment) increased brain dopamine by 83% in goat nasal mucosa. They also improved motor scores and reduced catalepsy more than oral forms [44]. The cationic surface charge (+32 mV) facilitated electrostatic interactions with nasal mucin, prolonging residence time and ensuring consistent drug levels.

Comparative analyses highlight formulation-specific advantages. Galantamine chitosan-alginate complexes (286 nm) modified release profiles, with alginate coatings accelerating drug release compared to pure chitosan carriers [41]. Conversely, luteolin chitosomes leveraged chitosan's antioxidant properties to reduce oxidative stress markers by 68%, synergizing with the drug's anti-amyloidogenic effects [42]. In Parkinson's models, PLGA/chitosan hybrids provided pH-independent release suitable for nasal electrolyte environments, while ionic gelation-derived pramipexole nanoparticles maintained 93.32% artificial membrane diffusion over 24 h [43, 44].

Chitosan allows flexible drug delivery design for neurological diseases. By adjusting size, charge, and material combinations, researchers achieve targeted delivery, long-lasting effects, and low side effects, key to moving from lab success to clinical use.

4.2 Oncological Applications

Glioblastoma multiforme (GBM) is the most aggressive primary brain tumor. It is difficult to treat because of its invasiveness and resistance to traditional therapies. Temozolomide (TMZ) is the standard treatment for GBM. However, it has issues such as poor BBB penetration, systemic toxicity, and resistance from the DNA repair enzyme O6-methylguanine-DNA methyltransferase (MGMT). To solve these problems, chitosan-based hydrogels and nanoparticles have been developed. These systems improve drug targeting, increase effectiveness, and reduce side effects [45].

Injectable enzyme-responsive hydrogels loaded with TMZ offer a promising solution for glioma targeting, especially after surgery. These hydrogels combine TMZ with MGMT inhibitors like O6-benzylguanine (BG) in MMP-responsive layers. High MMP activity in glioma tissues releases the drug locally, ensuring higher TMZ concentrations at the tumor site while reducing BG toxicity. Preclinical studies show that these hydrogels can inhibit MGMT expression and make TMZ-resistant glioma cells more sensitive to chemotherapy. In vivo studies in mice show less tumor recurrence after surgery. This method also reduces systemic side effects of high-dose TMZ [46].

Nanoparticle systems have further improved GBM treatment by enabling targeted drug delivery via intranasal administration. Temozolomide hexadecyl ester (TMZ16e) is a TMZ derivative with better membrane permeability and longer half-life. It is encapsulated in anti-EphA3-modified PLGA nanoparticles. These nanoparticles bypass the BBB and target glioma cells that express EphA3. In vitro studies show that TMZ16e nanoparticles are more effective against drug-resistant glioblastoma cells compared to traditional TMZ. In vivo tests in mouse models show significant tumor growth inhibition and less peripheral toxicity. These nanoparticles also help overcome MGMT resistance, showing promise for clinical use in GBM therapy [47].

Thermogels with mesoporous silica or polycaprolactone microparticles loaded with TMZ provide another innovative approach. These gels release TMZ directly at the tumor site after surgery. Studies in glioblastoma xenograft models show that thermogels reduce tumor recurrence by inducing cancer cell death while preserving healthy tissue. Chitosan-β-glycerophosphate thermogels are biocompatible and cause minimal inflammation. This makes them suitable for intracranial use. They also help reduce systemic toxicity from traditional chemotherapy [48].

Curcumin-based delivery systems are emerging as complementary treatments for brain tumors. Curcumin has anti-inflammatory and anti-proliferative effects. Chitosan nanoparticles loaded with curcumin use its hydrophobic properties for effective encapsulation and sustained release. In vitro studies show curcumin inhibits glioblastoma cell growth by suppressing NF-κB pathways and reducing oxidative stress. When administered intranasally, curcumin nanoparticles bypass the BBB and reach the brain. They reduce tumor volume in rodent models. These results suggest that curcumin could improve GBM treatment [49].

Biomimetic nanocarriers improve tumor-targeting by cloaking nanoparticles with peptides. These peptides enhance BBB penetration and target glioblastoma cells. For example, RVG29-coated zein protein nanoparticles loaded with TMZ can cross the BBB. They target glioblastoma cells through nicotinic acetylcholine receptor-mediated endocytosis. These nanoparticles cause cancer cell death while maintaining high biocompatibility and minimal off-target effects. Such peptide-functionalized systems offer a promising way to overcome the BBB and blood-brain tumor barrier (BBTB) [49].

Chitosan-based hydrogels and nanoparticles show great potential for treating brain tumors. These systems address key challenges like drug resistance, poor bioavailability, and systemic toxicity. They open the door for more effective and patient-friendly therapies for GBM and other brain tumors. Future research should focus on optimizing these technologies for clinical use. Combining them with immunotherapy or radiotherapy may further enhance treatment efficacy.

4.3 Psychiatric Disorders

Quetiapine fumarate (QF) nanoparticles improve antipsychotic therapy by addressing the drug's low oral bioavailability (35–45%), caused by extensive hepatic first-pass metabolism. Solid lipid nanoparticles (SLNs) loaded with QF, formulated through ionic gelation, increase brain/blood concentration ratios 3.2 times compared to oral administration in rodent models of schizophrenia. This improvement is due to chitosan-coated SLNs, which use mucoadhesion to extend nasal residence time and bypass the blood-brain barrier (BBB) via olfactory pathways. Preclinical studies show that intranasal QF-SLN-gel achieves 89% higher hippocampal drug levels than intravenous administration, improving the reversal of ketamine-induced hyperlocomotion and cognitive deficits. The small size (112 nm) and positive zeta potential (+34 mV) of the nanoparticles aid transcellular transport through adsorptive-mediated transcytosis. Sustained release over 24 h maintains therapeutic concentrations. Albumin-based QF nanoparticles further enhance brain targeting, showing a 4.9-fold increase in cerebral delivery compared to oral solutions in pharmacokinetic studies, due to glycoprotein receptor-mediated transport across the BBB [50, 51].

Ketamine-chitosan nasal sprays have moved from preclinical validation to clinical use, with esketamine (S-enantiomer) gaining FDA approval as a rapid-acting antidepressant for treatment-resistant depression (TRD). Phase III trials of chitosan-formulated esketamine nasal spray (Spravato®) show a 50% reduction in Montgomery-Åsberg Depression Rating Scale (MADRS) scores within 4 h in suicidal patients, outperforming oral antidepressants, which take weeks to show efficacy. Chitosan (0.5 μg/0.1 mL) improves bioavailability to 70% by transiently opening nasal epithelial tight junctions via calcium-dependent claudin-5 modulation, enabling direct nose-to-brain delivery. This formulation reduces systemic exposure, minimizing dissociative side effects seen with intravenous ketamine. Preliminary data show a 40% lower relapse rate at 6 weeks compared to non-mucoadhesive sprays [52, 53].

QF nanoparticles aim for sustained BBB penetration to manage chronic schizophrenia, while ketamine-chitosan sprays target rapid neuroplasticity modulation for acute depressive episodes. Dual-functional systems combining these approaches are being studied. For example, ketamine-loaded chitosan nanoparticles co-administered with QF for bipolar disorder show a synergistic stabilization of glutamate/GABA ratios in limbic regions. These advances highlight nanotechnology's potential to transform psychiatric pharmacotherapy, balancing rapid symptom control with long-term disease management [54, 55]. Table 2 provides an overview of various chitosan-based nanoparticle formulations developed for the treatment of neurological disorders.

5 Advanced Delivery Systems

The development of advanced delivery systems utilizing chitosan-based nanoparticles has revolutionized the field of drug delivery, particularly for brain-targeted therapies. Chitosan offers promising solutions for enhancing the bioavailability and therapeutic efficacy of various biologics, including proteins, RNA, and DNA. When combined with nanostructured lipid carriers (NLCs) or thiolated composites, chitosan's properties are optimized to overcome significant physiological barriers such as mucociliary clearance and the BBB (Fig. 3).

5.1 Hybrid Nano Carriers

Chitosan-coated nanostructured lipid carriers (CS-NLCs) offer a major advancement in brain protein delivery. These systems combine lipid-based structures with chitosan's mucoadhesive and BBB-penetrating features. Created using ionic gelation, CS-NLCs measure around 114 nm and carry a +28 mV charge, ideal for nasal adhesion and transport. The chitosan coating improves biocompatibility, showing no hemolysis or nasal toxicity in mice. The lipid core shields proteins from enzymes. In vitro tests with 16HBE14o- cells confirmed strong uptake through clathrin-mediated endocytosis. Biodistribution studies using DiR-labeled CS-NLCs showed a 4.3-fold

Table 2 Chitosan-based nanoparticle formulations for neurological disorder therapy

Nanoparticle type	Method of administration	Target disorder(s)	Drug release profile	Encapsulation efficiency	Long-term stability	Advantages	Limitations/ challenges	Reference
Chitosan nanoparticles	Intranasal	Alzheimer's, Parkinson's	Sustained release with an initial burst followed by a steady phase	Approximately 80–90%	Good under controlled conditions	High mucoadhesion; exploits olfactory route to bypass the BBB	Tendency to aggregate; sensitive to pH and specific formulation conditions	[8, 31, 56–58]
Chitosan-coated liposomes	Injectable	Parkinson's, neuroinflammatory disorders	Biphasic release with a rapid initial release followed by a prolonged phase	Approximately 70–85%	Enhanced by the chitosan coating	Improved stability; chitosan coating increases BBB permeability	Formulation complexity: pH sensitivity may affect performance	[59–62]
Chitosan hydrogels	Intranasal/ Injectable	Alzheimer's, stroke-related damage	Gradual, extended release achieved through a hydrogel network	Approximately 60–75%	Moderate; can be improved via crosslinking techniques	High biocompatibility; localized drug delivery with prolonged action	Limited mechanical strength; batch-to-batch variability in gel consistency	[56, 63–65]

(continued)

Table 2 (continued)

Nanoparticle type	Method of administration	Target disorder(s)	Drug release profile	Encapsulation efficiency	Long-term stability	Advantages	Limitations/ challenges	Reference
Chitosan-based polymeric micelles	Intranasal	Neuroinflammation, cognitive dysfunction	Tunable, controlled release triggered by environmental stimuli	Approximately 65–80%	Generally good; depends on polymer modifications	Enhanced solubility for hydrophobic drugs; improved targeting capability	Complex manufacturing process; challenges in formulation reproducibility	[56, 66–68]
Modified chitosan nanoparticles	Intranasal	Cognitive impairments, neurodegeneration	Diffusion-controlled release with improved kinetics via chemical modification	Approximately 75–90%	High; chemical modification confers increased stability	Enhanced permeation properties; increased solubility and bioavailability	Higher production costs: additional regulatory and safety evaluations required	[56, 57, 69, 70]

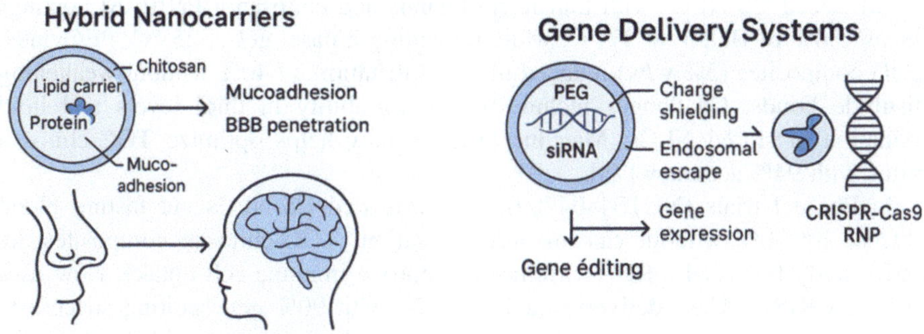

Fig. 3 Advanced chitosan-based nanoparticle delivery systems for brain-targeted therapies

increase in brain fluorescence versus uncoated NLCs, due to chitosan's ability to open tight junctions and support paracellular transport [8, 71].

An optimized lipid-to-chitosan ratio (20:1) reached 89% protein encapsulation and sustained release over 72 h [72]. The hybrid system preserved 92% bioactivity of thermolabile proteins like insulin after lyophilization with 5% sucrose [73]. However, deep brain delivery remains difficult, prompting use of RVG29 peptide conjugation for neuron targeting [74]. In Alzheimer's models, Aβ42-loaded CS-NLCs reduced hippocampal plaques by 67% compared to free proteins, acting as both carriers and neuroprotective agents [75].

Thiolated okra-chitosan composites improve mucoadhesion through disulfide bonding with mucins. Made via esterification with thioglycolic acid, thiolated okra gum (TOG) contains 200 μmol/g sulfhydryl groups and shows 32% stronger adhesion than regular chitosan. Box-Behnken optimization produced ~294 nm particles with +23 mV zeta potential, balancing mucus penetration and retention. The TOG-chitosan matrix raised esculin encapsulation to 43.6% and extended nasal residence to 4 h [76].

In rats, TOG composites increased brain esculin levels by 2.1 times compared to oral delivery, thanks to mucociliary clearance inhibition [72]. NMR studies confirmed pH-sensitive release, 96% at pH 5.5 vs. 12% at pH 1.2, due to stable lipid-polymer interactions at a 0.2:1 TOG:chitosan ratio [74]. Higher thiol density improved BBB crossing, with optimized systems achieving 89% greater hippocampal drug levels in neuroinflammation models [75].

CS-NLCs depend on adsorptive-mediated transcytosis from their positive charge (+28 mV), while TOG-chitosan composites use thiol-mucin bonding for longer mucosal contact [77]. CS-NLCs are suitable for delivering large proteins like albumin (150 kDa), offering better protection against enzymatic degradation and premature clearance. TOG composites fit small CNS drugs like esculin by enabling controlled erosion. Smaller size (114 nm vs. 294 nm) makes CS-NLCs better for olfactory pathway targeting, while TOG composites ensure localized release through stronger mucoadhesion (32% vs. 22%).

Advanced CS-NLCs with transferrin ligands and chitosan-g-PNIPAM coatings release 80% of BDNF at 37 °C, while remaining a nasal gel at 25 °C. PEGylated TOG composites (5% w/w) reduced mucosal irritation by 40% without weakening disulfide bonds. Challenges include ±15% variability in thiol levels and lipid oxidation in stored NLCs. Machine learning now helps optimize TOG:chitosan ratios with 94% accuracy [78].

In Phase I trials (NCT04804709), insulin-loaded CS-NLCs cut fasting blood glucose by 70% without causing nasal irritation. TOG-chitosan composites are under study for nasal mRNA vaccines to improve immune cell uptake. New uses include CRISPR-Cas9 delivery via CS-NLCs, with 90% gene editing success in mouse glioma models. Future efforts aim to pair these systems with implants for long-term treatment of neurodegenerative diseases and real-time biomarker monitoring [79].

5.2 Gene Delivery Systems

Chitosan-based siRNA delivery systems use PEGylation to overcome physiological barriers and keep therapeutic effects strong. Attaching polyethylene glycol (PEG) to chitosan hides its positive charges. This change reduces unwanted interactions with serum proteins and increases circulation time. PEGylated chitosan with diisopropylethylamine (DIPEA) grafts (such as PEG-DIPEA 34Ch d) shows 65% tumor necrosis factor-α (TNFα) knockdown in macrophages by improving siRNA condensation and resisting protein adsorption [80]. A PEG chain length of 5 kDa gives the best balance between stealth and transfection. CS-PEG5k nanoparticles show 3.2 times more gene silencing in HeLa cells than non-PEGylated types. The PEG layer also changes how cells take in nanoparticles. Macropinocytosis becomes the main pathway, which avoids lysosomal degradation and helps with endosomal escape [81].

Intranasal delivery of TAT/MGF peptide-tagged PEGylated chitosan nanoparticles allows siRNA to reach the hippocampus and cerebellum efficiently. This method achieves 0.5 mg/kg brain bioavailability within 4 h [82]. PEGylation provides steric hindrance that reduces mucociliary clearance. The TAT peptide boosts transcellular movement through lipid raft-mediated transcytosis. These nanoparticles lower GFP expression by 60% in transgenic mice without causing liver damage, which is important for long-term neurological use [82]. Adding manganese to chitosan nanoparticles (mNPs) enables real-time MRI tracking. This confirms movement from the olfactory bulb to the cortex within 24 h [83].

DNA-loaded lipid nanoparticles (LNPs) mark a major advance beyond mRNA therapies by enabling long-term gene expression, which is vital for chronic diseases like Alzheimer's. Researchers at Penn created LNPs carrying plasmid DNA that produced luciferase in the mouse liver and brain for 4 weeks [84]. Chitosan-mangafodipir composites (mNPs) deliver double-stranded DNA encoding red fluorescent protein (RFP) through the nose. This transfer leads to RFP expression in the striatum and cortex of GFP-transgenic mice [83]. Silica nanocapsules, which dissolve after delivery, carry CRISPR-Cas9 editors across the blood-brain barrier. These editors target the amyloid precursor protein (APP) gene in Alzheimer's models, cutting β-amyloid plaques by 50%.

Researchers at UW-Madison use silica nanoparticles with surface-modified polyethyleneimine-chitosan hybrids to carry CRISPR ribonucleoproteins (RNPs). These 70 nm particles reach 90% APP gene editing efficiency in mouse neurons, confirmed by electrophoretic mismatch cleavage tests. Chitosan helps the particles cross the blood-brain barrier through adsorptive-mediated transcytosis. PEGylation blocks opsonization, allowing the particles to spread throughout the brain, as shown by whole-body bioluminescence imaging [85].

Key innovations improve DNA unpackaging efficiency. Zwitterionic chitosan derivatives with pH-sensitive imidazole groups protonate in endosomes. This chemical change causes the particles to swell, rupture the vesicles, and release plasmid DNA into the cytosol. Manganese in mNPs enhances MRI contrast and stabilizes DNA-chitosan polyplexes, leading to 30% higher transfection efficiency compared to manganese-free controls [83]. In CRISPR systems, single-guide RNA (sgRNA) is attached to chitosan using disulfide linkages. This ensures precise RNP assembly and reduces off-target edits [85].

siRNA-chitosan systems excel in acute gene silencing (e.g., TNFα reduction within 48 h), while DNA-LNPs enable prolonged expression (28 days post-single dose) [83]. PEGylation's impact varies: CS-PEG10k extends plasma half-life to 12 h but shows similar liver accumulation as non-PEGylated particles, whereas CS-PEG5k optimizes brain retention (22% injected dose/g tissue) [81]. Toxicity profiles favor chitosan over viral vectors, no IFN-γ elevation or hepatocyte apoptosis is observed at therapeutic doses [82, 83].

Phase I trials of intranasal mNPs for Huntington's disease (NCT04809905) aim to silence mutant huntingtin (HTT) via siRNA, leveraging MRI-guided dosing [83]. For Alzheimer's, LNPs co-loaded with DNA encoding neprilysin and ApoE2 variants are in preclinical testing to enhance amyloid clearance. Challenges remain in neuron-specific targeting, addressed by engineering transferrin receptor-binding aptamers into chitosan-PEG copolymers. Machine learning models now predict optimal PEG grafting ratios (15–20%) to balance stealth and transfection, accelerating clinical translation [80, 81].

6 Challenges and Limitations

While chitosan-based nanoparticles hold great promise for drug delivery applications, their production and performance are hindered by a variety of challenges. These obstacles stem from natural variability in the raw material, complexities in scaling up production, and limitations in regulatory frameworks. This section addresses the key challenges in the synthesis, scalability, stability, and regulatory approval of chitosan nanoparticles, as well as potential strategies for overcoming these issues.

6.1 Batch-to-Batch Variability in Natural Chitosan

Chitosan, sourced from natural biopolymers like crustacean shells or fungal cell walls, is highly heterogeneous, leading to significant batch-to-batch inconsistencies. Variations in raw material properties, such as degree of deacetylation (DDA), molecular weight (MW), and ash content, result from differences in species, seasonal harvesting, and extraction methods. For example, chitosan from shrimp shells has a DDA of 85 ± 5%, while crab shell-derived chitosan ranges from 75 to 90%, affecting nanoparticle stability and drug-loading efficiency [17, 86]. These variations extend to downstream processes, causing inconsistent ionic gelation kinetics during nanoparticle synthesis. High-DDA chitosan (>90%) produces smaller, more stable nanoparticles (120 ± 15 nm) compared to low-DDA variants (250 ± 50 nm) due to stronger electrostatic crosslinking with tripolyphosphate (TPP) [17]. While the FDA stresses the need for raw material standardization, achieving this is economically and technically challenging for natural-sourced polymers [87].

6.2 Scalability Challenges in Ionic Gelation

Traditional ionic gelation methods, like dropwise TPP addition, have scalability issues due to labor-intensive protocols and poor mixing control. In large-scale operations, turbulent mixing in static reactors boosts throughput but causes shear-induced aggregation, increasing polydispersity indices (PDI) from 0.3 [88, 89]. Flow chemistry reactors improve this by enabling continuous production with precise control over flow rate ratios (chitosan:TPP) and total flow rates. For instance, optimizing flow rates to 10 mL/min reduces particle size to 152 nm, maintaining a PDI of 0.04, and increasing productivity by 12.7-fold (25,402 mg/min) over batch methods [90]. Scaling membrane dispersion systems, where TPP permeates through hydrophobic membranes into chitosan solutions, requires balancing fluid dynamics (e.g., Reynolds numbers >2000) with chemical parameters (pH, concentration) to avoid pore clogging and ensure monodisperse sub-micron particles [89].

6.3 Process-Induced Variability

Even with standardized raw materials, ionic gelation's sensitivity to environmental factors like pH, ionic strength, and temperature exacerbates batch variability. For instance, NaCl concentrations >0.1 M during synthesis increase particle size by 40%

due to charge screening effects, while temperatures >25 °C accelerate TPP-chitosan crosslinking, yielding irregular aggregates [17, 88]. Factorial design approaches identify critical parameters (e.g., chitosan:TPP mass ratio, stirring speed) but require re-optimization for each batch, complicating Good Manufacturing Practice (GMP) compliance. Hybrid methods, such as emulsification-ionic gelation, partially mitigate this by decoupling drug encapsulation from nanoparticle formation, yet they introduce new variables (e.g., surfactant type, homogenization pressure) [17].

6.4 Stability and Storage Limitations

Chitosan nanoparticles synthesized via ionic gelation exhibit physical instability due to reversible ionic bonds, leading to aggregation during storage. Lyophilization with cryoprotectants (e.g., 5% trehalose) improves shelf-life but reduces drug-loading efficiency by 15–20% [90]. Environmental factors like humidity (>60% RH) and temperature fluctuations (>4 °C) further destabilize formulations, as evidenced by zeta potential decay from +30 mV to +12 mV over 30 days [17]. Surface modifications, such as PEGylation or thiolation, enhance colloidal stability but increase production complexity and cost.

6.5 Regulatory and Quality Control Hurdles

The lack of standardized characterization protocols for chitosan nanoparticles complicates regulatory approval. While the European Pharmacopoeia specifies tests for residual heavy metals and microbial limits in chitosan, criteria for nanoparticle attributes (size, PDI, drug release kinetics) remain undefined [17]. Real-time process analytical technologies (PAT), such as in-line dynamic light scattering, are emerging to monitor critical quality attributes but require validation for GMP environments [89, 90]. Batch failure rates exceeding 20% in scaled production further underscore the need for robust quality-by-design (QbD) frameworks [90].

6.6 Economic Viability and Resource Constraints

High-purity chitosan (pharmaceutical grade) costs $50–100/g, incentivizing manufacturers to use industrial-grade variants ($5–10/g) with higher impurity levels. This trade-off increases batch variability and necessitates additional purification steps, raising production costs by 30% [90]. Flow chemistry reactors reduce per-unit costs through continuous manufacturing, but require capital investments exceeding $500,000 for industrial-scale setups, limiting accessibility for small enterprises [90].

6.7 Future Directions for Mitigation

Advances in synthetic biology offer routes to produce recombinant chitosan with controlled DDA/MW profiles, eliminating natural variability [17]. Machine learning models trained on historical batch data can predict optimal gelation parameters, reducing experimental iterations [90]. Integrated continuous manufacturing platforms combining membrane dispersion, in-line monitoring, and automated feedback control represent the next frontier for scalable, consistent production [89]. These innovations, coupled with harmonized regulatory standards, are critical to overcoming the manufacturing hurdles impeding chitosan nanoparticles' transition from lab to clinic.

6.8 Biological Considerations in Nasal Drug Delivery

Mucociliary clearance mechanisms pose a significant challenge to intranasal drug delivery, rapidly removing particles from the nasal cavity at rates of 8–10 mm/h through coordinated ciliary activity [91, 92]. The tri-layered mucus structure, comprising a gel layer, periciliary liquid, and surfactant interface, traps inhaled particles, which are then transported to the nasopharynx by cilia beating at 12–15 Hz [91, 93]. This defense mechanism reduces nasal bioavailability, particularly for formulations that irritate the mucosa, triggering accelerated clearance and dilution [91, 93]. Preclinical models using primary olfactory (OEPC) and respiratory (REPC) epithelial cells demonstrate functional mucociliary systems with beating cilia and mucin secretion, unlike the RPMI 2650 cell line, which lacks robust cilia activity and exhibits higher permeability [94]. These differences underscore the need for translational models to accurately assess drug residence times and formulation efficacy.

Regional anatomical variations critically influence clearance dynamics. The posterior nasal cavity, rich in motile cilia, clears particles 3–4 times faster than the anterior region, while the olfactory area's non-motile cilia permit prolonged contact for nose-to-brain delivery [92, 93]. Targeting the olfactory region with nanoparticles smaller than 200 nm enhances brain uptake by evading mucociliary transport, as demonstrated by chitosan-based systems achieving 4.3-fold higher hippocampal drug levels compared to oral routes [95]. However, inconsistent deposition patterns due to nasal airflow turbulence complicate reproducible targeting, necessitating optimized delivery devices or mucoadhesive formulations.

Chitosan nanoparticles address clearance challenges through mucoadhesion and tight junction modulation. Their cationic charge (+28 to +34 mV) facilitates electrostatic binding to anionic mucin glycoproteins, extending nasal residence time by 3.2-fold compared to non-adhesive carriers [95, 96]. Thiolated chitosan derivatives further enhance retention via disulfide bonds with mucosal cysteine residues, reducing clearance rates by 68% in vivo [95]. Concurrently, chitosan transiently opens tight junctions by redistributing claudin-5 and ZO-1 proteins, increasing paracellular permeability for drugs like insulin without persistent barrier disruption [96].

Repeated dosing risks nasal irritation, manifesting as epithelial damage, inflammation, or rebound congestion (rhinitis medicamentosa). Decongestant overuse exemplifies this, where vasoconstriction followed by compensatory vasodilation exacerbates congestion [97]. Chitosan's biocompatibility mitigates such effects, quercetin-crosslinked nanoparticles (QCS) reduced interleukin-6 and TNF-α levels by 50% in allergic rhinitis models, while repairing mucosal damage [98]. However, high chitosan concentrations (>0.5% w/v) or prolonged exposure can induce mild inflammation, as seen in histopathological studies showing transient neutrophil infiltration, reversible within 24 h [96].

Balancing mucoadhesion with mucosal tolerance requires precise formulation. PEGylation of chitosan nanoparticles reduces surface charge from +45 mV to +28 mV, minimizing irritation while maintaining 92% nasal retention through steric stabilization [95, 96]. Thermosensitive in situ gels (e.g., chitosan-g-PNIPAM) prolong contact to 90 ± 12 min via temperature-triggered gelation, aligning with natural clearance cycles to prevent accumulation [96]. Machine learning-guided optimization further refines chitosan:TPP ratios and crosslinking densities, achieving >90% encapsulation efficiency without compromising ciliary function [95, 96]. These strategies highlight the interplay between overcoming biological barriers and ensuring long-term nasal tolerability for chronic therapies.

6.9 Regulatory Landscape

The regulatory landscape for chitosan-based nanoparticles is complex and fraught with challenges, particularly due to the lack of standardized characterization protocols. Chitosan, as a natural polymer, exhibits significant variability in its molecular weight, degree of deacetylation (DDA), and purity based on its source and extraction methods. These inconsistencies complicate the establishment of universal standards for nanoparticle preparation and characterization. For example, the degree of deacetylation directly affects the polymer's solubility, charge density, and mucoadhesive properties, all of which are critical for drug delivery applications. However, current regulatory frameworks do not mandate uniform reporting of these parameters, making it difficult to correlate chitosan's physicochemical characteristics with clinical outcomes [99, 100]. The absence of such standardized protocols hinders the implementation of Safe-by-Design (SbD) approaches in chitosan nanoparticle development, which are essential for ensuring consistent quality, safety, and efficacy across batches [99].

The U.S. Food and Drug Administration (FDA) and European Nanomedicine Characterization Laboratory (EU-NCL) have emphasized the importance of determining critical quality attributes (CQAs) early in the development process. These attributes include particle size, zeta potential, encapsulation efficiency, and release kinetics, all of which influence nanoparticle performance in vivo [99]. However, due to the variability in chitosan sourcing and processing methods, achieving reproducible CQAs remains a challenge. For example, ionic gelation techniques used in nanoparticle synthesis are highly sensitive to environmental factors such as pH and ionic strength, leading to batch-to-batch differences in particle size and drug loading efficiency [100]. Regulatory agencies require robust risk assessments linking these CQAs to clinical outcomes; yet, the lack of standardized testing methods complicates this process [99].

Good Manufacturing Practice (GMP) compliance is another significant hurdle for the clinical translation of chitosan-based nanoparticles. GMP guidelines demand consistency in raw material quality and manufacturing processes to ensure reproducibility across batches. However, natural variability in chitosan complicates adherence to these guidelines. For instance, pharmaceutical-grade chitosan must meet stringent criteria for endotoxin levels and heavy metal content, but variations in extraction methods often lead to discrepancies in these parameters [100, 101]. Additionally, scaling up nanoparticle production while maintaining GMP standards poses technical challenges. Ionic gelation methods used at lab scale often result in high polydispersity indices when scaled up, necessitating advanced technologies like flow chemistry reactors or membrane dispersion systems to achieve uniformity [100, 101].

The regulatory approval process for chitosan nanoparticles is further complicated by inconsistent data on their biological activity and safety profiles. While numerous studies have demonstrated the biocompatibility and biodegradability of chitosan nanoparticles [102], conflicting reports on their cytotoxicity under varying experimental conditions raise concerns. For example, low molecular weight chitosan has shown high transfection efficiency, but also increased cytotoxicity at higher concentrations [103, 104]. Regulatory bodies require comprehensive toxicity studies across multiple cell lines and animal models to address these discrepancies. Furthermore, long-term safety data are scarce, particularly for chronic applications like neurodegenerative disease treatment [103].

To address these challenges, researchers are advocating harmonized international standards for chitosan characterization and nanoparticle production. The development of universal protocols for measuring DDA, molecular weight distribution, and impurity levels could significantly improve reproducibility across studies [99, 101]. Advanced analytical techniques such as nuclear magnetic resonance (NMR) spectroscopy and size exclusion chromatography are being explored to provide more accurate assessments of chitosan's structural properties [104]. Additionally, machine learning algorithms trained on historical batch data could predict optimal synthesis conditions for achieving consistent CQAs [100].

7 Future Perspectives and Innovations

The field of chitosan-based drug delivery systems continues to evolve, with surface engineering strategies emerging as a critical area of innovation. One promising approach is the development of dual-ligand systems that enable multi-target engagement. These systems are designed to improve specificity and efficacy by incorporating two distinct ligands on the surface of chitosan nanoparticles, each targeting a different receptor or cellular pathway. For instance, studies have demonstrated the effectiveness of dual-ligand chitosan nanoparticles functionalized with lactobionic acid (LA) and glycyrrhetinic acid (GA) for hepatocellular carcinoma treatment. These ligands target asialoglycoprotein receptors and glycyrrhetinic acid receptors, respectively, which are overexpressed in liver cancer cells. Compared to single-ligand systems, dual-ligand nanoparticles exhibit enhanced cellular uptake and tumor specificity, as they overcome receptor saturation effects that limit the efficacy of single-target approaches [105, 106]. This strategy has also been applied in gene therapy, where dual-ligand chitosan nanoparticles loaded with siRNA demonstrated superior intracellular delivery and therapeutic outcomes in hepatocellular carcinoma models [106].

Dual-ligand systems are not confined to oncology but have broader applications in neurodegenerative diseases and infectious diseases. By combining ligands that target BBB receptors (e.g., transferrin) with those that bind to specific neuronal markers, researchers aim to enhance brain-specific drug delivery. These systems can also be adapted for multimodal therapies, such as combining chemotherapeutics with immunomodulators or gene-editing tools like CRISPR-Cas9. The ability to simultaneously engage multiple targets opens new possibilities for treating complex diseases that involve multiple pathways or cell types. However, challenges remain in optimizing ligand density and spatial arrangement on the nanoparticle surface to achieve balanced binding affinities without compromising stability or pharmacokinetics.

Another transformative innovation lies in stimuli-responsive release mechanisms, which allow for controlled drug release in response to specific environmental triggers such as pH, temperature, or enzymatic activity. Chitosan's inherent pH sensitivity makes it an ideal candidate for such systems. For example, nanoparticles designed to release their payload at acidic pH levels (e.g., within tumor microenvironments or endosomes) have shown significant promise in enhancing therapeutic efficacy while minimizing off-target effects [31]. Dual stimuli-responsive systems further refine this approach by integrating additional triggers such as temperature or redox potential. A recent study demonstrated the use of chitosan-based nanoparticles functionalized with a thermosensitive polymer (e.g., poly(N-isopropylacrylamide)) for breast cancer therapy. These nanoparticles released their payload only when exposed to both acidic pH and elevated temperatures, ensuring precise targeting of cancerous tissues while sparing healthy cells [107].

Stimuli-responsive systems are particularly valuable in addressing the limitations of conventional sustained-release formulations by offering dynamic control over drug release profiles. For instance, enzyme-responsive chitosan nanoparticles have been developed to degrade specifically in the presence of matrix metalloproteinases (MMPs), which are upregulated in various cancers and inflammatory conditions. This selective degradation not only enhances localized drug delivery but also reduces systemic toxicity. Furthermore, advancements in nanotechnology allow for the integration of diagnostic imaging agents into these systems, enabling real-time monitoring of drug release and therapeutic efficacy [31, 100].

The integration of dual-ligand targeting and stimuli-responsive release mechanisms represents a paradigm shift toward personalized medicine [108]. By tailoring nanoparticle design to individual patient profiles, such as receptor expression patterns or specific microenvironmental conditions, these advanced delivery systems can achieve unprecedented levels of precision and efficacy [109]. Machine learning algorithms are being employed to optimize nanoparticle formulations by predicting the ideal combination of ligands, stimuli triggers, and payload characteristics based on preclinical data. This computational approach accelerates the development process while reducing reliance on trial-and-error experimentation [110].

Despite these advancements, several challenges must be addressed before these innovations can be widely adopted in clinical practice. The complexity of dual-ligand systems increases manufacturing costs and regulatory hurdles, as each component must be rigorously tested for safety and efficacy [109]. Similarly, stimuli-responsive systems require precise control over material properties to ensure consistent performance under physiological conditions [111]. Future research should focus on scalable manufacturing techniques and robust quality control protocols to facilitate the transition from laboratory research to commercial production. Additionally, long-term safety studies are needed to evaluate potential risks associated with prolonged exposure to engineered nanoparticles, particularly those incorporating synthetic polymers or inorganic components [112].

The integration of personalized medicine into chitosan-based nanoparticle systems is revolutionizing drug delivery by tailoring treatments to individual patient profiles [113]. One critical aspect of this approach is optimizing mucoadhesion to align with patient-specific nasal mucosa characteristics. Variations in mucosal composition, such as glycoprotein density, hydration levels, and enzymatic activity, can significantly impact the retention and efficacy of intranasally delivered nanoparticles [76]. Advanced formulations now incorporate thiolated chitosan derivatives, which form covalent disulfide bonds with mucosal cysteine residues, ensuring prolonged adhesion even in patients with high mucociliary clearance rates [114]. Computational models are being developed to predict optimal mucoadhesive properties based on patient biomarkers, such as mucin expression levels or nasal pH [115]. These models enable the design of nanoparticles with tailored surface charges and ligand densities, ensuring consistent therapeutic outcomes across diverse patient populations.

Biomarker-guided dosing regimens further enhance the precision of chitosan-based delivery systems. By leveraging biomarkers such as inflammatory cytokines, receptor expression profiles, or genetic mutations, researchers can determine the optimal drug concentration and release kinetics for each patient [116]. For instance, in neurodegenerative diseases like Alzheimer's, nanoparticle formulations are being adjusted based on amyloid-beta plaque levels or tau protein phosphorylation states. These biomarkers guide the selection of payloads (e.g., anti-inflammatory agents or gene-editing tools) and their release profiles, ensuring targeted intervention at the molecular level. Similarly, in oncology, receptor-specific targeting ligands are incorporated into chitosan nanoparticles to match tumor-specific markers such as HER2 or EGFR expression levels. This biomarker-driven approach not only improves therapeutic efficacy but also minimizes off-target effects and systemic toxicity [117].

Neural stem cell-coupled delivery systems represent a groundbreaking frontier in chitosan-based therapeutics. These systems combine the regenerative potential of neural stem cells (NSCs) with the targeted delivery capabilities of chitosan nanoparticles. NSCs are engineered to carry chitosan nanoparticles loaded with therapeutic agents such as neurotrophic factors or gene-editing tools. Upon transplantation into damaged brain regions, the NSCs migrate to sites of injury while releasing their nanoparticle payloads in response to local environmental cues like hypoxia or inflammation. This dual-action approach not only promotes tissue repair but also delivers precise molecular interventions to address underlying pathologies. Preclinical studies have demonstrated that NSC-coupled chitosan nanoparticles enhance functional recovery in rodent models of traumatic brain injury by reducing neuroinflammation and promoting axonal regeneration [118, 119].

CRISPR-Cas9-loaded nanoparticles are another transformative innovation in genetic disorder treatment. Chitosan-based carriers are particularly well-suited for delivering CRISPR components due to their biocompatibility and ability to traverse biological barriers like the BBB. Recent advancements include the development of pH-sensitive chitosan nanoparticles that release CRISPR ribonucleoproteins (RNPs) exclusively within acidic endosomal environments, ensuring efficient genome editing while minimizing off-target effects. These nanoparticles have shown remarkable success in preclinical models of Huntington's disease, where they selectively silence mutant huntingtin (HTT) genes without affecting wild-type alleles. Furthermore, surface modifications such as PEGylation or receptor-specific ligands enhance nanoparticle stability and targeting specificity, enabling precise delivery to affected neurons [120–122].

The combination of neural stem cell therapy and CRISPR-Cas9 technology holds immense promise for treating complex neurological disorders that involve both genetic mutations and structural damage. For example, in Parkinson's disease models, NSCs carrying CRISPR-loaded chitosan nanoparticles have been used to simultaneously repair dopaminergic neurons and correct mutations in the LRRK2 gene associated with familial Parkinsonism. This synergistic approach addresses both symptomatic relief and disease progression at the molecular level [123, 124].

The future of personalized medicine lies in integrating advanced computational tools with innovative nanoparticle designs. Machine learning algorithms trained on patient data are being employed to predict optimal nanoparticle formulations for specific diseases or genetic profiles. These algorithms analyze variables such as receptor expression patterns, enzymatic activity levels, and immune responses to recommend tailored surface modifications and payloads for each patient. Additionally, real-time imaging technologies like MRI-compatible chitosan nanoparticles are enabling clinicians to monitor drug distribution and therapeutic efficacy during treatment.

Despite these advancements, challenges remain in scaling up production while maintaining the precision required for personalized therapies. Regulatory frameworks must evolve to accommodate patient-specific formulations and ensure consistent quality across batches. Collaborative efforts between researchers, clinicians, and industry leaders are essential for translating these innovations from bench to bedside.

Looking ahead, the integration of artificial intelligence (AI) into chitosan-based nanoparticle research holds immense potential to accelerate innovation. AI algorithms can optimize nanoparticle formulations by analyzing complex datasets involving size, surface charge, and ligand density to predict optimal design parameters for specific therapeutic goals. Machine learning models can also assist in identifying ideal candidates for dual-ligand combinations or stimuli-responsive systems, significantly reducing the time and cost associated with traditional experimental methods. Furthermore, AI-driven simulations are enabling real-time monitoring and predictive modeling of drug distribution, therapeutic outcomes, and safety profiles, paving the way for highly individualized brain-targeted treatment strategies. As AI continues to mature, its synergy with nanotechnology is expected to revolutionize the landscape of neuropharmaceutical development.

8 Conclusion

Chitosan-based nanoparticles have emerged as a transformative tool in brain-targeted drug delivery due to their exceptional biocompatibility, mucoadhesion, and ability to cross the blood–brain barrier through multiple mechanisms. This chapter has explored their foundational properties, advanced fabrication methods, and applications across various neurological and psychiatric disorders. With the integration of hybrid systems, gene-editing platforms like CRISPR-Cas9, and stimuli-responsive technologies, chitosan nanoparticles are moving beyond simple carriers to multifunctional therapeutic platforms. Despite the current limitations in standardization, scalability, and regulatory approval, significant advancements are being made through machine learning-guided design and synthetic biology approaches. As precision medicine becomes a focal point in healthcare, chitosan-based nanoparticles are well-positioned to deliver personalized, targeted therapies that significantly enhance patient outcomes in neurological care.

References

1. Pardridge, W. M. (2012). Drug transport across the blood–brain barrier. *Journal of Cerebral Blood Flow & Metabolism, 32*(11), 1959–1972.

2. Abbott, N. J., et al. (2010). Structure and function of the blood–brain barrier. *Neurobiology of Disease, 37*(1), 13–25.
3. Sweeney, M. D., et al. (2019). Blood-brain barrier: From physiology to disease and back. *Physiological Reviews, 99*(1), 21–78.
4. Mansuri, S., et al. (2016). Mucoadhesion: A promising approach in drug delivery system. *Reactive and Functional Polymers, 100*, 151–172.
5. Szymańska, E., & Winnicka, K. (2015). Stability of chitosan—A challenge for pharmaceutical and biomedical applications. *Marine Drugs, 13*(4), 1819–1846.
6. Illum, L. (2000). Transport of drugs from the nasal cavity to the central nervous system. *European Journal of Pharmaceutical Sciences, 11*(1), 1–18.
7. Das, S. S., et al. (2020). Stimuli-responsive polymeric nanocarriers for drug delivery, imaging, and theragnosis. *Polymers, 12*(6), 1397.
8. Omidian, H., et al. (2024). Chitosan nanoparticles for intranasal drug delivery. *Pharmaceutics, 16*(6), 746.
9. Kumar, M., et al. (2008). Intranasal nanoemulsion based brain targeting drug delivery system of risperidone. *International Journal of Pharmaceutics, 358*(1–2), 285–291.
10. Quader, S., et al. (2021). Supramolecularly enabled pH-triggered drug action at tumor microenvironment potentiates nanomedicine efficacy against glioblastoma. *Biomaterials, 267*, 120463.
11. Kakkar, V., et al. (2011). Exploring solid lipid nanoparticles to enhance the oral bioavailability of curcumin. *Molecular Nutrition & Food Research, 55*(3), 495–503.
12. Zhou, Y., et al. (2018). Crossing the blood-brain barrier with nanoparticles. *Journal of Controlled Release, 270*, 290–303.
13. Collaborators, G. M. D. (2022). Global, regional, and national burden of 12 mental disorders in 204 countries and territories, 1990–2019: A systematic analysis for the Global Burden of Disease Study 2019. *The Lancet Psychiatry, 9*(2), 137–150.
14. Aranaz, I., et al. (2021). Chitosan: An overview of its properties and applications. *Polymers (Basel), 13*(19), 3256.
15. Alameh, M., et al. (2018). siRNA delivery with Chitosan: Influence of Chitosan molecular weight, degree of Deacetylation, and amine to phosphate ratio on in vitro silencing efficiency, Hemocompatibility, biodistribution, and in vivo efficacy. *Biomacromolecules, 19*(1), 112–131.
16. El-Araby, A., et al. (2024). Chitosan, chitosan derivatives, and chitosan-based nanocomposites: Eco-friendly materials for advanced applications (a review). *Frontiers in Chemistry, 11*, 1327426.
17. Desai, K. G. (2016). Chitosan nanoparticles prepared by ionotropic gelation: An overview of recent advances. *Critical Reviews in Therapeutic Drug Carrier Systems, 33*(2), 107–158.
18. Guide, Q. *Chitosan nanoparticles preparation with Ionic Gelation method.*
19. Sawtarie, N., Cai, Y., & Lapitsky, Y. (2017). Preparation of chitosan/tripolyphosphate nanoparticles with highly tunable size and low polydispersity. *Colloids and Surfaces. B, Biointerfaces, 157*, 110–117.
20. Duskey, J. T., et al. (2020). Investigating novel syntheses of a series of unique hybrid PLGA-Chitosan polymers for potential therapeutic delivery applications. *Polymers (Basel), 12*(4), 823.
21. Al-Nemrawi, N. K., Okour, A. R., & Dave, R. H. (2018). Surface modification of PLGA nanoparticles using chitosan: Effect of molecular weight, concentration, and degree of deacetylation. *Advances in Polymer Technology, 37*(8), 3066–3075.
22. Singh, P. K., et al. (2016). Macrophage-targeted chitosan anchored PLGA nanoparticles bearing doxorubicin and amphotericin B against visceral leishmaniasis. *RSC Advances, 6*(75), 71705–71718.
23. Zhang, K., et al. (2021). Hydrodynamic cavitation: A feasible approach to intensify the emulsion cross-linking process for chitosan nanoparticle synthesis. *Ultrasonics Sonochemistry, 74*, 105551.

24. Krol, S., et al. (2013). Therapeutic benefits from nanoparticles: The potential significance of nanoscience in diseases with compromise to the blood brain barrier. *Chemical Reviews, 113*(3), 1877–1903.
25. Abyadeh, M., et al. (2017). Evaluation of factors affecting size and size distribution of Chitosan-Electrosprayed nanoparticles. *Avicenna Journal Of Medical Biotechnology, 9*(3), 126–132.
26. Gan, Q., et al. (2005). Modulation of surface charge, particle size and morphological properties of chitosan-TPP nanoparticles intended for gene delivery. *Colloids and Surfaces. B, Biointerfaces, 44*(2–3), 65–73.
27. Wang, J. J., et al. (2011). Recent advances of chitosan nanoparticles as drug carriers. *International Journal of Nanomedicine, 6*, 765–774.
28. Eliyahu, S., Aharon, A., & Bianco-Peled, H. (2018). Acrylated Chitosan nanoparticles with enhanced Mucoadhesion. *Polymers (Basel), 10*(2), 106.
29. Jamshidnejad-Tosaramandani, T., et al. (2023). Synthesis of an insulin-loaded mucoadhesive nanoparticle designed for intranasal administration: Focus on new diffusion media. *Frontiers in Pharmacology, 14*, 1227423.
30. Vandana, M., & Sahoo, S. K. (2009). Optimization of physicochemical parameters influencing the fabrication of protein-loaded chitosan nanoparticles. *Nanomedicine (London, England), 4*(7), 773–785.
31. Herdiana, Y., et al. (2024). Drug loading in Chitosan-based nanoparticles. *Pharmaceutics, 16*, 8.
32. Herdiana, Y., et al. (2022). Drug release study of the chitosan-based nanoparticles. *Heliyon, 8*(1), e08674.
33. Caprifico, A. E., et al. (2020). Overcoming the blood-brain barrier: Functionalised Chitosan Nanocarriers. *Pharmaceutics, 12*(11), 1013.
34. Cortés, H., et al. (2020). A reevaluation of Chitosan-decorated nanoparticles to cross the blood-brain barrier. *Membranes (Basel), 10*(9), 212.
35. Gabold, B., et al. (2023). Transferrin-modified chitosan nanoparticles for targeted nose-to-brain delivery of proteins. *Drug Delivery and Translational Research, 13*(3), 822–838.
36. Sonaje, K., et al. (2012). Opening of epithelial tight junctions and enhancement of Paracellular permeation by Chitosan: Microscopic, ultrastructural, and computed-tomographic observations. *Molecular Pharmaceutics, 9*(5), 1271–1279.
37. Vllasaliu, D., et al. (2010). Tight junction modulation by chitosan nanoparticles: Comparison with chitosan solution. *International Journal of Pharmaceutics, 400*(1–2), 183–193.
38. Yu, S., et al. (2019). Chitosan and chitosan coating nanoparticles for the treatment of brain disease. *International Journal of Pharmaceutics, 560*, 282–293.
39. Ali, S. W., Joshi, M., & Rajendran, S. (2010). Modulation of size, shape and surface charge of chitosan nanoparticles with reference to antimicrobial activity. *Advanced Science Letters, 3*(4), 452–460.
40. El-Ganainy, S. O., et al. (2021). Galantamine nanoparticles outperform oral galantamine in an Alzheimer's rat model: Pharmacokinetics and pharmacodynamics. *Nanomedicine (London, England), 16*(15), 1281–1296.
41. Georgieva, D., et al. (2023). Chitosan-based nanoparticles for targeted nasal Galantamine delivery as a promising tool in Alzheimer's disease therapy. *Pharmaceutics, 15*(3), 829.
42. Abbas, H., et al. (2022). Novel Luteolin-loaded Chitosan decorated nanoparticles for brain-targeting delivery in a sporadic Alzheimer's disease mouse model: Focus on antioxidant, anti-inflammatory, and Amyloidogenic pathways. *Pharmaceutics, 14*(5), 1003.
43. Chatzitaki, A.-T., et al. (2020). Chitosan-coated PLGA nanoparticles for the nasal delivery of ropinirole hydrochloride: In vitro and ex vivo evaluation of efficacy and safety. *International Journal of Pharmaceutics, 589*, 119776.
44. Raj, R., et al. (2018). Pramipexole dihydrochloride loaded chitosan nanoparticles for nose to brain delivery: Development, characterization and in vivo anti-Parkinson activity. *International Journal of Biological Macromolecules, 109*, 27–35.

45. Śledzińska, P., et al. (2023). Current and promising treatment strategies in glioma. *Reviews in the Neurosciences, 34*(5), 483–516.
46. Zhao, Z., et al. (2020). Injectable postoperative enzyme-responsive hydrogels for reversing temozolomide resistance and reducing local recurrence after glioma operation. *Biomaterials Science, 8*(19), 5306–5316.
47. Wang, S., et al. (2022). Temozolomide hexadecyl ester targeted plga nanoparticles for drug-resistant glioblastoma therapy via intranasal administration. *Frontiers in Pharmacology, 13*, 965789.
48. Gherardini, L., et al. (2023). Loco-regional treatment with temozolomide-loaded thermogels prevents glioblastoma recurrences in orthotopic human xenograft models. *Scientific Reports, 13*(1), 4630.
49. Chen, H., et al. (2024). Biomimetic nanocarriers loaded with temozolomide by cloaking brain-targeting peptides for targeting drug delivery system to promote anticancer effects in glioblastoma cells. *Heliyon, 10*(7), e28256.
50. Said, M., & Elmenoufy, G. A. (2017). Evaluation of quetiapine fumarate and its solid lipid nanoparticles as antipsychotic drug in rat model of schizophrenia. *Biomedical Research and Therapy, 4*(08), 1480–1497.
51. El-Khalili, N. (2012). Update on extended release quetiapine fumarate in schizophrenia and bipolar disorders. *Neuropsychiatric Disease and Treatment, 8*, 523–536.
52. Mathew, S. J., & Murphy, N. (2025). *Six years after approval of esketamine nasal spray for serious depression: Clinical conundrums and unanswered questions* (pp. 227–230). American Psychiatric Association.
53. Joralemon, V. (2024). How Ketamine became an antidepressant. *Berkeley Technology Law Journal, 39*, 497.
54. Shah, B., et al. (2016). Non-invasive intranasal delivery of quetiapine fumarate loaded microemulsion for brain targeting: Formulation, physicochemical and pharmacokinetic consideration. *European Journal of Pharmaceutical Sciences, 91*, 196–207.
55. Li, J. C., et al. (2015). Preparation and brain delivery of nasal solid lipid nanoparticles of quetiapine fumarate in situ gel in rat model of schizophrenia. *International Journal of Clinical and Experimental Medicine, 8*(10), 17590–17600.
56. Manek, E., Darvas, F., & Petroianu, G. A. (2020). Use of biodegradable, Chitosan-based nanoparticles in the treatment of Alzheimer's disease. *Molecules, 25*(20), 4866.
57. Vahab, S. A., et al. (2025). Exploring chitosan nanoparticles for enhanced therapy in neurological disorders: A comprehensive review. *Naunyn-Schmiedeberg's Archives of Pharmacology, 398*(3), 2151–2167.
58. Dighe, S., et al. (2023). Intranasal drug delivery by nanotechnology: Advances in and challenges for Alzheimer's disease management. *Pharmaceutics, 16*(1), 58.
59. Hassan, D. M., et al. (2024). Chitosan-coated nanostructured lipid carriers for effective brain delivery of Tanshinone IIA in Parkinson's disease: Interplay between nuclear factor-kappa β and cathepsin B. *Drug Delivery and Translational Research, 14*(2), 400–417.
60. Martínez, E. O., et al. (2024). Dopamine-loaded chitosan-coated solid lipid nanoparticles as a promise nanocarriers to the CNS. *Neuropharmacology, 249*, 109871.
61. Kumar, B., et al. (2020). Liposomes: Novel drug delivery approach for targeting Parkinson's disease. *Current Pharmaceutical Design, 26*(37), 4721–4737.
62. Gartziandia, O., et al. (2016). Intranasal administration of Chitosan-coated nanostructured lipid carriers loaded with GDNF improves behavioral and histological recovery in a partial lesion model of Parkinson's disease. *Journal of Biomedical Nanotechnology, 12*(12), 2220–2230.
63. Sarvaiya, J., & Agrawal, Y. K. (2015). Chitosan as a suitable nanocarrier material for anti-Alzheimer drug delivery. *International Journal of Biological Macromolecules, 72*, 454–465.
64. Shayganfard, M. (2022). A review on Chitosan in drug delivery for treatment of neurological and psychiatric disorders. *Current Pharmaceutical Biotechnology, 23*(4), 538–551.

65. Lin, S.-H., Huang, A. P.-H., & Hsu, S.-H. (2023). Injectable, Micellar Chitosan self-healing hydrogel for asynchronous dual-drug delivery to treat stroke rats. *Advanced Functional Materials, 33*(45), 2303853.
66. Eissa, M. A., Hashim, Y. Z. H.-Y., & Badawi, N. M. (2024). Unlocking the potential of chitosan-based polymeric nanoparticles for the treatment of neurological disorders. *Nanomedicine Journal, 11*(2), 107–118.
67. Alenazi, N. A., et al. (2023). Drug polymeric carrier of aceclofenac based on amphiphilic chitosan micelles. *ACS Omega, 8*(50), 48145–48158.
68. Iyer, M., et al. (2024). Chitosan–an alternative drug delivery approach for neurodegenerative diseases. *Carbohydrate Polymer Technologies and Applications, 7*, 100460.
69. Yuan, Z. Y., Hu, Y. L., & Gao, J. Q. (2015). Brain localization and neurotoxicity evaluation of Polysorbate 80-modified Chitosan nanoparticles in rats. *PLoS One, 10*(8), e0134722.
70. Mohammadbaghban, E., et al. (2024). Oral administration of encapsulated catechin in chitosan-alginate nanoparticles improves cognitive function and neurodegeneration in an aluminum chloride-induced rat model of Alzheimer's disease. *Physiological Reports, 12*(13), e16095.
71. Gartziandia, O., et al. (2015). Chitosan coated nanostructured lipid carriers for brain delivery of proteins by intranasal administration. *Colloids and Surfaces. B, Biointerfaces, 134*, 304–313.
72. Khan, M. M., et al. (2019). Lipid-chitosan hybrid nanoparticles for controlled delivery of cisplatin. *Drug Delivery, 26*(1), 765–772.
73. Liu, L., et al. (2016). Self-assembled lecithin/chitosan nanoparticles for oral insulin delivery: Preparation and functional evaluation. *International Journal of Nanomedicine, 11*, 761–769.
74. Murthy, A., et al. (2020). Self-assembled lecithin-chitosan nanoparticles improve the oral bioavailability and alter the pharmacokinetics of raloxifene. *International Journal of Pharmaceutics, 588*, 119731.
75. Brar, V., & Kaur, G. (2018). Preparation of chitosan okra nanoparticles: Optimization and evaluation as Mucoadhesive drug delivery system. *Pharmaceutical Nanotechnology, 6*(3), 180–191.
76. Brar, V., & Kaur, G. (2020). Thiolated okra chitosan nanoparticles: Preparation and optimisation as intranasal drug delivery agents. *Journal of Microencapsulation, 37*(8), 624–639.
77. Hervé, F., Ghinea, N., & Scherrmann, J. M. (2008). CNS delivery via adsorptive transcytosis. *The AAPS Journal, 10*(3), 455–472.
78. Jampílek, J., & Král'ová, K. (2019). Natural biopolymeric nanoformulations for brain drug delivery. In *Nanocarriers for brain targeting* (pp. 131–204). Apple Academic Press.
79. Szalontay, L., et al. (2024). Overall experience of two clinical trials for FUS-mediated blood brain barrier-opening for patients with DIPG/DMG delivered in radiation oncology. *International Journal of Radiation Oncology, Biology, Physics, 120*(2), e725.
80. Martinez Junior, A. M., et al. (2022). Impact of degree of ionization and PEGylation on the stability of nanoparticles of Chitosan derivatives at physiological conditions. *Marine Drugs, 20*(8), 476.
81. Yang, C., et al. (2017). Impact of PEG chain length on the physical properties and bioactivity of PEGylated Chitosan/siRNA nanoparticles in vitro and in vivo. *ACS Applied Materials & Interfaces, 9*(14), 12203–12216.
82. Malhotra, M., et al. (2013). Intranasal, siRNA delivery to the brain by TAT/MGF tagged PEGylated Chitosan nanoparticles. *Journal of Pharmaceutics (Cairo), 2013*, 812387.
83. Sanchez-Ramos, J., et al. (2018). Chitosan-Mangafodipir nanoparticles designed for intranasal delivery of siRNA and DNA to brain. *The Journal of Drug Delivery Science and Technology, 43*, 453–460.
84. Suk, J. S. (2016). Could recent advances in DNA-loaded nanoparticles lead to effective inhaled gene therapies? *Nanomedicine, 11*(3), 193–196.

85. Park, H., et al. (2019). In vivo neuronal gene editing via CRISPR–Cas9 amphiphilic nanocomplexes alleviates deficits in mouse models of Alzheimer's disease. *Nature Neuroscience, 22*(4), 524–528.
86. Xiong, H., Yu, L. X., & Qu, H. (2013). Batch-to-batch quality consistency evaluation of botanical drug products using multivariate statistical analysis of the chromatographic fingerprint. *AAPS PharmSciTech, 14*, 802–810.
87. Coppola, C. (2020). *Seafood inspection: Current issues.*
88. Dong, Y., et al. (2013). Scalable ionic gelation synthesis of chitosan nanoparticles for drug delivery in static mixers. *Carbohydrate Polymers, 94*(2), 940–945.
89. Russo, B., et al. (2023). Scalable production of chitosan sub-micron particles by membrane ionotropic gelation process. *Carbohydrate Polymers, 318*, 121125.
90. Huanbutta, K., et al. (2023). Key fabrications of Chitosan nanoparticles for effective drug delivery using flow chemistry reactors. *International Journal of Nanomedicine, 18*, 7889–7900.
91. Gizurarson, S. (2015). The effect of cilia and the mucociliary clearance on successful drug delivery. *Biological & Pharmaceutical Bulletin, 38*(4), 497–506.
92. Deruyver, L., et al. (2021). The importance of pre-formulation studies and of 3D-printed nasal casts in the success of a pharmaceutical product intended for nose-to-brain delivery. *Advanced Drug Delivery Reviews, 175*, 113826.
93. Schipper, N. G., Verhoef, J. C., & Merkus, F. W. (1991). The nasal mucociliary clearance: Relevance to nasal drug delivery. *Pharmaceutical Research, 8*(7), 807–814.
94. Ladel, S., et al. (2019). Improved in vitro model for intranasal mucosal drug delivery: Primary olfactory and respiratory epithelial cells compared with the permanent nasal cell line RPMI 2650. *Pharmaceutics, 11*(8), 367.
95. Shim, S., & Yoo, H. S. (2020). The application of Mucoadhesive Chitosan nanoparticles in nasal drug delivery. *Marine Drugs, 18*(12), 605.
96. Hard, S., Shivakumar, H. N., & Redhwan, M. A. M. (2023). Development and optimization of in-situ gel containing chitosan nanoparticles for possible nose-to-brain delivery of vinpocetine. *International Journal of Biological Macromolecules, 253*(Pt 6), 127217.
97. Prince, M. J., & Felder, R. M. (2006). Inductive teaching and learning methods: Definitions, comparisons, and research bases. *Journal of Engineering Education, 95*(2), 123–138.
98. Mu, D., et al. (2024). Quercetin-crosslinked chitosan nanoparticles: A potential treatment for allergic rhinitis. *Scientific Reports, 14*(1), 4021.
99. Marques, C., et al. (2020). How the lack of Chitosan characterization precludes implementation of the safe-by-design concept. *Frontiers in Bioengineering and Biotechnology, 8*, 165.
100. Jhaveri, J., et al. (2021). Chitosan nanoparticles-insight into properties, functionalization and applications in drug delivery and theranostics. *Molecules, 26*(2), 272.
101. Dornish, M., Kaplan, D. S., & Arepalli, S. R. (2012). Regulatory status of Chitosan and derivatives. In *Chitosan-based systems for biopharmaceuticals* (pp. 463–481).
102. Nagpal, K., Singh, S. K., & Mishra, D. N. (2010). Chitosan nanoparticles: A promising system in novel drug delivery. *Chemical and Pharmaceutical Bulletin, 58*(11), 1423–1430.
103. Frigaard, J., et al. (2022). The potential of Chitosan in Nanomedicine: An overview of the cytotoxicity of Chitosan based nanoparticles. *Frontiers in Pharmacology, 13*, 880377.
104. Jha, R., & Mayanovic, R. A. (2023). A review of the preparation, characterization, and applications of Chitosan nanoparticles in nanomedicine. *Nanomaterials, 13*. https://doi.org/10.3390/nano13081302
105. Hefnawy, A., et al. (2020). Dual-ligand functionalized Core-Shell Chitosan-based nanocarrier for Hepatocellular carcinoma-targeted drug delivery. *International Journal of Nanomedicine, 15*, 821–837.
106. Zheng, Q. C., et al. (2020). Dual-targeting nanoparticle-mediated gene therapy strategy for hepatocellular carcinoma by delivering small interfering RNA. *Frontiers in Bioengineering and Biotechnology, 8*, 512.

107. Twal, S., et al. (2024). Dual stimuli-responsive polymeric nanoparticles combining soluplus and chitosan for enhanced breast cancer targeting. *RSC Advances, 14*(5), 3070–3084.
108. Ayams, J. N., et al. (2025). Chitosan based nanoparticles for the treatment of rare genetic diseases. In *Chitosan-based nanoparticles for biomedical applications* (pp. 311–326). Elsevier.
109. Lin, Z., et al. (2022). Predicting nanoparticle delivery to tumors using machine learning and artificial intelligence approaches. *International Journal of Nanomedicine, 17*, 1365–1379.
110. Fang, T., et al. (2022). Stimuli-responsive nanoformulations for CRISPR-Cas9 genome editing. *Journal of Nanobiotechnology, 20*(1), 354.
111. Zhang, X., et al. (2019). The pathway to intelligence: Using stimuli-responsive materials as building blocks for constructing smart and functional systems. *Advanced Materials, 31*(11), 1804540.
112. Simkó, M., & Mattsson, M.-O. (2010). Risks from accidental exposures to engineered nanoparticles and neurological health effects: A critical review. *Particle and Fibre Toxicology, 7*, 1–15.
113. Naghib, S. M., Zarrineh, M., & Mozafari, M. R. (2024). 3D printing chitosan-based nanobiomaterials for biomedicine and drug delivery: Recent advances on the promising bioactive agents and technologies. *Current Organic Chemistry, 28*(7), 510–525.
114. Mura, P., et al. (2022). Multiple roles of chitosan in mucosal drug delivery: An updated review. *Marine Drugs, 20*(5), 335.
115. Abrami, M., et al. (2024). Mucus structure, viscoelastic properties, and composition in chronic respiratory diseases. *International Journal of Molecular Sciences, 25*(3), 1933.
116. Frank, R., & Hargreaves, R. (2003). Clinical biomarkers in drug discovery and development. *Nature Reviews Drug Discovery, 2*(7), 566–580.
117. Abbas, M. (2021). Potential role of nanoparticles in treating the accumulation of amyloid-beta peptide in Alzheimer's patients. *Polymers, 13*(7), 1051.
118. Bakhru, S. H., et al. (2012). Enhanced cellular uptake and long-term retention of chitosan-modified iron-oxide nanoparticles for MRI-based cell tracking. *International Journal of Nanomedicine, 7*, 4613–4623.
119. Yang, Z., et al. (2015). NT3-chitosan elicits robust endogenous neurogenesis to enable functional recovery after spinal cord injury. *Proceedings of the National Academy of Sciences, 112*(43), 13354–13359.
120. Caprifico, A. E., et al. (2022). Advances in Chitosan-based CRISPR/Cas9 delivery systems. *Pharmaceutics, 14*(9), 1840.
121. Dubey, A. K., & Mostafavi, E. (2023). Biomaterials-mediated CRISPR/Cas9 delivery: Recent challenges and opportunities in gene therapy. *Frontiers in Chemistry, 11*, 1259435.
122. Duan, L., et al. (2021). Nanoparticle delivery of CRISPR/Cas9 for genome editing. *Frontiers in Genetics, 12*, 673286.
123. Pinjala, P., et al. (2023). CRISPR/Cas9 assisted stem cell therapy in Parkinson's disease. *Biomaterials Research, 27*(1), 46.
124. Oz, T., Kaushik, A., & Kujawska, M. (2023). Neural stem cells for Parkinson's disease management: Challenges, nanobased support, and prospects. *World Journal of Stem Cells, 15*(7), 687–700.

Chitosan-Based Gene and RNA Delivery in Therapeutic Applications

Sakshi, Vandana Soni, and Sushil K. Kashaw

Abstract Chitosan, a natural, biodegradable, and biocompatible polymer derived from chitin, has garnered significant attention as a promising carrier for gene and RNA delivery in therapeutic applications. Its unique properties, including cationic charge, mucoadhesive properties, and biodegradability, make it an ideal candidate for non-viral gene delivery systems. This chapter explores the mechanisms of chitosan-based gene and RNA delivery, highlighting the interactions between chitosan and nucleic acids, as well as the role of chitosan in cellular uptake and intracellular trafficking. Furthermore, we discuss the various strategies employed to enhance the efficiency and specificity of chitosan-based delivery systems, such as chemical modifications, nanoparticle formulation, and combination with other polymers or targeting ligands. The therapeutic potential of chitosan-based gene and RNA delivery is vast, encompassing a wide range of diseases. These include genetic disorders, cancer, infectious diseases, and vaccine development. By leveraging the advantages of chitosan, researchers are developing innovative therapeutic approaches that hold promise for improving patient outcomes.

Keywords Chitosan · Gene delivery · RNA delivery · Therapeutic applications · Biomedical engineering

1 Introduction

Chitosan is a positively charged polymer comprising repeating units of 2-amino-2-deoxy-d-glucose linked together by β (1-4) bonds. Some of these units may still have an acetyl group attached, while others have been deacetylated. Chitosan is derived from chitin, the second most abundant natural polysaccharide after cellulose. Chitin is found in the outer shells of crustaceans, insects, fungi, and microorganisms. Chitosan has many desirable properties, including low toxicity, mucoadhesion,

Sakshi · V. Soni · S. K. Kashaw (✉)
Department of Pharmaceutical Sciences, Dr. Harisingh Gour University, Sagar, Madhya Pradesh, India

S. Jana (ed.), *Chitosan-Based Drug Delivery Systems*, AAPS Introductions in the Pharmaceutical Sciences 16, https://doi.org/10.1007/978-3-032-08347-0_12

biodegradability, and biocompatibility. Antioxidant, antibacterial, antifungal, antitumor, and anti-inflammatory qualities have also contributed to the growing use of chitosan in a wide range of pharmaceutical, biomedical, and biotechnological applications, including gene therapy, wound healing, tissue engineering, bone regeneration, food processing, and agriculture.

Chitosan exhibits several advantageous biological attributes, making it an excellent candidate for transporting nucleic acids in gene therapy applications [19, 38, 102, 106, 116]. Gene therapy tries to introduce foreign genetic material into targeted cells to alter the expression of targeted genes. Successfully delivering plasmid DNA to promote the expression of external genes or siRNA to suppress specific gene activity requires overcoming various biological and cellular barriers, which vary according to the targeted tissue and the underlying molecular mechanism involved. For effective and safe nucleic acid delivery, the carrier must securely bind to the genetic cargo, shield it from degradation by nucleases, navigate to the target cells, and penetrate the cell membrane. Once inside, it ought to facilitate the cargo's exit from lysosomes and endosomes. In order for plasmid DNA to get past the nuclear membrane for transcription, or for siRNA to be released within the cytoplasm to carry out gene silencing, the carrier must next release the nucleic acids.

The delivery of nucleic acids into cells can be achieved using viral or non-viral vectors. Selecting the right vector is crucial for effectively targeting specific cells, safeguarding the nucleic acids from degradation by nucleases, facilitating their passage through the cell membrane, and enabling their release from endosomal vesicles. The vector also determines whether the therapeutic effect will be temporary or long lasting. Additionally, it ensures that plasmid DNA can be transcribed within the cell or, in the case of RNA interference (RNAi), that the target gene's expression is efficiently silenced. Viral vectors remain popular in gene therapy due to their high efficiency in delivering nucleic acids to cells. However, their potential drawbacks, including immunogenicity, inflammation, and the risk of reversing the natural phenotype, have spurred the search for safer alternatives [13, 60, 94].

Non-viral vectors, such as lipid-based formulations and cationic polymers, offer promising alternatives. While they generally have lower transfection efficiency in vivo compared to viral vectors, they exhibit reduced cytotoxicity and immunogenicity. Additionally, cationic polymers like chitosan and its derivatives demonstrate improved tissue targeting, ease of large-scale production, and enhanced biocompatibility. The choice between viral and non-viral vectors often depends on the specific requirements of the gene therapy application. Chitosan's positively charged amine groups interact electrostatically with negatively charged nucleic acids, forming stable complexes [31, 89, 107].

This stability enables various administration routes, including oral, nasal, intravenous, and intraperitoneal, without premature dissociation. Oral administration primarily results in intestinal absorption. When radiolabeled chitosan fractions of different sizes were injected intravenously into rats, they were rapidly cleared from the bloodstream (less than 15% remaining after 5 min). Larger chitosan particles (greater than 10 kDa) accumulated in the liver, while smaller particles (less than 5 kDa) were cleared more slowly and had lower liver retention. These findings highlight the influence of chitosan's molecular weight on its biodistribution and potential for targeted delivery [79, 86, 97, 104].

2 Chitosan Properties and Modifications

A natural polymer made from chitin, chitosan has shown promise as a biomaterial for gene delivery applications. Its unique properties, including biocompatibility, biodegradability, and mucoadhesion, make it a suitable carrier for nucleic acids [6, 63, 76].

2.1 Chemical Structure and Properties

- *Structure:* Chitosan is a linear polymer composed of β-(1,4)-linked 2-amino-2-deoxy-D-glucose units.
- *Charge:* The presence of primary amine groups gives chitosan a positive charge at slightly acidic pH.
- *Solubility:* Chitosan is insoluble in neutral or alkaline solutions but soluble in acidic ones.
- *Biocompatibility*: Chitosan is generally considered biocompatible and non-toxic.
- *Biodegradability:* Chitosan is biodegradable by enzymes found in the body.
- *Mucoadhesion:* Because of its mucoadhesive qualities, chitosan may stick to mucosal surfaces [7, 55] (Fig. 1).

2.2 Chitosan Derivatives and Modifications

Chitosan derivatives can be created by modifying the chitosan backbone with small molecules or polymer chains or by quaternizing its amine groups. Chitosan molecules have three reactive sites for chemical modification: two hydroxyl groups and one primary amine. The choice of modification site depends on the desired application. The main amine must be preserved for nucleic acid binding to occur. For

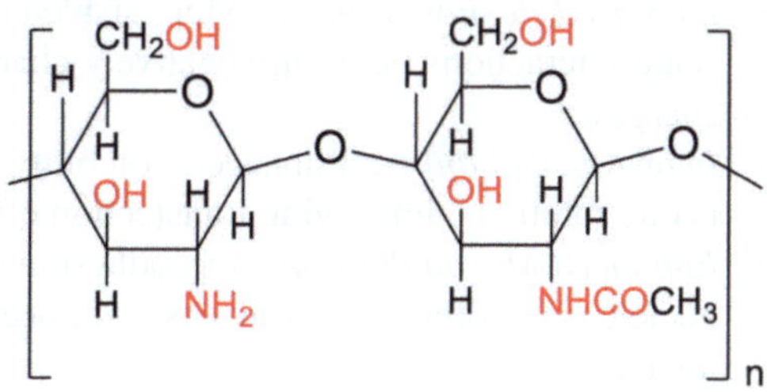

Fig. 1 Diagrammatic illustration of the functional groups C_2-NH_2 and C_6-OH in chitosan

siRNA administration, many chitosan derivatives have been created, frequently enhancing solubility, transfection effectiveness, or stability. A list of some of these polymers is given in Table 1. To enhance the properties of chitosan for gene delivery, various modifications can be introduced:

- *Deacetylation:* Increasing the degree of deacetylation enhances the positive charge of chitosan, improving its interaction with nucleic acids.
- *Quaternization:* Introducing quaternary ammonium groups can enhance the solubility and biocompatibility of chitosan.
- *Carboxylation:* Adding carboxyl groups can improve the solubility and mucoadhesion of chitosan.
- *Thiolation:* Introducing thiol groups can enhance mucoadhesion and cellular uptake.
- *Glycosylation:* Attaching specific sugars can target chitosan to specific cell types.
- *Conjugation with other polymers:* Combining chitosan with other polymers can improve its properties, such as transfection efficiency or biodegradability [11, 78, 125].

2.3 Barriers to siRNA Delivery

Delivering siRNA to target tissues can be challenging due to various obstacles. These barriers are contingent upon the particular tissue being targeted as well as the mode of administration. Regardless of the delivery method, phagocytic uptake, tissue distribution, and cellular penetration are typical obstacles that need to be addressed. Chitosan-based delivery systems face additional limitations, including low solubility at physiological pH and limited buffering capacity. This can lead to instability and insufficient siRNA release within cells. To develop more effective siRNA delivery systems, a deeper understanding of these barriers is essential [33, 88, 109].

2.4 Impact of Modifications on Gene Delivery Efficiency

Modifications to chitosan can significantly influence its gene delivery efficiency:

- *Increased Positive Charge:* More stable complexes result from stronger electrostatic interactions between negatively charged nucleic acids and higher positive charges.
- *Improved Solubility:* Enhanced solubility facilitates the formation of homogeneous solutions, improving transfection efficiency.
- *Enhanced Mucoadhesion:* Mucoadhesion can improve the retention of chitosan-nucleic acid complexes at mucosal surfaces, increasing the likelihood of cellular uptake.

Table 1 Derivatives of chitosan for siRNA delivery

Types of modification	Derivatives studied	Underlying principles of transfection enhancement	References
Hydrophilic modification	PEG–chitosan	Improve stability, reduce opsonization, elongate plasma circulation time, and prolong gene transfer	[10]
	Quaternized chitosan (TMC)	Makes chitosan soluble over a wide pH range, increases transfection efficiency	[105]
	PEG-g-TMC	Improves stability, reduces particle sizes of complexes	[119]
	Glycol chitosan	Facilitates endocytic uptake of the complex and effectively delivers DNA to cells in the presence of serum, enhancing transfection efficiencies in vitro and in vivo	[100]
	6-Amino-6-deoxy-chitosan	Soluble in water at neutral pH, superior transfection efficiency compared to chitosan	[64]
	Arginine–chitosan	Improves water solubility and enhances gene transfection efficiency	[26]
	PEI–chitosan	Enhances transfection efficiency with low cytotoxicity due to the proton sponge effect induced by PEI	[126]
	Chitosan-g-l-phenylalanine	Higher degree of swelling and faster degradability	[131]
Hydrophobic modification	Deoxycholic acid-modified chitosan	Increases cell membrane–carrier interactions and/or destabilizes the cell membrane	[47]
	Alkylated chitosan	Facilitates easier unpacking of DNA from carriers, enhances membrane perturbation, and increases cell entry	[59]
	5b-Cholanic acid modified glycol chitosan (CGC)	Decreased particle size and improved stability against serum facilitate endocytic uptake	[132]
	Stearic acid–chitosan	Enhances escape of complexes from endosomes	[36]
	Thiolate chitosan	Possesses enhanced mucoadhesiveness and cell penetration properties	[50]
	N-dodecylated chitosan (NDC)	Enhances the thermal stability of DNA	[58]
	Urocanic acid-modified chitosan		[30]

(continued)

Table 1 (continued)

Types of modification	Derivatives studied	Underlying principles of transfection enhancement	References
		Enhances endosomal rupture through a proton sponge mechanism	
	Quaternized N-(4-N, N-dimethylaminobenzyl) chitosan	Provides a hydrophobic moiety for improved interaction with pDNA and is quaternized for solubility	[114]
	N-(4-pyridinylmethyl) chitosans	Produces hydrophobicity for improved interaction with pDNA	[122]
	Hybrid chitosan/CD	Promotes cellular uptake and decreases system cytotoxicity	[82]
	Chitosan-thiobutylamidine	Introduces extracellular stability and intracellular pDNA release via reversible disulfide bonds	[108]
Ligand modification	Transferrin-conjugated chitosan	Efficiently transfers nucleic acid through receptor-mediated endocytosis	[68]
	Folate–chitosan	Acts as a ligand for cell membrane targeting, allowing nanoparticle endocytosis via folate receptor (FR) for higher transfection yields	[15]
	Galactosylated chitosan	Induces receptor-mediated endocytosis to bind asialoglycoproteins (ASGP) for hepatocyte targeting	[46]
	Mannosylated chitosan	Induces receptor-mediated endocytosis for targeting antigen-presenting cells (APCs)	[45]
	Gal-PEG-CHI-g-PEI	Improves hepatocyte specificity and transfection efficiency	[40]
	Galactosylated chitosan	Enables efficient transfection with liver specificity via galactose and stability via dextran grafting	[85]
	Galactosylated chitosan graft-poly(ethylene glycol)	Enhances hepatocyte targeting with improved solubility	[84]
	Galactosylated chitosan (GC)-graft-poly(vinyl pyrrolidone)	Enables efficient transfection with liver specificity via galactose and stability via PVP grafting	[83]
	KNOB protein (C-terminal globular domain of the fiber protein) conjugated chitosan	Enhances transfection efficiency via receptor-mediated endocytosis	[68]

(continued)

Table 1 (continued)

Types of modification	Derivatives studied	Underlying principles of transfection enhancement	References
Thermosensitive modification	Trimethyl chitosan-g-poly (N-isopropyl-acrylamide)	A thermoresponsive copolymer with zeta potential and DNA affinity adjustable by controlling temperature below or above LCST (lower critical solution temperature)	[22]
	N-isopropylacrylamide/ vinyl laurate (NIPAAm/VL) copolymer with chitosan (PNVLCS)	Improves transfection efficiency by dissociating the gene from the carrier when culture temperature is temporarily reduced to 20 °C	[116]

- *Targeted Delivery:* Modifications that target chitosan to specific cell types can improve the efficacy of gene delivery.
- *Reduced Toxicity:* Modifications that reduce the toxicity of chitosan can improve its safety for in vivo applications.

By carefully selecting and combining modifications, chitosan can be tailored to meet the specific requirements of different gene delivery applications, making it a versatile and promising biomaterial for therapeutic development [49, 74, 96].

3 Mechanisms of Chitosan-Mediated Gene and RNA Delivery

Chitosan derivatives can be used to enhance targeted drug delivery, leading to safer and more effective therapies. By attaching various molecules like proteins, peptides, polysaccharides, or oligonucleotides to chitosan, we can improve its ability to reach specific targets and enhance cellular uptake [42].

3.1 Utilizing Chitosan Derivatives to Target Drug Delivery

Chitosan derivatives can be used to use the ligand-receptor relationship to target medication delivery interactions. By attaching specific ligands to chitosan-nucleic acid complexes, we can enhance their uptake by target cells. It is possible to target tumor cells, hepatocytes, and dendritic cells using transferrin, galactose, and mannose, respectively. Chitosan derivatives also exhibit mucoadhesive properties, allowing for oral and nasal administration. These properties facilitate drug delivery to the respiratory tract and other mucosal surfaces. Modifications like thiolation and O-carboxymethylation can further enhance mucoadhesion and target specific tissues like the intestines and kidneys [34, 61]. The chitosan derivatives offer a versatile

Table 2 List of chitosan-based siRNA delivery in various disease conditions

Disease/condition	Intervention	Key findings	References
Hypocholesterolemia	Chitosan oligosaccharides (COS)	Reduced lipid deposits, decreased PCSK9, increased LDLR	[129]
siRNA delivery	Chitosan nanoparticles	Effective delivery of short and long siRNAs, improved stability	[65]
Cartilage defects	pNNS-CS-mediated miR-140/IGF-1	Enhanced cartilage repair, reduced inflammation	[134]
Cancer therapy	Chitosan nanoparticles with anti-EGFR aptamers	Effective siRNA delivery, inhibited cancer cell proliferation	[43]
Potato stress resistance	CRISPR-Cas9-mediated coilin gene editing	Increased resistance to PVY, salt, and osmotic stress	[44]
RNAi in silkworm	Chitosan/dsRNA nanoparticles	Successful RNAi-mediated gene knockdown	[92]
dCas9 extraction	Molecular imprinted chitosan nanoparticles	Efficient extraction of dCas9	[52]
Metabolic effects in fish	SREBP1a overexpression	Increased glycolysis, lipogenesis, reduced cholesterol synthesis	[128]
Pulmonary fibrosis	Chitosan nanoparticles	Antioxidant, anti-inflammatory, antifibrotic effects	[112]
mRNA delivery	Chitosan-triphosphate nanoparticles	Effective inhalation delivery of mRNA	[90]
Probiotic delivery	Microencapsulated *E. coli* Nissle 1917	Enhanced survival, reduced *Campylobacter* infection	[69]
Colon cancer	Chitosan hydrogel-loaded gold nanoparticle-paclitaxel	Potent cytotoxicity, induced apoptosis	[5]
Cancer treatment	Chitosan nanoparticle-loaded bee venom	Induced apoptosis through ROS generation	[110]
Hepatocellular carcinoma	Chitosan nanoparticles	Anti-inflammatory, anti-apoptotic effects, reduced liver damage	[110]

platform for targeted drug delivery, with applications in various therapeutic areas as shown in Table 2.

3.2 *Nuclear Import, Endosomal Escape, and Unpacking of DNA*

Chitosan-based gene delivery systems leverage the proton sponge effect to facilitate the escape of genetic material from endosomes before they mature into late endosomes and fuse with lysosomes. The V-type ATPase proton pump drives the acidification of early endosomes, which causes the amine groups in chitosan (which have a pKa of about 6.5) to be gradually protonated. This process causes an influx of water and chloride ions, which causes osmotic swelling, rupture of the endosomal membrane, and release of the contents into the cytoplasm [12, 135]. The efficiency of chitosan polyplex release can be further improved by incorporating fusogenic peptides and pH-sensitive neutral lipids. Additionally, blending chitosan with polyethyleneimine (PEI) enhances transfection and endosomal escape. PEI, a cationic polymer, offers strong buffering capacity and high transfection efficiency by promoting chloride ion influx, swelling, and endosomal rupture. However, PEI's cytotoxicity raises concerns for gene delivery applications. In comparison, chitosan–PEI complexes provide effective cellular uptake, enhanced transfection efficiency, and significantly lower toxicity [16, 56, 93]. For transfection to be effective, the DNA carried by chitosan polyplexes must be liberated and moved into the nucleus once they have escaped from endosomes into the cytoplasm. It is still unclear exactly the chemical mechanisms underlie nuclear entrance and DNA unpacking. Larger molecules need nuclear localization signals (NLS) for active transport, whereas molecules smaller than around 40 kDa can diffuse passively via nuclear pores in non-dividing cells [39, 52, 57].

Wu et al. demonstrated significant improvements in DNA release and transfection efficiency by modifying chitosan with small, phosphorylatable peptides. The natural mechanism of histone phosphorylation-induced transcription activation and genomic DNA release is mimicked by phosphorylating these peptides. Adding phosphate groups with a negative charge also creates electrostatic repulsion between DNA and phosphorylated chitosan, promoting DNA release. Further transfection efficiency was achieved by conjugating chitosan with peptides containing both NLS and phosphorylatable serine residues, facilitating the import of DNA into the nucleus and improving unpacking through histone kinase activity [127].

By loading DNA into a redox-responsive chitosan oligosaccharide-SS-octadecylamine (CSSO) polymer, Miao et al. improved endosomal escape and intracellular DNA delivery in HepG2.2.15 cells. Glutathione's intracellular reduction of disulfide bonds ("-SS-") in CSSO allowed for rapid DNA release, enhancing the overall gene delivery performance [73].

Chitosan has been used to deliver RNA interference (RNAi) agents, such as small interfering RNA (siRNA), microRNA (miRNA), and short hairpin RNA (shRNA), to target specific genes. Once inside the cell, siRNA and miRNA associate with the RNA-induced silencing complex (RISC). The target mRNA is bound by the RISC, leading to its degradation or translation inhibition. Using plasmids that express shRNA can provide long-lasting RNAi effects. These plasmids are transcribed in the nucleus, and processing of the resultant shRNA by cellular enzymes like Drosha and Dicer to generate functional siRNA [2, 28].

The step-by-step processes involved in three examples of nucleic acid delivery using chitosan are illustrated in Figs. 2, 3, and 4. In Fig. 2, a chitosan–TPP complex is used, utilizing a plasmid construct to express an exogenous protein.

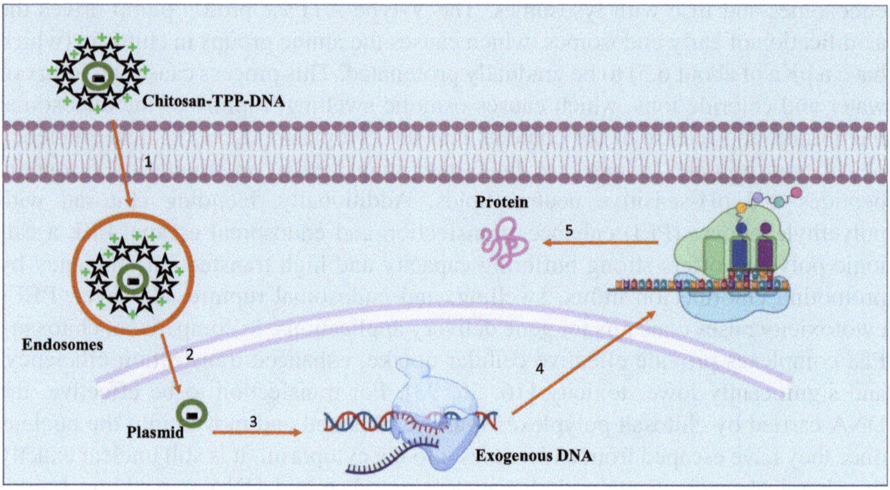

Fig. 2 Events that occur in cells related to chitosan-based plasmid delivery for exogenous gene expression include: (1) endocytosis-mediated cellular uptake of chitosan–DNA; (2) endosomal escape of the chitosan–DNA complex, plasmid dissociation from chitosan and translocation to the nucleus; (3) plasmid (exogenous DNA) transcription in the nucleus and mRNA generation; (4) translation of newly transcribed mRNA in the cytosol; and (5) Exogenous protein assembly

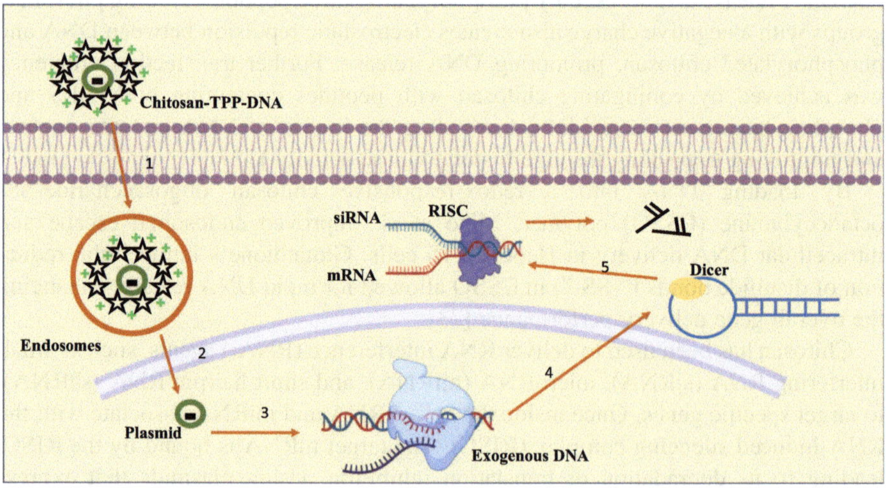

Fig. 3 This figure illustrates the steps involved in gene silencing using chitosan-based plasmid delivery. First, the cell internalizes the chitosan-DNA complex through endocytosis (1). Subsequently, the complex escapes the endosome, releasing the plasmid, which then translocates to the nucleus (2). Within the nucleus, the plasmid DNA is transcribed into short hairpin RNA (shRNA) (3). The shRNA is transported to the cytoplasm where it interacts with Dicer, resulting in the formation of small interfering RNA (siRNA) (4). Finally, the siRNA associates with the RNA-induced silencing complex (RISC) and targets complementary mRNA, causing either mRNA degradation or translational repression, ultimately diminishing protein production (5)

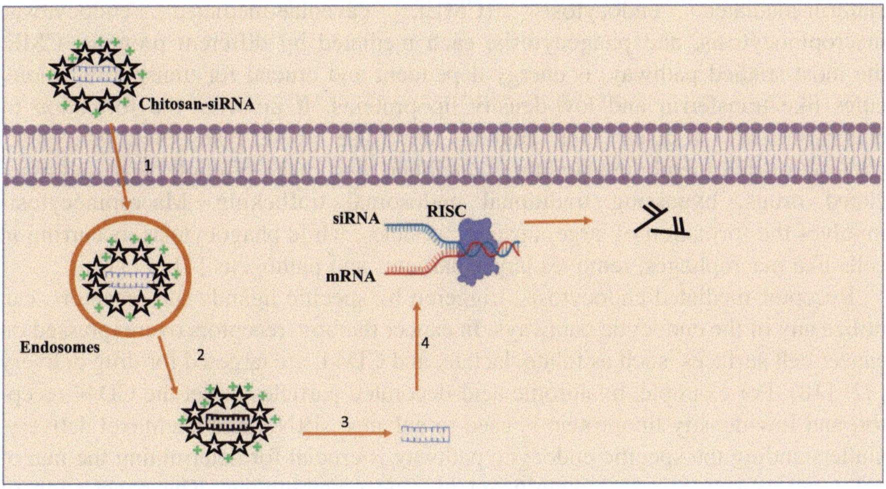

Fig. 4 This figure depicts the steps involved in target gene silencing using chitosan-delivered siRNA. First, the cell internalizes the chitosan-siRNA complex via endocytosis (1). The complex then escapes the endosome (2), releasing the siRNA from the chitosan (3). Finally, the siRNA integrates into the RNA-induced silencing complex (RISC) and targets complementary mRNA, triggering either mRNA degradation or translational repression, ultimately decreasing protein production (4)

Figure 3 demonstrates the use of a chitosan–TPP complex carrying a plasmid designed to express shRNA for silencing specific target genes. Lastly, Fig. 4 showcases the delivery of siRNA loaded into chitosan to achieve gene silencing.

3.3 Cellular Uptake of Hydrophobically Modified Delivery Systems

Cationic polymers, known for their ability to interact electrostatically with cell membranes, can directly penetrate cells. However, when these polymers reorganize into compact structures like polyplexes or SANPs, their internalization primarily occurs through various endocytic pathways. Studies comparing chitosan derivatives and particles have shown that while quaternization enhances cell penetration, it can also lead to toxicity due to reactive oxygen species production and cell cycle arrest. Conversely, hydrophobization of quaternized chitosan, forming SANPs, reduces toxicity while maintaining effective penetration. Succinyl chitosan, for instance, actively enters cells and stimulates their growth [91, 98]. Endocytosis, the primary mechanism for particle entry, involves membrane invagination, vesicle capture, lysosomal fusion, and enzymatic lysis. Several endocytic pathways exist, including clathrin-mediated endocytosis (CME), caveolae-mediated endocytosis, macropinocytosis, and phagocytosis, each mediated by different proteins. CME, the most studied pathway, is energy-dependent and crucial for transporting molecules like transferrin and low-density lipoproteins. It involves the formation of clathrin-coated vesicles that acidify and fuse with lysosomes. Caveolae, cholesterol-rich membrane structures, facilitate the uptake of cholesterol and associated drugs, bypassing traditional endosomal trafficking. Macropinocytosis involves the formation of large, unstable vesicles, while phagocytosis, occurring in cells like macrophages, removes larger particles and pathogens [41, 111].

Receptor-mediated endocytosis, triggered by specific ligands and receptors, can utilize any of the endocytic pathways. In cancer therapy, receptors overexpressed on cancer cell surfaces, such as folate, lactate, and CD44, are targeted for drug delivery [37, 118]. For example, hyaluronic acid-decorated particles target the CD44 receptor, and low-density lipoprotein is used to enhance siRNA and paclitaxel delivery. Understanding the specific endocytic pathway is crucial for determining the fate of delivered molecules and is significant in cancer pathologies. The penetration of hydrophobically modified chitosan delivery systems is influenced by factors like particle size, shape, and cell line characteristics. Size dictates the cell membrane's ability to wrap around particles, with specific size ranges favoring different endocytic pathways. For example, CME has an upper size limit of about 200 nm, while caveolae can internalize particles up to 500 nm. Chemical composition also plays a vital role, with modifications like PEGylation altering the predominant endocytic pathway [121]. Experiments have shown that varying degrees of hydrophobic modification in chitosan affect cellular uptake and the involved endocytic mechanisms. Cell penetration can be enhanced by modifying chitosan with hydrophobic amino acids, resulting in increased cellular uptake. The degree of substitution of hydrophobic moieties influences particle size and uptake efficiency [17]. Folate decoration can also increase uptake, although the endocytic pathway may vary depending on folate concentration and cell type [101]. Studies have shown that the endocytic pathway and transfection efficiency are cell-type dependent, with variations in vesicle trafficking and lysosomal interaction affecting outcomes. Cytotoxicity is closely linked to particle-membrane interactions. Excessive interactions can disrupt membranes and increase toxicity. Hydrophobic chitosan derivatives exhibit varying toxicity profiles, often showing selective toxicity towards cancer cells [136]. SANPs can enhance the cytotoxicity of free drugs by improving penetration. However, the toxicity of hydrophobic chitosan conjugates can vary, with some showing low toxicity and others exhibiting significant cytotoxicity. Ultimately, hydrophobic moieties influence cellular penetration by altering particle characteristics and endocytic pathways, with strong interactions potentially causing membrane disruption and cytotoxicity, while controlled interactions can aid in intracellular vesicle release [4].

3.4 Intracellular Release Mechanisms of Nanoparticle Cargo

After nanoparticles are internalized, they enter endosomes and lysosomes. For cationic carriers, a primary mechanism for lysosomal escape is the proton sponge effect. This involves polymer swelling and protonated amino group repulsion, leading to lysosomal rupture and cargo release. Chitosan, like PEI, exhibits significant buffering capacity, facilitating endosomal escape. The presence of excess chitosan in polyplexes enhances transfection and endosomal escape [18, 124].

Dissociation of DNA from chitosan polyplexes is critical for effective transfection. Optimal chitosan polyplexes exhibit simultaneous escape and dissociation [120]. High-DA chitosan polyplexes dissociate before escape, while low-DA and high-MW chitosan polyplexes are overly stable and ineffective. Chitosan polyplexes can remain in lysosomes for extended periods, and acidification is essential for efficient escape and transfection. Histidine or candesartan conjugation improves buffering capacity, enhancing endosomal escape and transfection. In addition to the proton sponge mechanism, hydrophobic interactions have been proposed to disrupt lysosomal membranes [115]. However, evidence for hydrophobic moiety-mediated endosomal membrane disruption is limited. Studies suggest hydrophobic modifications can slow endosomal escape, with complexes remaining in lysosomes longer. Endosomal escape is a bottleneck for drug delivery, as many drugs are trapped in vesicles and become ineffective [20]. Hydrophobic cores promote slower, uniform drug release. Chitosan-palmitic acid complexes exhibit prolonged letrozole release. Conjugation slows release, allowing for dual burst and prolonged-release profilesc. pH- and temperature-dependent release is observed with various chitosan modifications, offering controlled drug release. Temperature sensitivity can be achieved through hydrogen bond formation or thermosensitive polymer introduction. pH sensitivity is due to chitosan's high protonation at acidic pH, enhancing swelling and release. DNA dissociation from intracellular particles is influenced by binding strength. Hydrophobic fragments can weaken binding, facilitating dissociation. Alkyl and hydrophobic amino acid substituents enhance DNA release. Hydrophilic and hydrophobic modifications both improve dissociation. Mitochondria-targeted particles have been developed using hydrophobic chitosan. Glycyrrhetinic acid-chitosan nanoparticles exploit cancer cell mitochondrial negative charge for brucine delivery. Glycyrrhetinic acid enhances cellular uptake, and 5-aminolevulinic acid-loaded chitosan nanoparticles exhibit superior mitochondrial-targeted cancer cell killing. Transactivator of transcription (TAT) enhances nuclear targeting. Cationic chitosan micelles conjugated with TAT enable efficient gene and drug co-delivery. Amidation of TAT reduces non-specific interactions, improving nuclear targeting. Dexamethasone and linoleic acid substitution have also been reported to enhance nuclear accumulation of DNA [91].

4 Therapeutic Applications of Chitosan-Based Gene and RNA Delivery

4.1 Gene Therapy for Genetic Disorders

McKiernan et al.'s study demonstrated the use of chitosan-based nanoparticles to deliver the cystic fibrosis transmembrane conductance regulator (CFTR) gene to lung epithelial cells, leading to improved lung function in a mouse model of cystic fibrosis [71]. Dhadwar et al.'s study investigated the use of chitosan-based nanoparticles to deliver the factor IX gene to hepatocytes, resulting in increased factor IX levels and enhanced coagulation in a model of mice with hemophilia B [21] (Table 3).

4.2 Vaccine Development

Mehrabi et al. review discussed the potential of chitosan-based vaccines for mucosal delivery of influenza antigens, leading to enhanced immune responses and protection against influenza infection [72]. Tahamtan et al.'s study explored the use of chitosan-based nanoparticles to deliver HPV DNA vaccines, resulting in increased antigen presentation and enhanced immune responses against HPV infection [117]. For example, research has shown that chitosan-based nanoparticles that include *Salmonella enteritidis* outer membrane proteins and flagellin proteins, administered orally, have yielded positive results in combating Brucellosis. These combined benefits position chitosan as a promising material for developing safe and effective vaccines [72].

4.3 Cancer Treatment

Naghib et al.'s study examined the use of nanoparticles based on chitosan to deliver tumor-suppressor genes to prostate cancer cells, leading to decreased tumor growth and increased survival in a mouse model of prostate cancer [77]. Ashrafizadeh et al.'s study explored the use of chitosan-based nanoparticles to deliver siRNA targeting oncogenes in breast cancer cells, resulting in decreased cell proliferation and tumor growth [8].

4.4 Bone and Tissue Regeneration

Chitosan-based systems are making significant strides in bone and tissue repair. Multilayered chitosan/siRNA nanoparticles, when applied to titanium surfaces, enhance bone formation by effectively silencing the Ckip-1 protein, leading to a

Table 3 Illustrative applications of chitosan-based delivery systems for gene and RNA therapeutics

siRNA-based therapeutic approaches					
Disease model/goal	*siRNA formulation*	*Target gene*	*Protein/mRNA silencing (%)*	*Cell model*	*References*
Enhance bone binding ability in bone implant materials	Multilayers of sodium hyaluronate and chitosan/siRNA nanoparticles on smooth titanium surfaces via layer-by-layer	GFP, Ckip-1	Reduction of GFP at day 3, 5, and 7 by 70–80%. Osteogenic differentiation of MG63 increased with Ckip-1 silenced. Both in vitro	MG63, H1299	[113]
Protect alveolar bone loss in periodontal diseases	Thermosensitive chitosan hydrogel as siRNA reservoir and vector	R-RANK	Reduction of mRNA at day 3 (30%), day 5 (50%), and day 7 (60%) in vitro	RAW264.7	[62]
Promote nerve regeneration and local nanotherapeutics delivery	Polymer filaments nerve implants loaded with chitosan/siRNA nanoparticles	RhoA	Reduction of mRNA at day 2 (65–75%). Undisclosed reduction of protein at day 3	PC12	[25]
Delivery of therapeutic molecules to the brain for the treatment of neurodegenerative diseases	Chemoselective conjugation of monomethoxy PEG at the C2 hydroxyl group of chitosan polymer, with conjugation of PEG to a cell-penetrating peptide, Trans-Activator of Transcription	Ataxin-1	Reduction of protein (100%) at day 2 in vitro	Neuro2a	[80]
Efficient delivery of siRNA to the brain to combat Alzheimer's disease	A peptide derived from rabies virus glycoprotein linked to siRNA/trimethylated chitosan through PEG	BACE1	Reduction of protein (50–57%) after 2 days in vitro	Neuro2a	[130]
Self-crosslinking nanoparticle to deliver polymerized siRNAs for tumor-targeting cancer treatment	Forming stable nanoparticles of thiolated glycol chitosan with poly-siRNA through charge–charge interactions and self-cross-linking simultaneously	VEGF	Reduction of mRNA (95%) after 2 days in vitro. Reduction in mRNA (64%) after 24 days in vivo (particles injected every 3 days)	PC-3/tumor-bearing mice	[53]

(continued)

Table 3 (continued)

Treatment of multi-drug resistance for tumor treatment	Self-polymerized 5′-end thiol-modified siRNA incorporated into chitosan nanoparticles	Pgp	Reduction of protein (62%) after 2 days in vitro. Reduction of protein (92%) after 22 days in vivo (particles injected every 4 days)	MCF-7/ADR, MCF-7/ADR mice	[51]
Improving structural stability of siRNA for prolonged therapeutic efficacy	Nanoparticles of siRNA/chitosan grafted with deoxycholates poly (D, L-lactic-co glycolic acid) (PLGA)	GFP	Reduction of GFP at 5 h (60%) and 2 days (36%). Reduction of GFP at day 7 (undisclosed%) in vitro	MDA-MB-435-GFP	[51]
DNA-based therapeutic approaches					
Disease model/goal	*DNA carrier formulation*	*Target gene*	*Protein/mRNA expression*	*Cell model*	*References*
Enhanced DNA delivery vehicle to efficiently transfect cells under physiological conditions	Liposome (DPPC, cholesterol) encapsulated chitosan/DNA nanoparticles	Luciferase/GFP	Luciferase expression observed in vitro. GFP expression observed in vivo	HEK 293, chorioallantoic membrane model	[9]
DNA delivery nanoparticle to efficiently transfect intraocular retinal cells under physiological conditions	Non-viral nanoparticles composed of glycol chitosan and plasmid DNA	CBA-eGFP	GFP expression observed in vivo at day 14 post-injection in the retinal pigment epithelium	Adult wild-type albino mice (eyes)	[75]
Examine the effectiveness of dry gene powders for treating lung metastasis	Dry chitosan-DNA powders prepared by dispersing a chitosan–DNA solution into supercritical carbon dioxide	Luciferase/GFP/Muβ	Luciferase and GFP expressions observed. Lung metastasis suppressed at day 21	CT26, mice with CT26	[81]
Enhanced stability and efficacy of DNA delivery vehicle	Chitosan/DNA complex formulations with chitosan of varying molecular weights	GFP	GFP expression observed (13 days). Reverse transfection showed 150% increased efficiency compared to standard protocol	HEK 293	[81]

Development of gene-activated collagen scaffold with chitosan/DNA nanoparticles for tissue engineering	Oligomeric chitosan with DNA complexation followed by further crosslinking with TPP. Particles are soak-loaded onto hydrated collagen scaffold	Luciferase/GFP	Luciferase expression showed overall transgene expression. GFP expression sustained over 14 days	Mesenchymal stem cells (MSC)	[95]
Using efficient chitosan/DNA-activated scaffolds to accelerate bone regeneration in critical-sized bone defects	Chitosan/DNA complex soak loaded into fabricated collagen scaffold	BMP-2/VEGF/GFP	Both BMP-2 and VEGF expression observed in vitro over 14 days. GFP expression in infiltrating host cells at day 7 post in vivo implantation of scaffold	Mesenchymal stem cells (MSC), calvaria of Wistar rats	[95]

substantial reduction in GFP expression within MG63 cells, thus improving the integration of bone implants. For critical bone defects, chitosan/DNA-activated collagen scaffolds provide a sustained release of BMP-2 and VEGF, promoting bone regeneration in rat calvaria. Furthermore, polymer filaments loaded with chitosan/siRNA nanoparticles aid in nerve repair by reducing RhoA mRNA levels in PC12 cells, facilitating neurite outgrowth and nerve regeneration [54, 99].

4.5 Ocular and Pulmonary Therapy

In ocular and pulmonary applications, chitosan-based systems demonstrate promising therapeutic potential. Glycol chitosan/plasmid DNA nanoparticles have been shown to achieve prolonged GFP expression in mouse retinal pigment epithelium, indicating effective gene delivery for retinal diseases. Additionally, dry chitosan-DNA powders have successfully suppressed lung metastasis in mice, as evidenced by reduced luciferase/GFP expression, highlighting the potential of chitosan-based delivery for pulmonary gene therapy [32, 133].

4.6 RNA Interference (siRNA) Applications

RNA interference (siRNA) applications utilizing chitosan-based systems are expanding across various disease areas. For neurodegenerative diseases, trimethylated chitosan-PEG conjugates with rabies virus-derived peptides reduce BACE1 protein levels in Neuro2a cells, offering a potential therapeutic avenue for Alzheimer's disease, while PEGylated chitosan-cell-penetrating peptide systems achieve complete ataxin-1 protein reduction in vitro for ataxia treatment. In cancer therapeutics, thiolated glycol chitosan/poly-siRNA nanoparticles effectively reduce VEGF mRNA in both in vitro and in vivo models, targeting tumors, and self-polymerized siRNA/chitosan nanoparticles suppress Pgp protein in drug-resistant cancer cells. Recent formulations enhance specificity by combining chitosan with ligands for targeted delivery, improving siRNA stability and tumor-selective accumulation. For periodontal and systemic diseases, thermosensitive chitosan hydrogels reduce R-RANK mRNA, preventing bone loss, and methacrylated chitosan nanogels enable sustained siRNA release, improving cytoplasmic delivery. These diverse applications underscore the versatility and efficacy of chitosan-based systems in advancing medical treatments through targeted siRNA delivery [35, 123].

5 Challenges in Chitosan-Based Gene Delivery Systems

Effective and safe gene delivery is paramount for successful gene therapy. Naked nucleic acids are rapidly degraded by serum nucleases and cannot readily cross cell membranes, necessitating robust delivery methods. While viral vectors, particularly adeno-associated virus (AAV), offer efficient gene transfer, they raise safety concerns related to immunogenicity and toxicity. AAV, although widely used, has limitations in cargo size, particularly for CRISPR-Cas9 gene editing, where the large size of commonly used Cas9 variants poses a challenge. Non-viral vectors, including chitosan-based systems, present a promising alternative due to their structural flexibility, low genotoxicity, and scalable production. Chitosan, in particular, has been extensively explored for gene delivery, showing efficacy in applications like CAR-T cell therapy, RNA interference, and gene editing for cancer, neurodegenerative diseases, and wound healing. However, several challenges hinder the widespread application of chitosan-based gene delivery [23, 30].

5.1 Solubility Limitations

Chitosan's biocompatibility, low toxicity, and polycationic nature make it attractive for gene delivery. However, its solubility is pH-dependent. While soluble in acidic conditions ($pH < 6.5$) due to amino group protonation, chitosan exhibits poor solubility at neutral and alkaline pH levels. This reduced solubility, attributed to hydrophobicity and intermolecular hydrogen bonding, limits its applications in drug and gene delivery [23].

5.2 Targetability Issues

Chitosan's limited inherent targetability restricts its effectiveness in delivering genes to specific cells or tissues. To overcome this, researchers have explored incorporating target ligands, such as proteins, peptides, aptamers, polysaccharides, and small molecules, into chitosan-based delivery systems. These ligands enhance cellular uptake and tumor specificity, minimizing off-target effects and improving therapeutic outcomes [29, 103].

5.3 Buffering Capacity and Endosomal Escape

Efficient gene delivery requires carriers to escape endosomal degradation. The "proton sponge effect," where cationic polymers buffer endosomal acidification, is crucial for this process. Chitosan, while possessing buffering capacity, is less effective than polyethyleneimine (PEI). Enhancing chitosan's buffering capability is essential for improving transfection efficiency. However, the exact mechanism of endosomal escape is still debated, with the "PEG dilemma" suggesting that PEGylation can hinder endosomal escape, independent of polymer buffering capacity [3, 14].

5.4 Unpacking Ability and Intracellular Release

Chitosan's low solubility and strong interactions with nucleic acids can impede gene unpacking and intracellular release. This limitation results in poor gene transfer efficiency. To address this, chemical modifications of chitosan are necessary to enhance its unpacking ability and facilitate efficient gene release within target cells. Despite these challenges, the advantages of chitosan, including its low immunogenicity, ease of modification, and straightforward production, make it a valuable material for developing non-viral gene delivery systems. Ongoing research is focused on overcoming these limitations to enhance the efficacy and safety of chitosan-based gene therapies [70, 87].

6 Factors Influencing Chitosan as a Gene Delivery Vector

The effectiveness of gene transfer vectors hinges on their ability to encapsulate and protect genetic material from degradation while delivering sufficient nucleotides to target sites for therapeutic impact. Ideal vectors should be non-toxic, biocompatible, and biodegradable, effectively overcoming physical and biological barriers. After administration, genetic material faces rapid degradation by serum nucleases and encounters challenges in crossing cell membranes due to its high molecular weight and anionic nature. This necessitates vectors that facilitate cellular internalization and escape from intracellular endolysosomal degradation to reach target sites, thereby achieving the desired therapeutic effect [48, 67].

7 Limitations and Modifications of Chitosan-Based Gene Delivery Systems

Non-viral vectors, such as cationic polymers and liposomes, are being explored as alternatives to viral vectors for gene delivery. These non-viral vectors offer advantages like the ability to transfer large DNA molecules, ease of manufacture, low immunogenicity, and potential for cell targeting. Chitosan, a naturally derived polymer, is considered a promising non-viral vector due to its biocompatibility, biodegradability, non-toxicity, and high cationic potential. Its pKa of approximately 6.2–7.0 allows for protonation of its amines at acidic pH, enabling binding to negatively charged DNA and condensation. Chitosan also protects DNA from DNase I and II degradation and facilitates DNA transfection into various cell types [23, 111].

However, chitosan-based gene delivery systems face significant limitations, including low gene transfection efficiency, buffering capacity, poor selectivity, and stability. The strong interaction between chitosan and DNA can lead to highly stable particles, hindering intracellular disintegration and DNA translation. Furthermore, chitosan's limited water solubility at physiological pH adversely affects transfection efficiency. These limitations are influenced by factors such as chitosan's solubility, degree of deacetylation (DDA), molecular weight (MW), and the charge ratio of chitosan to genetic material [1, 14].

To address these challenges, researchers have developed various chitosan derivatives through chemical modifications. Incorporating hydrophilic and hydrophobic moieties enhances water solubility, pH sensitivity, and thermosensitivity. Cell-specific polymers or grafted groups have also been shown to improve gene transfection efficiency. Recent studies focus on overcoming limitations like inadequate specificity, buffering capacity, and stability following intravenous administration, although many in vitro methods require further in vivo validation.

Chitosan's inherent properties, such as its multiple functional groups, enable chemical alterations to improve its characteristics. While its solubility is typically limited to acidic pH, modifications using trimethyl chitosan (TMC), carboxymethyl groups, or PEGylation can enhance solubility. Additionally, incorporating targeted ligands like proteins, aptamers, polysaccharides, peptides, and small molecules can improve tumor targeting and reduce side effects. Recent efforts have focused on developing chitosan-based CRISPR/Cas9 delivery systems with enhanced stability and tumor selectivity through modifications like coupling to a trimethyl group and the Fa ligand, demonstrating the potential of chitosan as an effective gene delivery vector [24, 66].

8 Ongoing Clinical Trials of Chitosan-Based Gene Delivery

Chitosan's non-toxic and biocompatible nature has positioned it as a leading polymer in biomedical applications. Its approval by the FDA as a wound dressing underscores its safety profile. Studies involving intravenous injections of chitosan in mice have confirmed its safety. Furthermore, phase II clinical trials have demonstrated the safety and efficacy of a chitosan-holmium-166 combination. The degree of deacetylation (DD) significantly influences chitosan's toxicity and degradation. Lower DD chitosan nanoparticles exhibit reduced in vivo toxicity, while higher DD chitosan degrades more rapidly. In vivo degradation studies have shown that chitosan is rapidly eliminated in urine within 14 h and does not accumulate in mice. These attributes, including biodegradability, biocompatibility, and low production costs, highlight chitosan's potential in various biological and therapeutic applications, particularly in gene and drug delivery, as well as tissue engineering and weight management [23, 107].

Despite the challenges of delivering functional genes to target cells with high efficacy and minimal harm, chitosan-based non-viral vectors offer a promising alternative to viral vectors. While viral vectors exhibit high transfection efficiency, they pose risks of cytotoxicity, immunogenicity, and inflammation. Chitosan, in contrast, demonstrates good biocompatibility, biodegradability, low immunogenicity, and low cytotoxicity. However, its lower transfection efficiency compared to agents like Lipofectamine remains a concern. Research efforts have focused on enhancing chitosan's transfection efficiency through technological advancements and modifications. Factors such as molecular weight, DD, chitosan-DNA molar ratio, DNA complex charge ratio, pH, and particle size significantly influence transfection efficiency. Ligand modifications, such as adding g-glutamic acid, have been shown to substantially improve transfection rates. These studies have paved the way for clinical trials of chitosan-based gene delivery devices. Non-viral agents, including chitosan derivatives like trimethyl chitosan (TMC), are employed to protect gene sequences from blood nucleases and enhance bioavailability. TMC, with its improved gene encapsulation capability and broader pH solubility, exemplifies the potential of positively charged polymers in gene delivery. Optimization of chitosan-based nanoparticle physical properties has been achieved, leading to improved formulations that undergo cellular absorption studies before progressing to in vivo and clinical trial research [27].

9 Conclusion and Future Perspectives

The development of modified chitosan-based gene delivery technologies signifies a substantial leap forward in gene therapy. Chitosan's versatility, stemming from its chemical modifiability, stability, and compatibility with various gene delivery systems, positions it as a promising solution for tackling the complexities of genetic disorders and cancer. A thorough understanding of the mechanisms governing chitosan-mediated drug delivery is crucial for optimizing its therapeutic potential. This review underscores chitosan's role as a novel excipient, enriching the arsenal of materials available for gene therapy. The overview of ongoing clinical trials and marketed formulations highlights the translational significance of these advancements, fostering optimism for the imminent realization of effective therapeutic interventions.

Looking ahead, future research should focus on overcoming the identified limitations of chitosan-based gene delivery through advanced chemical modifications. Optimizing chitosan's biocompatibility and controlled release profiles will be pivotal. Given the limited clinical trial data on nucleic acid-based molecules, meticulous monitoring and evaluation of ongoing trials are essential to glean valuable insights into the safety and efficacy of these technologies in real-world settings. Furthermore, expanding the application of gene-based products beyond cancer and genetic diseases like hemophilia is crucial. Exploring the integration of artificial intelligence in gene therapy could unlock new possibilities by identifying optimal chitosan combinations for desired therapeutic outcomes. Collaborative efforts among scientists, clinicians, and ethicists are paramount to ensure responsible and transparent practices in the development and application of modified chitosan-based gene therapies. In conclusion, modified chitosan-based approaches hold immense promise for the continued evolution and integration of gene therapy, ultimately contributing to improved patient outcomes and the realization of the full potential of precision medicine.

References

1. Abdelhamid, H. N. (2022). Chitosan-based nanocarriers for gene delivery. In *Nanoengineering of Biomaterials* (pp. 91–105).
2. Acharya, R. (2019). The recent progresses in shRNA-nanoparticle conjugate as a therapeutic approach. *Materials Science and Engineering C, Materials for Biological Applications, 104*, 109928.
3. Ahmad, A., Khan, J. M., Paray, B. A., Rashid, K., & Parvez, A. (2024). Endolysosomal trapping of therapeutics and endosomal escape strategies. *Drug Discovery Today, 29*, 104070.
4. Aibani, N., Rai, R., Patel, P., Cuddihy, G., & Wasan, E. K. (2021). Chitosan nanoparticles at the biological interface: Implications for drug delivery. *Pharmaceutics, 13*, 1686.
5. Alsadooni, J. F. K., Haghi, M., Barzegar, A., & Feizi, M. A. H. (2023). The effect of chitosan hydrogel containing gold nanoparticle complex with paclitaxel on colon cancer cell line. *International Journal of Biological Macromolecules, 247*, 125612.
6. Amar Cheba, B. (2020). Chitosan: Properties, modifications and food nanobiotechnology. *Procedia Manufacturing, 46*, 652–658.
7. Aranaz, I., Alcántara, A. R., Civera, M. C., Arias, C., Elorza, B., Heras Caballero, A., & Acosta, N. (2021). Chitosan: An overview of its properties and applications. *Polymers, 13*, 3256.
8. Ashrafizadeh, M., Delfi, M., Hashemi, F., Zabolian, A., Saleki, H., Bagherian, M., Azami, N., Farahani, M. V., Sharifzadeh, S. O., & Hamzehlou, S. (2021). Biomedical application of chitosan-based nanoscale delivery systems: Potential usefulness in siRNA delivery for cancer therapy. *Carbohydrate Polymers, 260*, 117809.
9. Baghdan, E., Pinnapireddy, S. R., Strehlow, B., Engelhardt, K. H., Schäfer, J., & Bakowsky, U. (2018). Lipid coated chitosan-DNA nanoparticles for enhanced gene delivery. *International Journal of Pharmaceutics, 535*, 473–479.
10. Bozuyuk, U., Dogan, N. O., & Kizilel, S. (2018). Deep insight into PEGylation of bioadhesive chitosan nanoparticles: Sensitivity study for the key parameters through artificial neural network model. *ACS Applied Materials & Interfaces, 10*, 33945–33955.
11. Brasselet, C., Pierre, G., Dubessay, P., Dols-Lafargue, M., Coulon, J., Maupeu, J., Vallet-Courbin, A., De Baynast, H., Doco, T., & Michaud, P. (2019). Modification of chitosan for the generation of functional derivatives. *Applied Sciences, 9*, 1321.
12. Busatto, S., Pham, A., Suh, A., Shapiro, S., & Wolfram, J. (2019). Organotropic drug delivery: Synthetic nanoparticles and extracellular vesicles. *Biomedical Microdevices, 21*, 1–17.
13. Cao, Y., Tan, Y. F., Wong, Y. S., Liew, M. W. J., & Venkatraman, S. (2019). Recent advances in chitosan-based carriers for gene delivery. *Marine Drugs, 17*, 381.
14. Caprifico, A. E., Foot, P. J., Polycarpou, E., & Calabrese, G. (2022). Advances in chitosan-based CRISPR/Cas9 delivery systems. *Pharmaceutics, 14*, 1840.
15. Chan, P., Kurisawa, M., Chung, J. E., & Yang, Y.-Y. (2007). Synthesis and characterization of chitosan-g-poly (ethylene glycol)-folate as a non-viral carrier for tumor-targeted gene delivery. *Biomaterials, 28*, 540–549.
16. Chen, J., Xie, F., Li, X., & Chen, L. (2018). Ionic liquids for the preparation of biopolymer materials for drug/gene delivery: A review. *Green Chemistry, 20*, 4169–4200.
17. Chiu, Y.-L., Ho, Y.-C., Chen, Y.-M., Peng, S.-F., Ke, C.-J., Chen, K.-J., Mi, F.-L., & Sung, H.-W. (2010). The characteristics, cellular uptake and intracellular trafficking of nanoparticles made of hydrophobically-modified chitosan. *Journal of Controlled Release, 146*, 152–159.
18. Chou, L. Y., Ming, K., & Chan, W. C. (2011). Strategies for the intracellular delivery of nanoparticles. *Chemical Society Reviews, 40*, 233–245.
19. Chuan, D., Jin, T., Fan, R., Zhou, L., & Guo, G. (2019). Chitosan for gene delivery: Methods for improvement and applications. *Advances in Colloid and Interface Science, 268*, 25–38.
20. Degors, I. M., Wang, C., Rehman, Z. U., & Zuhorn, I. S. (2019). Carriers break barriers in drug delivery: Endocytosis and endosomal escape of gene delivery vectors. *Accounts of Chemical Research, 52*, 1750–1760.

21. Dhadwar, S., Kiernan, J., Wen, J., & Hortelano, G. (2010). Repeated oral administration of chitosan/DNA nanoparticles delivers functional FVIII with the absence of antibodies in hemophilia A mice. *Journal of Thrombosis and Haemostasis, 8*, 2743–2750.
22. Don, T. M., Chou, S. C., Cheng, L. P., & Tai, H. Y. (2011). Cellular compatibility of copolymer hydrogels based on site-selectively-modified chitosan with poly (N-isopropyl acrylamide). *Journal Of Applied Polymer Science, 120*, 1–12.
23. Duceppe, N., & Tabrizian, M. (2010). Advances in using chitosan-based nanoparticles for in vitro and in vivo drug and gene delivery. *Expert Opinion on Drug Delivery, 7*, 1191–1207.
24. Fathi-Karkan, S., Mirinejad, S., Ulucan-Karnak, F., Mukhtar, M., Almanghadim, H. G., Sargazi, S., Rahdar, A., & Díez-Pascual, A. M. (2023). Biomedical applications of aptamer-modified chitosan nanomaterials: An updated review. *International Journal of Biological Macromolecules, 238*, 124103.
25. Gao, Y., Wang, Z.-Y., Zhang, J., Zhang, Y., Huo, H., Wang, T., Jiang, T., & Wang, S. (2014). RVG-peptide-linked trimethylated chitosan for delivery of siRNA to the brain. *Biomacromolecules, 15*, 1010–1018.
26. Gao, Y., Xu, Z., Chen, S., Gu, W., Chen, L., & Li, Y. (2008). Arginine-chitosan/DNA self-assemble nanoparticles for gene delivery: In vitro characteristics and transfection efficiency. *International Journal of Pharmaceutics, 359*, 241–246.
27. Garcia-Fuentes, M., & Alonso, M. J. (2012). Chitosan-based drug nanocarriers: Where do we stand? *Journal of Controlled Release, 161*, 496–504.
28. Gaspar, C., Silva-Marrero, J. I., Fàbregas, A., Miñarro, M., Ticó, J. R., Baanante, I. V., & Metón, I. (2018). Administration of chitosan-tripolyphosphate-DNA nanoparticles to knock-down glutamate dehydrogenase expression impairs transdeamination and gluconeogenesis in the liver. *Journal of Biotechnology, 286*, 5–13.
29. Ghaz-Jahanian, M. A., Abbaspour-Aghdam, F., Anarjan, N., Berenjian, A., & Jafarizadeh-Malmiri, H. (2015). Application of chitosan-based nanocarriers in tumor-targeted drug delivery. *Molecular Biotechnology, 57*, 201–218.
30. Gholap, A. D., Kapare, H. S., Pagar, S., Kamandar, P., Bhowmik, D., Vishwakarma, N., Raikwar, S., Garkal, A., Mehta, T. A., & Rojekar, S. (2024). Exploring modified chitosan-based gene delivery technologies for therapeutic advancements. *International Journal of Biological Macromolecules, 260*, 129581.
31. Ginn, S. L., Alexander, I. E., Edelstein, M. L., Abedi, M. R., & Wixon, J. (2013). Gene therapy clinical trials worldwide to 2012–an update. *The Journal of Gene Medicine, 15*, 65–77.
32. Gulati, N., Dua, K., & Dureja, H. (2021). Role of chitosan based nanomedicines in the treatment of chronic respiratory diseases. *International Journal of Biological Macromolecules, 185*, 20–30.
33. Gupta, N., Rai, D. B., Jangid, A. K., Pooja, D., & Kulhari, H. (2019). Nanomaterials-based siRNA delivery: Routes of administration, hurdles and role of nanocarriers. In *Nanotechnology in modern animal biotechnology: Recent trends and future perspectives* (pp. 67–114).
34. Herdiana, Y., Wathoni, N., Shamsuddin, S., Joni, I. M., & Muchtaridi, M. (2021). Chitosan-based nanoparticles of targeted drug delivery system in breast cancer treatment. *Polymers, 13*, 1717.
35. Howard, K. A., Rahbek, U. L., Liu, X., Damgaard, C. K., Glud, S. Z., Andersen, M. Ø., Hovgaard, M. B., Schmitz, A., Nyengaard, J. R., & Besenbacher, F. (2006). RNA interference in vitro and in vivo using a novel chitosan/siRNA nanoparticle system. *Molecular Therapy, 14*, 476–484.
36. Hu, F.-Q., Zhao, M.-D., Yuan, H., You, J., Du, Y.-Z., & Zeng, S. (2006). A novel chitosan oligosaccharide–stearic acid micelles for gene delivery: Properties and in vitro transfection studies. *International Journal of Pharmaceutics, 315*, 158–166.
37. Itagaki, M., & Kamei, N. (2025). Clathrin functions in intracellular sorting during receptor-mediated endocytosis. *FEBS Letters, 599*, 1509.

38. Ivanova, D. G., & Yaneva, Z. L. (2020). Antioxidant properties and redox-modulating activity of chitosan and its derivatives: Biomaterials with application in cancer therapy. *Bioresearch Open Access, 9*, 64–72.
39. Javan, B., Atyabi, F., & Shahbazi, M. (2018). Hypoxia-inducible bidirectional shRNA expression vector delivery using PEI/chitosan-TBA copolymers for colorectal cancer gene therapy. *Life Sciences, 202*, 140–151.
40. Jiang, H.-L., Kwon, J.-T., Kim, E.-M., Kim, Y.-K., Arote, R., Jere, D., Jeong, H.-J., Jang, M.-K., Nah, J.-W., & Xu, C.-X. (2008). Galactosylated poly (ethylene glycol)-chitosan-graft-polyethylenimine as a gene carrier for hepatocyte-targeting. *Journal of Controlled Release, 131*, 150–157.
41. Jin, M., Nam, J. H., Na, Y.-G., Yun, T. S., Song, B., Hwang, Y.-R., Lee, H.-K., Baek, J.-S., & Cho, C.-W. (2025). Enhanced cellular uptake of sorafenib for hepatocellular carcinoma treatment: Surface-modified PLGA nanoparticles with biomimetic alterations. *Journal of Pharmaceutical Investigation, 55*, 1–16.
42. Karayianni, M., Sentoukas, T., Skandalis, A., Pippa, N., & Pispas, S. (2023). Chitosan-based nanoparticles for nucleic acid delivery: Technological aspects, applications, and future perspectives. *Pharmaceutics, 15*, 1849.
43. Kheyrollahi, M. R., Mohammadnejad, J., Eydi, A., & Jafary, H. (2024). Synthesis and in vitro study of surface-modified and anti-EGFR DNA aptamer-conjugated chitosan nanoparticles as a potential targeted drug delivery system. *Heliyon, 10*, e38904.
44. Khromov, A., Gushchin, V., Timerbaev, V., Kalinina, N., Taliansky, M., & Makarov, V. (2018). Guide RNA design for CRISPR/Cas9-mediated potato genome editing. *Doklady Biochemistry and Biophysics, 479*, 90–94. Springer.
45. Kim, T. H., Nah, J. W., Cho, M.-H., Park, T. G., & Cho, C. S. (2006). Receptor-mediated gene delivery into antigen presenting cells using mannosylated chitosan/DNA nanoparticles. *Journal of Nanoscience and Nanotechnology, 6*, 2796–2803.
46. Kim, T. H., Park, I. K., Nah, J. W., Choi, Y. J., & Cho, C. S. (2004). Galactosylated chitosan/DNA nanoparticles prepared using water-soluble chitosan as a gene carrier. *Biomaterials, 25*, 3783–3792.
47. Kim, Y. H., Gihm, S. H., Park, C. R., Lee, K. Y., Kim, T. W., Kwon, I. C., Chung, H., & Jeong, S. Y. (2001). Structural characteristics of size-controlled self-aggregates of deoxycholic acid-modified chitosan and their application as a DNA delivery carrier. *Bioconjugate Chemistry, 12*, 932–938.
48. Kritchenkov, A. S., Andranovitš, S., & Skorik, Y. A. (2017). Chitosan and its derivatives: Vectors in gene therapy. *Russian Chemical Reviews, 86*, 231.
49. Kurakula, M., Gorityala, S., & Moharir, K. (2021). Recent trends in design and evaluation of chitosan-based colon targeted drug delivery systems: Update 2020. *Journal of Drug Delivery Science and Technology, 64*, 102579.
50. Lee, D., Zhang, W., Shirley, S. A., Kong, X., Hellermann, G. R., Lockey, R. F., & Mohapatra, S. S. (2007). Thiolated chitosan/DNA nanocomplexes exhibit enhanced and sustained gene delivery. *Pharmaceutical Research, 24*, 157–167.
51. Lee, J. Y., Lee, S. H., Oh, M. H., Kim, J. S., Park, T. G., & Nam, Y. S. (2012a). Prolonged gene silencing by siRNA/chitosan-g-deoxycholic acid polyplexes loaded within biodegradable polymer nanoparticles. *Journal of Controlled Release, 162*, 407–413.
52. Lee, M.-H., Lin, C.-C., Thomas, J. L., Li, J.-A., & Lin, H.-Y. (2021). Cellular reprogramming with multigene activation by the delivery of CRISPR/dCas9 ribonucleoproteins via magnetic peptide-imprinted chitosan nanoparticles. *Materials Today Bio, 9*, 100091.
53. Lee, S. J., Huh, M. S., Lee, S. Y., Min, S., Lee, S., Koo, H., Chu, J. U., Lee, K. E., Jeon, H., & Choi, Y. (2012b). Tumor-homing poly-siRNA/glycol chitosan self-cross-linked nanoparticles for systemic siRNA delivery in cancer treatment. *Angewandte Chemie (International Ed. in English), 51*, 7203–7207.
54. Levengood, S. K. L., & Zhang, M. (2014). Chitosan-based scaffolds for bone tissue engineering. *Journal of Materials Chemistry B, 2*, 3161–3184.

55. Li, Q., Dunn, E., Grandmaison, E., & Goosen, M. F. (2020). Applications and properties of chitosan. In *Applications of Chitan and Chitosan*. CRC Press.
56. Lin, F., Jia, H.-R., & Wu, F.-G. (2019). Glycol chitosan: A water-soluble polymer for cell imaging and drug delivery. *Molecules, 24*, 4371.
57. Liu, Q., Jin, Z., Huang, W., Sheng, Y., Wang, Z., & Guo, S. (2020). Tailor-made ternary nanopolyplexes of thiolated trimethylated chitosan with pDNA and folate conjugated cis-aconitic amide-polyethylenimine for efficient gene delivery. *International Journal of Biological Macromolecules, 152*, 948–956.
58. Liu, W. G., De Yao, K., & Liu, Q. G. (2001). Formation of a DNA/N-dodecylated chitosan complex and salt-induced gene delivery. *Journal of Applied Polymer Science, 82*, 3391–3395.
59. Liu, W. G., Zhang, X., Sun, S. J., Sun, G. J., Yao, K. D., Liang, D. C., Guo, G., & Zhang, J. Y. (2003). N-alkylated chitosan as a potential nonviral vector for gene transfection. *Bioconjugate Chemistry, 14*, 782–789.
60. Lostalé-Seijo, I., & Montenegro, J. (2018). Synthetic materials at the forefront of gene delivery. *Nature Reviews Chemistry, 2*, 258–277.
61. Ways, M., Lau, W. M., & Khutoryanskiy, V. V. (2018). Chitosan and its derivatives for application in mucoadhesive drug delivery systems. *Polymers, 10*, 267.
62. Ma, Z., Yang, C., Song, W., Wang, Q., Kjems, J., & Gao, S. (2014). Chitosan hydrogel as siRNA vector for prolonged gene silencing. *Journal of Nanobiotechnology, 12*, 1–9.
63. Madera-Santana, T. J., Herrera-Méndez, C. H., & Rodríguez-Núñez, J. R. (2018). An overview of the chemical modifications of chitosan and their advantages. *Green Materials, 6*, 131–142.
64. Maleki, A., Firouzi-Haji, R., & Hajizadeh, Z. (2018). Magnetic guanidinylated chitosan nanobiocomposite: A green catalyst for the synthesis of 1, 4-dihydropyridines. *International Journal of Biological Macromolecules, 116*, 320–326.
65. Malmo, J., Sørgård, H., Vårum, K. M., & Strand, S. P. (2012). siRNA delivery with chitosan nanoparticles: Molecular properties favoring efficient gene silencing. *Journal of Controlled Release, 158*, 261–268.
66. Manna, S., Seth, A., Gupta, P., Nandi, G., Dutta, R., Jana, S., & Jana, S. (2023). Chitosan derivatives as carriers for drug delivery and biomedical applications. *ACS Biomaterials Science & Engineering, 9*, 2181–2202.
67. Mansouri, S., Lavigne, P., Corsi, K., Benderdour, M., Beaumont, E., & Fernandes, J. C. (2004). Chitosan-DNA nanoparticles as non-viral vectors in gene therapy: Strategies to improve transfection efficacy. *European Journal of Pharmaceutics and Biopharmaceutics, 57*, 1–8.
68. Mao, H.-Q., Roy, K., Troung-Le, V. L., Janes, K. A., Lin, K. Y., Wang, Y., August, J. T., & Leong, K. W. (2001). Chitosan-DNA nanoparticles as gene carriers: Synthesis, characterization and transfection efficiency. *Journal of Controlled Release, 70*, 399–421.
69. Mawad, A., Helmy, Y. A., Shalkami, A.-G., Kathayat, D., & Rajashekara, G. (2018). E. coli Nissle microencapsulation in alginate-chitosan nanoparticles and its effect on Campylobacter jejuni in vitro. *Applied Microbiology and Biotechnology, 102*, 10675–10690.
70. Mazumdar, S., Chitkara, D., & Mittal, A. (2021). Exploration and insights into the cellular internalization and intracellular fate of amphiphilic polymeric nanocarriers. *Acta Pharmaceutica Sinica B, 11*, 903–924.
71. Mckiernan, P. J., Cunningham, O., Greene, C. M., & Cryan, S.-A. (2013). Targeting miRNA-based medicines to cystic fibrosis airway epithelial cells using nanotechnology. *International Journal of Nanomedicine, 8*, 3907–3915.
72. Mehrabi, M., Montazeri, H., Mohamadpour Dounighi, N., Rashti, A., & Vakili-Ghartavol, R. (2018). Chitosan-based nanoparticles in mucosal vaccine delivery. *Archives of Razi Institute, 73*, 165–176.
73. Miao, J., Yang, X.-Q., Gao, Z., Li, Q., Meng, T.-T., Wu, J.-Y., Yuan, H., & Hu, F.-Q. (2019). Redox-responsive chitosan oligosaccharide-SS-Octadecylamine polymeric carrier for efficient anti-Hepatitis B Virus gene therapy. *Carbohydrate Polymers, 212*, 215–221.

74. Mikušová, V., & Mikuš, P. (2021). Advances in chitosan-based nanoparticles for drug delivery. *International Journal of Molecular Sciences, 22*, 9652.
75. Mitra, R. N., Han, Z., Merwin, M., Al Taai, M., Conley, S. M., & Naash, M. I. (2014). Synthesis and characterization of glycol chitosan DNA nanoparticles for retinal gene delivery. *ChemMedChem, 9*, 189–196.
76. Mourya, V., & Inamdar, N. N. (2008). Chitosan-modifications and applications: Opportunities galore. *Reactive and Functional Polymers, 68*, 1013–1051.
77. Naghib, S. M., Ahmadi, B., Kangarshahi, B. M., & Mozafari, M. (2024). Chitosan-based smart stimuli-responsive nanoparticles for gene delivery and gene therapy: Recent progresses on cancer therapy. *International Journal of Biological Macromolecules, 278*, 134542.
78. Negm, N. A., Hefni, H. H., Abd-Elaal, A. A., Badr, E. A., & Abou Kana, M. T. (2020). Advancement on modification of chitosan biopolymer and its potential applications. *International Journal of Biological Macromolecules, 152*, 681–702.
79. Neves, S., Faneca, H., Bertin, S., Konopka, K., Düzgüneş, N., Pierrefite-Carle, V., Simoes, S., & Pedroso De Lima, M. (2009). Transferrin lipoplex-mediated suicide gene therapy of oral squamous cell carcinoma in an immunocompetent murine model and mechanisms involved in the antitumoral response. *Cancer Gene Therapy, 16*, 91–101.
80. Nguyen, J., & Szoka, F. C. (2012). Nucleic acid delivery: The missing pieces of the puzzle? *Accounts of Chemical Research, 45*, 1153–1162.
81. Okamoto, H., Shiraki, K., Yasuda, R., Danjo, K., & Watanabe, Y. (2011). Chitosan–interferon-β gene complex powder for inhalation treatment of lung metastasis in mice. *Journal of Controlled Release, 150*, 187–195.
82. Opanasopit, P., Sajomsang, W., Ruktanonchai, U., Mayen, V., Rojanarata, T., & Ngawhirunpat, T. (2008). Methylated N-(4-pyridinylmethyl) chitosan as a novel effective safe gene carrier. *International Journal of Pharmaceutics, 364*, 127–134.
83. Park, I.-K., Ihm, J.-E., Park, Y., Choi, Y., Kim, S., Kim, W., Akaike, T., & Cho, C. (2003). Galactosylated chitosan (GC)-graft-poly (vinyl pyrrolidone)(PVP) as hepatocyte-targeting DNA carrier: Preparation and physicochemical characterization of GC-graft-PVP/DNA complex (1). *Journal of Controlled Release, 86*, 349–359.
84. Park, I., Kim, T., Park, Y., Shin, B., Choi, E., Chowdhury, E., Akaike, T., & Cho, C. (2001). Galactosylated chitosan-graft-poly (ethylene glycol) as hepatocyte-targeting DNA carrier. *Journal of Controlled Release, 76*, 349–362.
85. Park, Y., Park, Y., Shin, B., Choi, E., Park, Y., Akaike, T., & Cho, C. (2000). Galactosylated chitosan–graft–dextran as hepatocyte-targeting DNA carrier. *Journal of Controlled Release, 69*, 97–108.
86. Patil, S., Gao, Y.-G., Lin, X., Li, Y., Dang, K., Tian, Y., Zhang, W.-J., Jiang, S.-F., Qadir, A., & Qian, A.-R. (2019). The development of functional non-viral vectors for gene delivery. *International Journal of Molecular Sciences, 20*, 5491.
87. Pavlovic, I., Thakor, D. T., Vargas, J. R., Mckinlay, C. J., Hauke, S., Anstaett, P., Camuña, R. C., Bigler, L., Gasser, G., & Schultz, C. (2016). Cellular delivery and photochemical release of a caged inositol-pyrophosphate induces PH-domain translocation in cellulo. *Nature Communications, 7*, 10622.
88. Peptu, C., Humelnicu, A. C., Rotaru, R., Fortuna, M. E., Patras, X., Teodorescu, M., Tamba, B. I., & Harabagiu, V. (2019). Chitosan-based drug delivery systems. In *Chitin and Chitosan: Properties and applications* (pp. 259–289).
89. Picanço-Castro, V., Pereira, C. G., Covas, D. T., Porto, G. S., Athanassiadou, A., & Figueiredo, M. L. (2020). Emerging patent landscape for non-viral vectors used for gene therapy. *Nature Biotechnology, 38*, 151–157.
90. Pohlit, H., Bellinghausen, I., Frey, H., & Saloga, J. (2017). Recent advances in the use of nanoparticles for allergen-specific immunotherapy. *Allergy, 72*, 1461–1474.
91. Poshina, D. N., Rakshina, A. D., & Skorik, Y. A. (2025). Hydrophobic Chitosan derivatives for gene and drug delivery in cancer therapies. *Polysaccharides, 6*, 11.

92. Pugsley, C. E., Isaac, R., Warren, N. J., & Cayre, O. J. (2021). Recent advances in engineered nanoparticles for RNAi-mediated crop protection against insect pests. *Frontiers in Agronomy, 3*, 652981.
93. Qiu, J., Kong, L., Cao, X., Li, A., Wei, P., Wang, L., Mignani, S., Caminade, A.-M., Majoral, J.-P., & Shi, X. (2018). Enhanced delivery of therapeutic siRNA into glioblastoma cells using dendrimer-entrapped gold nanoparticles conjugated with β-cyclodextrin. *Nanomaterials, 8*, 131.
94. Raftery, R., O'brien, F. J., & Cryan, S.-A. (2013). Chitosan for gene delivery and orthopedic tissue engineering applications. *Molecules, 18*, 5611–5647.
95. Raftery, R. M., Castaño, I. M., Chen, G., Cavanagh, B., Quinn, B., Curtin, C. M., Cryan, S. A., & O'brien, F. J. (2017). Translating the role of osteogenic-angiogenic coupling in bone formation: Highly efficient chitosan-pDNA activated scaffolds can accelerate bone regeneration in critical-sized bone defects. *Biomaterials, 149*, 116–127.
96. Rahmani, S., Hakimi, S., Esmaeily, A., Samadi, F. Y., Mortazavian, E., Nazari, M., Mohammadi, Z., Tehrani, N. R., & Tehrani, M. R. (2019). Novel chitosan based nanoparticles as gene delivery systems to cancerous and noncancerous cells. *International Journal of Pharmaceutics, 560*, 306–314.
97. Ramamoorth, M., & Narvekar, A. (2015). Non viral vectors in gene therapy-an overview. *Journal of Clinical and Diagnostic Research: JCDR, 9*, GE01.
98. Ren, W., Zhu, H., Shang, D., Zhou, X., Zhang, H., Zhang, J., Liu, G., Wei, Y., Wang, K., & Jiang, C. (2025). Folate-targeting and reduction-responsive nano-delivery system based on curcumin polymer to enhance the anti-cancer activity. *Journal of Drug Delivery Science and Technology, 105*, 106616.
99. Ressler, A. (2022). Chitosan-based biomaterials for bone tissue engineering applications: A short review. *Polymers, 14*, 3430.
100. Ryu, J. H., Yoon, H. Y., Sun, I. C., Kwon, I. C., & Kim, K. (2020). Tumor-targeting glycol chitosan nanoparticles for cancer heterogeneity. *Advanced Materials, 32*, 2002197.
101. Sabharanjak, S., & Mayor, S. (2004). Folate receptor endocytosis and trafficking. *Advanced Drug Delivery Reviews, 56*, 1099–1109.
102. Sacco, P., Cok, M., Scognamiglio, F., Pizzolitto, C., Vecchies, F., Marfoglia, A., Marsich, E., & Donati, I. (2020). Glycosylated-chitosan derivatives: A systematic review. *Molecules, 25*, 1534.
103. Sachdeva, B., Sachdeva, P., Negi, A., Ghosh, S., Han, S., Dewanjee, S., Jha, S. K., Bhaskar, R., Sinha, J. K., & Paiva-Santos, A. C. (2023). Chitosan nanoparticles-based cancer drug delivery: Application and challenges. *Marine Drugs, 21*, 211.
104. Saffari, M., Moghimi, H. R., & Dass, C. R. (2016). Barriers to liposomal gene delivery: From application site to the target. *Iranian Journal of Pharmaceutical Research, 15*, 3.
105. Sahariah, P., Cibor, D., Zielińska, D., Hjálmarsdóttir, M. Á., Stawski, D., & Másson, M. (2019). The effect of molecular weight on the antibacterial activity of N, N, N-trimethyl chitosan (TMC). *International Journal of Molecular Sciences, 20*, 1743.
106. Samar, M. M., El-Kalyoubi, M., Khalaf, M., & Abd El-Razik, M. (2013). Physicochemical, functional, antioxidant and antibacterial properties of chitosan extracted from shrimp wastes by microwave technique. *Annals of Agricultural Sciences, 58*, 33–41.
107. Santos-Carballal, B., Fernández Fernández, E., & Goycoolea, F. M. (2018). Chitosan in non-viral gene delivery: Role of structure, characterization methods, and insights in cancer and rare diseases therapies. *Polymers, 10*, 444.
108. Schmitz, T., Bravo-Osuna, I., Vauthier, C., Ponchel, G., Loretz, B., & Bernkop-Schnürch, A. (2007). Development and in vitro evaluation of a thiomer-based nanoparticulate gene delivery system. *Biomaterials, 28*, 524–531.
109. Serrano-Sevilla, I., Artiga, Á., Mitchell, S. G., De Matteis, L., & De La Fuente, J. M. (2019). Natural polysaccharides for siRNA delivery: Nanocarriers based on chitosan, hyaluronic acid, and their derivatives. *Molecules, 24*, 2570.

110. Sharaf, M., Zahra, A. A., Alharbi, M., Mekky, A. E., Shehata, A. M., Alkhudhayri, A., Ali, A. M., Al Suhaimi, E. A., Zakai, S. A., & Al Harthi, N. (2024). Bee chitosan nanoparticles loaded with apitoxin as a novel approach to eradication of common human bacterial, fungal pathogens and treating cancer. *Frontiers in Microbiology, 15*, 1345478.
111. Sharma, P., Pasha, T. Y., Aparna, T. N., Singaram, G., Shirsat, M. K., Kumar, V., & Tiwari, G. (2025). Inulin: A versatile polymer for drug delivery system. In *Inulin for pharmaceutical applications: A versatile biopolymer*. Springer.
112. Singh, S., & Wairkar, S. (2023). Long-circulating thiolated chitosan nanoparticles of nintedanib with N-acetyl cysteine for treating idiopathic pulmonary fibrosis: In vitro assessment of cytotoxicity, antioxidant, and antifibrotic potential. *International Journal of Pharmaceutics, 644*, 123322.
113. Song, W., Song, X., Yang, C., Gao, S., Klausen, L. H., Zhang, Y., Dong, M., & Kjems, J. (2015). Chitosan/siRNA functionalized titanium surface via a layer-by-layer approach for in vitro sustained gene silencing and osteogenic promotion. *International Journal of Nanomedicine, 10*, 2335–2346.
114. Sripetthong, S., Nalinbenjapun, S., Basit, A., & Ovatlarnporn, C. (2024). Synthesis of Quarternized Chitosans and their potential applications in the solubility enhancement of indomethacin by solid dispersion. *AAPS PharmSciTech, 25*, 179.
115. Strand, S. P., Lelu, S., Reitan, N. K., De Lange Davies, C., Artursson, P., & Vårum, K. M. (2010). Molecular design of chitosan gene delivery systems with an optimized balance between polyplex stability and polyplex unpacking. *Biomaterials, 31*, 975–987.
116. Sun, M., Wang, T., Pang, J., Chen, X., & Liu, Y. (2020). Hydroxybutyl chitosan centered biocomposites for potential curative applications: A critical review. *Biomacromolecules, 21*, 1351–1367.
117. Tahamtan, A., Barati, M., Tabarraei, A., Mohebbi, S. R., Shirian, S., Gorji, A., & Ghaemi, A. (2018). Antitumor immunity induced by genetic immunization with chitosan nanoparticle formulated adjuvanted for HPV-16 E7 DNA vaccine. *Iranian Journal of Immunology, 15*, 269–280.
118. Tashima, T. (2025). Non-invasive delivery of CRISPR/Cas9 ribonucleoproteins (Cas9 RNPs) into cells via nanoparticles for membrane transport. *Pharmaceutics, 17*, 201.
119. Tekade, M., Maheshwari, N., Youngren-Ortiz, S. R., Pandey, V., Chourasiya, Y., Soni, V., Deb, P. K., & Sharma, M. C. (2019). Thiolated-chitosan: A novel mucoadhesive polymer for better-targeted drug delivery. In *Biomaterials and bionanotechnology*. Elsevier.
120. Thibault, M., Nimesh, S., Lavertu, M., & Buschmann, M. D. (2010). Intracellular trafficking and decondensation kinetics of chitosan–pDNA polyplexes. *Molecular Therapy, 18*, 1787–1795.
121. Tian, Y., & Shi, Y. (2025). Mechanisms of targeted drug delivery for liver cancer: Active, passive, and subcellular strategies. *Journal of Biosciences and Medicines, 13*, 369–384.
122. Trivedi, M., & Kumar, S. (2022). Potential of chitosan-based nanocomposites for biomedical application in gene therapy. In *Chitosan-based nanocomposite materials: Fabrication, characterization and biomedical applications*. Springer.
123. Vauthier, C., Zandanel, C., & Ramon, A. L. (2013). Chitosan-based nanoparticles for in vivo delivery of interfering agents including siRNA. *Current Opinion in Colloid & Interface Science, 18*, 406–418.
124. Von Haartman, E., Lindberg, D., Prabhakar, N., & Rosenholm, J. M. (2016). On the intracellular release mechanism of hydrophobic cargo and its relation to the biodegradation behavior of mesoporous silica nanocarriers. *European Journal of Pharmaceutical Sciences, 95*, 17–27.
125. Wang, W., Meng, Q., Li, Q., Liu, J., Zhou, M., Jin, Z., & ZHAO, K. (2020). Chitosan derivatives and their application in biomedicine. *International Journal of Molecular Sciences, 21*, 487.
126. Wong, K., Sun, G., Zhang, Dai, H., Liu, Y., He, C., & Leong, K. W. (2006). PEI-g-chitosan, a novel gene delivery system with transfection efficiency comparable to polyethylenimine in vitro and after liver administration in vivo. *Bioconjugate Chemistry, 17*, 152–158.

127. Wu, Y., Rashidpour, A., Almajano, M. P., & Metón, I. (2020). Chitosan-based drug delivery system: Applications in fish biotechnology. *Polymers, 12*, 1177.
128. Xu, Y.-H., Tan, X.-Y., Xu, Y.-C., Zhao, T., Zhang, L.-H., & Luo, Z. (2019). Novel insights for SREBP-1 as a key transcription factor in regulating lipogenesis in a freshwater teleost, grass carp Ctenopharyngodon idella. *British Journal of Nutrition, 122*, 1201–1211.
129. Yang, X., Zhang, J., Chen, L., Wu, Q., & Yu, C. (2018). Chitosan oligosaccharides enhance lipid droplets via down-regulation of PCSK9 gene expression in HepG2 cells. *Experimental Cell Research, 366*, 152–160.
130. Yhee, J. Y., Song, S., Lee, S. J., Park, S.-G., Kim, K.-S., Kim, M. G., Son, S., Koo, H., Kwon, I. C., & Jeong, J. H. (2015). Cancer-targeted MDR-1 siRNA delivery using self-cross-linked glycol chitosan nanoparticles to overcome drug resistance. *Journal Of Controlled Release, 198*, 1–9.
131. Yoksan, R., & Akashi, M. (2009). Low molecular weight chitosan-gl-phenylalanine: Preparation, characterization, and complex formation with DNA. *Carbohydrate Polymers, 75*, 95–103.
132. Yoo, H. S., Lee, J. E., Chung, H., Kwon, I. C., & Jeong, S. Y. (2005). Self-assembled nanoparticles containing hydrophobically modified glycol chitosan for gene delivery. *Journal of Controlled Release, 103*, 235–243.
133. Zacaron, T. M., Silva, M. L. S. E., Costa, M. P., Silva, D. M. E., Silva, A. C., Apolônio, A. C. M., Fabri, R. L., Pittella, F., Rocha, H. V. A., & Tavares, G. D. (2023). Advancements in chitosan-based nanoparticles for pulmonary drug delivery. *Polymers, 15*, 3849.
134. Zhao, R.-L., Zhang, X.-M., Jia, L.-N., Song, W., Sun, Y.-L., Meng, X.-Y., & Peng, X.-X. (2019). pNNS-conjugated Chitosan mediated IGF-1 and miR-140 overexpression in articular chondrocytes improves cartilage repair. *BioMed Research International, 2019*, 2761241.
135. Zhou, Y., Zhu, F., Liu, Y., Zheng, M., Wang, Y., Zhang, D., Anraku, Y., & Zou, Y. (2020). Blood-brain barrier–penetrating siRNA nanomedicine for Alzheimer's disease therapy. *Science Advances, 6*, eabc7031.
136. Zubareva, A., Shagdarova, B., Varlamov, V., Kashirina, E., & Svirshchevskaya, E. (2017). Penetration and toxicity of chitosan and its derivatives. *European Polymer Journal, 93*, 743–749.

Chitosan-Based Liposomes in Cancer Therapy

Govindaraj Sabarees, Sivapregassame Vishvaja, Dhamodharan Priyadharshini, Veerachamy Alagarsamy, and Viswas Raja Solomon

Abstract The ongoing search for effective cancer treatments has spotlighted nanotechnology-based drug delivery systems as a means to enhance therapeutic outcomes and reduce systemic toxicity. Among these, liposomes have gained prominence for their ability to encapsulate diverse therapeutic agents and modulate drug release profiles. However, conventional liposomes face challenges such as limited stability, rapid clearance, and suboptimal tumor targeting. To address these issues, chitosan, a biocompatible, biodegradable, and mucoadhesive natural polymer, has been integrated with liposomal systems to create chitosan-based liposomes. These hybrid nanocarriers exhibit improved colloidal stability, prolonged circulation time, enhanced cellular uptake, and pH-responsive release, making them particularly suited for targeting the acidic tumor microenvironment. Chitosan-based liposomes thus combine the strengths of both components, offering improved drug delivery for chemotherapy, gene therapy, and siRNA applications. Their tunability and potential for targeted functionalization make them strong candidates for precision cancer treatment. This chapter explores the design, formulation, therapeutic applications, and current progress of chitosan-based liposomes in cancer therapy, while highlighting the challenges and future directions for clinical translation.

Keywords Chitosan · Liposomes · Cancer therapy · Nanocarriers · Targeted drug delivery · Biopolymers

---

G. Sabarees
Department of Pharmaceutical Chemistry, Saveetha College of Pharmacy, Saveetha Institute of Medical and Technical Sciences, Chennai, India

S. Vishvaja · D. Priyadharshini
Shri Venkateshwara College of Pharmacy, Ariyur, Puducherry, India

V. Alagarsamy (✉) · V. R. Solomon (✉)
Medicinal Chemistry Research Laboratory, MNR College of Pharmacy, Greater Hyderabad, Sangareddy, India

S. Jana (ed.), *Chitosan-Based Drug Delivery Systems*, AAPS Introductions in the Pharmaceutical Sciences 16, https://doi.org/10.1007/978-3-032-08347-0_13

1 Introduction

Cancer remains one of the leading causes of morbidity and mortality worldwide, characterized by the uncontrolled proliferation and spread of abnormal cells [1]. Despite advances in diagnostic tools and therapeutic modalities, the global burden of cancer continues to rise, with the World Health Organization reporting nearly 10 million cancer-related deaths annually [2]. Conventional treatment strategies such as surgery, chemotherapy, radiotherapy, and immunotherapy have improved survival rates for certain cancer types, but they often fall short due to critical limitations, including systemic toxicity, non-specific targeting, drug resistance, and poor pharmacokinetics (Fig. 1) [3]. Chemotherapy, a mainstay of cancer treatment, suffers from a narrow therapeutic index. Most chemotherapeutic agents lack the ability to distinguish between cancerous and healthy cells, leading to significant off-target effects such as bone marrow suppression, gastrointestinal toxicity, and cardiotoxicity [4]. In addition, tumor heterogeneity and the evolution of drug-resistant cancer cell clones pose major obstacles to long-term therapeutic success. Multi-drug resistance (MDR), often mediated by efflux transporters such as P-glycoprotein, further limits the efficacy of many anticancer agents [5]. Radiotherapy, while localized, can still cause collateral damage to surrounding healthy tissues and often fails to address metastatic lesions. Similarly, immunotherapy and targeted therapies, though promising, are limited by high costs, immune-related adverse effects, and variable patient response [6]. Moreover, the tumor microenvironment itself presents biological barriers to effective therapy, including hypoxia, high interstitial fluid pressure, and an abnormal vasculature that restricts drug penetration [7]. These challenges highlight the urgent need for advanced drug delivery systems that can improve therapeutic index, overcome biological barriers, and enable site-specific release of anticancer agents [8].

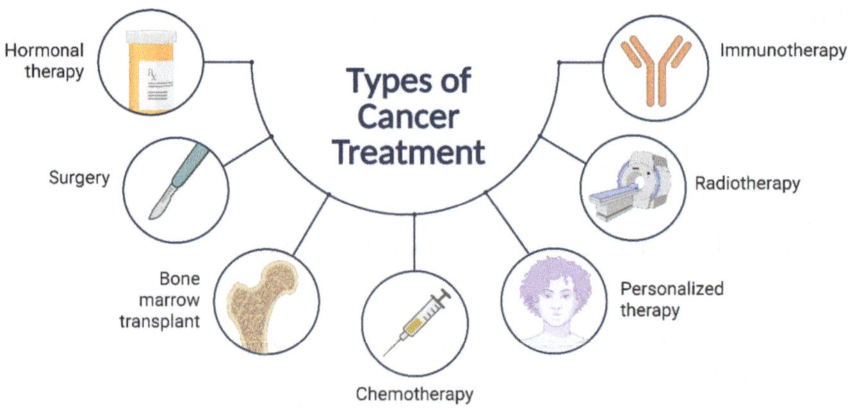

Fig. 1 Conventional cancer treatment approaches

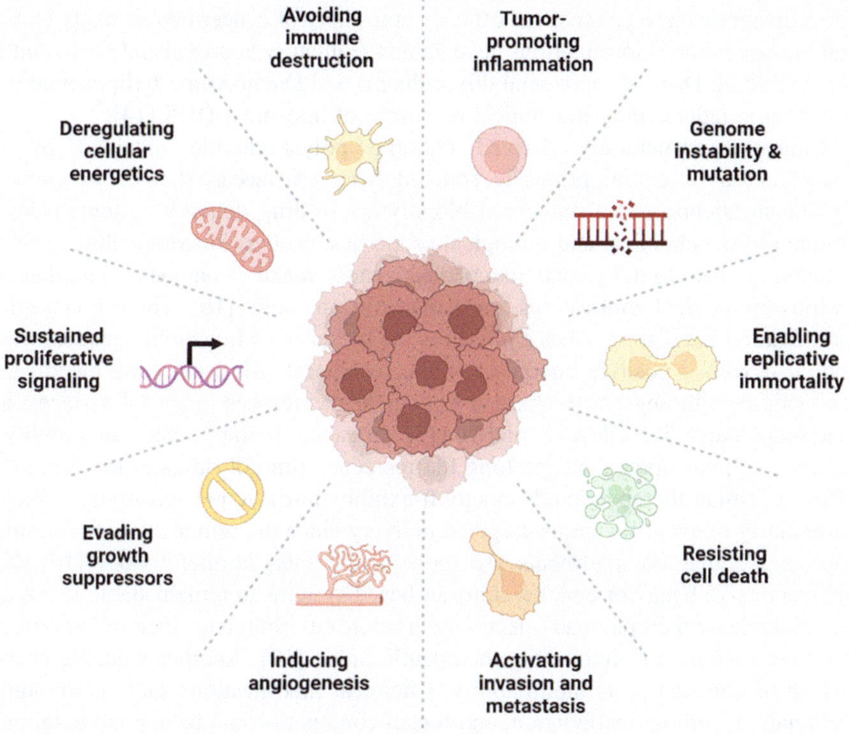

Fig. 2 Hallmarks of cancer and enabling characteristics

The hallmarks of cancer are key biological traits driving tumor progression, including sustained proliferative signaling, evasion of growth suppressors, angiogenesis, invasion and metastasis, resistance to cell death, and replicative immortality. Enabling factors such as deregulated cellular energetics and immune evasion further support these traits. Emerging hallmarks, like tumor-promoting inflammation and genome instability, contribute to cancer evolution [9]. Understanding these features is vital for developing targeted cancer therapies (Fig. 2).

Liposomes have emerged as a promising drug delivery system for cancer therapy due to their ability to encapsulate both hydrophilic and lipophilic drugs, enhancing their stability, bioavailability, and targeting efficiency [10]. These nanosized vesicles are composed of phospholipid bilayers that can be modified to improve their specificity for cancer cells, either by incorporating targeting ligands or altering surface properties [11]. The ability of liposomes to passively accumulate in tumor tissues through the enhanced permeability and retention (EPR) effect, combined with their capacity for controlled and sustained release of therapeutic agents, makes them particularly effective in reducing systemic toxicity and improving therapeutic outcomes [12]. Moreover, liposomal formulations can be used for a range of cancer treatments, including chemotherapy, gene therapy, and immunotherapy, offering a versatile approach to enhance the efficacy and safety of cancer treatments [13]. Several liposome-based formulations have gained regulatory approval and are in clinical use, including Doxil® (liposomal doxorubicin) and DaunoXome® (liposomal daunorubicin), underscoring the clinical relevance of liposomal DDS [14].

Chitosan, a naturally derived cationic polysaccharide obtained by the deacetylation of chitin, primarily sourced from crustacean shells, has garnered significant attention as a functional biopolymer in drug delivery systems [15]. Its unique physicochemical and biological properties, such as biocompatibility, biodegradability, low toxicity, and mucoadhesiveness make it an ideal candidate for various biomedical applications, particularly in oncology [16]. The primary structural feature that distinguishes chitosan is the presence of free amino groups, which endow it with a positive charge at acidic to neutral pH, facilitating electrostatic interactions with negatively charged biological membranes, mucosal surfaces, and macromolecules like DNA or siRNA [17]. In cancer therapy, chitosan's ability to enhance cellular uptake and prolong the residence time of drugs at the tumor site offers a distinct therapeutic advantage. It exhibits intrinsic pH-sensitivity, which is particularly relevant for tumor-targeted delivery since the tumor microenvironment and intracellular compartments are more acidic than normal tissues [18]. This pH-responsive behavior enables chitosan-based systems to remain stable in circulation and release their payload selectively in acidic environments, thereby minimizing systemic toxicity and improving therapeutic index [19]. Another valuable characteristic of chitosan is its modifiability. Chemical modifications such as thiolation, PEGylation, carboxymethylation, and ligand conjugation can be applied to tailor its solubility, targeting specificity, and drug release profile [20]. For instance, PEGylation enhances circulation time by reducing recognition by the reticuloendothelial system (RES), while ligand conjugation can enable active targeting of overexpressed tumor cell receptors such as folate, transferrin, or epidermal growth factor receptors (EGFR) [21]. Chitosan has also demonstrated potential as a gene delivery vector due to its ability to form stable polyplexes with nucleic acids and facilitate endosomal escape. Its use in combination therapies, co-delivery systems, and synergistic formulations further amplifies its value in cancer treatment [22]. Moreover, being a naturally sourced and generally regarded as safe (GRAS) material, chitosan aligns with the growing demand for sustainable and clinically translatable biomaterials [23].

In this context, nanotechnology-based approaches, particularly liposomal drug carriers, have emerged as a promising solution. When combined with biopolymers like chitosan, these systems offer a versatile platform capable of enhancing drug solubility, stability, bioavailability, and tumor-targeting efficiency [24]. Thus, the development of chitosan-based liposomes represents a significant stride toward addressing the current limitations of cancer therapy [25]. The efficacy of cancer treatment is highly dependent on the precise delivery of therapeutic agents to malignant tissues while sparing normal cells. Conventional drug delivery methods often result in systemic exposure, leading to dose-limiting toxicities and suboptimal therapeutic outcomes [26]. In response, the field of oncology has increasingly embraced drug delivery systems (DDS) designed to improve pharmacokinetics, enhance tumor specificity, and minimize adverse effects [27]. In particular, biopolymer-based coatings, including chitosan, offer advantages like increased mucoadhesion, pH sensitivity, and improved cellular uptake [28]. Incorporating such modifications into liposomal systems holds great promise for overcoming current drug delivery barriers in cancer therapy and achieving more effective, targeted treatment regimens [29, 30]. However, to enhance their performance, liposomes can be further modified with biopolymers such as chitosan, which confer mucoadhesiveness, pH sensitivity, and enhanced interaction with negatively charged cancer cell membranes [31]. Chitosan-coated liposomes combine the structural advantages of phospholipid vesicles with the functional benefits of chitosan. The cationic nature of chitosan enables strong electrostatic interactions with the negatively charged phospholipid surfaces of liposomes, forming a stable coating that improves colloidal stability, prevents aggregation, and reduces drug leakage [32]. This coating can also shield the encapsulated drug from enzymatic degradation and promote prolonged circulation time in vivo [33]. Importantly, the mucoadhesive and bioadhesive properties of chitosan enhance the retention of liposomes at the site of administration and promote their absorption across biological barriers, such as the gastrointestinal mucosa, blood-brain barrier, or tumor vasculature [34]. Furthermore, chitosan-coated liposomes exhibit improved cellular uptake due to their positive surface charge, facilitating endocytosis and intracellular drug delivery. These features are particularly advantageous for targeting cancer cells, which often display negatively charged membranes and acidic microenvironments that favor chitosan's pH-sensitive behavior [35]. Additionally, chitosan-based liposomes can be further modified to include active targeting moieties such as folic acid, peptides, or antibodies enabling receptor-mediated delivery to tumor cells and enhancing therapeutic specificity [36]. They can also be engineered for stimuli-responsive drug release, triggered by pH, redox potential, or enzymatic activity within the tumor microenvironment, ensuring that the drug payload is released precisely where and when it is needed [37]. The versatility of chitosan in functionalization and the well-established safety profile of liposomes support the clinical translation of this hybrid system. Taken together, chitosan-based liposomes present a synergistic platform that addresses multiple challenges in cancer drug delivery such as targeting, stability, and bioavailability, making them a compelling candidate for next-generation oncologic nanomedicine [38]. This chapter offers a novel perspective by integrating the unique properties of chitosan with liposomal drug delivery for cancer therapy. It highlights recent advances, including stimuli-responsive systems, gene delivery applications, and hybrid nanocarriers, while addressing translational challenges and clinical potential [39]. By bridging fundamental science with emerging innovations, this work provides a timely and forward-looking overview to guide future development of chitosan-based liposomes in oncology [40].

2 Chitosan

2.1 *Source and Chemical Structure*

Chitosan is a linear polysaccharide derived from the deacetylation of chitin, a naturally abundant biopolymer found in the exoskeletons of crustaceans, insects, and fungal cell walls [41]. Structurally, it consists of repeating units of β-(1→4)-linked *D*-glucosamine and *N*-acetyl-*D*-glucosamine. The degree of deacetylation (DD)—a critical parameter typically ranging from 50% to 95%—influences its solubility, charge density, and biological activity (Fig. 3) [42]. Chitosan is soluble in mildly acidic aqueous solutions (pH < 6.5) due to the protonation of its amino groups, which gives it a polycationic nature. This cationic property distinguishes chitosan from most other natural polymers and is central to its ability to interact with negatively charged molecules, including DNA, proteins, and cell membranes [43]. The molecular weight of chitosan also varies significantly, typically ranging from 50 to over 1000 kDa. This parameter affects its viscosity, film-forming ability, degradation rate, and bioactivity, and is therefore often tailored depending on the intended biomedical application [44]. Low-molecular-weight chitosan is favored for drug and gene delivery, while higher molecular weights are preferred in wound healing and tissue engineering due to enhanced mechanical properties [45].

2.2 *Biocompatibility, Biodegradability, and Mucoadhesive Properties*

Chitosan is widely regarded as biocompatible and biodegradable, making it suitable for in vivo applications. Upon administration, it undergoes enzymatic degradation primarily by lysozymes and chitinases, producing non-toxic, absorbable oligosaccharides [46]. Its degradation products are excreted naturally, reducing the risk of accumulation and long-term toxicity [47]. One of the most significant attributes of

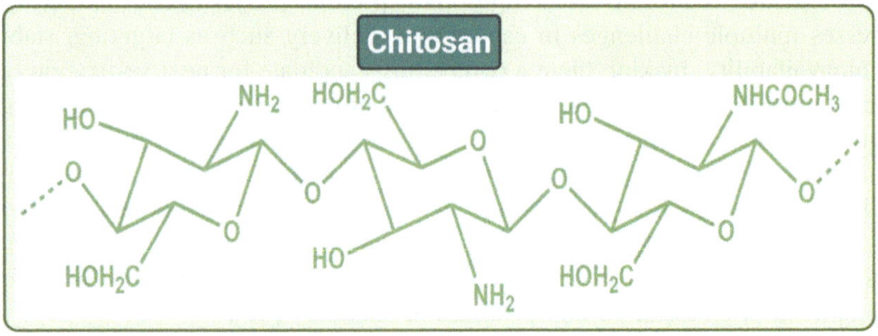

Fig. 3 Chemical structure of chitosan

chitosan is its mucoadhesiveness. The positively charged amino groups of chitosan interact electrostatically with negatively charged sialic acid residues and glycoproteins present in mucus, facilitating prolonged retention at mucosal surfaces [48]. This property is particularly beneficial in drug delivery across mucosal barriers such as the nasal, pulmonary, gastrointestinal, and vaginal routes. Prolonged residence time enhances local drug concentration and absorption, thereby improving bioavailability [49]. Chitosan has also demonstrated immunomodulatory and hemostatic effects, further supporting its role in biomedical applications. Its excellent safety profile has led to approval by regulatory agencies such as the FDA for use in dietary supplements, wound dressings, and drug delivery systems [50].

2.3 pH-Sensitivity and Tumor-Targeting Advantages

Chitosan exhibits pH-responsive solubility and swelling behavior, which is critical for site-specific drug delivery in oncology. It remains stable and less soluble at physiological pH (7.4) but becomes protonated and more soluble under mildly acidic conditions (pH 5.0–6.5), such as those found in the tumor microenvironment, endosomes, or inflamed tissues [51]. This pH sensitivity can be leveraged to trigger drug release selectively in acidic tumor regions, improving therapeutic outcomes while minimizing systemic toxicity [52].

Additionally, the positive surface charge of chitosan promotes interaction with negatively charged cell membranes, facilitating enhanced cellular uptake via adsorptive endocytosis. This electrostatic affinity can be further optimized to exploit the characteristic features of tumor vasculature and extracellular matrices [53]. These interactions support chitosan's ability to promote deep tumor penetration, enhance drug accumulation at the site of action, and improve therapeutic index [54].

2.4 Functionalization Strategies

To overcome limitations such as poor solubility at neutral pH and rapid in vivo clearance, chitosan can be chemically modified through various functionalization strategies. Thiolation, which introduces thiol groups onto the chitosan backbone, enhances mucoadhesiveness by forming disulfide bonds with cysteine-rich domains in mucus and also improves drug permeation across epithelial barriers [55]. PEGylation, the attachment of polyethylene glycol (PEG) chains, increases chitosan's hydrophilicity, reduces its immunogenicity, and prolongs systemic circulation by minimizing recognition and clearance by macrophages—making it especially advantageous for intravenous administration and stealth liposome applications [56]. Ligand conjugation involves linking targeting moieties such as folic acid, transferrin, RGD peptides, or monoclonal antibodies to chitosan, facilitating receptor-mediated uptake by cancer cells and enhancing site-specific drug delivery while minimizing off-target effects [57]. These functionalization approaches enable precise control over chitosan's physicochemical and biological properties, establishing it as a highly versatile platform for the development of tumor-targeted drug delivery systems [58].

3 Liposomes

3.1 General Structure and Composition

Liposomes are spherical vesicles composed of one or more phospholipid bilayers encapsulating an aqueous core (Fig. 4). Their architecture closely mimics biological membranes, which contributes to their excellent biocompatibility and low toxicity [59]. Depending on the method of preparation, liposomes can vary in size—from small unilamellar vesicles (SUVs, 20–100 nm) to multilamellar vesicles (MLVs, 100 nm to several microns)—and in lamellarity (number of lipid bilayers) [60]. The typical components of liposomes include natural or synthetic phospholipids (e.g., phosphatidylcholine, phosphatidylserine) and cholesterol, which stabilizes the bilayer by reducing its permeability and enhancing rigidity [61]. The amphiphilic nature of lipids allows the formation of closed bilayer structures in aqueous environments, providing a versatile system capable of carrying both hydrophilic (in the core) and hydrophobic (within the bilayer) drugs [62]. This dual-loading capacity, combined with the ability to modify surface properties and encapsulation efficiency, makes liposomes a highly customizable platform for drug delivery, especially in cancer treatment, where localized delivery and controlled release are crucial [63].

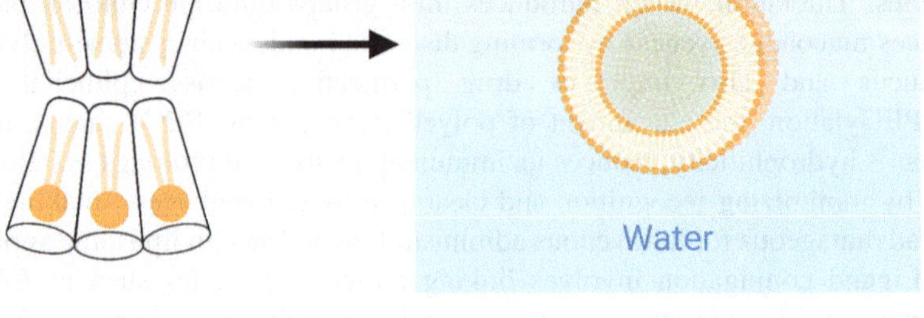

Fig. 4 Liposomes structure

3.2 Types of Liposomes

Liposomes can be classified based on their surface characteristics, composition, and targeting strategies, each tailored to optimize therapeutic efficacy. Conventional liposomes, composed primarily of phospholipids and cholesterol, are simple to prepare but suffer from rapid clearance by the mononuclear phagocyte system (MPS), limiting their circulation time [64]. To overcome this, PEGylated liposomes—also known as stealth liposomes—are modified with polyethylene glycol (PEG), which forms a hydrophilic steric barrier that reduces immune recognition and significantly extends systemic circulation, enhancing passive accumulation in tumors via the enhanced permeation and retention (EPR) effect [65]. Targeted liposomes are further engineered by conjugating ligands such as antibodies, peptides, or aptamers that bind selectively to receptors overexpressed on cancer cells, such as HER2, EGFR, or folate receptors, thereby facilitating active targeting and improved cellular uptake [66]. Additionally, stimuli-responsive liposomes are designed to release their cargo in response to specific internal stimuli like acidic pH, enzymatic activity, or redox conditions, or external triggers such as heat, light, or ultrasound [67]. These smart systems enable precise, site-specific drug release, minimizing systemic toxicity. Moreover, liposomes can be co-loaded with multiple therapeutic agents or imaging molecules, supporting combination therapy and real-time diagnostic monitoring. This diversity in liposomal design allows for fine-tuning based on therapeutic needs and target site characteristics [68].

3.3 Advantages in Cancer Therapy

Liposomes present numerous advantages for cancer drug delivery, owing to their biocompatibility, structural versatility, and capacity to enhance the pharmacokinetics and biodistribution of anticancer agents [69]. Liposomes release therapeutic agents through various mechanisms that can be tailored based on their composition and the target environment. One common mechanism is passive diffusion, where small and lipophilic drugs slowly leak through the lipid bilayer into the surrounding tissue. In tumor microenvironments, pH-sensitive liposomes exploit the acidic conditions to destabilize their membranes, triggering rapid drug release. Additionally, enzymes such as phospholipases present in the body can degrade the liposomal structure, leading to controlled drug release. Thermosensitive liposomes are designed to release their payload when exposed to elevated temperatures, such as those used in localized hyperthermia therapy. Another effective mechanism involves the fusion of liposomes with cancer cell membranes, facilitating direct drug delivery into the cytoplasm. These mechanisms collectively enhance the therapeutic efficiency of liposome-based drug delivery systems while minimizing off-target effects (Fig. 5). Their nanoscale size facilitates passive targeting through the enhanced permeability and retention (EPR) effect, allowing preferential accumulation in

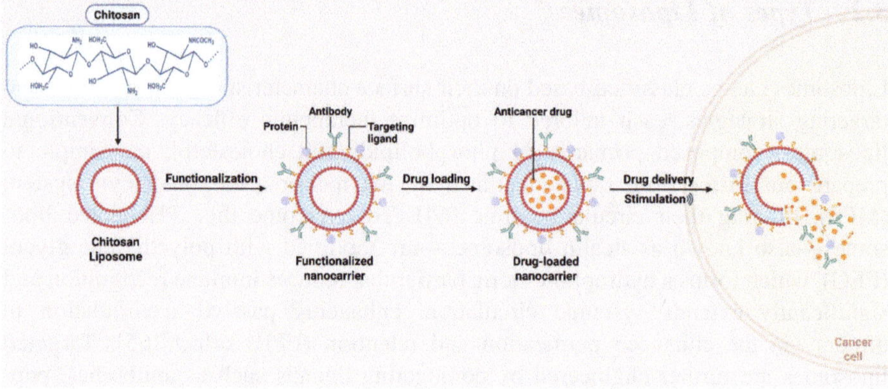

Fig. 5 Anticancer mechanisms of chitosan-based liposomes

tumor tissues characterized by leaky vasculature while minimizing impact on healthy cells [70]. Beyond passive targeting, liposomes can be engineered for active targeting by incorporating surface ligands that specifically bind to tumor-associated antigens or receptors, thereby improving selective uptake by malignant cells and reducing systemic exposure and side effects [71]. Encapsulation of cytotoxic drugs within the liposomal lipid bilayer also helps mitigate systemic toxicity by shielding healthy tissues from direct exposure and enabling controlled, sustained drug release at the tumor site [72]. These features collectively result in a higher therapeutic index, better patient tolerability, and lower incidence of drug resistance. Furthermore, liposomes serve as adaptable platforms for multifunctional delivery systems, capable of co-delivering multiple therapeutic agents, incorporating imaging probes for theranostic applications, and responding to tumor-specific stimuli such as pH, enzymes, or redox conditions for on-demand drug release [73]. This multifunctionality positions liposomes as powerful tools in the advancement of personalized and precision oncology. Several liposomal formulations have received FDA approval and are in clinical use, providing robust evidence of liposomes' translational potential in cancer therapy [74] (Table 1).

4 Chitosan-Based Liposomes: Synthesis and Characterization

4.1 Methods of Preparation

Several advanced techniques have been established for the synthesis of chitosan-coated liposomes, each offering specific advantages in terms of particle size control, drug encapsulation efficiency, and scalability [75]. The widely used thin film hydration method involves dissolving lipids in an organic solvent, forming a thin

Table 1 FDA-approved liposomal formulations for cancer therapy

Product name	Drug encapsulated	Liposome type	Approved indications	Key advantages
Doxil®	Doxorubicin	PEGylated liposome	Ovarian cancer, multiple myeloma, Kaposi's sarcoma	Prolonged circulation, reduced cardiotoxicity
DaunoXome®	Daunorubicin	Conventional liposome	HIV-associated Kaposi's sarcoma	Enhanced tumor delivery, reduced systemic toxicity
Marqibo®	Vincristine	Conventional liposome	Philadelphia chromosome–negative acute lymphoblastic leukemia (ALL)	Extended systemic exposure, improved efficacy
Onivyde®	Irinotecan	PEGylated liposome	Metastatic pancreatic cancer (in combination therapy)	Improved tumor accumulation, reduced gastrointestinal toxicity
Vyxeos®	Daunorubicin and Cytarabine (5:1 molar ratio)	Liposomal co-formulation	Therapy-related AML (t-AML), AML with myelodysplasia-related changes (AML-MRC)	Synergistic drug ratio, improved survival outcomes

film upon solvent evaporation, and subsequently hydrating it with an aqueous drug solution. Chitosan coating is achieved by incubating the formed liposomes with a chitosan solution, enabling electrostatic adsorption [76]. Although simple and versatile, this method may result in heterogeneous liposome populations. In contrast, the reverse-phase evaporation method, which emulsifies lipids with aqueous drug solutions in a water-in-oil system followed by solvent removal under reduced pressure, generates large unilamellar vesicles (LUVs) with enhanced encapsulation efficiency—especially for hydrophilic drugs [77]. The ethanol injection method, wherein lipids dissolved in ethanol are rapidly introduced into an aqueous phase under stirring, yields small-sized liposomes quickly and reproducibly, although it is less suitable for hydrophilic payloads [78]. Emerging microfluidic platforms now enable precise control over liposome size, lamellarity, and polydispersity, with chitosan integration occurring either in-line or post-formation through secondary mixing, offering high reproducibility and industrial scalability [79].

4.2 *Chitosan Coating and Surface Modification*

Chitosan coating is primarily mediated by electrostatic interaction between positively charged chitosan chains and negatively charged phospholipid headgroups. Coating efficiency is influenced by parameters such as chitosan's molecular weight, degree of deacetylation, pH, and ionic strength [80]. Further surface modifications, including crosslinking with agents like genipin, PEGylation, ligand conjugation, or layer-by-layer assembly with polyanions such as alginate, can enhance stability, targeting specificity, and controlled release properties. Overall, chitosan-functionalized liposomes present a robust, tunable platform for tumor-targeted drug delivery with improved bioavailability and therapeutic efficacy [81].

4.3 Physicochemical Characterization

A comprehensive physicochemical characterization of chitosan-based liposomes is critical for assessing their suitability as drug delivery systems. Among the most important parameters is particle size, typically measured by dynamic light scattering (DLS), which also provides the polydispersity index (PDI)—a measure of size uniformity that influences biodistribution, cellular uptake, and pharmacokinetics [82]. Zeta potential analysis is used to evaluate surface charge, where successful chitosan coating is reflected in a shift toward positive values (generally +20 to +50 mV), indicating effective surface modification and enhanced colloidal stability [83]. Morphological assessment using transmission electron microscopy (TEM), scanning electron microscopy (SEM), and atomic force microscopy (AFM) reveals insights into liposome shape, structural integrity, and surface texture. Chitosan-coated liposomes typically exhibit a slightly roughened surface compared to their uncoated counterparts, consistent with polymer adsorption [84]. Drug encapsulation efficiency (EE%) and loading capacity (DL%) are key indicators of formulation success, with quantification typically performed via ultracentrifugation followed by HPLC or UV–visible spectroscopy. These metrics determine how much of the therapeutic agent is retained within the carrier and its proportion relative to the total formulation mass [85]. In vitro drug release studies, commonly conducted under dialysis conditions that simulate both physiological (pH 7.4) and tumor microenvironments (pH ~5.5), demonstrate that chitosan-coated liposomes often provide sustained and pH-sensitive release profiles. Such behavior underscores their potential for site-specific drug release within the acidic tumor milieu, reducing systemic side effects and enhancing therapeutic efficacy [86].

5 Mechanisms of Anticancer Action

5.1 Enhanced Permeation and Retention (EPR) Effect

The Enhanced Permeation and Retention (EPR) effect is a cornerstone principle of passive tumor targeting exploited by nanoscale drug delivery systems. In solid tumors, the vasculature is often irregular, dilated, and fenestrated, with gaps ranging from 100 to 800 nm, due to rapid and aberrant angiogenesis [87]. Simultaneously, lymphatic drainage is inefficient or absent in tumor tissues, limiting the clearance of macromolecules and nanoparticles. Chitosan-based liposomes, typically ranging from 50 to 200 nm in size, can extravasate through these leaky vessels and preferentially accumulate in the tumor interstitium [88]. Moreover, the mucoadhesive nature of chitosan, resulting from its cationic amino groups, facilitates stronger interaction with negatively charged tumor cell membranes and extracellular matrix components, promoting longer residence time and enhanced therapeutic efficacy. These properties collectively make chitosan-coated liposomes highly suitable for exploiting the EPR effect in oncological applications [89].

5.2 *Tumor-Targeted Delivery (Passive and Ligand-Mediated Active Targeting)*

While passive targeting relies solely on the EPR effect, active targeting introduces a layer of molecular specificity. Chitosan-based liposomes can be modified with targeting ligands that recognize overexpressed receptors on cancer cells. Common ligands include folic acid (for folate receptor-positive cancers), transferrin (targeting transferrin receptors), RGD peptides (integrin receptors), antibodies (e.g., anti-HER2, anti-EGFR), and nucleic acid aptamers [90]. These ligands are covalently attached to the chitosan backbone or incorporated into the liposomal membrane, facilitating receptor-mediated endocytosis once the nanocarrier reaches the tumor site. This strategy enhances selective uptake by malignant cells, improves intracellular drug accumulation, and reduces systemic toxicity [91]. When combined with the bioadhesive properties of chitosan, active targeting further amplifies tumor specificity and therapeutic precision [92].

5.3 *Controlled and Sustained Drug Release in the Tumor Microenvironment*

Chitosan-based liposomes exhibit pH-responsive behavior due to the protonation of amino groups in acidic environments. The tumor microenvironment typically presents a lower extracellular pH (~6.5) compared to normal tissues (~7.4), and intracellular compartments such as endosomes and lysosomes are even more acidic (~5.0–5.5). In such environments, chitosan undergoes structural changes—swelling, increased solubility, or partial disintegration—that facilitate drug release [93]. This pH-sensitive mechanism ensures that drug release is minimal in the bloodstream, reducing systemic side effects, and is maximized at the tumor site, enhancing therapeutic outcomes. Advanced formulations also incorporate dual or multi-stimuli responsiveness, such as redox or enzyme-triggered systems, to further refine spatiotemporal control over drug release [94].

5.4 Cellular Uptake, Endosomal Escape, and Intracellular Trafficking

Effective anticancer therapy depends not only on tumor accumulation but also on efficient intracellular delivery. Chitosan-coated liposomes are taken up predominantly via adsorptive endocytosis, driven by electrostatic interactions between the positively charged chitosan and the negatively charged cell membranes [95]. In ligand-functionalized systems, receptor-mediated endocytosis becomes the dominant pathway, providing both specificity and enhanced uptake efficiency. After internalization, one major barrier is the endosomal entrapment of the therapeutic payload [96]. Chitosan facilitates endosomal escape through the "proton sponge" effect: the accumulation of protons within endosomes leads to osmotic swelling and eventual rupture, releasing the drug into the cytoplasm. This mechanism is particularly important for the delivery of nucleic acids (e.g., siRNA, mRNA) or protein-based drugs that require cytosolic or nuclear localization for their therapeutic activity [97].

5.5 Apoptosis Induction, Anti-Angiogenic, and Antiproliferative Mechanisms

The culmination of these delivery strategies is the activation of cellular pathways that inhibit tumor growth and survival. Encapsulated chemotherapeutics or gene therapeutics, once released intracellularly, initiate apoptosis via multiple mechanisms—mitochondrial membrane depolarization, reactive oxygen species (ROS) generation, and caspase cascade activation [98]. Chitosan-based formulations also demonstrate anti-angiogenic effects by inhibiting vascular endothelial growth factor (VEGF) signaling, thereby limiting tumor vascularization and nutrient supply. Additionally, these nanocarriers suppress cell proliferation by arresting the cell cycle at specific checkpoints (e.g., G2/M phase) and inhibiting key proliferative signaling pathways such as PI3K/Akt, MAPK/ERK, and NF-κB [99]. Intriguingly, chitosan itself has been reported to exert mild antitumor activity, possibly via immune modulation, inhibition of metalloproteinases, or interaction with toll-like receptors (TLRs), further enhancing the synergistic therapeutic outcome when combined with active agents [100].

6 Preclinical and Clinical Studies

Chitosan-based liposomes have emerged as promising nanocarriers for targeted cancer therapy due to their biocompatibility, biodegradability, and mucoadhesive properties. These hybrid systems combine the structural advantages of liposomes

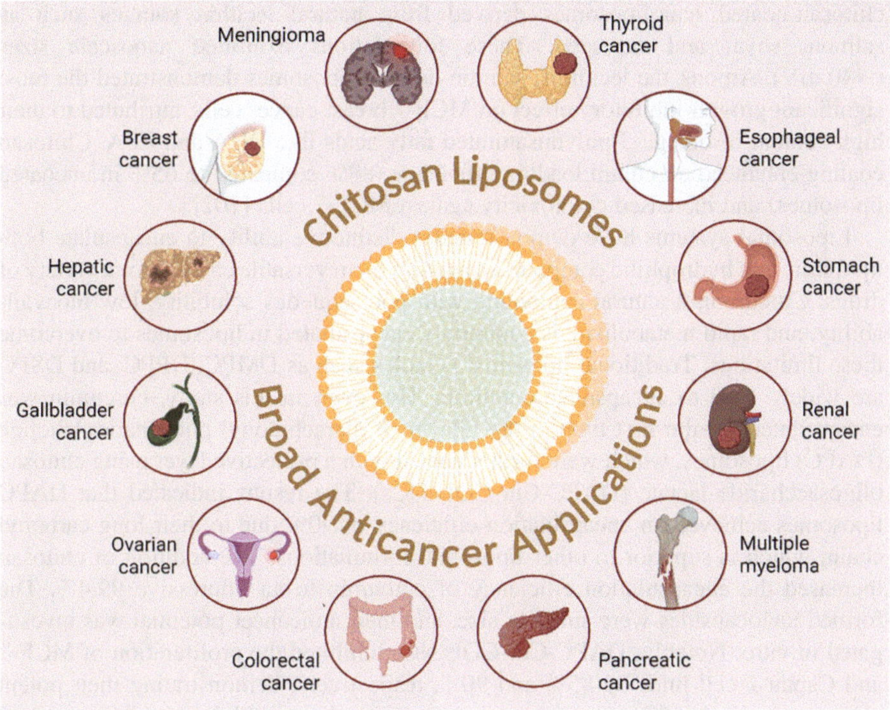

Fig. 6 Schematic representation of chitosan-based liposomes and their applications in various cancer therapies

with the functional versatility of chitosan. Preclinical studies have demonstrated improved drug delivery, tumor targeting, and reduced systemic toxicity. While clinical translation is still limited, early findings suggest strong potential for future therapeutic applications [101]. Chitosan-based liposomes have been explored for the treatment of multiple cancer types (Fig. 6).

6.1 *Breast Cancer*

Breast cancer remains a leading cause of cancer-related mortality among women worldwide. Despite advancements in treatment modalities, challenges such as poor drug solubility, limited bioavailability, and systemic toxicity persist. Nanotechnology-based delivery systems, including liposomes and nanofibers, have emerged as promising strategies to overcome these limitations by enhancing targeted delivery and controlled release of anticancer agents. Curcumin, a polyphenolic compound derived from turmeric, exhibits potent anticancer properties but suffers from poor bioavailability. To address this, curcumin was encapsulated in chitosan-coated nanoliposomes derived from natural lecithin sources such as salmon, soya, and rapeseed. These formulations exhibited nanoscale sizes (−40 mV). Among the lecithins, salmon-derived liposomes demonstrated the most significant growth-inhibitory effect on MCF-7 breast cancer cells, attributed to their high content of omega-3 polyunsaturated fatty acids like DHA and EPA. Chitosan coating enhanced curcumin loading efficiency (88% compared to 65% in uncoated liposomes) and increased cytotoxicity against MCF-7 cells [102].

Liposomal systems have demonstrated a distinctive ability to encapsulate both lipophilic and hydrophilic compounds, making them versatile carriers for a variety of drugs. Curcumin, a natural compound with poor aqueous solubility, low bioavailability, and rapid metabolism, is commonly encapsulated in liposomes to overcome these limitations. Traditional liposomal vehicles such as DMPC, DPPC, and DSPC are widely used to encapsulate curcumin. However, in this study, curcumin was encapsulated for the first time in the bilayer of diarachidonyl phosphatidylcholine (DAPC) liposomes, which were further coated with a protective layer using chitosan oligosaccharide lactate (DAPC-Cur-COL NCs). The results indicated that DAPC liposomes achieved an encapsulation efficiency of 90% due to their long carbonyl chain, which is superior to other liposomal formulations. The addition of chitosan increased the encapsulation efficiency of curcumin to an impressive 99.4%. The formed nanocapsules were small in size, and their anticancer potential was investigated in vitro. Notably, DAPC-Cur-COL NCs inhibited the proliferation of MCF-7 and Capan-1 cell lines by 85% and 90%, respectively, demonstrating their potent anticancer activity. To further improve curcumin delivery, thiolated chitosan-coated liposomal hydrogels were developed. These thermosensitive hydrogels remained fluid at room temperature and rapidly gelled at physiological temperature (37 °C), allowing for in situ injection. The hydrogels exhibited a sustained release profile, with approximately 31.6% cumulative curcumin release over 12 h. In vitro studies showed significant cytotoxicity against MCF-7 cells after 72 h, while in vivo experiments demonstrated inhibition of breast cancer recurrence post-surgical resection and promoted tissue repair [103].

Paclitaxel-loaded liposome-incorporated chitosan/poly(ε-caprolactone) (PCL) nanofibers were engineered to enhance breast cancer treatment. These nanofibers provided sustained drug release over 30 days, following the Korsmeyer–Peppas kinetic model. In vitro assays revealed high cytotoxicity against MCF-7 cells (85% ± 2.5%) while maintaining over 98% viability in normal L929 fibroblasts. In vivo studies showed a significant reduction in tumor weight, highlighting the potential of these formulations for effective breast cancer therapy [103].

A novel oral delivery system was developed by co-encapsulating exemestane (EXE) and genistein (GEN) into chitosan-coated nanoliposomes using the ethanol injection method. The optimized formulation exhibited particle sizes around 120 nm and a positive zeta potential (+22.4 mV), indicating successful chitosan coating. Encapsulation efficiencies ranged between 70% and 80%. In vitro release studies followed the Korsmeyer–Peppas model, indicating sustained release. Ex vivo studies demonstrated enhanced permeation, and cytotoxicity assays showed improved efficacy against MCF-7 cells with an IC50 of 7.253 ± 0.34 μM. The formulation also exhibited superior antioxidant activity (94.56%) and was deemed safe in animal toxicity studies. Chitosan-coated liposomes delivery systems offer significant advantages in breast cancer therapy by improving drug solubility, bioavailability, and targeted delivery while minimizing systemic toxicity. These innovative platforms demonstrate enhanced therapeutic efficacy in both in vitro and in vivo models, indicating their potential for clinical application in breast cancer treatment [104].

The application of liposomal drug delivery systems has gained significant attention in recent years, particularly for enhancing the bioavailability and therapeutic efficacy of poorly water-soluble drugs. Curcumin (Cur), a hydrophobic compound with established anticancer potential, suffers from poor solubility and rapid systemic elimination, which limit its clinical utility. To address this challenge, curcumin-loaded liposomes (Cur-Lip) were developed and subsequently coated with thiolated chitosan (CSSH) to form an advanced thermosensitive hydrogel system (CSSH/Cur-Lip gel). These hydrogels demonstrated sol-to-gel transition behavior, remaining fluidic at ambient temperatures and rapidly forming a gel at physiological temperature (37 °C), making them suitable for in situ injectable applications. The cumulative release profile of curcumin from the CSSH/Cur-Lip gel at a concentration of 200 μM showed a controlled release pattern, with $31.57 \pm 1.34\%$ released at 12 h, effectively mitigating the typical burst release associated with conventional formulations. Mechanical testing revealed that the hydrogels retained compressive strength even after multiple deformation cycles, indicating structural resilience. In vitro cytotoxicity assessments confirmed the biocompatibility of the blank gel; however, curcumin-loaded hydrogels significantly suppressed the viability of MCF-7 breast cancer cells after 72 h, suggesting potent antitumor activity. Moreover, in vivo studies involving breast cancer recurrence models demonstrated that post-resection application of CSSH/Cur-Lip gel not only inhibited tumor recurrence but also supported tissue regeneration at the surgical site. These findings highlight the dual therapeutic and regenerative capabilities of the formulation, proposing it as a promising local drug delivery system for breast cancer treatment [105].

Triple-negative breast cancer (TNBC) is characterized by its aggressive nature and lack of specific therapeutic targets, resulting in poor clinical outcomes. CD44, a cell surface adhesion receptor implicated in tumor progression and metastasis, is overexpressed in TNBC and presents a viable target for selective drug delivery. A novel approach involved the fabrication of chitosan oligosaccharide (CO)-coated liposomes, incorporating a photosensitizer (Photochlor, HPPH) and a hypoxia-activated prodrug (evofosfamide, TH-302). This dual-loaded nanocarrier system enabled both diagnostic imaging via HPPH-mediated fluorescence and therapeutic action through combined photodynamic therapy (PDT) and chemotherapy. The targeted liposomes exhibited superior tumor-specific accumulation, confirmed by enhanced imaging capabilities in vivo. When compared to non-targeted controls, the CO-coated liposomes showed increased cytotoxicity and more effective tumor inhibition, both in vitro and in murine TNBC models. Importantly, this formulation maintained a favorable safety profile, with minimal systemic toxicity. The integration of targeted delivery, multimodal therapy, and diagnostic functionality underscores the translational potential of this system for personalized management of TNBC and other CD44-overexpressing malignancies [106].

6.2 Colorectal Cancer

Colorectal cancer (CRC) remains a significant global health issue, necessitating the development of advanced therapeutic strategies. One promising approach involves the formulation of chitosan-coated flexible liposomes, referred to as chitosomes, to enhance the targeted delivery of anticancer agents such as 5-fluorouracil (5-FU). In a recent study, 5-FU-loaded flexible liposomal formulations (F1, F2, and F3) were prepared using the thin-film hydration method. These liposomes were subsequently coated with chitosan through electrostatic deposition, resulting in chitosomal formulations (F4, F5, and F6). The prepared chitosomes exhibited particle sizes ranging from 212 to 271 nm and possessed positive zeta potentials between +6.1 and +14.7 mV. In comparison, the uncoated liposomes showed smaller particle sizes (108 to 234 nm) and negative surface charges ranging from −2.3 to −16.3 mV. Morphological analysis revealed that the F3 and F6 formulations had a spherical shape with a rough surface, indicating structural modifications due to chitosan coating. In vitro drug release studies demonstrated that chitosomes provided a more controlled and sustained release of 5-FU compared to the free drug solution and uncoated liposomes. Furthermore, cytotoxicity evaluations against HT-29 colon cancer cells indicated that the chitosomal formulations exhibited enhanced anticancer efficacy, likely due to improved drug retention and prolonged exposure to cancer cells. These findings support the potential of chitosan-coated liposomal systems as effective nanocarriers for colorectal cancer therapy [107].

In addition to chemotherapeutic applications, attention has been drawn to the role of certain microbial strains, particularly *Escherichia coli*, in CRC progression. To address this aspect, a co-delivery system encapsulating usnic acid (UA) and ceftazidime (CAZ) within liposomes was developed. The liposomes were fabricated using the lipid film hydration technique, followed by sonication and chitosan coating via a drip method to improve their oral delivery properties. Characterization results showed that the chitosan-coated liposomes had particle sizes ranging from approximately 116.5 $\pm$ 5.3 nm to 240.3 $\pm$ 3.5 nm and displayed zeta potentials between +16.4 $\pm$ 0.6 and + 28 $\pm$ 0.8 mV. Drug encapsulation efficiency was notably high for both agents, with UA reaching 99.94 $\pm$ 0.1% and CAZ showing 51.5 $\pm$ 0.2%. The antimicrobial potential of the formulations was tested against various *E. coli* strains, including ATCC 25922, NCTC 13846, and H10407. Both minimum inhibitory concentration (MIC) and minimum bactericidal concentration (MBC) values were determined using the microdilution method. Additionally, biofilm formation and disruption were assessed using the crystal violet assay. The chitosan-coated liposomal formulations, including Lipo-CAZ-Chi, Lipo-UA-Chi, and their combination (either co-encapsulated or mixed post-formulation), exhibited significant antibacterial activity and effectively inhibited both the formation and persistence of *E. coli* biofilms. These results suggest a dual-function potential of the chitosan-coated liposomal system: enhancing chemotherapeutic efficacy while simultaneously addressing bacterial factors that may contribute to CRC development. This integrated approach holds promise for future oral therapeutic interventions and warrants further in vivo investigation to validate its clinical applicability in the management of colorectal cancer [108].

6.3 Colon Cancer

A significant hurdle in the treatment of colon cancer using 5-fluorouracil (5-FU) is its premature release and degradation in the upper gastrointestinal tract, particularly in the stomach and small intestine. This premature drug loss limits its availability at the target site—the colon—where its therapeutic activity is needed most. To address this challenge, innovative drug delivery systems utilizing liposomal carriers have been developed, with a particular focus on achieving colon-specific release and enhancing drug stability under varying pH conditions. In a recent study, researchers explored the potential of chitosan succinate (CSSC), known for its protective and colon-targeting properties, in the formulation of polyelectrolyte liposomes (PEL). These systems were designed using chitosan and CSSC to create a dual-layered delivery platform capable of resisting acidic degradation and promoting sustained release in the alkaline pH of the colon. The central composite design method was employed to optimize the 5-FU-loaded chitosomal formulation, which was initially prepared using the thin-film hydration technique. Subsequent coating with CSSC introduced an anionic character due to its abundant carboxylic acid groups, which electrostatically interacted with the cationic amino groups of chitosan, forming a stable polyelectrolyte complex. Comprehensive characterization of the optimized formulations revealed favorable attributes, including high drug entrapment efficiency, uniform and small particle size, and controlled drug release profiles. Particularly, the PELs demonstrated superior ability to limit drug release under acidic conditions, attributed to the deprotonation of carboxylate ions in CSSC, leading to electrostatic repulsion and swelling of the polymer matrix. These features enabled a prolonged and site-specific release of 5-FU in colonic environments. In vitro cytotoxicity studies using HT-29 colon cancer cells showed a significant reduction in cell viability upon PEL treatment, underscoring their enhanced anticancer activity. Furthermore, in vivo pharmacokinetic evaluation indicated that PELs markedly improved 5-FU absorption and bioavailability at the target site compared to conventional liposomes and chitosomes. These findings suggest that polyelectrolyte-based liposomal systems could serve as promising platforms for colon-targeted chemotherapeutic delivery [108].

6.4 Lung Cancer

Lung cancer remains the leading cause of cancer-related mortality worldwide. Despite advances in therapy, traditional chemotherapeutic approaches often fail due to poor tumor targeting, insufficient drug accumulation at tumor sites, and systemic toxicity. To address these limitations, several novel liposome-based delivery systems have been developed, particularly using chitosan and its derivatives to enhance therapeutic efficacy.

A recent study introduced chitosan oligosaccharide (CSO)-modified liposomes to improve drug accumulation in lung tumors. CSO was conjugated with Pluronic P123 to create polymers with varying CSO grafting ratios, designated as CP50 and CP20. These were used to formulate CSO-modified liposomes (CP50-LSs and CP20-LSs), which demonstrated significantly enhanced cellular uptake in A549 lung cancer cells and superior tumor accumulation in vivo compared to non-modified liposomes (P-LSs). The increased tumor affinity was attributed to the strong interaction between CSO-modified liposomes and collagen, abundantly present in lung tumors. In vivo, intravenous administration of paclitaxel-loaded CP50-LSs every 3 days for 21 days yielded an impressive tumor inhibition rate of 86.4%, indicating the strong therapeutic potential of CSO-based liposomal systems for lung cancer treatment [109]. Another study explored the potential of Mallow (*Malva* spp.) plant extracts encapsulated in alginate/chitosan-based liposomes. Two types of extracts were processed and entrapped into liposomal structures. FTIR analysis confirmed successful encapsulation. The encapsulation efficiency and drug loading content were determined to be 87.64% and 3.94% for one formulation, and 91.56% and 5.24% for the other, respectively. A sustained release profile was observed, with 90% of the encapsulated extract released over approximately 40 h. In vitro cytotoxicity tests using the A549 cell line showed that the IC50 values of free extracts ranged from 103 to 147 μg/mL at 24 h, whereas encapsulated extracts exhibited improved efficacy, with IC50 values reduced to 41 and 39 μg/mL, respectively. These formulations demonstrated dose- and time-dependent cytotoxicity, suggesting that encapsulation significantly enhances the anti-proliferative activity of natural plant extracts [109]. To overcome resistance mechanisms in epidermal growth factor receptor (EGFR)-mutated lung cancer, a novel approach involving anti-EGFR aptamer (Apt)-conjugated chitosan (Apt-Cs) anchored liposomes was employed to deliver erlotinib, an EGFR tyrosine kinase inhibitor (TKI). These liposomal constructs (Apt-CL-E) were engineered to enhance targeting and uptake by EGFR-mutant cancer cells. The Apt-CL-E complex displayed excellent biostability and specificity for EGFR, leading to enhanced cellular internalization. Cytotoxicity assays revealed that Apt-CL-E induced cell cycle arrest and apoptosis more effectively than non-targeted liposomes, demonstrating a promising targeted therapy platform for TKI-resistant lung cancer [110]. In a different combinatorial therapeutic strategy, liposomes coated with hyaluronic acid (HA) and chitosan (CS) were used to deliver a combination of berberine (BER) and doxorubicin (DOX). Characterization via FTIR, XRD, SEM, and TEM confirmed the successful formation of the

HA-CS@BER&DOX-L complex with appropriate physicochemical properties. The formulation demonstrated sustained release of both drugs under various physiological conditions. In vitro studies on A549 cells indicated significantly enhanced cytotoxic effects, with an IC50 value of 89.19 μg/300 μL for the HA-CS@BER&DOX liposomes. This formulation effectively penetrated A549 cells, induced apoptosis, and demonstrated improved therapeutic performance compared to free drugs or unmodified liposomes, establishing its potential as a robust lung cancer treatment modality [111].

Beyond lung cancer, innovative liposomal systems have also been explored for targeting hepatocellular carcinoma (HCC). A dual-drug delivery system was developed by incorporating niacin into chitosan-stabilized lecithin liposomes and curcumin into the outer liposomal shell. Folic acid-conjugated chitosan was added to facilitate tumor-targeted delivery. This configuration improved niacin loading efficiency, minimized premature drug release at physiological pH (7.4), and promoted selective uptake in cancerous tissues. In vitro studies with HePG2 cells revealed significant growth inhibition after 48 h of treatment: 91% inhibition for pure niacin, 55% for pure curcumin, 83% for niacin nanoparticles, and 51% for curcumin-niacin nanoparticles. Additionally, gene expression analysis showed that mRNA levels of mTOR and p62—both linked to autophagy—were significantly modulated. These findings underline the therapeutic promise of natural biomaterials such as niacin and curcumin delivered via chitosan-based liposomal carriers for effective cancer therapy [112].

6.5 Cervical Cancer

Cervical cancer remains one of the most prevalent malignancies affecting women globally. Among the various emerging therapeutic strategies, photodynamic therapy (PDT) has garnered considerable attention as a non-invasive and selective modality for treating neoplastic diseases, including cervical cancer. PDT involves the administration of a photosensitizing agent followed by exposure to light of a specific wavelength, generating reactive oxygen species (ROS) that induce cellular apoptosis and necrosis. Despite its therapeutic promise, a key limitation of many photosensitizers is their poor aqueous solubility, which hampers their bioavailability and clinical efficacy. Consequently, the development of an efficient delivery system that can enhance the solubility, stability, and tumor-targeting capacity of photosensitizers is of paramount importance.

To address this limitation, researchers have synthesized composite films incorporating chitosan, polyvinylpyrrolidone (PVP), and liposomes encapsulating zinc phthalocyanine (ZnPc). These biopolymeric films were characterized using UV–visible and fluorescence spectroscopy to assess the photophysical properties of ZnPc. Additionally, the films were evaluated for their swelling behavior, mechanical integrity, and water vapor permeability. The in vitro photodynamic efficacy of these films was tested using HeLa cervical cancer cells, revealing a significant decrease in cell viability exceeding 95% upon irradiation. These findings suggest that ZnPc-liposome-loaded chitosan/PVP films possess strong potential as photoactive therapeutic platforms for localized cervical cancer treatment [113].

In another innovative approach, a chitosan-coated liposomal formulation of ursolic acid (UA)—referred to as CS-UA-L—was developed to enhance tumor-targeted drug delivery. The CS-UA-L nanoparticles were spherical with an average diameter of approximately 130 nm, facilitating enhanced permeability and retention (EPR) in tumor tissues. The surface modification with chitosan imparted a positive charge to the liposomes, enabling electrostatic interaction with negatively charged tumor cell membranes and improving cellular uptake. Moreover, the formulation exhibited pH-responsive behavior, releasing UA more effectively in the acidic tumor microenvironment (pH 5.0) compared to physiological pH (7.4). In vitro studies demonstrated significant anti-proliferative effects on HeLa cells (76.46% inhibition), while in vivo experiments in U14 cervical tumor-bearing mice showed a notable reduction in tumor growth (61.26% inhibition). Histopathological analysis of treated tumor tissues revealed extensive apoptosis, necrosis, and minimal cell proliferation. These outcomes underscore the potential of CS-UA-L as a precision nanocarrier system capable of minimizing systemic toxicity while enhancing anticancer efficacy [114].

Further advancements in liposome-based nanocarriers involve the incorporation of quaternized N,O-oleoyl chitosan (QCS) into liposomal vesicles, forming a novel system termed Lip-QCS. These liposomes demonstrated a narrow size distribution (171.5 ± 0.8 nm), positive surface charge (zeta potential of 11.7 ± 0.7 mV), and mucoadhesive properties, enhancing their retention in mucosal tissues such as the cervix. Cellular uptake studies confirmed that Lip-QCS exhibited time-dependent internalization, with significant uptake observed after 24 h. When loaded with cisplatin, a widely used chemotherapeutic agent, the system achieved a drug loading efficiency of $27.45 \pm 0.78\%$. The therapeutic efficacy of cisplatin-loaded Lip-QCS was evaluated using a three-dimensional (3D) spheroid model of SiHa cervical cancer cells. The results demonstrated marked cytotoxicity and apoptosis induction, indicating that this system provides controlled drug release and improved penetration into tumor-like structures. Collectively, these findings highlight the potential of chitosan-based and surface-modified liposomal systems as versatile and effective nanocarriers for cervical cancer therapy. By enhancing drug solubility, facilitating targeted delivery, and enabling controlled release, these nanocarrier platforms offer promising avenues for improving clinical outcomes while minimizing adverse effects [115].

6.6 Gastric Cancer

This study aimed to enhance autophagic activity in hepatocellular carcinoma cells to promote cellular degradation and improve therapeutic efficacy. To achieve this, liposomes incorporating chitosan in their core were developed to enhance lecithin stability and increase the loading efficiency of niacin. Curcumin, a hydrophobic compound, was entrapped within the liposomal bilayer and functioned as a surface layer to control and reduce the release of niacin at physiological pH (7.4). Furthermore, folic acid-conjugated chitosan was employed to facilitate targeted delivery of the liposomes specifically to cancer cells. Successful liposomal formation and high encapsulation efficiency were confirmed using transmission electron microscopy (TEM), UV–visible spectroscopy, and Fourier-transform infrared spectroscopy (FTIR). In vitro cytotoxicity studies conducted on HePG2 cells revealed significant growth inhibition following 48 h of incubation at a concentration of 100 μg/mL. The inhibition rates were recorded as 91 ± 1% (pure niacin, $P \leq 0.002$), 55 ± 3% (pure curcumin, $P \leq 0.001$), 83 ± 1.5% (niacin-loaded nanoparticles, $P \leq 0.001$), and 51 ± 1.5% (curcumin-niacin-loaded nanoparticles, $P \leq 0.0001$), compared to untreated controls. Gene expression analysis further revealed a significant upregulation of mTOR mRNA levels: 0.72 ± 0.08 ($P \leq 0.001$) for pure niacin, 1.0 ± 0.1 ($P \leq 0.001$) for pure curcumin, 0.5 ± 0.07 ($P \leq 0.01$) for niacin nanoparticles, and 1.3 ± 0.02 ($P \leq 0.001$) for curcumin-niacin nanoparticles, relative to the control group (0.3 ± 0.08). Similarly, the expression of p62 mRNA was significantly elevated: 0.92 ± 0.07 ($P \leq 0.05$) for pure niacin, 1.7 ± 0.07 ($P \leq 0.0001$) for pure curcumin, 0.72 ± 0.08 ($P \leq 0.5$) for niacin nanoparticles, and 2.1 ± 0.1 ($P \leq 0.0001$) for curcumin-niacin nanoparticles, compared to the control expression level of 0.72 ± 0.08. Collectively, these findings underscore the therapeutic potential of biomaterial-based nanoformulations derived from natural compounds, offering a promising alternative to conventional chemotherapies in the treatment of gastric and liver cancers [112].

6.7 *Wound Cancer*

This study focused on the development of a novel chemowrap patch composed of hydrogel-forming chitosan-based microneedles (HFM) loaded with 5-fluorouracil (5-FU)-encapsulated PEGylated liposomes (LPs), aiming to enhance the therapeutic efficacy of 5-FU for the treatment of pre-carcinogenic and carcinogenic skin lesions. The resulting HFM chemowrap patch demonstrated excellent physical and mechanical properties, including complete penetration capability, making it suitable for effective transdermal drug delivery. In vivo skin permeation studies revealed that the 5-FU-PEGylated LPs-loaded HFM patch achieved the highest level of 5-FU skin permeation (42.06 ± 11.82%) and skin deposition (75.90 ± 1.13%) compared to other formulations. Moreover, the formulation exhibited a marked improvement in wound healing, with a significant percentage of complete wound closure observed in vivo (47.00 ± 5.77% on day 7) and in normal human fibroblast (NHF) cells (92.79 ± 7.15% at 48 h). Importantly, the 5-FU-PEGylated LPs-loaded HFM patch demonstrated potent anticancer activity while preserving the viability of normal cells, indicating its selective cytotoxicity. Apoptotic analysis further confirmed that the novel chemowrap formulation induced a higher rate of apoptosis than the conventional 5-FU solution, underscoring its enhanced therapeutic potential. The developed 5-FU-PEGylated LPs-loaded HFM chemowrap patch offers a promising and safe drug delivery strategy for the effective treatment of both pre-cancerous and cancerous skin conditions, combining efficient drug penetration, enhanced therapeutic action, and selective cytotoxicity [116].

6.8 Melanoma

This study explores the development of an innovative drug delivery platform utilizing liposomes coated with multifunctional polymers for tumor-targeted therapy. The research focuses on the synthesis and characterization of novel Folic Acid–Cysteine-Thiolated Chitosan (FTC) derivatives, which possess both active targeting ability and redox responsiveness. These FTC derivatives were employed to fabricate FTC-coated liposomes, which were then characterized for their size, surface charge, and drug encapsulation efficiency before and after coating. The formation of the coating layer on the liposomal surface was confirmed by observing a slight increase in particle size and changes in zeta potential, indicating successful coating. The FTC-coated liposomes exhibited a redox-dependent drug release profile, demonstrating good stability under physiological conditions, with rapid release of the entrapped calcein in the presence of glutathione, a characteristic of the reductive microenvironment typical of tumor cells. The uptake and cytotoxicity of doxorubicin-loaded FTC-coated liposomes were evaluated in murine B16-F10 and human SKMEL2 melanoma cancer cells. The results showed a significant enhancement in both cellular uptake and antitumor efficacy of FTC-coated liposomes compared to control chitosan-coated liposomes. This improved performance is attributed to increased cellular uptake via folate receptor-mediated endocytosis and the triggered release of the drug in the reductive tumor microenvironment. The findings suggest that these novel FTC-coated liposomes have great potential as targeted nanocarriers for cancer therapy, offering a promising strategy for improving the efficacy of drug delivery to melanoma cells [117].

6.9 Pancreatic Cancer

Lawsone, a natural bioactive compound, was encapsulated in liposomal nanoparticles coated with chitosan–folate (L-LNPs-CF), creating a novel drug delivery system. The antioxidant, anticancer, and proapoptotic activities of L-LNPs-CF were evaluated through cell viability assays, acridine orange/propidium iodide (AO/PI) staining, and flow cytometry. Additionally, gene profiling of key apoptotic markers, including caspase 3, caspase 9, and Bax, was performed using real-time PCR. The results revealed that L-LNPs-CF had a particle size of 111.98 ± 2.15 nm and a surface charge of 20.89 mV, with lawsone encapsulation efficiency of 83.8%. The cellular and molecular experiments demonstrated that L-LNPs-CF exhibited strong free radical scavenging activity, with respective DPPH and ABTS values of 40.9% and 84.14%. Importantly, L-LNPs-CF significantly inhibited the proliferation of pancreatic cancer cells (IC50: 171.94 μg/mL), and notably upregulated the expression of caspase 3, caspase 9, and Bax, which are key genes involved in apoptosis. These findings suggest that L-LNPs-CF enhances cellular uptake and represents a promising therapeutic system for pancreatic cancer [118].

6.10 Liver Cancer

In the context of liver cancer, polyethylene glycol (PEG)-coated magnetic polymeric liposome (MPL) nanoparticles were developed for in vivo liver targeting. These MPLs were assembled from octadecyl quaternized carboxymethyl chitosan (OQC), PEGylated OQC, cholesterol, and magnetic nanoparticles, and functionalized with epithelial growth factor receptor (EGFR) peptides. The two-step liver targeting strategy, which combined magnetic force and EGFR peptide conjugation, was evaluated in a subcutaneous hepatocellular carcinoma model in nude mice. The results showed that EGFR-conjugated MPLs accumulated in the liver due to the magnetic force and could also diffuse into tumor cells through EGFR-mediated targeting. Moreover, paclitaxel (PTX) was incorporated into small EGFR-conjugated MPLs (102.0 $\pm$ 0.7 nm), resulting in spherical particles with high drug encapsulation efficiency (>90%). Magnetic targeting further enhanced the transport of PTX-loaded MPLs to the tumor site, as confirmed by PTX level detection. These findings suggest that PTX-loaded EGFR-conjugated MPLs could serve as an effective drug delivery system for targeted liver cancer therapy [119].

6.11 Leukemia

In the field of leukemia treatment, researchers have explored the effectiveness of various nanofiber formulations for controlled irinotecan (CPT) release. Specifically, poly(ε-caprolactone) (PCL)/chitosan (CS) core–shell nanofibers were developed, including PCL/CS (core)/irinotecan (CPT) nanofibers (termed PCL-CS10 CPT), PCL/chitosan/irinotecan (core)/PCL/chitosan (shell) nanofibers (termed CS/CPT/PCL/CS), and irinotecan-loaded liposome-incorporated PCL/chitosan-chitosan nanofibers (termed CPT@Lipo/CS/PCL/CS). The primary goal of these formulations was to investigate their potential for controlled irinotecan release in the treatment of leukemia. The fabricated nanofibers were thoroughly characterized using techniques such as Fourier transform infrared (FTIR) analysis, transmission electron microscopy (TEM), scanning electron microscopy (SEM), dynamic light scattering (DLS), zeta potential measurement, and polydispersity index (PDI).

The results indicated that irinotecan was released in a controlled manner from the liposome-loaded nanofibers over a period of 30 days. Additionally, cell viability assays demonstrated that the fabricated nanofibers exhibited minimal toxicity to human umbilical vein endothelial cells (HUVECs), with viability remaining >98% ± 1% after 168 h. Notably, the CPT@Lipo/CS/PCL/CS nanofibers showed a significant cytotoxic effect on K562 leukemia cancer cells, achieving a maximal cytotoxicity of 85% ± 2.5%. The release profile of the CPT@Lipo/CS/PCL/CS nanofibers followed a three-stage drug release pattern, further supporting their potential for effective and controlled drug delivery. These findings highlight the promise of liposome-incorporated core–shell nanofibers as a valuable platform for future leukemia therapy [120].

7 Challenges and Limitations

The development of chitosan-based liposomes for cancer therapy faces multiple challenges, particularly in terms of stability, scale-up, regulatory approval, and long-term safety. Formulation stability is a significant hurdle that can directly impact the effectiveness of liposomal drug delivery systems [121]. Physical stability issues arise when chitosan-coated liposomes aggregate or fuse, especially under suboptimal storage conditions, leading to a shift in size distribution and a loss in encapsulation efficiency [122]. The cationic nature of chitosan can also cause undesirable interactions with formulation components, compromising the structural integrity of liposomes. Chemical stability is another concern, as chitosan is susceptible to degradation over time, particularly in the presence of moisture or acidic environments. This degradation can cause premature release of encapsulated drugs or reduced stability of the liposomal membrane, which may affect therapeutic efficacy [123]. Scale-up challenges are crucial when transitioning from laboratory-scale synthesis to large-scale manufacturing. The consistency of size, surface charge, and encapsulation efficiency must be maintained during mass production, which often requires process optimization. Techniques like thin-film hydration or reverse-phase evaporation may need substantial fine-tuning to ensure reproducibility, which is vital for commercial-scale production [124]. Additionally, variability in the quality of raw materials, such as chitosan derived from different sources (e.g., shrimp or crab shells), can result in batch-to-batch inconsistencies that complicate standardization during manufacturing [125]. Regulatory considerations add complexity to the approval process for chitosan-based liposomes. While some liposomal formulations, such as Doxil®, have received FDA approval, regulatory bodies may require extensive safety and efficacy data for each new chitosan-based liposome formulation [126]. The biopolymer's natural characteristics—such as varying molecular weight or degree of deacetylation—could lead to concerns regarding consistency, requiring comprehensive evaluations of toxicity, immunogenicity, and long-term biocompatibility [127]. Immunogenicity and long-term safety are further factors that need careful evaluation. Despite chitosan's generally favorable biocompatibility and biodegradability, its cationic charge may induce a mild immune response, leading to inflammation or complement activation [128]. The long-term accumulation of liposomes, particularly in organs like the liver and spleen, may cause tissue damage or toxicity over time. Consequently, optimizing surface modifications to reduce immunogenicity, such as PEGylation, and assessing the long-term effects in clinical trials are essential steps for ensuring patient safety [129]. The cost-effectiveness of chitosan-based liposomes is another significant challenge, as the cost of raw materials and specialized manufacturing equipment can be prohibitively high. Scaling up production methods while maintaining high encapsulation efficiency and uniformity remains a major concern [130]. To mitigate these challenges, researchers are exploring alternative production methods, such as microfluidics or electrohydrodynamic atomization, which offer the potential for producing more uniform liposomes at a lower cost. Additionally, engaging early with regulatory agencies to navigate the approval process can facilitate smoother transitions to clinical applications, ensuring that chitosan-based liposomes meet the necessary safety and efficacy standards [131]. Overcoming these hurdles will pave the way for the successful commercialization and clinical adoption of chitosan-based liposomal formulations [132].

8 Future Prospects and Emerging Trends

The future of chitosan-based liposomes in cancer therapy is poised to be shaped by several emerging trends and technological advancements that promise to significantly improve their efficacy, precision, and personalized approach to treatment. Among the most promising developments are stimuli-responsive chitosan-liposomes, which respond to specific cues from the tumor microenvironment to trigger controlled drug release. pH-responsive liposomes take advantage of the acidic nature of tumors (pH ~6.5–7.0), enabling the liposomes to release their payload specifically at the tumor site, minimizing premature drug release and reducing toxicity to healthy tissues [133]. Redox-responsive liposomes capitalize on the altered redox states of cancer cells, with increased levels of reactive oxygen species (ROS) and reduced glutathione (GSH). By incorporating disulfide bonds or other redox-sensitive materials, chitosan-liposomes can be designed to release their encapsulated drugs upon encountering the oxidative environment of the tumor. Another exciting development is enzyme-responsive liposomes, where chitosan-coated liposomes are functionalized with enzyme-sensitive linkers, such as those cleaved by matrix metalloproteinases (MMPs), which are overexpressed in the tumor microenvironment [134]. These systems enable targeted release of drugs, further enhancing the precision of drug delivery and reducing off-target effects. Such stimuli-responsive systems not only improve the therapeutic index but also hold the potential to significantly reduce side effects by ensuring that drugs are released only at the desired site [135]. In addition to stimuli-responsive systems, smart hybrid nanocarriers, combining chitosan-based liposomes with other nanomaterials, are gaining attention for their ability to enhance the therapeutic potential of cancer therapies. For instance, polymer-liposome hybrids integrate chitosan with polymers like polyethylene glycol (PEG) or poly(lactic-co-glycolic acid) (PLGA), which help improve liposome stability, reduce aggregation, and enhance targeted delivery [136]. These hybrids can be designed to encapsulate hydrophobic drugs and provide controlled release properties while also targeting specific cancer cell receptors. Multi-functional hybrids take this a step further by combining liposomes with nanostructures such as magnetic nanoparticles, quantum dots, or gold nanoparticles, which can provide additional functionalities. For example, magnetic liposome hybrids can be directed to tumor sites using an external magnetic field, while quantum dots can be used for real-time imaging and monitoring of drug release, allowing for better treatment tracking [137]. As cancer therapy becomes more personalized, patient-specific nanocarriers are expected to play a crucial role in improving treatment outcomes. By analyzing genetic profiles and tumor markers, chitosan-liposomes can be tailored to target specific mutations or biomarkers unique to an individual's cancer. This personalized approach ensures that the therapeutic agents are delivered with greater specificity, thereby reducing off-target effects [138]. Genetic profiling can guide the design of chitosan-liposomes functionalized with specific ligands, such as antibodies or peptides, that target mutations like epidermal growth factor receptor (EGFR) mutations in non-small-cell lung cancer (NSCLC) or HER2 overexpression in breast cancer [139]. Moreover, patient-specific nanocarriers can be engineered to deliver drugs directly to tumors based on the specific genetic characteristics of the patient's cancer, revolutionizing the way cancer therapies are developed and administered [140]. One of the most ground-breaking advancements in the field is the use of CRISPR/Cas9 gene-editing technology for cancer treatment. Chitosan-liposomes are emerging as promising delivery vehicles for CRISPR/Cas9 components, such as Cas9 proteins and guide RNAs, enabling precise gene editing directly within tumor cells [141]. This system allows for the targeted modification of oncogenes, repair of tumor suppressor genes, or induction of apoptosis in cancer cells. The integration of CRISPR/Cas9 with chitosan-liposomes can revolutionize cancer therapy by offering a non-invasive method for genetic modifications within tumor cells, bypassing the need for traditional therapies like chemotherapy and radiation [142]. Targeted gene editing using chitosan-liposomes functionalized with tumor-specific ligands can ensure that CRISPR components are delivered selectively to the tumor, further enhancing the precision of gene therapy while minimizing off-target effects [143]. Lastly, the integration of artificial intelligence (AI) and nanorobotics with chitosan-based liposomes has the potential to revolutionize drug delivery systems. AI algorithms can be utilized to predict the pharmacokinetic behavior, stability, and drug release profiles of chitosan-liposomes, optimizing their design and ensuring they are tailored for maximum therapeutic benefit [144]. AI can also assist in personalizing treatment plans based on patient data, ensuring that the most appropriate drug formulations are selected. Nanorobots, on the other hand, can be designed to carry therapeutic agents within chitosan-liposomes and autonomously navigate the body, detecting and targeting tumor cells [145]. These nanorobots could release drugs in a highly controlled manner upon detecting specific tumor markers, providing a more precise and adaptive approach to cancer treatment. By combining AI and nanorobotics with chitosan-liposome technology, cancer therapies could become more intelligent, adaptive, and responsive to the dynamic changes in tumor biology, offering the potential for highly effective, personalized, and targeted treatments [146]. In summary, the future prospects of chitosan-based liposomes in cancer therapy are bright, with numerous emerging trends and technologies that promise to enhance their efficacy, specificity, and versatility [147]. The combination of stimuli-responsive systems, hybrid nanocarriers, personalized medicine, CRISPR/Cas9 gene-editing, and cutting-edge AI and nanorobotics will undoubtedly play a pivotal role in transforming cancer treatment paradigms, making therapies more precise, effective, and tailored to individual patient needs [148].

9 Conclusion

Chitosan-based liposomes represent a transformative approach in cancer therapy, combining the advantageous properties of chitosan, such as its biocompatibility, biodegradability, and mucoadhesion, with the efficient drug delivery capabilities of liposomes. These nanocarriers offer enhanced stability, targeted delivery, and controlled release of therapeutic agents, making them a promising platform for reducing the side effects of chemotherapy while improving drug efficacy. Innovations such as stimuli-responsive systems, hybrid nanocarriers, and gene-editing delivery are pushing the boundaries of what chitosan-liposomes can achieve, offering personalized and precise treatment options. While challenges related to stability, scalability, and regulatory approval persist, the future of chitosan-based liposomes in clinical oncology remains promising, with continued research likely to lead to breakthroughs in cancer treatment.

References

1. Cooper, G. M., & Hausman, R. E. (2007). The development and causes of cancer. In *The cell: A molecular approach* (p. 743). Sinauer Associates.
2. Thun, M. J., DeLancey, J. O., Center, M. M., Jemal, A., & Ward, E. M. (2009). The global burden of cancer: Priorities for prevention. *Carcinogenesis, 31*, 100–110. https://doi.org/10.1093/carcin/bgp263
3. Bravo, M., Fortuni, B., Mulvaney, P., Hofkens, J., Uji-i, H., Rocha, S., & Hutchison, J. A. (2024). Nanoparticle-mediated thermal cancer therapies: Strategies to improve clinical translatability. *Journal of Controlled Release, 372*, 751–777. https://doi.org/10.1016/j.jconrel.2024.06.055
4. Nakamura, H., & Maeda, H. (2013). Cancer chemotherapy. In *Fundamentals of pharmaceutical nanoscience* (pp. 401–427). https://doi.org/10.1007/978-1-4614-9164-4_15
5. Bin Emran, T., Shahriar, A., Mahmud, A. R., Rahman, T., Abir, M. H., Siddiquee, M. F. R., Ahmed, H., Rahman, N., Nainu, F., Wahyudin, E., Mitra, S., Dhama, K., Habiballah, M. M.,

Haque, S., Islam, A., & Hassan, M. M. (2022). Multidrug resistance in cancer: Understanding molecular mechanisms, immunoprevention and therapeutic approaches. *Frontiers in Oncology, 12*, 891652. https://doi.org/10.3389/fonc.2022.891652
6. Majeed, H., & Gupta, V. (2020). *Adverse effects of radiation therapy*. StatPearls.
7. Jain, R. K. (2013). Normalizing tumor microenvironment to treat cancer: Bench to bedside to biomarkers. *Journal of Clinical Oncology, 31*, 2205–2218. https://doi.org/10.1200/JCO.2012.46.3653
8. Ezike, T. C., Okpala, U. S., Onoja, U. L., Nwike, C. P., Ezeako, E. C., Okpara, O. J., Okoroafor, C. C., Eze, S. C., Kalu, O. L., Odoh, E. C., Nwadike, U. G., Ogbodo, J. O., Umeh, B. U., Ossai, E. C., & Nwanguma, B. C. (2023). Advances in drug delivery systems, challenges and future directions. *Heliyon, 9*, e17488. https://doi.org/10.1016/j.heliyon.2023.e17488
9. Hanahan, D., & Weinberg, R. A. (2011). Hallmarks of cancer: The next generation. *Cell, 144*, 646–674. https://doi.org/10.1016/j.cell.2011.02.013
10. Nsairat, H., Khater, D., Sayed, U., Odeh, F., Al Bawab, A., & Alshaer, W. (2022). Liposomes: Structure, composition, types, and clinical applications. *Heliyon, 8*, e09394. https://doi.org/10.1016/j.heliyon.2022.e09394
11. Rani, V., Venkatesan, J., & Prabhu, A. (2022). Liposomes- a promising strategy for drug delivery in anticancer applications. *Journal of Drug Delivery Science and Technology, 76*, 103739. https://doi.org/10.1016/j.jddst.2022.103739
12. Chen, J., Hu, S., Sun, M., Shi, J., Zhang, H., Yu, H., & Yang, Z. (2024). Recent advances and clinical translation of liposomal delivery systems in cancer therapy. *European Journal of Pharmaceutical Sciences, 193*, 106688. https://doi.org/10.1016/j.ejps.2023.106688
13. Rommasi, F., & Esfandiari, N. (2021). Liposomal Nanomedicine: Applications for drug delivery in cancer therapy. *Nanoscale Research Letters, 16*, 1–20. https://doi.org/10.1186/s11671-021-03553-8
14. Bulbake, U., Doppalapudi, S., Kommineni, N., & Khan, W. (2017). Liposomal formulations in clinical use: An updated review. *Pharmaceutics, 9*, 12. https://doi.org/10.3390/pharmaceutics9020012
15. Hisham, F., Maziati Akmal, M. H., Ahmad, F., Ahmad, K., & Samat, N. (2024). Biopolymer chitosan: Potential sources, extraction methods, and emerging applications. *Ain Shams Engineering Journal, 15*, 102424. https://doi.org/10.1016/j.asej.2023.102424
16. Manohar, M., Shetty, A., Lobo, C. L., Jain, P., Hebbar, S., Dhas, N., Sutar, K. P., Sukeewandhi, J., Perumalsamy, H., Balusamy, S. R., Jamous, Y. F., Dubey, A., & El-Zahaby, S. A. (2025). Emerging trends in chitosan based colloidal drug delivery systems: A translational journey from research to practice. *Carbohydrate Polymers, 360*, 123604. https://doi.org/10.1016/j.carbpol.2025.123604
17. Herdiana, Y., Wathoni, N., Shamsuddin, S., & Muchtaridi, M. (2022). Drug release study of the chitosan-based nanoparticles. *Heliyon, 8*, e08674. https://doi.org/10.1016/j.heliyon.2021.e08674
18. Herdiana, Y., Wathoni, N., Gozali, D., Shamsuddin, S., & Muchtaridi, M. (2023). Chitosan-based Nano-smart drug delivery system in breast cancer therapy. *Pharmaceutics., 15*, 879. https://doi.org/10.3390/pharmaceutics15030879
19. Karuppaiah, A., Babu, D., Selvaraj, D., Natrajan, T., Rajan, R., Gautam, M., Ranganathan, H., Siram, K., Nesamony, J., & Sankar, V. (2021). Building and behavior of a pH-stimuli responsive chitosan nanoparticles loaded with folic acid conjugated gemcitabine silver colloids in MDA-MB-453 metastatic breast cancer cell line and pharmacokinetics in rats. *European Journal of Pharmaceutical Sciences, 165*, 105938. https://doi.org/10.1016/j.ejps.2021.105938
20. Žigrayová, D., Mikušová, V., & Mikuš, P. (2024). Advances in Chitosan derivatives: Preparation, properties and applications in pharmacy and medicine. *Gels., 10*, 701. https://doi.org/10.3390/gels10110701

21. Yao, Y., Zhou, Y., Liu, L., Xu, Y., Chen, Q., Wang, Y., Wu, S., Deng, Y., Zhang, J., & Shao, A. (2020). Nanoparticle-based drug delivery in cancer therapy and its role in overcoming drug resistance. *Frontiers in Molecular Biosciences, 7*, 193. https://doi.org/10.3389/fmolb.2020.00193
22. Buschmann, M. D., Merzouki, A., Lavertu, M., Thibault, M., Jean, M., & Darras, V. (2013). Chitosans for delivery of nucleic acids. *Advanced Drug Delivery Reviews, 65*, 1234–1270. https://doi.org/10.1016/j.addr.2013.07.005
23. Kumar, S., Arora, A., Mathur, D., Chaudhary, A., Pant, V., Guchhait, S., & Singh, B. K. (2025). A review on chitosan and chitosan-based bionanocomposites: Promising biological macromolecules for sustainable corrosion inhibition. *International Journal of Biological Macromolecules, 301*, 140392. https://doi.org/10.1016/j.ijbiomac.2025.140392
24. Liu, Y., Liang, Y., Yuhong, J., Xin, P., Han, J. L., Zhu, R., Zhang, M., Chen, W., Ma, Y., Du, Y., & Yu, X. (2024). Advances in nanotechnology for enhancing the solubility and bioavailability of poorly soluble drugs. *Drug Design, Development and Therapy, 18*, 1469–1495. https://doi.org/10.2147/DDDT.S447496
25. Shokati Eshkiki, Z., Mansouri, F., Karamzadeh, A. R., Namazi, A., Heydari, H., Akhtari, J., Tabaeian, S. P., & Akbari, A. (2024). Chitosan and its derivative-based nanoparticles in gastrointestinal cancers: Molecular mechanisms of action and promising anticancer strategies. *Journal of Clinical Pharmacy and Therapeutics, 2024*, 1239661. https://doi.org/10.1155/2024/1239661
26. Agostinis, P., Berg, K., Cengel, K. A., Foster, T. H., Girotti, A. W., Gollnick, S. O., Hahn, S. M., Hamblin, M. R., Juzeniene, A., Kessel, D., Korbelik, M., Moan, J., Mroz, P., Nowis, D., Piette, J., Wilson, B. C., & Golab, J. (2011). Photodynamic therapy of cancer: An update. *CA: A Cancer Journal for Clinicians, 61*, 250–281. https://doi.org/10.3322/caac.20114
27. Li, K., Guo, B., Gu, J., Ta, N., Gu, J., Yu, H., Sun, M., & Han, T. (2025). Emerging advances in drug delivery systems (DDSs) for optimizing cancer complications. *Materials Today Bio, 30*, 101375. https://doi.org/10.1016/j.mtbio.2024.101375
28. Desai, N., Rana, D., Salave, S., Gupta, R., Patel, P., Karunakaran, B., Sharma, A., Giri, J., Benival, D., & Kommineni, N. (2023). Chitosan: A potential biopolymer in drug delivery and biomedical applications. *Pharmaceutics., 15*, 1313. https://doi.org/10.3390/pharmaceutics15041313
29. Amin, M., Seynhaeve, A. L. B., Sharifi, M., Falahati, M., & ten Hagen, T. L. M. (2022). Liposomal drug delivery systems for cancer therapy: The Rotterdam experience. *Pharmaceutics, 14*, 2165. https://doi.org/10.3390/pharmaceutics14102165
30. Al-Thani, A. N., Jan, A. G., Abbas, M., Geetha, M., & Sadasivuni, K. K. (2024). Nanoparticles in cancer theragnostic and drug delivery: A comprehensive review. *Life Sciences, 352*, 122899. https://doi.org/10.1016/j.lfs.2024.122899
31. Vosoughi, P., Naghib, S. M., Jafari, T., & Kangarshahi, B. M. (2025). Chitosan-encapsulated lipid-based nanovesicles for therapeutic applications and tissue engineering: A review. *Carbohydrate Polymer Technologies and Applications, 10*, 100805. https://doi.org/10.1016/j.carpta.2025.100805
32. Sebaaly, C., Trifan, A., Sieniawska, E., & Greige-Gerges, H. (2021). Chitosan-coating effect on the characteristics of liposomes: A focus on bioactive compounds and essential oils: A review. *PRO, 9*, 1–47. https://doi.org/10.3390/pr9030445
33. Wang, T., Zhang, D., Sun, D., & Gu, J. (2020). Current status of in vivo bioanalysis of nano drug delivery systems. *Journal of Pharmaceutical Analysis, 10*, 221–232. https://doi.org/10.1016/j.jpha.2020.05.002
34. Mura, P., Maestrelli, F., Cirri, M., & Mennini, N. (2022). Multiple roles of Chitosan in mucosal drug delivery: An updated review. *Marine Drugs, 20*, 335. https://doi.org/10.3390/md20050335
35. Grewal, A. K., & Salar, R. K. (2024). Chitosan nanoparticle delivery systems: An effective approach to enhancing efficacy and safety of anticancer drugs. *Nano TransMed, 3*, 100040. https://doi.org/10.1016/j.ntm.2024.100040

36. Geethakumari, D., Bhaskaran Sathyabhama, A., Raji Sathyan, K., Mohandas, D., Somasekharan, J. V., & Thavarool Puthiyedathu, S. (2022). Folate functionalized chitosan nanoparticles as targeted delivery systems for improved anticancer efficiency of cytarabine in MCF-7 human breast cancer cell lines. *International Journal of Biological Macromolecules, 199*, 150–161. https://doi.org/10.1016/j.ijbiomac.2021.12.070
37. Thomas, R. G., Surendran, S. P., & Jeong, Y. Y. (2020). Tumor microenvironment-stimuli responsive nanoparticles for anticancer therapy. *Frontiers in Molecular Biosciences, 7*, 610533. https://doi.org/10.3389/fmolb.2020.610533
38. Biswas, R., Mondal, S., Ansari, M. A., Sarkar, T., Condiuc, I. P., Trifas, G., & Atanase, L. I. (2025). Chitosan and its derivatives as Nanocarriers for drug delivery. *Molecules, 30*, 1297. https://doi.org/10.3390/molecules30061297
39. Ashrafizadeh, M., Delfi, M., Zarrabi, A., Bigham, A., Sharifi, E., Rabiee, N., Paiva-Santos, A. C., Kumar, A. P., Tan, S. C., Hushmandi, K., Ren, J., Zare, E. N., & Makvandi, P. (2022). Stimuli-responsive liposomal nanoformulations in cancer therapy: Pre-clinical & clinical approaches. *Journal of Controlled Release, 351*, 50–80. https://doi.org/10.1016/j.jconrel.2022.08.001
40. Bhattacharya, T., Preetam, S., Ghosh, B., Chakrabarti, T., Chakrabarti, P., Samal, S. K., & Thorat, N. (2023). Advancement in biopolymer assisted cancer theranostics. *ACS Applied Bio Materials, 6*, 3959–3983. https://doi.org/10.1021/acsabm.3c00458
41. Elieh-Ali-Komi, D., & Hamblin, M. R. (2016). Chitin and Chitosan: Production and application of versatile biomedical nanomaterials. *International Journal of Advanced Research, 4*, 411–427.
42. Younes, I., & Rinaudo, M. (2015). Chitin and chitosan preparation from marine sources. Structure, properties and applications. *Marine Drugs, 13*, 1133–1174. https://doi.org/10.3390/md13031133
43. Yadav, M., Kaushik, B., Rao, G. K., Srivastava, C. M., & Vaya, D. (2023). Advances and challenges in the use of chitosan and its derivatives in biomedical fields: A review. *Carbohydrate Polymer Technologies and Applications, 5*, 100323. https://doi.org/10.1016/j.carpta.2023.100323
44. Ul-Islam, M., Alabbosh, K. F., Manan, S., Khan, S., Ahmad, F., & Ullah, M. W. (2024). Chitosan-based nanostructured biomaterials: Synthesis, properties, and biomedical applications. *Advanced Industrial and Engineering Polymer Research, 7*, 79–99. https://doi.org/10.1016/j.aiepr.2023.07.002
45. Singha, I., & Basu, A. (2022). Chitosan based injectable hydrogels for smart drug delivery applications. *Sensors Int., 3*, 100168. https://doi.org/10.1016/j.sintl.2022.100168
46. Yadav, H., Malviya, R., & Kaushik, N. (2024). Chitosan in biomedicine: A comprehensive review of recent developments. *Carbohydrate Polymer Technologies and Applications, 8*, 100551. https://doi.org/10.1016/j.carpta.2024.100551
47. Dimassi, S. N., Hahladakis, J. N., Yahia, M. N. D., Ahmad, M. I., Sayadi, S., & Al-Ghouti, M. A. (2022). Degradation-fragmentation of marine plastic waste and their environmental implications: A critical review. *Arabian Journal of Chemistry, 15*, 104262. https://doi.org/10.1016/j.arabjc.2022.104262
48. Prabakaran, A., & Alexander, A. (2024). Biophysical analysis on molecular interactions between chitosan-coated sinapic acid loaded liposomes and mucin. *Biochimica Et Biophysica Acta - General Subjects, 1868*, 130517. https://doi.org/10.1016/j.bbagen.2023.130517
49. Ghadiri, M., Young, P. M., & Traini, D. (2019). Strategies to enhance drug absorption via nasal and pulmonary routes. *Pharmaceutics, 11*, 113. https://doi.org/10.3390/pharmaceutics11030113
50. Kantak, M. N., & Bharate, S. S. (2022). Analysis of clinical trials on biomaterial and therapeutic applications of chitosan: A review. *Carbohydrate Polymers, 278*, 118999. https://doi.org/10.1016/j.carbpol.2021.118999
51. Shikuku, R., Hasnat, M. A., Mashrur, S. B. A., Haque, P., Rahman, M. M., & Khan, M. N. (2024). Chitosan-based pH-sensitive semi-interpenetrating network nanoparticles as a

sustained release matrix for anticancer drug delivery. *Carbohydrate Polymer Technologies and Applications, 7*, 100515. https://doi.org/10.1016/j.carpta.2024.100515
52. Chu, S., Shi, X., Tian, Y., & Gao, F. (2022). pH-responsive polymer nanomaterials for tumor therapy. *Frontiers in Oncology, 12*, 855019. https://doi.org/10.3389/fonc.2022.855019
53. Jhaveri, J., Raichura, Z., Khan, T., Momin, M., & Omri, A. (2021). Chitosan nanoparticles-insight into properties, functionalization and applications in drug delivery and theranostics. *Molecules, 26*, 272. https://doi.org/10.3390/molecules26020272
54. Herdiana, Y., Wathoni, N., Shamsuddin, S., Joni, I. M., & Muchtaridi, M. (2021). Chitosan-based nanoparticles of targeted drug delivery system in breast cancer treatment. *Polymers (Basel)., 13*, 1717. https://doi.org/10.3390/polym13111717
55. Jain, S., Nuwal, K., Mahmood, A., Piplani, M., Chander, S., Dubey, S. K., & Singhvi, G. (2021). Thiolated chitosan as an improved bioadhesive polymer in drug delivery. In *Chitosan in drug delivery* (pp. 247–276). https://doi.org/10.1016/B978-0-12-819336-5.00013-3
56. Suk, J. S., Xu, Q., Kim, N., Hanes, J., & Ensign, L. M. (2016). PEGylation as a strategy for improving nanoparticle-based drug and gene delivery. *Advanced Drug Delivery Reviews, 99*, 28–51. https://doi.org/10.1016/j.addr.2015.09.012
57. Meyer, R. A., Neshat, S. Y., Green, J. J., Santos, J. L., & Tuesca, A. D. (2022). Targeting strategies for mRNA delivery. *Materials Today Advances, 14*. https://doi.org/10.1016/j.mtadv.2022.100240
58. Yan, S., Na, J., Liu, X., & Wu, P. (2024). Different targeting ligands-mediated drug delivery systems for tumor therapy. *Pharmaceutics, 16*, 248. https://doi.org/10.3390/pharmaceutics16020248
59. Lombardo, D., & Kiselev, M. A. (2022). Methods of liposomes preparation: Formation and control factors of versatile nanocarriers for biomedical and nanomedicine application. *Pharmaceutics, 14*, 543. https://doi.org/10.3390/pharmaceutics14030543
60. Sleytr, U. B., Schuster, B., Egelseer, E. M., Pum, D., Horejs, C. M., Tscheliessnig, R., & Ilk, N. (2011). Nanobiotechnology with S-layer proteins as building blocks. *Progress in Molecular Biology and Translational Science, 103*, 277–352. https://doi.org/10.1016/B978-0-12-415906-8.00003-0
61. Çetin, M., Aytekin, E., Yavuz, B., & Bozda-Pehlivan, S. (2017). Nanoscience in targeted brain drug delivery. In *Nanotechnology methods for neurological diseases and brain tumors drug delivery across the blood-brain barrier* (pp. 117–147). https://doi.org/10.1016/B978-0-12-803796-6.00007-1
62. Fanani, M. L., Nocelli, N. E., & de las M. Zulueta Díaz, Y. (2022). What can we learn about amphiphile-membrane interaction from model lipid membranes? *Biochimica Et Biophysica Acta, 1864*, 183781. https://doi.org/10.1016/j.bbamem.2021.183781
63. Obalola, A. A., Abrahamse, H., & Dhilip Kumar, S. S. (2025). Enhanced therapeutic precision using dual drug-loaded nanomaterials for targeted cancer photodynamic therapy. *Biomedicine & Pharmacotherapy, 184*, 117909. https://doi.org/10.1016/j.biopha.2025.117909
64. Maja, L., Željko, K., & Mateja, P. (2020). Sustainable technologies for liposome preparation. *Journal of Supercritical Fluids, 165*, 104984. https://doi.org/10.1016/j.supflu.2020.104984
65. Immordino, M. L., Dosio, F., & Cattel, L. (2006). Stealth liposomes: Review of the basic science, rationale, and clinical applications, existing and potential. *International Journal of Nanomedicine, 1*, 297–315.
66. Luiz, M. T., Dutra, J. A. P., Tofani, L. B., de Araújo, J. T. C., Di Filippo, L. D., Marchetti, J. M., & Chorilli, M. (2022). Targeted liposomes: A nonviral gene delivery system for cancer therapy. *Pharmaceutics., 14*, 821. https://doi.org/10.3390/pharmaceutics14040821
67. Zangabad, P. S., Mirkiani, S., Shahsavari, S., Masoudi, B., Masroor, M., Hamed, H., Jafari, Z., Taghipour, Y. D., Hashemi, H., Karimi, M., & Hamblin, M. R. (2018). Stimulus-responsive liposomes as smart nanoplatforms for drug delivery applications. *Nanotechnology Reviews, 7*, 95–122. https://doi.org/10.1515/ntrev-2017-0154

68. Wang, X., Li, C., Wang, Y., Chen, H., Zhang, X., Luo, C., Zhou, W., Li, L., Teng, L., Yu, H., & Wang, J. (2022). Smart drug delivery systems for precise cancer therapy. *Acta Pharmaceutica Sinica B, 12*, 4098–4121. https://doi.org/10.1016/j.apsb.2022.08.013
69. Olusanya, T. O. B., Ahmad, R. R. H., Ibegbu, D. M., Smith, J. R., & Elkordy, A. A. (2018). Liposomal drug delivery systems and anticancer drugs. *Molecules, 23*, 907. https://doi.org/10.3390/molecules23040907
70. Nakamura, Y., Mochida, A., Choyke, P. L., & Kobayashi, H. (2016). Nanodrug delivery: Is the enhanced permeability and retention effect sufficient for curing cancer? *Bioconjugate Chemistry, 27*, 2225–2238. https://doi.org/10.1021/acs.bioconjchem.6b00437
71. Riaz, M. K., Riaz, M. A., Zhang, X., Lin, C., Wong, K. H., Chen, X., Zhang, G., Lu, A., & Yang, Z. (2018). Surface functionalization and targeting strategies of liposomes in solid tumor therapy: A review. *International Journal of Molecular Sciences, 19*, 195. https://doi.org/10.3390/ijms19010195
72. Yingchoncharoen, P., Kalinowski, D. S., & Richardson, D. R. (2016). Lipid-based drug delivery systems in cancer therapy: What is available and what is yet to come. *Pharmacological Reviews, 68*, 701–787. https://doi.org/10.1124/pr.115.012070
73. Petcov, T. E., Silberschmidt, V. V., Pandele, M. A., Chiticaru, E. A., Ioniță, M., & Manole, M. (2025). Nanostructures: An efficient drug delivery platform for therapy of multiple myeloma. *Eur. J. Med. Chem. Reports., 14*, 100263. https://doi.org/10.1016/j.ejmcr.2025.100263
74. Nikolova, M. P., Kumar, E. M., & Chavali, M. S. (2022). Updates on responsive drug delivery based on liposome vehicles for cancer treatment. *Pharmaceutics., 14*, 2195. https://doi.org/10.3390/pharmaceutics14102195
75. Cunha, J., Latocheski, E., Fidalgo, A. C. D., Gerola, A. P., Marin, d. C. F. F., & Ribeiro, A. J. (2025). Core-shell hybrid liposomes: Transforming imaging diagnostics and therapeutic strategies. *Colloids and Surfaces B: Biointerfaces, 251*, 114597. https://doi.org/10.1016/j.colsurfb.2025.114597
76. Huang, R., Song, H., Li, S., & Guan, X. (2025). Selection strategy for encapsulation of hydrophilic and hydrophobic ingredients with food-grade materials: A systematic review and analysis. *Food Chemistry X, 25*, 102149. https://doi.org/10.1016/j.fochx.2024.102149
77. Musielak, E., Feliczak-Guzik, A., & Nowak, I. (2022). Synthesis and potential applications of lipid nanoparticles in medicine. *Materials (Basel), 15*, 682. https://doi.org/10.3390/ma15020682
78. Shaker, S., Gardouh, A., & Ghorab, M. (2017). Factors affecting liposomes particle size prepared by ethanol injection method. *Research in Pharmaceutical Sciences, 12*, 346–352. https://doi.org/10.4103/1735-5362.213979
79. Bi, Y., Xie, S., Li, Z., Dong, S., & Teng, L. (2025). Precise nanoscale fabrication technologies, the "last mile" of medicinal development. *Acta Pharmaceutica Sinica B*. https://doi.org/10.1016/j.apsb.2025.03.040
80. Jochelavicius, K., Pereira, A. R., Fiamingo, A., Nobre, T. M., Campana-Filho, S. P., & Oliveira, O. N. (2022). Chitosan effects on monolayers of zwitterionic, anionic and a natural lipid extract from E. coli at physiological pH. *Colloids and Surfaces B: Biointerfaces, 209*, 112146. https://doi.org/10.1016/j.colsurfb.2021.112146
81. Kurl, S., Kumar, A., Reena, Mittal, N., Singh, D., Bassi, P., & Kaur, G. (2023). Challenges, opportunities, and future prospects of polysaccharide-based nanoparticles for colon targeting: A comprehensive review. *Carbohydrate Polymer Technologies and Applications, 6*, 100361. https://doi.org/10.1016/j.carpta.2023.100361
82. Sreeharsha, N., Prasanthi, S., Rao, G. S. N. K., Gajula, L. R., Biradar, N., Goudanavar, P., Naveen, N. R., Shiroorkar, P. N., Meravanige, G., Telsang, M., Asif, A. H., & Sreenivasalu, P. K. P. (2025). Formulation optimization of chitosan surface coated solid lipid nanoparticles of griseofulvin: A Box-Behnken design and in vivo pharmacokinetic study. *European Journal of Pharmaceutical Sciences, 204*, 106951. https://doi.org/10.1016/j.ejps.2024.106951

83. Moraru, C., Mincea, M., Menghiu, G., & Ostafe, V. (2020). Understanding the factors influencing chitosan-based nanoparticles-protein corona interaction and drug delivery applications. *Molecules, 25*, 4758. https://doi.org/10.3390/molecules25204758
84. Baxa, U. (2018). Imaging of liposomes by transmission electron microscopy. *Methods in Molecular Biology, 1682*, 73–88. https://doi.org/10.1007/978-1-4939-7352-1_8
85. Singh, G., & Pai, R. S. (2014). Optimization (central composite design) and validation of HPLC method for investigation of emtricitabine loaded poly(lactic-co-glycolic acid) nanoparticles: In vitro drug release and in vivo pharmacokinetic studies. *Scientific World Journal, 2014*, 583090. https://doi.org/10.1155/2014/583090
86. Kazemi-Andalib, F., Mohammadikish, M., Divsalar, A., & Sahebi, U. (2022). Hollow microcapsule with pH-sensitive chitosan/polymer shell for in vitro delivery of curcumin and gemcitabine. *European Polymer Journal, 162*, 110887. https://doi.org/10.1016/j.eurpolymj.2021.110887
87. Vagena, I. A., Malapani, C., Gatou, M. A., Lagopati, N., & Pavlatou, E. A. (2025). Enhancement of EPR effect for passive tumor targeting: Current status and future perspectives. *Applied Sciences, 15*, 3189. https://doi.org/10.3390/app15063189
88. Barua, S., & Mitragotri, S. (2014). Challenges associated with penetration of nanoparticles across cell and tissue barriers: A review of current status and future prospects. *Nano Today, 9*, 223–243. https://doi.org/10.1016/j.nantod.2014.04.008
89. Ways, T. M. M., Lau, W. M., & Khutoryanskiy, V. V. (2018). Chitosan and its derivatives for application in mucoadhesive drug delivery systems. *Polymers, 10*, 267. https://doi.org/10.3390/polym10030267
90. Hong, L., Li, W., Li, Y., & Yin, S. (2023). Nanoparticle-based drug delivery systems targeting cancer cell surfaces. *RSC Advances, 13*, 21365–21382. https://doi.org/10.1039/d3ra02969g
91. Kesharwani, P., Chadar, R., Sheikh, A., Rizg, W. Y., & Safhi, A. Y. (2022). CD44-targeted nanocarrier for cancer therapy. *Frontiers in Pharmacology, 12*, 800481. https://doi.org/10.3389/fphar.2021.800481
92. Kurczewska, J. (2023). Chitosan-based nanoparticles with optimized parameters for targeted delivery of a specific anticancer drug—A comprehensive review. *Pharmaceutics., 15*, 503. https://doi.org/10.3390/pharmaceutics15020503
93. Liu, Y., Si, L., Jiang, Y., Jiang, S., Zhang, X., Li, S., Chen, J., & Hu, J. (2025). Design of pH-responsive nanomaterials based on the tumor microenvironment. *International Journal of Nanomedicine, 20*, 705–721. https://doi.org/10.2147/IJN.S504629
94. Imtiaz, S., Ferdous, U. T., Nizela, A., Hasan, A., Shakoor, A., Zia, A. W., & Uddin, S. (2025). Mechanistic study of cancer drug delivery: Current techniques, limitations, and future prospects. *European Journal of Medicinal Chemistry, 290*, 117535. https://doi.org/10.1016/j.ejmech.2025.117535
95. Imran, H. M., Tang, Y., Wang, S., Yan, X., Liu, C., Guo, L., Wang, E., & Xu, C. (2024). Optimized DOX drug deliveries via Chitosan-mediated nanoparticles and stimuli responses in cancer chemotherapy: A review. *Molecules, 29*, 31. https://doi.org/10.3390/molecules29010031
96. Pei, D., & Buyanova, M. (2019). Overcoming endosomal entrapment in drug delivery. *Bioconjugate Chemistry, 30*, 273–283. https://doi.org/10.1021/acs.bioconjchem.8b00778
97. Antoniou, V., Mourelatou, E. A., Galatou, E., Avgoustakis, K., & Hatziantoniou, S. (2024). Gene therapy with Chitosan nanoparticles: Modern formulation strategies for enhancing cancer cell transfection. *Pharmaceutics., 16*, 868. https://doi.org/10.3390/pharmaceutics16070868
98. Ghazizadeh, Y., Sharifi-Ardani, S. E., Tajik, N., Mirzaei, R., & Pourahmad, J. (2025). Exploring the potential of mitochondria-targeted drug delivery for enhanced breast cancer therapy. *International Journal of Breast Cancer, 2025*, 3013009. https://doi.org/10.1155/ijbc/3013009

99. Li, Y., Wang, W., Zhang, Y., Wang, X., Gao, X., Yuan, Z., & Li, Y. (2019). Chitosan sulfate inhibits angiogenesis: Via blocking the VEGF/VEGFR2 pathway and suppresses tumor growth in vivo. *Biomaterials Science, 7*, 1584–1597. https://doi.org/10.1039/c8bm01337c
100. Li, X., Min, M., Du, N., Gu, Y., Hode, T., Naylor, M., Chen, D., Nordquist, R. E., & Chen, W. R. (2013). Chitin, chitosan, and glycated chitosan regulate immune responses: The novel adjuvants for cancer vaccine. *Clinical & Developmental Immunology, 2013*, 387023. https://doi.org/10.1155/2013/387023
101. Shaikh, M. A. J., Afzal, O., Almalki, W. H., Kazmi, I., Alzarea, S. I., Jafar, M., Alfawaz Altamimi, A. S., Jakhmola, V., Anand, K., Singh, S. K., Dua, K., & Gupta, G. (2023). Harnessing chitosan-adorned liposomes for enhanced drug delivery in cancer. *Journal of Drug Delivery Science and Technology, 85*, 104619. https://doi.org/10.1016/j.jddst.2023.104619
102. Hasan, M., Elkhoury, K., Belhaj, N., Kahn, C., Tamayol, A., Barberi-Heyob, M., Arab-Tehrany, E., & Linder, M. (2020). Growth-inhibitory effect of chitosan-coated liposomes encapsulating curcumin on MCF-7 breast cancer cells. *Marine Drugs, 18*. https://doi.org/10.3390/md18040217
103. El Kurdi, R., Mesmar, J., Estephan, M., Badran, A., Baydoun, E., & Patra, D. (2022). Anticancer activity of Diarachidonyl Phosphatidyl Choline Liposomal Curcumin coated with Chitosan against breast and pancreatic cancer cells. *BioNanoScience, 12*, 1158–1165. https://doi.org/10.1007/s12668-022-01019-4
104. Sharma, S., Gupta, P., Kawish, S. M., Ahmad, S., Iqbal, Z., Vohora, D., & Kohli, K. (2024). Novel Chitosan-coated liposomes Coloaded with Exemestane and Genistein for an effective breast cancer therapy. *ACS Omega, 9*, 9735–9752. https://doi.org/10.1021/acsomega.3c09948
105. Li, R., Lin, Z., Zhang, Q., Zhang, Y., Liu, Y., Lyu, Y., Li, X., Zhou, C., Wu, G., Ao, N., & Li, L. (2020). Injectable and in situ-formable Thiolated Chitosan-coated liposomal hydrogels as Curcumin carriers for prevention of in vivo breast cancer recurrence. *ACS Applied Materials & Interfaces, 12*, 17936–17948. https://doi.org/10.1021/acsami.9b21528
106. Ding, Y., Yang, R., Yu, W., Hu, C., Zhang, Z., Liu, D., An, Y., Wang, X., He, C., Liu, P., Tang, Q., & Chen, D. (2021). Chitosan oligosaccharide decorated liposomes combined with TH302 for photodynamic therapy in triple negative breast cancer. *Journal of Nanobiotechnology, 19*. https://doi.org/10.1186/s12951-021-00891-8
107. Alomrani, A., Badran, M., Harisa, G. I., Shehry, M. A. L., Alhariri, M., Alshamsan, A., & Alkholief, M. (2019). The use of chitosan-coated flexible liposomes as a remarkable carrier to enhance the antitumor efficacy of 5-fluorouracil against colorectal cancer. *Saudi Pharmaceutical Journal, 27*, 603–611. https://doi.org/10.1016/j.jsps.2019.02.008
108. de Souza, J. B., de Lacerda Coriolano, D., dos Santos Silva, R. C., da Costa Júnior, S. D., de Almeida Campos, L. A., Cavalcanti, I. D. L., de Lira Nogueira, M. C. B., Pereira, V. R. A., Brelaz-de-Castro, M. C. A., & Cavalcanti, I. M. F. (2024). Ceftazidime and Usnic acid encapsulated in Chitosan-coated liposomes for oral administration against colorectal cancer-inducing Escherichia coli. *Pharmaceuticals, 17*, 802. https://doi.org/10.3390/ph17060802
109. Miao, Y. Q., Chen, M. S., Zhou, X., Guo, L. M., Zhu, J. J., Wang, R., Zhang, X. X., & Gan, Y. (2021). Chitosan oligosaccharide modified liposomes enhance lung cancer delivery of paclitaxel. *Acta Pharmacologica Sinica, 42*, 1714–1722. https://doi.org/10.1038/s41401-020-00594-0
110. Li, F., Mei, H., Xie, X., Zhang, H., Liu, J., Lv, T., Nie, H., Gao, Y., & Jia, L. (2017). Aptamer-conjugated Chitosan-anchored liposomal complexes for targeted delivery of Erlotinib to EGFR-mutated lung cancer cells. *The AAPS Journal, 19*, 814–826. https://doi.org/10.1208/s12248-017-0057-9
111. Uma Maheswari, R. T., Ajithkumar, V., Varalakshmi, P., & Rajan, M. (2023). CD44 tagged hyaluronic acid - chitosan liposome carrier for the delivery of berberine and doxorubicin into lung cancer cells. *International Journal of Biological Macromolecules, 253*, 126599. https://doi.org/10.1016/j.ijbiomac.2023.126599

112. Hanafy, N. A. N., Sheashaa, R. F., Moussa, E. A., & Mahfouz, M. E. (2023). Potential of curcumin and niacin-loaded targeted chitosan coated liposomes to activate autophagy in hepatocellular carcinoma cells: An in vitro evaluation in HePG2 cell line. *International Journal of Biological Macromolecules, 245*. https://doi.org/10.1016/j.ijbiomac.2023.125572
113. Miretti, M., Prucca, C. G., Baumgartner, M. T., & Martinelli, M. (2023). Combining ZnPc-liposomes and chitosan on a hybrid matrix for enhanced photodynamic therapy. *International Journal of Biological Macromolecules, 253*, 127544. https://doi.org/10.1016/j.ijbiomac.2023.127544
114. Wang, M., Zhao, T., Liu, Y., Wang, Q., Xing, S., Li, L., Wang, L., Liu, L., & Gao, D. (2017). Ursolic acid liposomes with chitosan modification: Promising antitumor drug delivery and efficacy. *Materials Science and Engineering: C, 71*, 1231–1240. https://doi.org/10.1016/j.msec.2016.11.014
115. Saesoo, S., Bunthot, S., Sajomsang, W., Gonil, P., Phunpee, S., Songkhum, P., Laohhasurayotin, K., Wutikhun, T., Yata, T., Ruktanonchai, U. R., & Saengkrit, N. (2016). Phospholipid-chitosan hybrid nanoliposomes promoting cell entry for drug delivery against cervical cancer. *Journal of Colloid and Interface Science, 480*, 240–248. https://doi.org/10.1016/j.jcis.2016.06.071
116. Suriyaamporn, P., Dechsri, K., Charoenying, T., Ngawhirunpat, T., Rojanarata, T., Patrojanasophon, P., Opanasopit, P., & Pamornpathomkul, B. (2024). Multiple strategies approach: A novel crosslinked hydrogel forming chitosan-based microneedles chemowrap patch loaded with 5-fluorouracil liposomes for chronic wound cancer treatment. *International Journal of Biological Macromolecules, 279*. https://doi.org/10.1016/j.ijbiomac.2024.134973
117. Mazzotta, E., Marazioti, A., Mourtas, S., Muzzalupo, R., & Antimisiaris, S. G. (2024). Liposomes coated with novel synthetic Bifunctional Chitosan derivatives as potential carriers of anticancer drugs. *Pharmaceutics, 16*. https://doi.org/10.3390/pharmaceutics16030319
118. Ghafaripour, H., Homayouni Tabrizi, M., Karimi, E., & Barati Naeeni, N. (2023). Lawsone encapsulated polylactic-co-glycolic acid nanoparticles modified with chitosan-folic acid successfully inhibited cell growth and triggered apoptosis in Panc-1 cancer cells. *IET Nanobiotechnology, 17*, 425–437. https://doi.org/10.1049/nbt2.12139
119. Łukasiewicz, S., Szczepanowicz, K., Błasiak, E., & Dziedzicka-Wasylewska, M. (2015). Biocompatible polymeric nanoparticles as promising candidates for drug delivery. *Langmuir, 31*, 6415–6425. https://doi.org/10.1021/acs.langmuir.5b01226
120. Govindasamy, C., El Newehy, A. S., Hussein-Al-Ali, S. H., Arulselvan, P., Bharathi, M., & Parthasarathy, S. (2024). Investigation of antiproliferative efficacy and apoptosis induction in leukemia cancer cells using irinotecan-loaded liposome-embedded nanofibers constructed from chitosan. *International Journal of Biological Macromolecules, 270*, 132284. https://doi.org/10.1016/j.ijbiomac.2024.132284
121. Elumalai, K., Srinivasan, S., & Shanmugam, A. (2024). Review of the efficacy of nanoparticle-based drug delivery systems for cancer treatment. *Biomedical Technology, 5*, 109–122. https://doi.org/10.1016/j.bmt.2023.09.001
122. Zhuang, J., Ping, Q., Song, Y., Qi, J., & Cui, Z. (2010). Effects of chitosan coating on physical properties and pharmacokinetic behavior of mitoxantrone liposomes. *International Journal of Nanomedicine, 5*, 407–416. https://doi.org/10.2147/ijn.s10189
123. Sawant, R. R., & Torchilin, V. P. (2012). Challenges in development of targeted liposomal therapeutics. *The AAPS Journal, 14*, 303–315. https://doi.org/10.1208/s12248-012-9330-0
124. Eugster, R., & Luciani, P. (2025). Liposomes: Bridging the gap from lab to pharmaceuticals. *Current Opinion in Colloid & Interface Science, 75*, 101875. https://doi.org/10.1016/j.cocis.2024.101875
125. Chouhan, D., & Mandal, B. B. (2020). Silk biomaterials in wound healing and skin regeneration therapeutics: From bench to bedside. *Acta Biomaterialia, 103*, 24–51. https://doi.org/10.1016/j.actbio.2019.11.050

126. Liu, P., Chen, G., & Zhang, J. (2022). A review of liposomes as a drug delivery system: Current status of approved products, regulatory environments, and future perspectives. *Molecules, 27*, 1372. https://doi.org/10.3390/molecules27041372
127. Kyriakides, T. R., Raj, A., Tseng, T. H., Xiao, H., Nguyen, R., Mohammed, F. S., Halder, S., Xu, M., Wu, M. J., Bao, S., & Sheu, W. C. (2021). Biocompatibility of nanomaterials and their immunological properties. *Biomedical Materials, 16*. https://doi.org/10.1088/1748-605X/abe5fa
128. Masimov, R., & Wasan, E. K. (2024). Chitosan non-particulate vaccine delivery systems. *Journal of Pharmacy & Pharmaceutical Sciences, 27*, 12921. https://doi.org/10.3389/jpps.2024.12921
129. Inglut, C. T., Sorrin, A. J., Kuruppu, T., Vig, S., Cicalo, J., Ahmad, H., & Huang, H. C. (2020). Immunological and toxicological considerations for the design of liposomes. *Nanomaterials (Basel), 10*, 190. https://doi.org/10.3390/nano10020190
130. Kang, B. R., Park, J. S., Ryu, G. R., Jung, W. J., Choi, J. S., & Shin, H. M. (2022). Effect of Chitosan coating for efficient encapsulation and improved stability under loading preparation and storage conditions of bacillus Lipopeptides. *Nanomaterials, 12*, 4189. https://doi.org/10.3390/nano12234189
131. Yu, B., Lee, R. J., & Lee, L. J. (2009). Microfluidic methods for production of liposomes. *Methods in Enzymology, 465*, 129–141. https://doi.org/10.1016/S0076-6879(09)65007-2
132. Sangnim, T., Dheer, D., Jangra, N., Huanbutta, K., Puri, V., & Sharma, A. (2023). Chitosan in Oral drug delivery formulations: A review. *Pharmaceutics, 15*, 2361. https://doi.org/10.3390/pharmaceutics15092361
133. Izadiyan, Z., Misran, M., Kalantari, K., Webster, T. J., Kia, P., Basrowi, N. A., Rasouli, E., & Shameli, K. (2025). Advancements in liposomal nanomedicines: Innovative formulations, therapeutic applications, and future directions in precision medicine. *International Journal of Nanomedicine, 20*, 1213–1262. https://doi.org/10.2147/IJN.S488961
134. Chiang, Y. T., Yen, Y. W., & Lo, C. L. (2015). Reactive oxygen species and glutathione dual redox-responsive micelles for selective cytotoxicity of cancer. *Biomaterials, 61*, 150–161. https://doi.org/10.1016/j.biomaterials.2015.05.007
135. Li, J., Wang, Q., Xia, G., Adilijiang, N., Li, Y., Hou, Z., Fan, Z., & Li, J. (2023). Recent advances in targeted drug delivery strategy for enhancing oncotherapy. *Pharmaceutics, 15*, 2233. https://doi.org/10.3390/pharmaceutics15092233
136. Naghib, S. M., Amiri, S., & Mozafari, M. R. (2024). Stimuli-responsive chitosan-based nanocarriers for drug delivery in wound dressing applications: A review. *Carbohydrate Polymer Technologies and Applications, 7*, 100497. https://doi.org/10.1016/j.carpta.2024.100497
137. Sailor, M. J., & Park, J. H. (2012). Hybrid nanoparticles for detection and treatment of cancer. *Advanced Materials, 24*, 3779–3802. https://doi.org/10.1002/adma.201200653
138. Jadhav, L. A., & Mandlik, S. K. (2025). Nanocarriers in skin cancer treatment: Emerging drug delivery approaches and innovations. *Nano TransMed, 4*, 100068. https://doi.org/10.1016/j.ntm.2024.100068
139. Akbarzadeh Khiavi, M., Safary, A., Barar, J., Ajoolabady, A., Somi, M. H., & Omidi, Y. (2020). Multifunctional nanomedicines for targeting epidermal growth factor receptor in colorectal cancer. *Cellular and Molecular Life Sciences, 77*, 997–1019. https://doi.org/10.1007/s00018-019-03305-z
140. Zhu, J., Lee, H. J., Huang, R., Zhou, J., Zhang, J., Yang, X., Zhou, W., Jiang, W., & Chen, S. (2024). Harnessing nanotechnology for cancer treatment. *Frontiers in Bioengineering and Biotechnology, 12*, 1514890. https://doi.org/10.3389/fbioe.2024.1514890
141. Aljabali, A. A. A., El-Tanani, M., & Tambuwala, M. M. (2024). Principles of CRISPR-Cas9 technology: Advancements in genome editing and emerging trends in drug delivery. *Journal of Drug Delivery Science and Technology, 92*, 105338. https://doi.org/10.1016/j.jddst.2024.105338

142. Nujoom, N., Koyakutty, M., Biswas, L., Rajkumar, T., & Nair, S. V. (2024). Emerging gene-editing nano-therapeutics for cancer. *Heliyon, 10*, e39323. https://doi.org/10.1016/j.heliyon.2024.e39323
143. Yin, X., Harmancey, R., McPherson, D. D., Kim, H., & Huang, S. L. (2023). Liposome-based carriers for CRISPR genome editing. *International Journal of Molecular Sciences, 24*, 12844. https://doi.org/10.3390/ijms241612844
144. Haider, A., Khan, S., Iqbal, D. N., Shrahili, M., Haider, S., Mohammad, K., Mohammad, A., Rizwan, M., Kanwal, Q., & Mustafa, G. (2024). Advances in chitosan-based drug delivery systems: A comprehensive review for therapeutic applications. *European Polymer Journal, 210*, 112983. https://doi.org/10.1016/j.eurpolymj.2024.112983
145. Bhange, M., & Telange, D. (2025). Convergence of nanotechnology and artificial intelligence in the fight against liver cancer: A comprehensive review. *Discover Oncology, 16*, 77. https://doi.org/10.1007/s12672-025-01821-y
146. Kong, X., Gao, P., Wang, J., Fang, Y., & Hwang, K. C. (2023). Advances of medical nanorobots for future cancer treatments. *Journal of Hematology & Oncology, 16*, 74. https://doi.org/10.1186/s13045-023-01463-z
147. Kumar, A., & Kumar, A. (2024). Chitosan-based drug conjugated nanocomposites: Advances and innovation in cancer therapy. *Regenerative Engineering and Translational Medicine, 10*, 1–8. https://doi.org/10.1007/s40883-023-00310-4
148. Fang, T., Cao, X., Ibnat, M., & Chen, G. (2022). Stimuli-responsive nanoformulations for CRISPR-Cas9 genome editing. *Journal of Nanobiotechnology, 20*, 354. https://doi.org/10.1186/s12951-022-01570-y

Chitosan-Based System in 3D/4D (Bio) Printing for Drug Delivery and Tissue Engineering Applications

Hani Nasser Abdelhamid

Abstract This chapter discusses a chitosan-based system for three-dimensional (3D)/four-dimensional (4D) (bio)printing for drug delivery and tissue-engineered scaffolds. These methodologies have advanced biomedical research into the digital manufacturing age by facilitating high-throughput and reproducibility, offering single-step assembly of a complex system that can be applied to heterogeneous components, and producing complex cellularized scaffolds. Several efforts have boosted control over scaffold construction, mechanical characteristics, and dynamic responsiveness to external stimuli by integrating chitosan into 3D and 4D printing platforms. These biofabrication techniques facilitate the development of intricate, cellular structures with exceptional precision and reproducibility, meeting the requirements of personalized treatment. Furthermore, the capacity to integrate therapeutic molecules directly into printable biomaterials enables localized, sustained drug delivery, thereby mitigating systemic toxicity and enhancing treatment efficacy. This chapter emphasizes the revolutionary influence of chitosan-based printable systems in advancing tissue engineering and drug delivery within the realm of digital, high-throughput biomedical manufacturing.

Keywords Chitosan · Printing · Drug delivery · Tissue engineering

1 Introduction

Advanced biomedical technologies, such as drug delivery [7] and tissue engineering [20, 22, 44], have advanced health and improved medical treatment for several diseases. These applications have been improved over the past two decades, driven by rapid technological progress and a deeper understanding of complex biological systems. A notable milestone in this evolution has been the emergence of three-dimensional (3D) or four-dimensional (4D) printing and bioprinting technologies,

H. N. Abdelhamid (✉)
Department of Chemistry, College of Science, Imam Mohammad Ibn Saud Islamic University (IMSIU), Riyadh, Saudi Arabia
e-mail: hnabdelhamid@imamu.edu.sa; hani.nasserabdelhamid@gmail.com

S. Jana (ed.), *Chitosan-Based Drug Delivery Systems*, AAPS Introductions in the Pharmaceutical Sciences 16, https://doi.org/10.1007/978-3-032-08347-0_14

which have revolutionized the fabrication of hydrogel-based constructs [16]. 3D printing was also known as additive manufacturing (AM), rapid prototyping (RP), robocasting direct ink writing (DIW), and robocast-assisted deposition (robocasting) [2, 36, 48]. 4D printing objects offer dynamic responses under external stimuli [9]. These approaches have advanced biomedical research into the digital manufacturing era by enabling high-throughput, reproducible, and single-step assembly of structurally intricate, heterogeneous, and cellularized scaffolds [61].

Polysaccharides are promising materials for several applications, including 3D printing [30, 31, 35, 50, 60]. They have advanced fields, including biomedical technologies [47]. Among the vast array of natural and synthetic polysaccharides, chitosan-based materials have garnered significant attention due to their inherent biocompatibility, bioactivity, and generally low cost [1, 3, 4, 19]. Chitosan-based materials offered several reactive functional groups; e.g., carboxyl, hydroxyl, and amine moieties, that facilitate the chemical modification to fulfill the requirements, broadening their applicability across diverse biomedical fields [49]. 3D/4D printed chitosan-based materials were applied for several applications, including microfluidics-based lab/organ-on-a-chip [49], and drug delivery [56].

This chapter discusses and analyzes chitosan-based inks for 3D and 4D printing, particularly in tissue engineering and drug delivery applications. It outlines the leading 3D printing platforms suitable for chitosan processing and the key criteria for selecting compatible hydrogel inks. It provides a detailed overview of chitosan and its derivatives, examining how they are formulated into functional inks, individually or with other materials, e.g., biopolymers and nanomaterials. It highlights their synthesis pathways and the physicochemical properties influencing their performance in printing applications. It focuses on applying 3D/4D printed chitosan-based materials in drug delivery and tissue engineering. It presents a critical discussion on challenges and future directions for enhancing the utility of chitosan-based hydrogels in biomedical manufacturing.

2 Chitosan

Chitosan is a linear, semi-crystalline polysaccharide derived from the deacetylation of chitin, a naturally occurring structural biopolymer found abundantly in the exoskeletons of marine organisms such as crabs and shrimp [40]. Its molecular structure comprises β-(1→4)-linked D-glucosamine and N-acetyl-D-glucosamine units, randomly distributed along the polymer chain. The extent of deacetylation, defined as the ratio of D-glucosamine to total glucosamine units, governs its solubility and functional characteristics. Chitosan becomes soluble in mildly acidic aqueous solutions when its degree of deacetylation exceeds 50%, distinguishing it from its parent compound, chitin. The source material and extraction technique significantly influence chitosan's physicochemical and mechanical properties, including degradation rate, water solubility, and structural strength. However, crosslinking strategies often stabilize chitosan-based materials under physiological conditions (pH 7.4 and 37 °C).

These strategies preserve the inherent biocompatibility and structural fidelity of the material.

The functional groups in chitosan can undergo chemical reactions for covalent modification. For example, Diels–Alder chemistry method was reported to crosslink a chitosan–gelatin blend with maleimide-functionalized star-polyethylene glycol (PEG) and furan-modified biopolymers [29]. This method, devoid of harmful reagents and byproducts, facilitated the modulation of scaffold dynamics and sustained U87 cell viability for up to 6 days [29]. Numerous chemical modifications of chitosan have resulted in derivatives with enhanced and diversified properties:

1. *Carboxymethyl Chitosan (CMC):* CMC can be synthesized via carboxymethylation at the hydroxyl and/or amine groups. CMC exhibits improved water solubility at neutral pH and enhanced biological activities, including antimicrobial and anticancer effects [33, 45]. Despite its ampholytic nature and compatibility for forming biopolymer complexes, additional crosslinking is necessary to improve scaffold stability.
2. *Chitosan Methacrylate (CS-MA):* This derivative incorporates photocross-linkable methacrylic groups, enabling the formation of hydrogels with tunable properties such as swelling behavior, mechanical strength, and photopolymerization kinetics [62]. CS-MA has proven valuable in biofabrication techniques such as stereolithography and extrusion-based bioprinting.
3. *Glycol Chitosan (GC):* By conjugating chitosan with ethylene glycol, GC becomes water-soluble and suitable for bioprinting. Its amine groups allow further functionalization, including methacrylation for photocrosslinking. GC is often combined with other polymers to form stable hydrogels without post-processing.
4. *Maleic Anhydride Modified Chitosan (MAC):* The modification introduces anionic carboxyl groups that enhance biofunctionality, particularly in bone tissue engineering. However, MAC hydrogels are weak and require blending with other polymers to improve their mechanical properties and crosslinking efficiency.
5. *Succinylated Chitosan* (Chi-S): It offers improved water solubility due to succinyl group insertion, making it an excellent candidate for bioinks in 3D cell-laden printing. Its residual reactive groups also allow for subsequent functionalization, expanding its applicability.
6. *Hydroxybutyl Chitosan:* This derivative exhibits thermoresponsive behavior with a sol-gel transition above ~28 °C, making it useful for injectable hydrogels. Its weak mechanical strength is reinforced by blending with other biomaterials or ionic treatment. Furthermore, it can be modified with methacrylic groups to create dual cross-linkable systems.
7. *Catechol-Conjugated Chitosan:* This derivative benefits from intrinsic adhesion to biological tissues and does not require light-induced crosslinking. Despite its cytocompatibility and ease of scaffold formation, it suffers from poor mechanical strength, which is commonly addressed by introducing metal-ion-mediated crosslinking. A mussel-inspired strategy was employed to develop a catechol-conjugated chitosan bioink capable of rapid gelation upon extrusion into standard

culture media [26]. Improving the adhesive properties of catechol and amine groups, the bioink formed stable 3D constructs through immediate complexation with serum proteins. Incorporating vanadyl ions (VO^{2+}) at a low metal-to-catechol ratio significantly enhanced covalent crosslinking via metal–catechol coordination, improving mechanical strength and printability. This system supported in situ bioprinting with approximately 90% cell viability, demonstrating its potential as a biocompatible, stimuli-free crosslinking bioink for 3D tissue fabrication [26].

3 3D Printing of Chitosan-Based Materials

Developing customized hydrogel precursor inks is central to the success of 3D/4D printing fabrication techniques. Ideally, chitosan-based inks should exhibit excellent printability and precision [49]. Among the vast array of natural and synthetic hydrogel systems, chitosan-based inks with other additives derived from materials such as alginate, hyaluronic acid, cellulose, and dextran have garnered significant attention due to their inherent biocompatibility, bioactivity, and low cost. Additionally, these molecules can be chemically bonded to the reactive functional groups, such as carboxyl, hydroxyl, and amine moieties, present in the chitosan backbone. Covalent bonds between chitosan and other derivatives enable broadening their applicability across diverse biomedical fields.

Chitosan-based inks have emerged as an up-and-coming class of biomaterials, attracting increasing interest due to their unique combination of physicochemical, mechanical, and antibacterial properties. Moreover, chitosan's abundant hydroxyl and primary amine groups allow for extensive chemical modification, enabling the creation of advanced derivatives with improved printability, functionality, and biological performance. The chemical structure of chitosan enables the formation of a hydrogel that can be applied as an effective ink for 3D printing [61].

Chitosan and its derivatives represent a highly adaptable category of biopolymers, exhibiting a wide array of physicochemical properties and processing capabilities. They offer high diversity that facilitates multiple tactics in developing bioinks and biomaterials inks. For clarity, these can be classified into three categories: (i) single-component inks having a sole biopolymer, (ii) polymer mix inks consisting of two or more biopolymers, and (iii) composite inks that integrate multiphase components or fillers. Each category presents unique benefits for printability, structural integrity, mechanical performance, and biological function. Key parameters affecting the printing of categories are summarized in Table 1. They can be classified into three steps; the First step for ink preparation, the second step represents printing, and the third step for treatment after printing. Characterization of the 3D printing process by analyzing the effect of ink properties, i.e., rheological properties and solvent evaporation [54]. Process-related printing parameters should be evaluated, i.e., nozzle diameter, robot velocity, and applied pressure (Table 1).

Table 1 Key parameters that affect the 3D printing procedure and their impact

Parameter	Details/options	Impact
Chitosan concentration	Weight percentage (1–20%)	Affects viscosity, printability, mechanical strength, and degradation rate
Solvent	Acetic acid (2–50% v/v)	Facilitates chitosan dissolution
Printing temperature	Room temperature (RT, 25 °C) to 37 °C	Maintains material consistency during extrusion
Printing needle size	50–250 μm (inner diameter, conical)	Controls strand thickness and resolution
Printing pressure	1–5 bar	Influences extrusion flow and structural fidelity
Printing speed	1–10 mm/s	Affects line width, accuracy, and scaffold geometry
Scaffold geometry	Layer-by-layer	Determines mechanical behavior and pore structure
Pore size	Adjustable 0.5–5 mm	Influences mechanical strength, cell infiltration, and nutrient diffusion
Drying methods	Air drying, warm drying, vacuum drying	Affects mechanical strength and dimensional stability
Crosslinking agent	Sodium hydroxide (5% w/v) Glutaraldehyde	Enables hydrogel formation via ionic and hydrogen bonding
Crosslinking conditions	RT	Critical for scaffold integrity
Post-crosslinking neutralization	Washing with distilled water and drying at 37 °C, 65% humidity	Removes excess NaOH and stabilizes the scaffold
Mechanical testing	Elastic modulus (dependent on drying and pore size)	Determines scaffold stiffness and load-bearing capability
Pore interconnectivity	Verified by optical microscopy or electron microscopy	Supports cell migration and nutrient exchange
Degradation and swelling	Higher with lower chitosan concentration and larger pores	Affects scaffold lifespan and behavior in vivo

Single-component inks represent the simplest formulations and generally depend on chitosan's inherent gelation characteristics or physical crosslinking qualities. Sadeghianmaryan *et al.* formulated a 3D-printable chitosan-based ink at concentrations of 8%, 10%, and 12% (w/v), achieved by dissolving chitosan in 50% (v/v) acetic acid while stirring for 24 h at ambient temperature (Fig. 1) [41]. Employing a 3D-Bioplotter, they constructed 10-layer cubic scaffolds ($20 \times 20 \times 5$ mm^3) featuring a 0–90° deposition pattern with pore dimensions of 2, 3, and 4 mm, printed at 37 °C utilizing a 250 μm conical needle, 3.5 bar pressure, and a velocity of 10 mm/s. The impact of drying and crosslinking methods on scaffolds' mechanical and biological properties was methodically examined. Three drying techniques, e.g., air drying (room temperature, 24 h), warm drying (60 °C, 12 h), and vacuum drying (50 °C, 12 h), were utilized before crosslinking in 5% (w/v) sodium hydroxide for 2 h at 25 °C. The pH-induced crosslinking transformed NH_3^+ into NH_2 groups, enhancing

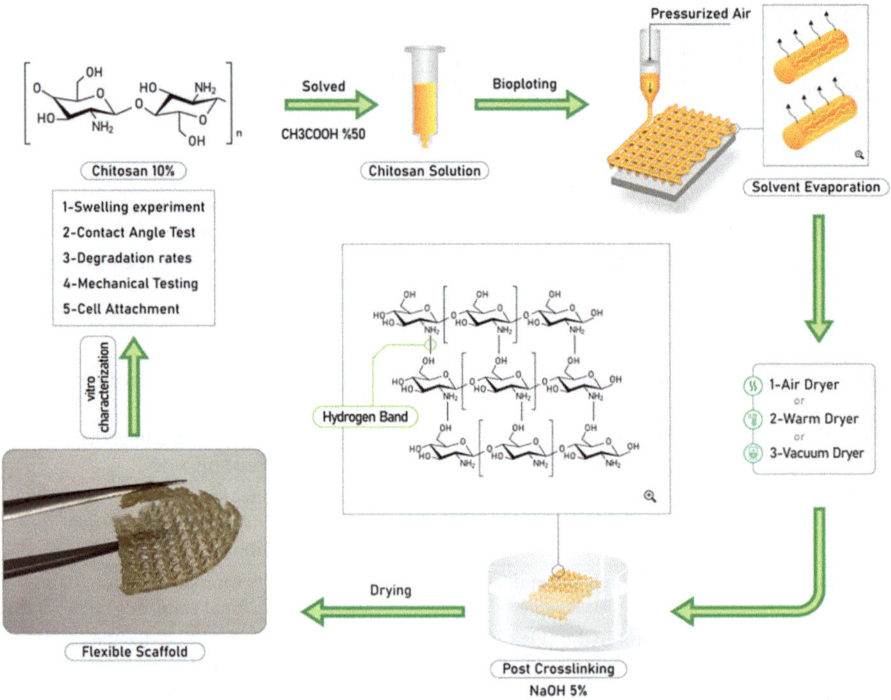

Fig. 1 3D printing procedure for chitosan scaffolds, including steps for solution preparation, chemical reaction, solvent evaporation, post-crosslinking, and drying. (Figure reprinted with permission from Ref. [41] © 2020 Elsevier B.V. All rights reserved)

physical interactions like hydrogen bonding and hydrophobic connections. Neutralization was accomplished by rinsing with distilled water and drying at 37 °C with 65% humidity. Of the evaluated formulations, 10% chitosan exhibited superior printability and was chosen for additional examination. Mechanical testing indicated that air-dried scaffolds, especially those with 2 mm pore diameters, demonstrated the highest elastic modulus. The printed objects exhibit distinct morphology and interconnected porous structures, but a little reduction in pore size was noted relative to the original design. Biocompatibility investigations utilizing ATDC5 chondrocyte cells demonstrated superior cell adhesion and survivability, particularly on air-dried scaffolds with smaller pores, where cells preserved their rounded morphology and mainly localized at the scaffold peripheries. Scaffolds with reduced chitosan concentrations and increased pore diameters exhibited high degradation and swelling ratios. These findings highlight the substantial impact of drying methods and pore structure on the efficacy of chitosan-based scaffolds. Air-dried, 10% chitosan scaffolds with 2 mm pores exhibit notable mechanical strength and cytocompatibility, offering a promising option for future cartilage tissue engineering applications [41].

Integrating thermosensitive substances like β-glycerophosphate, potassium phosphate, and sodium bicarbonate into chitosan solutions improves gelation at physiological conditions (pH 7.4, 37 °C) [23]. Ku *et al.* selected sodium bicarbonate ($NaHCO_3$), demonstrating enhanced gelation kinetics, extrusion performance, and biocompatibility. These formulations facilitated direct cell encapsulation while maintaining high cell viability during and after bioprinting. A custom-built 3D bioprinting system was utilized to print polycaprolactone (PCL) at 65 °C as a supporting framework, while a syringe-based apparatus deposited the cell-laden chitosan hydrogel. Periodontal ligament stem cells (PDLSCs) were encapsulated at a concentration of 5×10^5 cells/mL in hydrogel, thereafter transferred to sterile syringes, and printed in an alternating configuration using PCL to create 8-layer lattice structures (2 mm in thickness). Gelation was initiated after printing through incubation at 37 °C. The authors investigated different chitosan-based bioinks, concluding that the kind of solvent (acetic, lactic, or hydrochloric acid) had little impact on gelation, although acetic acid demonstrated superior biocompatibility. The selection of a gelling agent exerted a more pronounced influence. $NaHCO_3$-based inks exhibited accelerated gel stiffening (increase in G'), more irregular porosity architectures, and improved cell adhesion relative to β-GP and K_2HPO_4. Although the inks exhibited comparable overall performance in sustaining cell-laden structures, $NaHCO_3$ exhibited the greatest fit for bioprinting. These findings highlighted the promise of $NaHCO_3$-infused chitosan hydrogels for application in tailored, patient-specific regenerative therapies [23]. Due to its biocompatibility and biodegradability, chitosan is appropriate for these uses. Thermosensitive hydrogels, dependent on non-covalent interactions, reduce cytotoxicity and preserve cell viability without chemical crosslinkers [23].

Combining chitosan with other biopolymers is a common approach to modify the mechanical, rheological, and biological characteristics of bioinks. Chitosan and its derivatives (e.g., carboxymethyl chitosan, CMC) have been effectively integrated with polymers like alginate, gelatin, hyaluronic acid, polyethylene glycol (PEG), and keratin. Puertas–Bartolomé *et al.* formulated a reactive hydrogel bioink consisting of carboxymethyl chitosan (CMCh) and partly oxidized hyaluronic acid (HAox), which rapidly self-crosslinks through Schiff base production [39]. A dual-syringe extrusion system with a static mixer facilitated in situ gelation during printing, allowing low-viscosity precursors to preserve good shape fidelity and structural integrity (Fig. 2) [39]. Post-printing crosslinking with Fe^{3+} ions improved long-term mechanical stability, markedly strengthening the printed constructions. The resultant scaffolds exhibited mild swelling, superior biocompatibility, and mechanical stability over a 14-day cell culture period. The reported procedure offered a customizable and scalable bioprinting technique providing a universal platform suitable for diverse cell types, materials, and bioactive agents in soft tissue regeneration applications [39].

Photopolymerization techniques, e.g., stereolithography (SLA) [11], are a commonly utilized technique for 3D printing of chitosan-based materials. You *et al.* synthesized a photo-crosslinkable bioink by conjugating methacrylated glycol chitosan (MeGC) with keratin methacrylate, resulting in a hybrid protein–polysaccharide hydrogel appropriate for 3D bioprinting. Upon exposure to blue light, the formulation generated resilient, bioactive scaffolds that facilitated stem cell growth

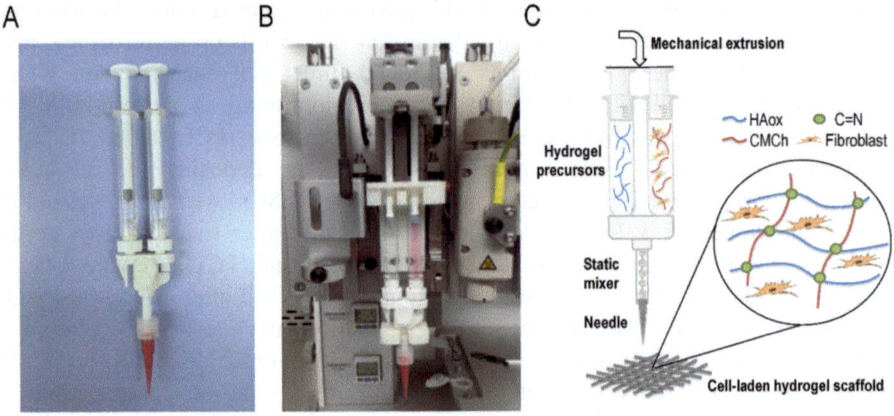

Fig. 2 (**a**) Picture of the static mixing tool, (**b**) Static mixing tool coupled to the 3D printer, and (**c**) Scheme of the bioprinting process using the mixing device coupled to the 3D printer. (Figure reprinted from Ref. [39]. © 2020 by the authors. Licensee MDPI, Basel, Switzerland. This article is an open-access article distributed under the terms and conditions of the Creative Commons Attribution (CC BY.4) license)

and maintained spheroid integrity [59]. Despite the poor mechanical strength of methacrylated chitosan alone, its combination with keratin methacrylate markedly improved the mechanical and biological efficacy of the construct. The 2% (w/v) MeGC and 2% (w/v) keratin methacrylate blend demonstrated the best swelling behavior, high cell survival, and enhanced cell proliferation among the evaluated formulations, thus being selected as the bioink for bioprinting trials. Using an extrusion-based approach, singular cells and spheroids were effectively manufactured and preserved in specified geometries. This study emphasizes an approach for developing bioactive, customizable hydrogels for prospective stem cell-derived tissue engineering uses [59]. Another study was reported for formulating a biomaterial ink consisting of maleic anhydride-modified chitosan (MCS), benzaldehyde-terminated PEG (PEGDF), and polyethylene glycol diacrylate (PEGDA) [8]. The resulting shear-thinning hydrogel, formed via Schiff-base reactions between MCS and PEGDF, could be injected and conform to irregular wound geometries, effectively preventing fluid leakage. The hydrogel rapidly solidified in situ upon UV exposure, enhancing adhesion and sealing capabilities. The hydrogel demonstrated strong stability even in harsh acidic environments (pH 1), indicating its potential for treating internal wounds such as gastric ulcers. Additionally, its applicability in in situ 3D printing enables use as adhesive wound dressings. The printing process offered UV-curable, shape-adaptive hydrogel, which offers significant promise as a tissue adhesive for wound closure and other biomedical applications [8]. Photo-assisted printing was also reported for succinylated chitosan (Chi-S) co-extrusion with 13% (w/v) cell-laden gelatin methacryloyl (GelMA) [52] and acidified chitosan and RHCMA (methacrylated collagen) [55].

Li *et al.* utilized a mixture of hydroxybutyl chitosan (HBC), oxidized chitosan (OCS), and human adipose-derived mesenchymal stem cells (MSCs) to fabricate inverse-geometry gels employing sacrificial Pluronic molds [27]. These creations preserved elevated cell viability and provoked a negligible macrophage response. A bioink including HBC and hyaluronic acid derivatives facilitated neural stem cell viability and proliferation for 7 days, preserving stemness and enhancing network development [27].

Composite inks incorporate functional fillers or secondary phases into chitosan-based matrices to enhance mechanical characteristics, bioactivity, and print resolution. Biopolymers such as cellulose nanocrystals (CNCs) were incorporated into chitosan inks, improving mechanical strength and cellular compatibility (Fig. 3) [32]. A 3D extrusion-based bioprinting method created constructions designed for bone tissue engineering and regeneration. Bioinks were developed utilizing thermogelling chitosan in conjunction with glycerophosphate, hydroxyethyl cellulose, and CNCs (Fig. 3). Formulations were optimized by varying the amounts of glycerophosphate (80–300 mM), hydroxyethyl cellulose (0–0.5 mg/mL), and CNCs (0–2% w/v) to achieve quick gelation (<7 s at 37 °C) and preserve structural integrity after printing. The impact of CNCs and pre-osteoblasts (MC3T3-E1 cells)

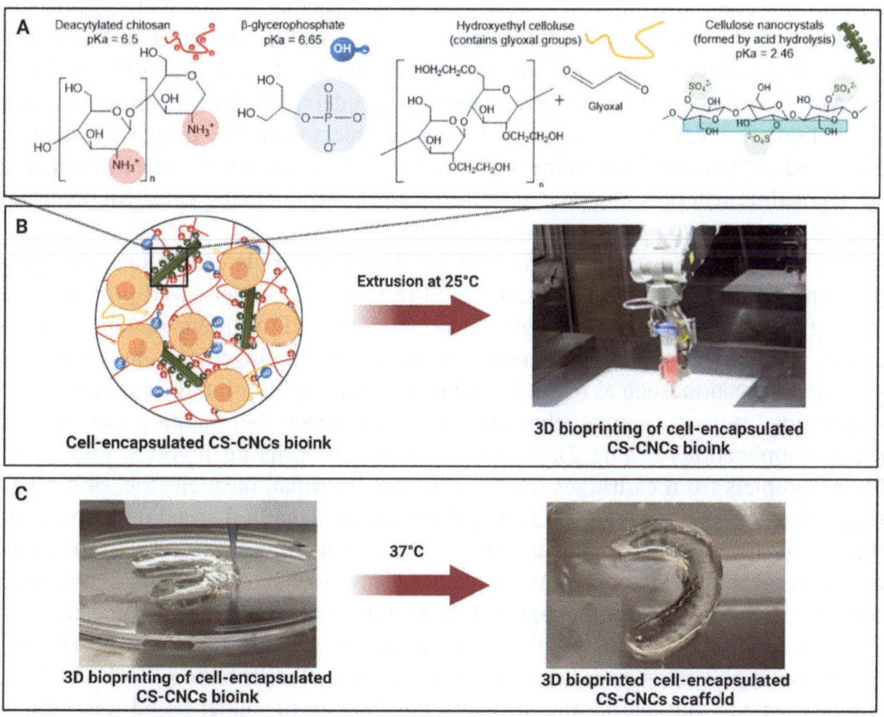

Fig. 3 3D printing of chitosan-CNCs scaffold. (Figure reprinted from Ref. [32]. Published 2021 by the American Chemical Society. This publication is licensed under CC-BY-NC-ND 4.0)

on rheological properties, printability, and scaffold mechanics was methodically evaluated. The integration of CNCs and cells (5 million cells/mL) markedly increased the bioink's viscosity and augmented the printed scaffolds' mechanical properties. The bioinks exhibited significant biocompatibility and printability within a pressure range of 12–20 kPa, maintaining cell viability. CNCs facilitated osteogenic development, as seen by high alkaline phosphatase activity, calcium deposition, and extracellular matrix production. The CNC-enhanced chitosan hydrogel is a promising and versatile bioink for 3D bioprinting in bone tissue engineering and other biomedical applications (Fig. 3) [32].

The properties of chitosan for gel formation enable the simple integration of nanoparticles. For example, iron oxide (Fe_3O_4) nanoparticles were integrated with chitosan–polyolefin forming a double-network hydrogel [18]. The interaction of Fe ions with chitosan, facilitated by ionic coordination and hydrogen bonding, imparted the hydrogel with superior mechanical strength, self-healing properties, and cytocompatibility. The hydrogel was printed into a saturated NaCl solution with ammonium persulfate, which enhanced stiffness by radical-induced crosslinking [18].

The 3D printing of chitosan-based materials can proceed using several methods, each presenting distinct advantages and problems. These methods fundamentally operate on a common principle: constructing a 3D structure incrementally, guided by a pre-designed model, usually depicted as a computer-aided design (CAD) file. The file is subsequently transformed into a standardized format, typically "STL," "3MF," or "AMF," which is acknowledged by most 3D printers. The drawing program, such as "Solidworks" can be also used to design the object [43]. The design can then be converted to "G-codes" by slicing software [43]. The printer constructs the structure by depositing materials in a predetermined pattern, guaranteeing excellent precision and the capability to produce intricate designs.

In general, printing methods for chitosan-based materials can be classified into three main techniques: (1) Extrusion-based printing, (2) Inkjet-based printing, and (3) Light-based printing. Although the principles of 3D printing remain uniform across various techniques, the specific technologies employed for chitosan-based scaffold fabrication, such as inkjet bioprinting, extrusion bioprinting, and light-based bioprinting, possess unique characteristics that affect their appropriateness for diverse applications (Table 2). Inkjet technology bioprinting involves dispensing bioink droplets from cartridges onto a substrate. Thermal, piezoelectric, or electromagnetic actuators can regulate the expulsion of droplets. In thermal inkjet printers, the ink is heated to generate a vapor bubble that expels the ink. Conversely, piezoelectric inkjet printers produce a pressure wave to eject the ink. It has numerous benefits, such as high printing speed and resolution (below 100 micrometers) and compatibility with various bioinks. Inkjet-based bioprinting is especially advantageous for applications necessitating control over cell distribution, such as tissue engineering. Nevertheless, the primary constraints of inkjet-based bioprinting encompass the requirement for low viscosity bioinks, which may limit material options, as well as issues about nozzle obstruction and inconsistent droplet size.

Table 2 Chitosan-based ink preparation methods for 3D/4D printing

Bioprinting type	Method	Composition	Nanomaterials	Preparation steps	Advantages	Disadvantages
Extrusion-based	Ionic gelation	Chitosan + TPP	None or nanoclay, AgNPs	1. Dissolve chitosan in acid2. Add TPP3. Stir for gelation	Mild, fast gelation, biocompatible	Weak mechanical properties
	pH-induced gelation	Chitosan + β-GP	Bioactive glass, GO	1. Acidic chitosan solution2. Add β-GP3. Neutralize	Injectable, cell-compatible	Slow gelation, weak strength
	Thermoresponsive gelation	Chitosan + pluronic F127	AuNPs, TiO_2	1. Dissolve chitosan2. Add thermoresponsive agent3. Adjust pH/temp	4D-compatible, injectable	Temperature-sensitive
	Enzymatic crosslinking	Chitosan + HRP + H_2O_2	HA nanoparticles, nanocellulose	1. Dissolve chitosan2. Add HRP and $H_2O_2$3. Allow crosslinking	Biocompatible, mild	Slow, costly enzymes
	Chemical crosslinking	Chitosan + genipin/ glutaraldehyde	Magnetic nanoparticles	1. Dissolve chitosan2. Add crosslinker3. React under mild heating	Strong networks	Toxic crosslinkers
	Freeze-drying	Chitosan gel	Retained from original ink	1. Print/cast gel2. Freeze3. Lyophilize	Porous structure	Time-consuming
Inkjet-based	Thermoresponsive gelation	Chitosan + β-GP + gelatin	AuNPs, TiO_2	1. Prepare solution2. Mix additives3. Adjust temp/pH	Responsive, soft tissues	Requires low viscosity
	Polyelectrolyte complexation	Chitosan + alginate	CNTs, silica	1. Mix polyelectrolytes2. Adjust conditions3. Form complex	Simple, tunable	Phase separation risk
	Composite/hybrid	Chitosan + collagen/ GO	GO, clay, MOFs	1. Mix materials2. Adjust rheology3. Homogenize	Customizable, functional	Aggregation risk
Light-based	Photopolymerization	Methacrylated chitosan + photoinitiator	Silica nanoparticles	1. Modify chitosan2. Add photoinitiator3. UV/light exposure	High resolution, tunable	UV risk, phototoxicity
	Composite/hybrid	Chitosan + photo-sensitive polymers	Nanodiamonds, GO	1. Mix with light-curable materials2. Homogenize3. Print under light	Improved biofunctionality	Complex formulation

Notes: TPP Tripolyphosphate, *GO* Graphene Oxide, *β-GP* Beta-Glycerophosphate, *Pluronic F127* A triblock copolymer used as a thermoresponsive gelator, *HRP* Horseradish Peroxidase, H_2O_2 Hydrogen Peroxide, *CNTs* Carbon Nanotubes, *HA* Hydroxyapatite or Hyaluronic Acid (depending on context), *AuNPs* Gold Nanoparticles, TiO_2 Titanium Dioxide Nanoparticles, *MOFs* Metal-Organic Frameworks, *ChiMA* Methacrylated Chitosan, *UV* Ultraviolet, *AgNPs* Silver Nanoparticles

Moreover, the solidification of droplets post-deposition should be transpire swiftly to prevent deformities in the printed structure, which may complicate the procedure.

Extrusion bioprinting is the most extensively researched and employed 3D printing methodology for chitosan-based scaffolds. The process involves extruding a continuous filament of bioink, often in hydrogel form, through a nozzle, propelled either mechanically or pneumatically. It is most efficacious for materials demonstrating shear-thinning properties, wherein the viscosity of the bioink diminishes under stress but rapidly recuperates upon removing that stress. It is exceptionally adaptable, adept at processing diverse materials and facilitating multi-material printing. It has been considered as optimal for fabricating scaffolds with diverse mechanical characteristics or functional strata. It can generate structures with high cell density, which is advantageous for tissue engineering applications. However, the resolution of extrusion-based bioprinting is typically lower than that of other technologies, with feature sizes greater than 100 micrometers, which can limit its precision. The capacity to regulate the exact configuration of cells within the scaffold is frequently constrained, and cell viability rates may vary from 40% to 86%, contingent upon the particular bioink employed. Notwithstanding these limitations, recent developments in extrusion-based bioprinting, such as coaxial and microfluidic bioprinting, offer advantages such as high resolution and greater control over cell patterning.

Photonic bioprinting, including technologies like stereolithography (SLA) and digital light processing (DLP), uses light to cure photo-crosslinkable chitosan-based bioinks layer by layer, selectively. In stereolithography, a laser is directed at a resin surface to cure the material, while digital light processing uses a digital projector to cure an entire layer of resin at once, significantly reducing printing time. These technologies are particularly advantageous for producing high-resolution structures with complex geometries, with layer thicknesses ranging from 25 to 100 micrometers. However, light-based bioprinting requires photo-crosslinkable materials that contain photoinitiators to initiate polymerization upon exposure to light. While this enables the creation of highly intricate and accurate structures, the requirement for suitable bioinks is a significant challenge, as limited materials can perform well in light-based bioprinting. Additionally, the toxicity of the photocuring agents, the need for support structures when printing complex geometries, and the material wastage due to excess bioink are notable disadvantages of this method. Nevertheless, the high accuracy and resolution offered by light-based bioprinting make it a promising technology for creating highly detailed scaffolds for advanced applications in tissue engineering.

3D printing technologies, e.g., inkjet-based bioprinting, extrusion-based bioprinting, and light-based bioprinting, offer distinct benefits and limitations when used for chitosan-based scaffold fabrication. The choice of technology depends mainly on the specific requirements of the application. Light-based bioprinting is ideal for high-resolution printing with complex geometries, but its material limitations and the need for support structures can hinder its use in some scenarios. On the other hand, extrusion-based bioprinting is cost-effective and versatile, supporting a wide range of bioinks and creating high-density cell structures. However, its lower resolution and challenges with cell patterning may make it less suitable for specific precision applications. Inkjet-based bioprinting offers high speed and resolution, with the added benefit of precise control over cell distribution, but the need for low-viscosity bioinks and the potential for nozzle clogging can limit its applicability. The method's advantages and drawbacks should be carefully considered based on the specific goals and constraints of the chitosan-based material fabrication process.

3D printing techniques could involve more than one techniques. For example, fused deposition modeling (FDM) and injection volume filling (IVF) could be combined to fabricate camptothecin-loaded chitosan micelles for colon-specific delivery [6].

4 3D Printed Chitosan-Based Materials for Drug Delivery

3D printing of chitosan-based materials can be used to deliver several pharmaceutical active components, such as anticancer, antibiotics, growth factors, enzymes, proteins, and gene-based therapeutic agents (Table 3) [34]. 3D printing chitosan-based materials enabled localized drug delivery [25]. It can be used for simultaneous control by drug delivery systems (DDSs) [25]. Thus, several drugs were reported, including doxorubicin [10, 11], antibiotic levofloxacin [38], ibuprofen (IBP) [56], camptothecin [6], and Vancomycin (Table 3) [28].

Two-photon 3D printing from a chitosan-based magnetic (superparamagnetic iron oxide nanoparticles) polymer nanocomposite was reported for the fabrication of a magnetically actuated, double-helical microswimmer, with a diameter of 6 μm and a length of 20 μm [10]. The printed object was applied to release chemotherapeutic drugs, e.g., anticancer drug Doxorubicin (DOX), on demand when stimulated by external light. The microswimmers are biocompatible and biodegradable for secure use in biological settings. Doxorubicin (DOX), the encapsulated drug, is conjugated through a photocleavable linker to amino groups on the surface of the microswimmer. These microswimmers exhibit accurate navigational control within a 10 mT rotating magnetic field. When UV radiation at 365 nm (0.34 W/cm^2), over 60% of the medication was released within 5 min. The release can be regulated via patterned light exposure, enabling temporal control over dosage. Moreover, under physiological settings, the microswimmers undergo substantial degradation after 204 h, resulting in non-toxic metabolites. Light-responsive drug delivery and magnetically powered mobility offered precision-targeted therapeutic systems necessitating multi-modal control [10]. 3D printing of chitosan/DOX using glutaraldehyde was also reported offering fabrication microneedles (MNs) for prolonged drug release time without drug-burst treatment is promising [11].

Hydrogel inks that are biocompatible and printable, made from chitosan, gelatin, and collagen, were developed and effectively 3D-printed via a pneumatic extrusion technique [38]. The printability of several chitosan–gelatin inks was assessed, indicating that inks with chitosan concentrations under 4% w/v exhibited insufficient

Table 3 Drug delivery using printing chitosan-based materials

Composition	Printing techniques	Drug	Cells	Efficiency	References
PCL ibuprofen Chitosan	FDM method combined with hot-melt extrusion (HME)	Ibuprofen	Mouse fibroblast L-929	100% after 12 h	[56]
Chitosan Methacrylamide chitosan (ChMA) Magnetic material	Two-photon-based 3D printing	Doxorubicin	SKBR3 breast Cancer cell	60% within 5 min	[10]
Chitosan Glutaraldehyde	Stereolithography (SLA)	Doxorubicin	Skin	56% after 3.5 h	[11]
Chitosan Gelatin	Extrusion-based 3D Bioprinter	Levofloxacin	Fibroblasts from human skin	55% after 15 min	[38]
Chitosan Polycaprolactone (PCL)	FDM	Vancomycin	ah-BM-MSCs	10 μg/mL after 3 h	[28]
Polylactic acid (PLA) *O*-methyl-*O*′-succinylpolyethylene glycol-chitosan-oleic acid (mPEG-CS-OA polymer)	FDM and IVF techniques	Camptothecin (CPT)	C2BBe1 [Caco-2 clone] (human colorectal adenocarcinoma)	70% after <30 min	[6]
Hydroxybutyl methacrylated chitosan (HBCMA) Rhodamine B	DLP	Rhodamine B	Agarose gel	70% after 170 h	[15]
PLA Chitosan	Extrusion	Levodopa	hAD-MSCs	100% after 14 days	[43]

Notes: *ah-BM-MSCs* Adult Human Bone Marrow-Derived Mesenchymal Stem Cells printability. Conversely, a formulation containing 4% w/v chitosan and 1% w/v gelatin (CG) exhibited favorable extrusion properties and structural integrity. Adding a minor quantity of collagen (0.1% w/v) to this improved formulation (CGC) did not adversely impact printability. Stabilizing post-printing scaffolds with potassium hydroxide (KOH) resulted in uniform morphology and structural integrity, surpassing the effectiveness of neutralization with bases like NaOH/ethanol or ammonia vapors, producing smaller pores and less stable structures. Studies on water evaporation revealed that KOH-neutralized scaffolds exhibited reduced evaporation rates due to robust intermolecular interactions that create a stable cross-linked network. The composition of chitosan enables the formation of polyelectrolyte complexes in Chitosan-Gel and Chitosan-Gel-Collagen blends, hence enhancing their mechanical stability. The swelling behavior at varying pH settings further validated structural integrity, with KOH-treated samples exhibiting minimal swelling and maximal durability. Following the optimization of the ink composition, the scaffolds were infused with 10 wt% of the antibiotic levofloxacin to assess drug delivery efficacy. The drug loading was around 4% for both GC and CGC formulations. CG Levo exhibited over 80% drug release within the initial hour, achieving complete release within 24 h. In contrast, the CGK Levo construct demonstrated a more gradual and regulated release, with 55% released in the first 15 min and a consistent profile after 4 h, indicating controlled diffusion from the expanded hydrogel matrix [38]. Biocompatibility was assessed by MTT assays on fibroblast cells, and antibacterial efficacy was verified against *E. coli* and *S. aureus*. The inhibitory zone against *E. coli* increased from 8 $\pm$ 0.4 mm (control) to 16.4 $\pm$ 0.37 mm using the KOH-treated CG Levo scaffold. The findings indicate that the engineered hydrogel inks, refined printing parameters, and post-processing methods facilitate the creation of structurally robust, drug-releasing scaffolds appropriate for tissue engineering and antibacterial purposes [38].

A FDM method was combined with hot-melt extrusion (HME) for creating personalized, biocompatible drug delivery implants (Table 3) [56]. Specifically, polycaprolactone (PCL) was used as a carrier material, ibuprofen (IBP) served as the model drug, and chitosan was incorporated to enable controlled and efficient drug release. The HME process allowed solvent-free preparation of drug-loaded PCL-CS filaments, addressing drug stability and biocompatibility requirements. These filaments were successfully 3D printed into implants using the FDM process. Mechanical testing showed that filament strength decreased with increasing drug and chitosan content, without significant degradation or chemical changes during processing. Drug release was highly efficient, reaching over 99% in 120 h, primarily governed by a diffusion–erosion mechanism. Cytocompatibility testing showed a cell viability of 75.3% after exposure to 100% implant extract for 72 h, confirming acceptable biocompatibility. This approach highlights a promising strategy for fabricating customized, implantable drug delivery systems with sustained release profiles [56].

Microneedles of chitosan-based biomaterials were fabricated using DLP-based 3D printing (Table 3) [15]. Hydroxybutyl methacrylated chitosan (HBCMA) was used as chitosan material with both temperature- and photo-responsive properties to produce microneedle structures. The DLP technique enabled rapid, high-resolution fabrication of HBCMA microneedles. These microneedles exhibit 4D behavior, dynamically altering their dimensions in response to temperature, which results in enhanced resolution, sharper tips, and improved mechanical strength. The microneedles demonstrated the capability to encapsulate and sustainably release small-molecule drugs while effectively penetrating soft tissue. Drug release performance was also validated using an agarose gel model, where RhoB fluorescence intensity increased over time after microneedle removal at 1, 12, and 48 h, confirming successful drug delivery. These results highlight the potential of HBCMA-based microneedles for non-dermal drug delivery applications, especially in soft tissue targeting [15].

3D-printed brain tissue scaffolds (pore dimensions of 100–200 μm) made from polylactic acid (PLA) and chitosan were reported for the regulated release of levodopa in managing Parkinson's disease (Table 3) [43]. Integrating chitosan into the PLA matrix enhanced the scaffolds' swelling capability and weight loss characteristics. Levodopa was released for a period of 14 days. Human adipose-derived mesenchymal stem cells (hAD-MSCs) were utilized to evaluate biocompatibility via MTT assays, fluorescence imaging, and scanning electron microscopy (SEM). Data analysis validated the scaffold's appropriateness for cellular proliferation. Authors of the study highlighted the promise of PLA/Chitosan 3D-printed scaffolds as an effective medium for targeted levodopa delivery in neural tissue engineering for Parkinson's disease treatment [43].

Incorporating pharmaceuticals or drugs directly into bioinks during the bioprinting process allows the resulting structures to function as localized, sustained-release drug delivery systems [57]. This approach provides several significant advantages, including precise targeting of therapeutic agents to specific disease sites, which enhances treatment efficacy and minimizes systemic side effects. The system enables controlled and prolonged release by embedding the drug within the printed scaffold, reducing the need for frequent dosing. Furthermore, bioprinting offers the flexibility to customize drug formulations and delivery profiles to suit individual patient needs, promoting personalized medicine. Integrating drug delivery with tissue engineering holds great promise for improving outcomes in regenerative therapies and localized treatments [57]. The formation of chitosan blend with other biopolymers such as starch (13 wt.%) exhibited good properties, including structural characteristics features [12].

5 Tissues Engineering

Tissue engineering has arisen as a revolutionary method to tackle the substantial clinical issues related to tissue and organ loss due to traffic accidents, trauma, burns, and congenital deformities. Conventional treatment modalities, including cell therapy, autologous and allogeneic transplantation, xenotransplantation, and prostheses, frequently encounter significant constraints such as donor shortages, immune rejection, inadequate integration with host tissues, and restricted long-term efficacy. While direct stem cell injection demonstrates potential, especially in muscle function restoration, this approach is limited by low cell viability post-injection, insufficient extracellular matrix (ECM) formation, and inadequate vascularization. Tissue engineering is a multidisciplinary approach that combines biology, materials science, and engineering to create functioning, biomimetic constructions to regenerate or replace injured tissues and organs. Three fundamental components of tissue engineering are seed cells (including embryonic stem cells, adult stem cells, and induced pluripotent stem cells), growth factors modulating cellular behavior and tissue development, and scaffolds replicating the natural extracellular matrix to facilitate cell attachment, proliferation, and differentiation. Contemporary scaffolds are frequently constructed from biocompatible and biodegradable polymers, including chitosan. Tissue engineering has the potential to transform regenerative medicine by surpassing the constraints of traditional therapies, facilitating the development of tailored, functioning tissues with improved integration and sustained therapeutic results. Tissue engineering components are summarized in Table 4.

Chitosan is a promising component in tissue engineering [17, 53]. 3D printing of hydrogel can be used for tissue engineering [5]. Ideally, chitosan-based inks should exhibit a supportive microenvironment conducive to 3D cell proliferation, differentiation, and organization into functional tissue analogues. They can be applied to cartilage tissue engineering (CTE) [35]. 3D printing procedure enabled a multi-scale procedure for processing scaffolds for tissue engineering [30].Chitosan is a promising material for tissue engineering and regenerative medicine owing to its advantageous characteristics, including hydrophilicity, biocompatibility, biodegradability, and antibacterial properties [42].

A photocurable chitosan-based bioink (chitosan-MA) was developed for application in digital light processing (DLP) 3D printing (Fig. 4) [46]. The chitosan-MA precursor was synthesized via a simple one-step N-acylation reaction to graft methacryloyl groups onto chitosan chains, enabling photo-crosslinking. The formulation was optimized by varying the concentration and the degree of substitution (DS) of chitosan-MA to fine-tune the rheological behavior, photocuring kinetics, and mechanical properties of the resulting hydrogels (Fig. 4). At a high DS of 33.6%, chitosan-MA enabled the formation of stable 150 μm-thick hydrogel layers in under 10 s, making it highly suitable for layer-by-layer DLP fabrication. The addition of Lithium phenyl-2,4,6-trimethylbenzoylphosphinate (LAP) as a photoinitiator facilitated rapid and efficient photocuring. Notably, the photo-crosslinking process and the resulting hydrogels exhibited excellent biocompatibility in cytotoxicity assays. The Chitosan-MA bioink was processed using DLP printing into intricate, high-resolution 3D hydrogel structures with superior fidelity and structural stability. These results highlight the promise of chitosan-MA as a versatile and efficient bioink for DLP and other light-based 3D bioprinting technologies [46]. The photocurable properties of chitosan-MA enable shaping into different forms that can recover to their original shape without residual deformity under external stimuli such as pressure, i.e., high elastic properties (Fig. 5) [21].

Table 4 Key components of tissue engineering

Element	Description	Key characteristics/examples	Challenges/ considerations
Seed cells	Stem cells with self-renewal and multi-lineage differentiation potential are used for tissue repair and regeneration	*Types*: Embryonic stem cells (ESCs), adult stem cells (ASCs), induced pluripotent stem cells (iPSCs)	Ethical issues Immune rejection Low survival post-injection Lack of structural guidance and ECM support Integration with biomaterials remains a challenge
Growth Factors (GFs)	Bioactive molecules (cytokines, polypeptides, hormones) that regulate cell proliferation, migration, and differentiation	*Functions*: stimulate cell signaling pathways for tissue growth *Types*: inflammatory signals, hormone cues *Delivery methods*: matrix-bound, scaffold-assisted release	High cost of recombinant production Protein instability Limited spatio-temporal control Non-specific hormone actions Need for scaffold-engineered delivery systems
Scaffolds	3D biodegradable structures that mimic the extracellular matrix (ECM), supporting cell adhesion, growth, and tissue regeneration	*Ideal properties*: Biocompatibility, high porosity, good mechanical stability *Materials*: hydrogels, alginate, chitosan, hyaluronic acid, collagen, gelatin, silk fibroin *Functions*: deliver GFs, support cell behavior	Designing ECM-mimetic architecture Controlled GF delivery Achieving long-term mechanical integrity Scaffold–cell integration and biological functionality

Ye *et al.* devised an expedited post-printing cross-linking technique utilizing isopropyl alcohol, which maintained scaffold shape and facilitated the viability of infrapatellar fat pad adipose-derived stem cells (IPFP-ASCs), rendering it appropriate for cartilage tissue creation [58]. IPFP-ASCs, recognized for their chondrogenic capabilities, were inoculated onto 3D-printed chitosan scaffolds and subjected to stimulation with TGFβ3 and BMP6. By the fourth week, a cartilage-like matrix developed, resulting in a pearlescent layer on the scaffold surface. Histological examination revealed chondrocyte-like morphology, with the extracellular matrix

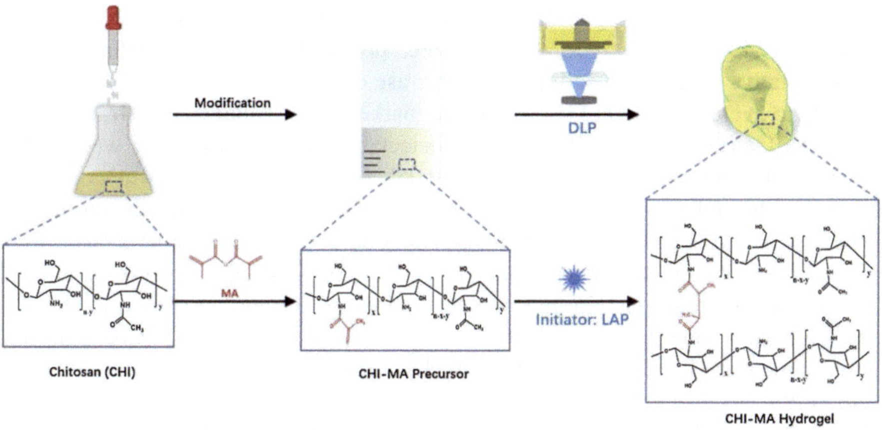

Fig. 4 Chitosan modification and 3D printing via DLP. (Figure reprinted with permission from Ref. [46] © 2020 Elsevier Ltd)

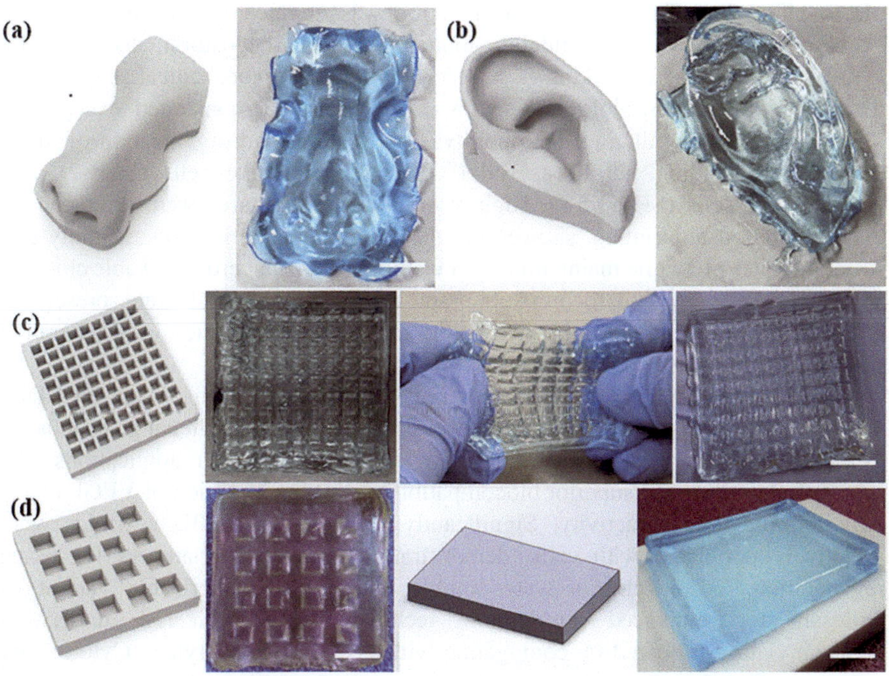

Fig. 5 printing procedures for CHIMA/PAM bioinks; (**a**, **b**) the CAD model for (**a**) nose and (**b**) ear auricle, (**c**, **d**) Lattice structures of CHIMA/PAM hydrogels. Bar represent 10 mm. (Figure reprinted from Open Access Ref. [21] © 2021 The Authors. Published by Elsevier Ltd)

exhibiting positive staining with toluidine blue. Immunohistochemistry verified the existence of collagen type II, cartilage proteoglycans, and collagen type I. Quantitative reverse transcription polymerase chain reaction (qRT-PCR) demonstrated the overexpression of chondrogenic markers, e.g., collagen type II, aggrecan, and SOX9. Collagen type II, initially undetectable, was expressed by week 2 and dramatically increased by week 4 ($p < 0.05$), but collagen type I was constantly expressed throughout the culture duration. These findings highlighted the promise of IPFP-ASCs and chitosan scaffolds for osteochondral transplant applications [58].

A chitosan-based bioink was created by integrating a dual cross-linking method that merges thermal gelation and photocrosslinking to improve structural stability and biological functioning for 3D bioprinting applications [51]. Chitosan was initially methacrylated via the reaction of chitosan solution (solubilized in 2% acetic acid (1.5% w/v)), with methacrylic anhydride for 4 h. The resultant methacrylated chitosan was subjected to dialysis and lyophilization after neutralization with sodium carbonate. To facilitate thermoresponsive properties and physiological compatibility, methacrylated chitosan was combined with β-glycerol phosphate, resulting in a sol-gel transition at physiological temperature (37 °C) and adjusting the formulation pH to about 7.4. The hydrogel formulation was engineered for initial thermal gelation to preserve print fidelity immediately post-deposition, followed by UV-triggered photocrosslinking ($\lambda = 365$ nm), enhancing mechanical stability and resistance in aqueous biological settings. The bioink demonstrated exceptional biocompatibility, facilitating the viability, proliferation, and organization of many cell types, including fibroblasts (NIH/3T3), osteoblast-like cells (Saos-2), and neuronal-like cells (SH-SY5Y). High-resolution 3D structures were produced by extrusion-based bioprinting, showcasing the material's superior processability and structural integrity while maintaining cell viability. This dual cross-linkable chitosan hydrogel is a resilient, cytocompatible, and printable bioink substrate, presenting considerable potential for fabricating intricate, cell-laden tissues in regenerative medicine and sophisticated tissue modeling [51].

Chitosan was modified with synthetic poly(ethylene glycol) diacrylate (PEGDA) and covalently cross-linked with a photosensitive vascular endothelial growth factor (VEGF) [24]. The printed materials were investigated for in vivo applications. The materials demonstrated superior biocompatibility, and the inclusion of VEGF markedly improved its bioactivity. Significantly, Chitosan-based 3D scaffolds were effectively populated with cells, demonstrating their appropriateness for tissue regeneration. To develop a water-soluble and photopolymerizable variant of chitosan for enhanced production techniques, the material underwent N-succinylation followed by conjugation with glycidyl methacrylate. Cytotoxicity assays verified that these chemical changes were non-toxic to human pulmonary microvascular endothelial cells (HPMECs). Rheological analyses revealed adjustable viscoelastic characteristics among several hydrogel formulations, providing opportunities for customized cell manipulation. Moreover, accurate chitosan microstructures with diverse topologies and porosities were produced using two-photon polymerization. These findings highlight the adaptability and potential of modified chitosan as a bioink appropriate for microstructuring and 3D printing applications in tissue engineering [24].

Chang *et al.* developed a dual-functionalized chitosan that integrates glycol and methacrylate groups to enhance solubility and mechanical strength [14]. The modified chitosan, methacrylated glycol chitosan (MeGC), was subjected to visible light-induced photocrosslinking with riboflavin as a photoinitiator, yielding hydrogels with improved rheological characteristics, structural integrity, and heightened MG63 cell proliferation. MeGC was chosen for its enhanced water compatibility and formulated into a bioink, including MG63 cells for 3D bioprinting. Optimal printing was attained with MeGC concentration (3 wt.%), riboflavin (12 μM), cured for 70 s under visible light, extruded via a 26-gauge nozzle at 120 kPa, and printed at a velocity of 6 mm/s. Quantitative tests revealed a tendency to decrease pore width with increased crosslinking: for MeGC-30 to MeGC-90, pore sizes in freeze-dried samples varied from approximately 193 to 151 μm, and in hydrated samples from approximately 218 to 168 μm. The resultant constructions exhibited superior printability, over 92% cellular viability, more than 96% cellular proliferation, and hemolysis rates below 2%, thereby endorsing their use as patient-specific scaffolds for bone regeneration. MeGC hydrogels exhibited interpenetrating porosity networks crucial for nutrition delivery and cellular proliferation. The structural variations, shaped by polymer shrinkage due to drying, impacted surface energy and protein adsorption. Although not statistically significant, protein binding seems to augment with a bigger pore size owing to the increased accessible surface area, thus further corroborating MeGC's appropriateness as a biomaterial for tissue engineering [14]. Another study reported on 3D printing of ChMA and GelMA, including nanohydroxyapatite (nano-HAp) [37]. The fabrication of resilient and bioactive scaffolds for tissue engineering. The hybrid polymer system first undergoes physical gelation by thermal procedures to provide immediate structural integrity after printing, followed by covalent photo-crosslinking of both ChMA and GelMA to ensure long-term mechanical stability. By adjusting the GelMA concentration, the hydrogel's rheological characteristics and the mechanical integrity of the printed structures were effectively optimized, resulting in improved shape fidelity and precise layer-by-layer deposition. The biological assessment indicated that the scaffolds facilitated stem cell adhesion, spreading, and proliferation following 7 days of in vitro culture. The addition of nano-HAp did not substantially affect the mechanical characteristics or cytocompatibility of the hydrogels, suggesting that bioactivity was improved without detriment to scaffold performance. This shear-thinning, structurally stable, and biocompatible hydrogel ink is a promising contender for 3D bioprinting in regenerative medicine [37]. 3D printing was also reported for methacrylated chitosan-laponite (MAC-Lp), methacrylated gelatin-laponite (MAG-Lp), MAC, and MAG were developed to evaluate their potential for bone tissue engineering [13]. Among these, MAC-Lp scaffolds exhibited superior biological performance, including enhanced cell viability, proliferation, and biomineral deposition, when compared to MAG-Lp counterparts. Fourier-transform infrared spectroscopy (FTIR) confirmed a higher presence of phosphate groups, indicative of biomineralization and more prominent amide peaks associated with ECM-like collagen formation on MAC-Lp scaffolds. Scanning electron microscopy (SEM) revealed a denser ECM-like tissue morphology in MAC-Lp scaffolds, while Alizarin Red staining demonstrated a greater abundance of mineralized nodules. X-ray absorption near-edge structure (XANES) analysis verified the formation of calcium-phosphate (Ca–P) species, further supporting the enhanced mineralization capacity of MAC-Lp constructs. These scaffolds were fabricated using robocast-assisted deposition (robocasting), a technique offering precise control over filament dimensions, pore architecture, and overall porosity. In vitro studies confirmed the potential of MAC-Lp scaffolds to promote osteoblast adhesion, growth, and differentiation more effectively than gelatin-based scaffolds. The findings highlight the promise of chitosan-laponite composites as functional biomaterials for bone regeneration, with a greater capacity to support biomineral formation and osteogenic activity [13].

6 Conclusion

Chitosan-based 3D and 4D (bio)printing techniques signify a substantial advancement in the design and production of tissue-engineered scaffolds and drug delivery devices. These technologies offer structural support for cellular proliferation and tissue regeneration while facilitating precise, regulated drug delivery customized to the specific requirements of patients. Incorporating chitosan into these digital manufacturing platforms improves scaffold performance by enhancing biocompatibility, biodegradability, and responsiveness to biological settings. The adaptability of chitosan facilitates the creation of intricate, multi-material structures essential for tackling the complexities of actual biological systems. As the discipline progresses, chitosan-based bioinks and printing methodologies are anticipated to be pivotal in advancing customized medicine and regenerative healthcare solutions.

References

1. Abdelhamid, H. N., El-Bery, H. M., Metwally, A. A., Elshazly, M., & Hathout, R. M. (2019). Synthesis of CdS-modified chitosan quantum dots for the drug delivery of Sesamol. *Carbohydrate Polymers [Internet], 214*, 90–99. Available from: https://linkinghub.elsevier.com/retrieve/pii/S0144861719302905.
2. Abdelhamid, H. N., Sultan, S., & Mathew, A. P. (2023). Three-dimensional printing of cellulose/covalent organic frameworks (CelloCOFs) for CO 2 adsorption and water treatment. *ACS Applied Materials & Interfaces [Internet]., 15*(51), 59795–59805. Available from: https://pubs.acs.org/doi/10.1021/acsami.3c13966.
3. Abdelhamid, H. N., & Wu, H.-F. (2013a). Multifunctional graphene magnetic nanosheet decorated with chitosan for highly sensitive detection of pathogenic bacteria. *Journal of Materials Chemistry B [Internet], 1*(32), 3950–3961. Available from: http://xlink.rsc.org/?DOI=c3tb20413h.

4. Abdelhamid, H. N., & Wu, H.-F. (2013b). Probing the interactions of chitosan capped CdS quantum dots with pathogenic bacteria and their biosensing application. *Journal of Materials Chemistry. B [Internet], 1*(44), 6094. Available from: http://xlink.rsc.org/?DOI=c3tb21020k.
5. Advincula, R. C., Dizon, J. R. C., Caldona, E. B., Viers, R. A., Siacor, F. D. C., Maalihan, R. D., et al. (2021). On the progress of 3D-printed hydrogels for tissue engineering. *MRS Communications [Internet], 11*(5), 539–553. Available from: https://link.springer.com/10.1557/s43579-021-00069-1.
6. Almeida, A., Linares, V., Mora-Castaño, G., Casas, M., Caraballo, I., & Sarmento, B. (2021). 3D printed systems for colon-specific delivery of camptothecin-loaded chitosan micelles. *European Journal of Pharmaceutics and Biopharmaceutics [Internet], 167*, 48–56. Available from: https://linkinghub.elsevier.com/retrieve/pii/S0939641121001879.
7. Beach, M. A., Nayanathara, U., Gao, Y., Zhang, C., Xiong, Y., Wang, Y., et al. (2024). Polymeric nanoparticles for drug delivery. *Chemical Reviews [Internet], 124*(9), 5505–5616. Available from: https://pubs.acs.org/doi/10.1021/acs.chemrev.3c00705.
8. Bian, S., Zheng, Z., Liu, Y., Ruan, C., Pan, H., & Zhao, X. (2019). A shear-thinning adhesive hydrogel reinforced by photo-initiated crosslinking as a fit-to-shape tissue sealant. *Journal of Materials Chemistry B [Internet], 7*(42), 6488–6499. Available from: https://xlink.rsc.org/?DOI=C9TB01521C.
9. Bodaghi, M., Wang, L., Zhang, F., Liu, Y., Leng, J., Xing, R., et al. (2024). 4D printing roadmap. *Smart Materials and Structures [Internet], 33*(11), 113501. Available from: https://iopscience.iop.org/article/10.1088/1361-665X/ad5c22.
10. Bozuyuk, U., Yasa, O., Yasa, I. C., Ceylan, H., Kizilel, S., & Sitti, M. (2018). Light-triggered drug release from 3D-printed magnetic chitosan microswimmers. *ACS Nano [Internet], 12*(9), 9617–9625. Available from: https://pubs.acs.org/doi/10.1021/acsnano.8b05997.
11. Camcı, Y., Türk, S., Gepek, E., İyibilgin, O., & Özsoy, M. İ. (2022). Fabrication and characterization of innovative chitosan/doxorubicin coated <scp>3D</scp> printed microneedle patch for prolonged drug delivery. *Journal of Applied Polymer Science [Internet], 139*-(32) Available from: https://onlinelibrary.wiley.com/doi/10.1002/app.52759.
12. Cardoso, S., Narciso, F., Monge, N., Bettencourt, A., & Ribeiro, I. A. C. (2023). Improving Chitosan hydrogels printability: A comprehensive study on printing Scaffolds for customized drug delivery. *International Journal of Molecular Sciences [Internet], 24*(2), 973. Available from: https://www.mdpi.com/1422-0067/24/2/973.
13. Cebe, T., Ahuja, N., Monte, F., Awad, K., Vyavhare, K., Aswath, P., et al. (2020). Novel 3D-printed methacrylated chitosan-laponite nanosilicate composite scaffolds enhance cell growth and biomineral formation in MC3T3 pre-osteoblasts. *Journal of Materials Research [Internet], 35*(1), 58–75. Available from: http://link.springer.com/10.1557/jmr.2018.260.
14. Chang, H. K., Yang, D. H., Ha, M. Y., Kim, H. J., Kim, C. H., Kim, S. H., et al. (2022). 3D printing of cell-laden visible light curable glycol chitosan bioink for bone tissue engineering. *Carbohydrate Polymers [Internet], 287*, 119328. Available from: https://linkinghub.elsevier.com/retrieve/pii/S0144861722002326.
15. Che, Q. T., Seo, J. W., Charoensri, K., Nguyen, M. H., Park, H. J., & Bae, H. (2024). 4D-printed microneedles from dual-sensitive chitosan for non-transdermal drug delivery. *International Journal of Biological Macromolecules [Internet], 261*, 129638. Available from: https://linkinghub.elsevier.com/retrieve/pii/S0141813024004410.
16. Cidonio, G., Costantini, M., Pierini, F., Scognamiglio, C., Agarwal, T., & Barbetta, A. (2021). 3D printing of biphasic inks: Beyond single-scale architectural control. *Journal of Materials Chemistry C [Internet], 9*(37), 12489–12508. Available from: https://xlink.rsc.org/?DOI=D1TC02117F.
17. Croisier, F., & Jérôme, C. (2013). Chitosan-based biomaterials for tissue engineering. *European Polymer Journal [Internet], 49*(4), 780–792. Available from: https://linkinghub.elsevier.com/retrieve/pii/S0014305712004181.
18. Gang, F., Yan, H., Ma, C., Jiang, L., Gu, Y., Liu, Z., et al. (2019). Robust magnetic double-network hydrogels with self-healing, MR imaging, cytocompatibility and 3D printability.

Chemical Communications [Internet], 55(66), 9801–9804. Available from: https://xlink.rsc.org/?DOI=C9CC04241E.

19. Gopal, J., Abdelhamid, H. N., Hua, P. –Y., & Wu, H. -F. (2013) Chitosan nanomagnets for effective extraction and sensitive mass spectrometric detection of pathogenic bacterial endotoxin from human urine. *Journal of Materials Chemistry. B [Internet]* . The Royal Society of Chemistry; [cited 2016 Jan 17], *1*(19), 2463. Available from: http://pubs.rsc.org/en/content/articlehtml/2013/tb/c3tb20079e.
20. Haider, A., Khan, S., Iqbal, D. N., Khan, S. U., Haider, S., Mohammad, K., et al. (2024). Chitosan as a tool for tissue engineering and rehabilitation: Recent developments and future perspectives—A review. *International Journal of Biological Macromolecules [Internet], 278*, 134172. Available from: https://linkinghub.elsevier.com/retrieve/pii/S0141813024049778.
21. He, Y., Wang, F., Wang, X., Zhang, J., Wang, D., & Huang, X. (2021). A photocurable hybrid chitosan/acrylamide bioink for DLP based 3D bioprinting. *Materials and Design [Internet], 202*, 109588. Available from: https://linkinghub.elsevier.com/retrieve/pii/S0264127521001416.
22. Khan, M. U. A., Stojanović, G. M., Abdullah, M. F. B., Dolatshahi-Pirouz, A., Marei, H. E., Ashammakhi, N., et al. (2024). Fundamental properties of smart hydrogels for tissue engineering applications: A review. *International Journal Of Biological Macromolecules [Internet], 254*, 127882. Available from: https://linkinghub.elsevier.com/retrieve/pii/S0141813023047815.
23. Ku, J., Seonwoo, H., Park, S., Jang, K.-J., Lee, J., Lee, M., et al. (2020). Cell-Laden thermosensitive Chitosan hydrogel bioinks for 3D bioprinting applications. *Applied Sciences [Internet], 10*(7), 2455. Available from: https://www.mdpi.com/2076-3417/10/7/2455.
24. Kufelt, O., El-Tamer, A., Sehring, C., Meißner, M., Schlie-Wolter, S., & Chichkov, B. N. (2015). Water-soluble photopolymerizable chitosan hydrogels for biofabrication via two-photon polymerization. *Acta Biomaterialia [Internet], 18*, 186–195. Available from: https://linkinghub.elsevier.com/retrieve/pii/S1742706115000963.
25. Lazaridou, M., Bikiaris, D. N., & Lamprou, D. A. (2022). 3D bioprinted Chitosan-based hydrogel Scaffolds in tissue engineering and localised drug delivery. *Pharmaceutics [Internet], 14*(9), 1978. Available from: https://www.mdpi.com/1999-4923/14/9/1978.
26. Lee, D., Park, J. P., Koh, M.-Y., Kim, P., Lee, J., Shin, M., et al. (2018). Chitosan-catechol: A writable bioink under serum culture media. *Biomaterials Science [Internet], 6*(5), 1040–1047. Available from: https://xlink.rsc.org/?DOI=C8BM00174J.
27. Li, C., Wang, K., Zhou, X., Li, T., Xu, Y., Qiang, L., et al. (2019). Controllable fabrication of hydroxybutyl chitosan/oxidized chondroitin sulfate hydrogels by 3D bioprinting technique for cartilage tissue engineering. *Biomedical Materials [Internet], 14*(2), 025006. Available from: https://iopscience.iop.org/article/10.1088/1748-605X/aaf8ed.
28. López-González, I., Hernández-Heredia, A. B., Rodríguez-López, M. I., Auñón-Calles, D., Boudifa, M., Gabaldón, J. A., et al. (2023). Evaluation of the in vitro antimicrobial efficacy against Staphylococcus aureus and epidermidis of a novel 3D-printed degradable drug delivery system based on Polycaprolactone/Chitosan/Vancomycin—Preclinical study. *Pharmaceutics [Internet], 15*(6), 1763. Available from: https://www.mdpi.com/1999-4923/15/6/1763.
29. Magli, S., Rossi, G. B., Risi, G., Bertini, S., Cosentino, C., Crippa, L., et al. (2020). Design and synthesis of Chitosan—Gelatin hybrid hydrogels for 3D printable in vitro models. *Frontiers in Chemistry [Internet], 8*. Available from: https://www.frontiersin.org/article/10.3389/fchem.2020.00524/full.
30. Mahendiran, B., Muthusamy, S., Sampath, S., Jaisankar, S. N., Popat, K. C., Selvakumar, R., et al. (2021). Recent trends in natural polysaccharide based bioinks for multiscale 3D printing in tissue regeneration: A review. *International Journal of Biological Macromolecules [Internet], 183*, 564–588. Available from: https://linkinghub.elsevier.com/retrieve/pii/S0141813021009521.

31. Mandal, S., Nagi, G. K., Corcoran, A. A., Agrawal, R., Dubey, M., & Hunt, R. W. (2023). Algal polysaccharides for 3D printing: A review. *Carbohydrate Polymers [Internet], 300*, 120267. Available from: https://linkinghub.elsevier.com/retrieve/pii/S0144861722011729.
32. Maturavongsadit, P., Narayanan, L. K., Chansoria, P., Shirwaiker, R., & Benhabbour, S. R. (2021). Cell-Laden Nanocellulose/Chitosan-based bioinks for 3D bioprinting and enhanced osteogenic cell differentiation. *ACS Applied Bio Materials [Internet], 4*(3), 2342–2353. Available from: https://pubs.acs.org/doi/10.1021/acsabm.0c01108.
33. Müller, W. E. G., Tolba, E., Schröder, H. C., Neufurth, M., Wang, S., Link, T., et al. (2015). A new printable and durable N,O-carboxymethyl chitosan–Ca 2+ –polyphosphate complex with morphogenetic activity. *Journal of Materials Chemistry B [Internet], 3*(8), 1722–1730. Available from: https://xlink.rsc.org/?DOI=C4TB01586J.
34. Naghib, S. M., Zarrineh, M., & Moepubi, M. R. (2024). 3D printing Chitosan-based Nanobiomaterials for biomedicine and drug delivery: Recent advances on the promising bioactive agents and technologies. *Current Organic Chemistry [Internet], 28*(7), 510–525. Available from: https://www.eurekaselect.com/228060/article.
35. Naranda, J., Bračič, M., Vogrin, M., & Maver, U. (2021). Recent advancements in 3D printing of polysaccharide hydrogels in cartilage tissue engineering. *Materials (Basel) [Internet], 14*(14), 3977. Available from: https://www.mdpi.com/1996-1944/14/14/3977.
36. Nasser Abdelhamid, H., Sultan, S., & Mathew, A. P. (2023). Binder-free Three-dimensional (3D) printing of cellulose-ZIF8 (CelloZIF-8) for water treatment and carbon dioxide (CO2) adsorption. *Chemical Engineering Journal [Internet], 468*, 143567. Available from: https://linkinghub.elsevier.com/retrieve/pii/S1385894723022982.
37. Osi, A. R., Zhang, H., Chen, J., Zhou, Y., Wang, R., Fu, J., et al. (2021). Three-dimensional-printable thermo/photo-cross-linked Methacrylated chitosan–gelatin hydrogel composites for tissue engineering. *ACS Applied Materials & Interfaces [Internet], 13*(19), 22902–22913. Available from: https://pubs.acs.org/doi/10.1021/acsami.1c01321.
38. Palamidi, A., Koumentakou, I., Michopoulou, A., Bikiaris, D. N., & Terzopoulou, Z. (2024). Optimization of chitosan-gelatin-based 3D-printed scaffolds for tissue engineering and drug delivery applications. *International Journal of Pharmaceutics [Internet], 666*, 124776. Available from: https://linkinghub.elsevier.com/retrieve/pii/S037851732401010X.
39. Puertas-Bartolomé, M., Włodarczyk-Biegun, M. K., del Campo, A., Vázquez-Lasa, B., & San Román, J. (2020). 3D printing of a reactive hydrogel bio-ink using a static mixing tool. *Polymers (Basel) [Internet], 12*(9), 1986. Available from: https://www.mdpi.com/2073-4360/12/9/1986.
40. Rinaudo, M. (2006). Chitin and chitosan: Properties and applications. *Progress in Polymer Science [Internet], 31*(7), 603–632. Available from: https://linkinghub.elsevier.com/retrieve/pii/S0079670006000530.
41. Sadeghianmaryan, A., Naghieh, S., Alizadeh Sardroud, H., Yazdanpanah, Z., Afzal Soltani, Y., Sernaglia, J., et al. (2020). Extrusion-based printing of chitosan scaffolds and their in vitro characterization for cartilage tissue engineering. *International Journal Of Biological Macromolecules [Internet], 164*, 3179–3192. Available from: https://linkinghub.elsevier.com/retrieve/pii/S0141813020342719.
42. Sahranavard, M., Zamanian, A., Ghorbani, F., & Shahrezaee, M. H. (2020). A critical review on three dimensional-printed chitosan hydrogels for development of tissue engineering. *Bioprinting [Internet], 17*, e00063. Available from: https://linkinghub.elsevier.com/retrieve/pii/S2405886619300016.
43. Saylam, E., Akkaya, Y., Ilhan, E., Cesur, S., Guler, E., Sahin, A., et al. (2021). Levodopa-loaded 3D-Printed Poly (lactic) acid/Chitosan neural tissue Scaffold as a promising drug delivery system for the treatment of Parkinson's disease. *Applied Sciences [Internet], 11*(22), 10727. Available from: https://www.mdpi.com/2076-3417/11/22/10727.
44. Servais, B., Mahmoudi, N., Gautam, V., Tong, W., Ibbotson, M. R., Nisbet, D. R., et al. (2024). Engineering brain-on-a-chip platforms. *Nature Reviews Bioengineering [Internet], 2*(8), 691–709. Available from: https://www.nature.com/articles/s44222-024-00184-3.

45. Shariatinia, Z. (2018). Carboxymethyl chitosan: Properties and biomedical applications. *International Journal of Biological Macromolecules [Internet], 120*, 1406–1419. Available from: https://linkinghub.elsevier.com/retrieve/pii/S0141813018337759.
46. Shen, Y., Tang, H., Huang, X., Hang, R., Zhang, X., Wang, Y., et al. (2020). DLP printing photocurable chitosan to build bio-constructs for tissue engineering. *Carbohydrate Polymers [Internet], 235*, 115970. Available from: https://linkinghub.elsevier.com/retrieve/pii/S0144861720301442.
47. Shokrani, H., Shokrani, A., Sajadi, S. M., Khodadadi Yazdi, M., Seidi, F., Jouyandeh, M., et al. (2022). Polysaccharide-based nanocomposites for biomedical applications: A critical review. *Nanoscale Horizons [Internet], 7*(10), 1136–1160. Available from: https://xlink.rsc.org/?DOI=D2NH00214K.
48. Sultan, S., Abdelhamid, H. N., Zou, X., & Mathew, A. P. (2019). CelloMOF: Nanocellulose enabled 3D printing of metal–organic frameworks. *Advanced Functional Materials [Internet], 29*(2), 1805372. Available from: http://doi.wiley.com/10.1002/adfm.201805372.
49. Taghizadeh, M., Taghizadeh, A., Yazdi, M. K., Zarrintaj, P., Stadler, F. J., Ramsey, J. D., et al. (2022). Chitosan-based inks for 3D printing and bioprinting. *Green Chemistry [Internet], 24*(1), 62–101. Available from: https://xlink.rsc.org/?DOI=D1GC01799C.
50. Tamay, D. G., Dursun Usal, T., Alagoz, A. S., Yucel, D., Hasirci, N., & Hasirci, V. (2019). 3D and 4D printing of polymers for tissue engineering applications. *Frontiers in Bioengineering and Biotechnology [Internet], 7*. Available from: https://www.frontiersin.org/article/10.3389/fbioe.2019.00164/full.
51. Tonda-Turo, C., Carmagnola, I., Chiappone, A., Feng, Z., Ciardelli, G., Hakkarainen, M., et al. (2020). Photocurable chitosan as bioink for cellularized therapies towards personalized scaffold architecture. *Bioprinting [Internet], 18*, e00082. Available from: https://linkinghub.elsevier.com/retrieve/pii/S2405886620300099.
52. Turner, P. R., Murray, E., McAdam, C. J., McConnell, M. A., & Cabral, J. D. (2020). Peptide Chitosan/Dextran Core/Shell vascularized 3D constructs for wound healing. *ACS Applied Materials & Interfaces [Internet], 12*(29), 32328–32339. Available from: https://pubs.acs.org/doi/10.1021/acsami.0c07212.
53. Wang, L., Xu, Z., Zhang, H., & Yao, C. (2023). A review on chitosan-based biomaterial as carrier in tissue engineering and medical applications. *European Polymer Journal [Internet], 191*, 112059. Available from: https://linkinghub.elsevier.com/retrieve/pii/S0014305723002422.
54. Wu, Q., Therriault, D., & Heuzey, M.-C. (2018). Processing and properties of Chitosan inks for 3D printing of hydrogel microstructures. *ACS Biomaterials Science & Engineering [Internet], 4*(7), 2643–2652. Available from: https://pubs.acs.org/doi/10.1021/acsbiomaterials.8b00415.
55. Yang, Y., Wang, Z., Xu, Y., Xia, J., Xu, Z., Zhu, S., et al. (2022a). Preparation of Chitosan/recombinant human collagen-based photo-responsive bioinks for 3D bioprinting. *Gels [Internet], 8*(5), 314. Available from: https://www.mdpi.com/2310-2861/8/5/314.
56. Yang, Y., Wu, H., Fu, Q., Xie, X., Song, Y., Xu, M., et al. (2022b). 3D-printed polycaprolactone-chitosan based drug delivery implants for personalized administration. *Materials and Design [Internet], 214*, 110394. Available from: https://linkinghub.elsevier.com/retrieve/pii/S0264127522000156.
57. Yao, Z., Feng, X., Wang, Z., Zhan, Y., Wu, X., Xie, W., et al. (2024). Techniques and applications in 3D bioprinting with chitosan bio-inks for drug delivery: A review. *International Journal of Biological Macromolecules [Internet], 278*, 134752. Available from: https://linkinghub.elsevier.com/retrieve/pii/S0141813024055570.
58. Ye, K., Felimban, R., Traianedes, K., Moulton, S. E., Wallace, G. G., Chung, J., et al. (2014). Chondrogenesis of infrapatellar fat pad derived adipose stem cells in 3D printed Chitosan Scaffold. Reilly G, editor. *PLoS One [Internet], 9*(6), e99410. Available from: https://dx.plos.org/10.1371/journal.pone.0099410.
59. Yu, K.-F., Lu, T.-Y., Li, Y.-C. E., Teng, K.-C., Chen, Y.-C., Wei, Y., et al. (2022). Design and synthesis of stem cell-laden Keratin/Glycol Chitosan methacrylate bioinks for 3D bioprinting.

Biomacromolecules [Internet], 23(7), 2814–2826. Available from: https://pubs.acs.org/doi/10.1021/acs.biomac.2c00191.
60. Zamboulis, A., Michailidou, G., Koumentakou, I., & Bikiaris, D. N. (2022). Polysaccharide 3D printing for drug delivery applications. *Pharmaceutics [Internet], 14*(1), 145. Available from: https://www.mdpi.com/1999-4923/14/1/145.
61. Zhang, X. N., Zheng, Q., & Wu, Z. L. (2022). Recent advances in 3D printing of tough hydrogels: A review. *Composites Part B: Engineering [Internet], 238*, 109895. Available from: https://linkinghub.elsevier.com/retrieve/pii/S1359836822002748.
62. Zhong, C., Wu, J., Reinhart-King, C. A., & Chu, C. C. (2010). Synthesis, characterization and cytotoxicity of photo-crosslinked maleic chitosan–polyethylene glycol diacrylate hybrid hydrogels. *Acta Biomaterialia [Internet], 6*(10), 3908–3918. Available from: https://linkinghub.elsevier.com/retrieve/pii/S174270611000200X.

Chitosan-Based Systems in Wound Dressing Applications

David Servín de la Mora López, Jesús Rubén Rodríguez-Núñez, Judith Fortiz-Hernandez, Itzel Y. López Peña, Rogerio R. Sotelo-Mundo, and Tomás J. Madera-Santana

Abstract Chitosan, a biopolymer produced from chitin, is gaining a lot of interest in biomedical research due to its intrinsic biocompatibility, biodegradability, and antimicrobial properties, making it suitable for wound dressing applications. This chapter provides a comprehensive overview of chitosan-based systems for wound treatment solutions. It begins with an analysis of chitosan's chemical composition and physicochemical properties, which support its biological functionality. The antimicrobial mechanism of chitosan is thoroughly investigated, including its capacity to disrupt microbial membranes, interact with DNA, and influence fungal cells. Chitosan's interaction with mammalian cells involved in wound healing is also examined, with a focus on its immunomodulatory and regenerative potential. The molecular weight, degree of deacetylation, pH, and origin of chitosan are examined as key parameters determining its antimicrobial activity. The wound healing process is divided into three main stages: inflammation, proliferation, and remodeling. Finally, this chapter presents recent advances in wound dressings formulated with chitosan in conjunction with natural or synthetic polymers and bioactive agents. These composite materials have improved antimicrobial, structural, and regenerative properties, making them potentially useful for the treatment of acute and chronic wounds.

D. S. de la Mora López
Departamento de Biotecnología y Ciencia de los Alimentos, Instituto Tecnológico de Sonora, Obregón, Sonora, Mexico

J. R. Rodríguez-Núñez
Programa de Biotecnología, División de Ciencias de la Salud e Ingenierías, Universidad de Guanajuato Campus Celaya-Salvatierra, Celaya, Guanajuato, Mexico

J. Fortiz-Hernandez · I. Y. López Peña · T. J. Madera-Santana (✉)
Coordinación de Tecnología de Alimentos de Origen Vegetal, Centro de Investigación en Alimentación y Desarrollo, Hermosillo, Sonora, Mexico
e-mail: madera@ciad.mx

R. R. Sotelo-Mundo
Coordinación de Tecnología de Alimentos de Origen Animal, Centro de Investigación en Alimentación y Desarrollo, Hermosillo, Sonora, Mexico

S. Jana (ed.), *Chitosan-Based Drug Delivery Systems*, AAPS Introductions in the Pharmaceutical Sciences 16, https://doi.org/10.1007/978-3-032-08347-0_15

Keywords Chitosan · Biodegradable · Biopolymer · Biocompatibility · Antimicrobial · Chitin · Natural polymer

1 Introduction

Wound healing is a complex, multi-phase biological process that includes hemostasis, inflammation, proliferation, and remodeling. Developing advanced wound dressings is critical to enabling this process, particularly for managing chronic wounds and preventing infections. Chitosan, a natural polysaccharide derived from chitin, has received significant attention in biomedical applications due to its biocompatibility, biodegradability, and inherent antimicrobial properties [67, 77]. The distinctive physicochemical properties of chitosan, including its cationic capacity and ability to form films, sponges, and hydrogels, make it an ideal candidate for wound dressing materials. These forms can maintain a moist wound environment, promote cell proliferation, and allow for the controlled release of therapeutic agents [86].

Chitosan's antimicrobial activity is attributed to its ability to disrupt microbial cell membranes, chelate metal ions, and inhibit DNA transcription. Its antimicrobial activity is influenced by molecular weight, degree of deacetylation, pH, and other bioactive compounds. Moreover, chitosan can modulate the immune response by stimulating cytokine production, therefore enhancing the wound-healing process [52].

Recent advancements have focused on developing chitosan-based composite dressings by incorporating natural and synthetic polymers to enhance mechanical strength, bioactivity, and healing efficiency. These composites aim to mimic the extracellular matrix, providing structural support and promoting tissue regeneration.

This chapter discusses the composition and properties of chitosan, its mechanisms of action against pathogenic microorganisms, factors influencing its antimicrobial effects, the stages of wound healing, and the development of chitosan-based wound dressings incorporating natural and synthetic polymers. By exploring these themes, we aim to provide a comprehensive understanding of chitosan's significance in modern wound care applications.

2 Composition and Properties

2.1 *Chitosan: Molecular and Structural Composition*

Chitin is a linear biopolymer composed of the monomers of N-acetyl-D-glucosamine and D-glucosamine residues linked by β-(1,4) glycosidic bonds. This has a high molecular weight and crystalline structure and is present in the exoskeletons of crustaceans, insects, and fungi. Its function is to provide support and shape to the exoskeletons of arthropods [65]. Regarding its beneficial characteristics, it is

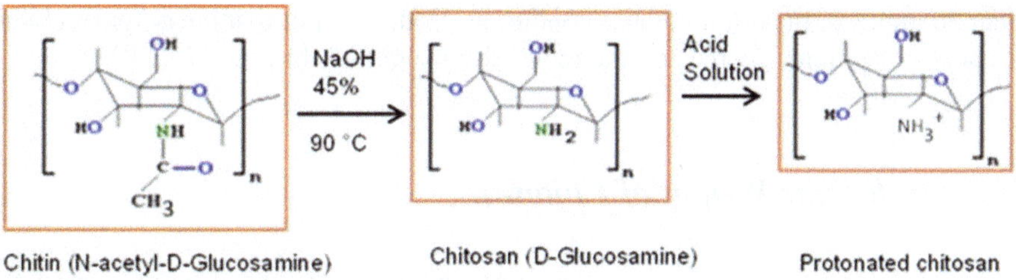

Fig. 1 Chemical deacetylation method for obtaining chitosan from chitin

non-toxic, biodegradable, and abundant in nature. However, it is highly insoluble in water, has low reactivity, and only dissolves in concentrated acids, fluoroalcohols, and 5% lithium chloride solutions, limiting its range of applications [70].

Chitosan is a linear polycationic polymer derived from chitin through a deacetylation process and composed of D-glucosamine units linked by β-(1,4) glycosidic bonds (Fig. 1). This biopolymer presents free functional groups as amino(-NH_2) and hydroxyl (-OH), which allows it to form new covalent bonds through esterification, reductive amination, and etherification reactions [25].

Chitosan is a non-toxic, biocompatible, and biodegradable polymer; however, it is insoluble in water but soluble in acidic solvents such as diluted acetic, lactic, and formic acid, as well as some solvents like dimethyl sulfoxide and p-toluene sulfonic acid. In acidic solutions—below pH 6.5—the free amine groups (-NH_2,with pKa 6.5) of chitosan are protonated, acquiring a positive formal charge (-NH_3^+), thus obtaining a cationic biopolymer that allows the ability to solubilize in acid solutions [47]. Other important chemical-structural parameters are the degree of deacetylation, the molecular weight, and the crystalline structure that chitosan presents, since these are related to the functional properties of chitosan, such as antimicrobial and antifungal capacity, antioxidant capacity, solubility, viscosity, and mechanical properties [11].

2.2 *Degree of Deacetylation of Chitosan*

The difference between chitin and chitosan refers to the degree of acetylation of the biopolymer. When it is less than 50%, it is called chitin; when it is above 50%, it is called chitosan (free –NH_2 groups). However, in practice, the term chitosan refers to polymers soluble in diluted acids and chitin when it is not soluble in diluted acids. Achieving a high degree of deacetylation depends on several factors, such as alkali concentration, particle size, treatment time, chitin density, the degree of chitin crystallinity, and even prior treatments. In practice, the maximum level of deacetylation achieved ranges from 75 to 85% with a single alkaline treatment process [25, 47]. The conventional deacetylation method is chemical, using strong concentrated bases (NaOH or KOH, 45%) at high temperatures (>60 °C) for more than 30 minutes. In addition, these conditions produce rupture of the polymer chains (depolymerization), decreasing the molecular weight of chitosan [70, 78].

2.3 Molecular Weight of Chitosan

The molecular weight of chitosan is an important parameter because it influences the physicochemical properties of chitosan, such as rheology, crystallinity, viscosity, antioxidant, antimicrobial, and antifungal capacity. Chitosan can be classified into three ranges based on molecular weight: high (310–375 kDa), medium (190–350 kDa), and low molecular weight (150–250 kDa). This range includes the chito-oligosaccharides ($\leq$10 monomer units of D-glucosamine) [83]. However, high molecular weight chitosan tends to intertwine and tangle, leading to increased intermolecular forces and reduced water solubility, while chitosan with low molecular weight enhances biological activities. For this reason, work is being done on methods to reduce the molecular weight of chitosan, and the most widely used is the chemical method through acid hydrolysis, using strong acids (HNO_3, HCl, and HF) and weak acids (CH_3COOH and H_3PO_4). The latter do not generate environmental problems and do not leave toxic residues. In addition, methods with oxidizing agents such as hydrogen peroxide (H_2O_2), electrochemical degradations, and ultraviolet light are being tested for depolymerization of chitosan chains [78, 83].

2.4 Chitosan Properties

Chitosan has multiple properties of interest in wound healing, but the most important is its antimicrobial, biodegradable, and biocompatible properties [80]. The antimicrobial properties of chitosan are mainly due to the presence of charged amino groups in the molecular structure of chitosan, which can interact with negatively charged compounds of the bacterial cell wall (Fig. 2). In Gram (+) bacteria, there are negatively charged components such as teichoic acids and peptidoglycans. In Gram (−) bacteria, the main compounds are phospholipids and lipopolysaccharides. In both cases, the amino groups of chitosan cause cell rupture by interacting electrostatically with the negatively charged groups of the bacteria, causing the release of intracellular material and consequently bacterial death [4, 66].

In addition, chitosan may be able to penetrate through the cell nucleus and interfere with the synthesis of mRNA and vital proteins [7]. Another inhibition mechanism that has been studied is the ability of chitosan to form chelates with other trace elements or compounds. This mechanism inhibits enzymatic activity due to interactions between chitosan and DNA that alter the mRNA synthesis [72]. Some authors have shown chitosan can provide antimicrobial activity against certain strains of Gram (+) bacteria such as *Staphylococcus aureus, S. epidermidis, Enterococcus faecalis, Listeria monocytogenes, Lactobacillos plantarum,* and *Bacillus*

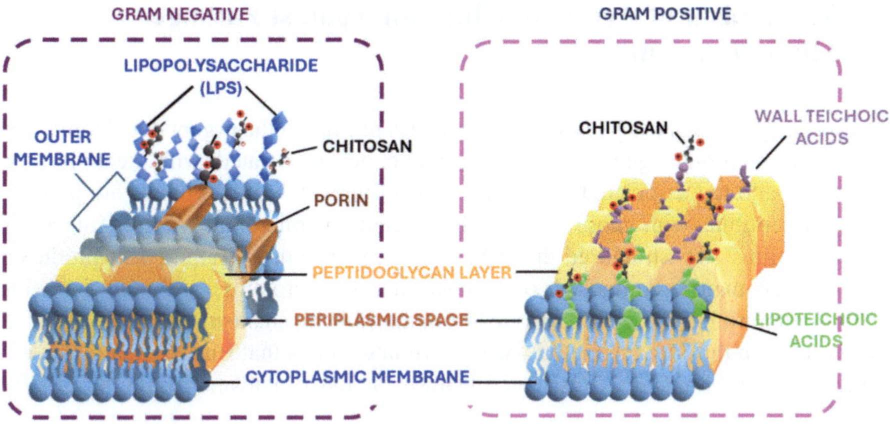

Fig. 2 Interaction of chitosan with Gram-positive and Gram-negative bacteria. Positively charged chitosan (in red) electrostatically bonds to lipopolysaccharides (Gram −) and teichoic acids (Gram +), leading cell disruption and potential rupture of the cytoplasmic membrane

megaterium. With respect to Gram (−) bacteria, there are some strains such as *Proteus mirabilis, Salmonella enteritidis, Enterobacter aerogenes, Pseudomonas aeruginosa, Escherichia coli,* and *Klebsiella pneumoniae*. In addition, chitosan can provide antimicrobial capacity against different fungi species, such as *Candida albicans* and *C. parapsilosis* [31, 40]. All this made the chitosan a material of interest in biomedicine for wound healing purposes.

The biodegradable properties of chitosan are associated with its affinity to be hydrolyzed by lysozyme (a lymphoid tissue enzyme), which first degrades the chitosan chain molecules in oligosaccharides, and subsequently into monomers of D-glucosamine and N-acetyl-D-glucosamine [20, 35, 71]. The latter are precursor metabolites of glycosaminoglycans that are present in the extracellular matrix of connective tissue. This factor widely contributes to the fact that chitosan is considered a material with biocompatible properties. However, the biocompatibility of chitosan is also due to its ability to properly interact with different cell types such as fibroblasts, keratinocytes, endothelial cells, chondrocytes, and hepatocytes [35].

In wound healing, chitosan may have different functions and biological properties of interest. It has been demonstrated that chitosan may be able to adhere to extracellular matrix proteins and stimulate the release of different types of growth factors that favor the healing process [71]. Likewise, chitosan can provide a provisional matrix that serves as a support to stimulate the adhesion, migration, and proliferation processes of different types of cells [35]. On the other hand, it has been revealed that chitosan can suppress the inflammatory process by accelerating the functionality of mRNA for the formation of TGF-β1 and type III collagen. This factor helps to significantly increase the rate of healing [9].

3 Mechanism of Action of Chitosan Against Pathogen Microorganisms

The global growth in drug-resistant bacteria has prompted the urgent development of novel antimicrobial agents due to the challenges associated with treating such infections and the associated healthcare expenditures. Chronic wounds are particularly susceptible, with more than 70% affected by microbial biofilm infections [49]. Biofilms are structured microbial aggregates encased in a self-produced exopolymer matrix composed of polysaccharides, proteins, and nucleic acids [36]. Their antibiotic resistance derives from multiple mechanisms, including limited drug diffusion through the matrix, survival of latent cells that survive treatment, and horizontal gene transfer of resistance features across organisms inside the biofilm [51, 76].

Quorum-sensing molecules—autoinducers that coordinate virulence factor expression—play a critical role in the regulation of biofilm-forming pathogens such as *Staphylococcus aureus* and *Pseudomonas aeruginosa*. These biofilms disrupt the wound-healing process by blocking immunological responses, inhibiting leukocyte activity, and preventing the penetration of antimicrobial agents [51]. They impair phagocytosis and the production of reactive oxygen species (ROS) by immune cells. As antimicrobial resistance grows, natural and synthetic antimicrobial agents have gained attention. The European Centre for Disease Prevention and Control reported that antibiotic-resistant infections cause roughly 33,000 fatalities in Europe alone each year [24]. Among promising candidates, antimicrobial polymers have emerged as a new generation of agents, with chitosan being especially noteworthy. Antimicrobial polymers, particularly chitosan, have emerged as potentially interesting candidates for a new generation of drugs.

Chitosan, a partially to fully deacetylated form of chitin, is derived from crustacean shells, fungi, and insect exoskeletons [70]. Commercially, seafood processing byproducts primarily yield chitin, thereby helping to reduce environmental pollution from discarded shell waste [46]. With global crustacean farming producing almost 7 million tons per year, this waste is a valuable source of chitin and other bioactive compounds [28, 33]. Chitosan's properties, such as solubility in acidic environments and polycationic nature, are determined by its degree of acetylation and molecular weight. These characteristics impact its bioactivity, viscosity, and biodegradability. Its noteworthy biological functions—biocompatibility, low toxicity, antimicrobial, antioxidant, and anticancer activities—make it highly versatile across pharmaceuticals, biomedicine, agriculture, food, cosmetics, and waste treatment applications [2, 22, 82].

Chitosan, a natural polymer, has been extensively studied for its antimicrobial properties since the 1980s. Early research identified both bacteriostatic and bactericidal effects, but it often did not distinguish between them. More recent research emphasizes its bacteriostatic activity, despite the exact mechanism remaining unclear [51]. However, the literature has proposed several generally accepted antimicrobial mechanisms, which are frequently cited to explain chitosan's action against various pathogens. This highlights chitosan's potential as an effective biopolymer in antimicrobial applications.

3.1 Cell Membrane or Cell Wall Disruption

One of the most frequently discussed mechanisms of chitosan's antimicrobial activity is the loss of cell membrane integrity via electrostatic interactions. The positively charged amino groups in chitosan interact with negatively charged microbial cell membranes, increasing permeability and cell lysis [10, 63]. Chitosan's antimicrobial properties are related to its protonated amino groups ($-NH_3^+$), which become active in acidic conditions below its pKa of 6.3, rendering the polymer soluble and positively charged [81]. This charge facilitates interaction with negatively charged microbial surfaces, such as teichoic acids in Gram-positive bacteria, lipopolysaccharides in Gram-negative bacteria, and similar structures in fungi and viruses. Fu et al. [29] showed that teichoic acids in *Staphylococcus aureus* interact with chitosan derivatives, resulting in considerable bacterial inhibition, up to nearly 100% in some cases. The membrane damage in *Escherichia coli* after a chitosan treatment was detected by detecting extracellular DNA and proteins via UV absorbance at 260 nm and 280 nm and by observing disrupted membranes using transmission electron microscopy [43]. Furthermore, the quaternized chitosan shows an enhanced positive charge that exhibits greater antibacterial activity against both Gram-positive and Gram-negative strains compared to native chitosan [34].

3.2 Interaction with the DNA of Microorganisms

Chitosan has been found to inhibit microbial activity through a variety of mechanisms, including interference with protein biosynthesis. Galván Márquez et al. [30] demonstrated this effect by using β-galactosidase expression assays in yeast cells. Their study revealed that at a chitosan concentration of 1.25 mg/mL, β-galactosidase activity was reduced by 13%, indicating a considerable impact on protein synthesis processes. This finding supports the idea that beyond membrane disruption, chitosan can affect intracellular functions critical for microbial survival.

3.3 Interactions with Fungal Cells

Chitosan has antifungal effects through multiple mechanisms that disrupt fungal growth and physiology. One possible mechanism involves the formation of a dense film on the surface of fungal cells, which may hinder nutrient intake and gas exchange and ultimately prevent fungal growth. This physical barrier may also result in host defense responses such as chitinase accumulation, callus and lignin synthesis, and production of proteinase inhibitors [17]. These metabolic changes are part of the plant's immune defense and may enhance resistance against fungal infection. Another theory is that fungi, particularly yeast cells, may adapt to or recover after extended exposure to chitosan. This adaptation could be related to the secretion of chitinases, which can degrade chitosan, particularly when deacetylated, lowering its antifungal activity [45].

Oliveira Junior et al. [62] studied the antifungal effects of chitosan by examining morphological changes in fungal mycelia. The study found significant antifungal activity against *Alternaria alternata* (500 μg/mL), *Botrytis cinerea* (1000 μg/mL), and *Penicillium expansum* (500 μg/mL) in chitosan samples with varying molar fractions of acetyl groups (FA 0.16 and 0.18) and degrees of polymerization (DP 1089 and 1242). The therapy reduced fungal growth and caused structural anomalies such as mycelial aggregation, increased branching, decreased hyphal size, and cell wall swelling. Additionally, Yien Ing et al. [84] discovered that *Candida albicans* was more sensitive to chitosan nanoparticles than other fungal species. This heightened sensitivity may be attributed to the strong electrostatic interactions between the positively charged chitosan and negatively charged components on the fungal cell surface. These findings collectively demonstrate that chitosan's antifungal action involves both structural and biochemical mechanisms. Physicochemical properties such as the degree of polymerization and acetylation, along with the fungal species targeted, influence its efficacy. The ability of fungi to degrade chitosan through enzymatic activity also presents a potential limitation to its long-term antifungal effectiveness.

3.4 Interactions with Mammalian Cells During Wound-Healing

In addition to its antimicrobial and antifungal activities, chitosan exhibits inherent biocompatibility and low cytotoxicity toward living cells and tissues. Its biocompatibility has been extensively evaluated in vitro using various cell types, including fibroblasts, keratinocytes, hepatocytes, as well as myocardial and endothelial cells.

While wounds generally heal spontaneously, some people, such as diabetics, may experience impaired wound healing, resulting in chronic, non-healing ulcers that result in prolonged discomfort and psychological distress. These disorders frequently need lengthier treatment times and contribute to higher healthcare expenditures. As a result, there is increased interest in natural medicinal agents that can successfully enhance wound healing while remaining accessible and affordable. Chitosan has emerged as a viable candidate due to its several bioactivities. It promotes wound healing by activating critical immunological and structural cells, such as inflammatory cells, macrophages, and fibroblasts, which accelerates the early inflammatory phase. This early activation shortens the inflammatory stage and expedites the transition to the proliferative phase of wound healing. Furthermore, chitosan has been shown to influence granulation tissue formation and angiogenesis, facilitating the orderly deposition of collagen fibers and promoting the regeneration of injured dermal structures [51, 74]. When assessing chitosan's wound-healing efficacy, its hemostatic and analgesic qualities should also be considered. Chitosan is unique in that it induces hemostasis independently of the typical coagulation pathways. Furthermore, it modulates the activity of neutrophils and macrophages, further contributing to re-epithelialization and fibroplasia during tissue repair.

Chitosan has structural similarities to glycosaminoglycans, which are important elements of the extracellular matrix (ECM), making it a promising candidate for skin tissue engineering. Its capacity to operate as a scaffold for the immobilization of bioactive compounds, such as cell-adhesive peptides, has the potential to enhance cellular responses and adherence. Karakeçili et al. [37] reported the covalent immobilization of murine epidermal growth factor (EGF) onto chitosan membranes. Because EGF plays an important role in boosting fibroblast proliferation, their findings showed that EGF-functionalized chitosan increased the proliferation of L929 murine fibroblast cells.

Although several studies have investigated chitosan's interactions with mammalian cells involved in wound healing, there is still a knowledge gap about how chitosan is recognized by the innate immune system and the nature of the resultant immune response. Polysaccharides, including chitosan, are identified by pattern recognition receptors; therefore, their size and molecular structure can influence immunogenicity. The absence of chitin and chitosan in mammalian cells makes these biopolymers visible to immune surveillance mechanisms. However, mammalian chitinases may degrade chito-oligosaccharides. A recent study has shown that chitosan—but not its acetylated form, chitin—can induce inflammatory responses by activating the NLRP3 inflammasome in a phagocytosis-dependent manner. This activity is considerably reduced when chitosan is converted to chitin by acetylation, demonstrating the relevance of its chemical structure in immunological activation [53]. Given chitosan's extensive use in biomedical applications, further investigation into its immunological interactions remains critical.

4 Factors Influencing Chitosan's Antimicrobial Effect

Various studies have demonstrated the antimicrobial activity of chitosan through inhibition tests on a wide variety of bacteria, both Gram-positive and Gram-negative (Fig. 3). In in vitro studies, the antimicrobial effect of chitosan has been reported on different species of Gram-negative bacteria (*Escherichia coli*, *Salmonella typhimurium*) and Gram-positive bacteria (*Bacillus cereus*, *Staphylococcus aureus*, *Listeria monocytogenes*) [70, 73]. However, there is debate about the mechanisms of the antimicrobial effect and the physicochemical parameters that affect the chitosan activity; among the most studied are the molecular weight, degree of acetylation,

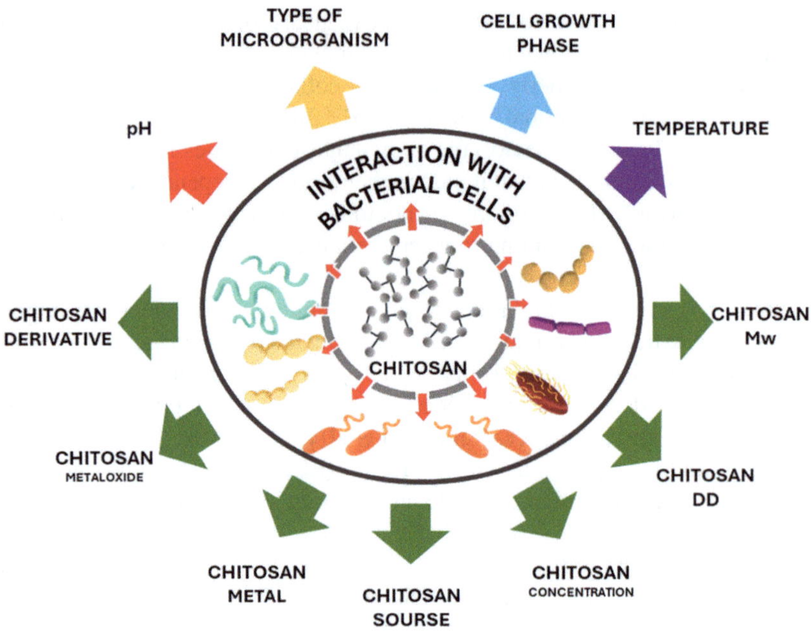

Fig. 3 Interactions between chitosan and bacterial cells depend on several factors that affect antimicrobial activity

degree of polymerization, positive charge content, and pH conditions, as well as the microbial structural characteristics and cell age [16, 70].

4.1 Influence of Molecular Weight and Degree of Deacetylation

Since the 90s, there has been a discussion about the effect of the molecular weight of chitosan on antimicrobial activity and, in summary, it has been reported that lower molecular weight chitosan has better antimicrobial activity than high molecular weight chitosan; however, the exact molecular weight ranges vary between different studies, reporting ranges from 1.5 to 300 kDa [44, 87], indicating a maximum antimicrobial activity between 50 and 90 kDa [50]. Some studies suggest that the molecular weight activity relationship varies among different bacterial species [61], and others mention that Gram-positive bacteria show a positive correlation between molecular weight and antimicrobial activity, while Gram-negative bacteria exhibit the opposite trend [50].

In this sense, Zheng and Zhu [87] reported that for *S. aureus* (Gram-positive bacteria), as the molecular weight of chitosan increased, the antimicrobial effect was enhanced, but in contrast, the effect on *E. coli* was weakened as the molecular weight increased. The main reason might be that chitosan of higher molecular weight forms a film around the membrane, which inhibits nutrient adsorption. Also, Abedian et al. [1] mentioned a higher antimicrobial activity with chitosan of high molecular weight (600–800 kDa) than chitosan with low molecular weight (100 kDa) against Gram-positive bacteria (*Streptococcus mutans* and *Streptococcus sobrinus*).

Several authors have reported the influence of the deacetylation degree on antimicrobial activity due to its relationship with solubility and the development of its net charge [3, 23]. Although it is established that an increased degree of deacetylation generally increases antimicrobial activity, this topic is not currently under debate because it is well-proven [50]. Increasing the degree of deacetylation (removal of acetyl groups from chitin) increases the number of free amino groups ($-NH_2$) in the chitosan structure, which are capable of being protonated ($-NH_3^+$), thus giving it a cationic character. This allows chitosan to interact with the negative charge of the cell surface of bacteria, causing a superficial agglutination that causes the rupture of the cell membrane [31].

4.2 *Influence of pH and Type of Bacteria on Antimicrobial Activity*

Some authors have conducted studies on the antimicrobial activity of chitosan, evaluating various parameters. Regarding the type of microorganism, it has been reported that chitosan has a greater antimicrobial effect against Gram-negative bacteria, and this is attributed to the fact that they have lipopolysaccharides with phosphate and pyrophosphate groups, which provide a negative charge density greater than that observed in Gram-positive bacteria. This property promotes interaction with the protonated amino groups (cationic character) of chitosan [31].

In this sense, Younes et al. [85] mentioned that the antibacterial activity of chitosan against Gram-negative bacteria (*Escherichia coli*, *Pseudomonas aeruginosa*, *Klebsiella pneumoniae,* and *Salmonella typhi*) increases with decreasing deacetylation degree, molecular weight, and pH. However, for Gram-positive bacteria (*Staphylococcus aureus*, *Bacillus cereus*, *Enterococcus faecalis,* and *Micrococcus luteus*), the influence of deacetylation degree and pH was lower than with Gram-negative bacteria; even for some of these bacteria, better efficiency was obtained with chitosan of high molecular weight. Additionally, Chang et al. [16] reported that chitosan exhibits good antimicrobial activity against *Escherichia coli* and *Staphylococcus aureus* at acidic pH, even when the molecular weight was high (>2.93 kDa), and this activity is independent of temperature and microorganism. Also, it was reported that at pH 5.0 and 6.0, chitosan exhibited good water solubility, and zeta potential decreased with the molecular weight, whereas the solubility and zeta potential decreased with increasing molecular weight at pH 7.0, which was attributed to the decrease in antimicrobial activity of chitosan at pH close to neutral. Furthermore, Másson [50] shows a summary of various studies where it is observed that most of the studies of antimicrobial capacity with chitosan are carried out at pH 6 or close to 6. This is because the cationic character of chitosan is maintained, and the best antimicrobial capacity is obtained.

5 Wound Healing: Concept and Phases

Wound healing requires a process that involves four overlapping phases: hemostasis, inflammation, proliferation, and cell maturation. In these stages, different factors will promote biological and enzymatic processes to stimulate healing [79]. The healing process begins when the tissue suffers trauma or injury, which compromises its physical integrity [39]. The generation of the trauma will result in a series of physiological and metabolic changes, with the main objective of avoiding significant blood loss. In the hemostasis stage, certain vascular cells in the smooth muscle will start the process of vasoconstriction [79]. At the same time, platelets arrive at the site of injury and adhere to the extracellular matrix of the damaged tissue; this stage also activates the coagulation factors to promote thrombogenesis. Here, platelets promote the release of fibrin, fibronectin, vitronectin, and thrombospondin for clot formation [54]. This blood clot is responsible for stopping bleeding and protecting the area of injury from the outside.

The second phase of healing is inflammation, which occurs shortly after the clot is generated. Here, platelets begin to release some chemotactic components to attract monocytes to the wound site. Some of the main chemotactic components are transforming growth factor β (TGF-β), complement systems C3a and C5a, and formyl methionyl peptides [21, 79]. When neutrophils reach the injury site, they cause the release of proteases, tumor necrosis factor α (TNF-α), oxygen free radicals, inducible nitrous oxide synthase (INOS), and proinflammatory cytokines (IL-1β, IL-6, IL-12). These factors together help induce the effects of bacterial phagocytosis, as well as apoptosis in monocytes. Proinflammatory cytokines have chemoattractant effects that allow macrophages to arrive at the site of injury so that they can affect the phagocytosis of apoptotic bodies. Macrophages in the lesion secrete some growth factors, such as fibroblast growth factor (FGF), TGF-α, and activating mechanisms of endothelial cells, fibroblasts, and keratinocytes [54]. These factors provide responses that initiate the cell proliferation stage [21].

In the proliferation phase, platelet-derived growth factors (PDGF) and TGF-β are produced, which stimulate the migration of fibroblasts and myofibroblasts to the wound bed. The cells proliferate, forming an extracellular matrix that is made up of fibrin, fibronectin, glycosaminoglycans, proteoglycans, hyalurans, and procollagens types I and III [5, 21, 27]. These components together give rise to granulation tissue, which is vital for the processes of cell migration and tissue restoration. Angiogenesis also occurs during the cell proliferation stage, where new capillary vessels originate from surrounding vessels that were not affected [5, 54]. The reconstructed vessels help facilitate the transport of other cells to reconstitute the tissues that were damaged. Subsequently, keratinocytes begin to migrate and proliferate into the granulation tissue matrix to reconstitute the epidermal layer and completely shrink the wound [15, 32].

The final phase of wound healing consists of cell maturation, where the new tissue that will form the scar develops. This stage is a process that can last months or even years. The activity of proinflammatory cells and epithelial cells decreases as the wound contracts. Over time, type III procollagen (synthesized by fibroblasts and myofibroblasts) matures into type I collagen through the action of proteases found in the tissue [38]. Type I collagen remodels and begins to acquire a better structural organization, which provides better tensile properties to the tissue [5, 32].

6 Wound Dressings Based on Chitosan, Natural and Synthetic Polymers

Conventional dressings serve to safeguard wounds against infections; the gauzes typically employed in wound therapy and cleansing are composed of textiles like cotton or synthetic fibers. However, these possess several disadvantages, including insufficient moisture regulation to avoid tissue death from dehydration, as well as potential mechanical injury resulting from dressing removal. Consequently, efforts have been directed toward the development of dressings that serve as a protective barrier against environmental factors, preventing microbial infections, while also possessing properties that promote tissue regeneration [77]. An ideal wound dressing must have biocompatibility, biodegradability, antibacterial qualities, mechanical characteristics similar to the target tissue, and a three-dimensional structure that mimics the extracellular matrix. There are different materials developed, such as films, hydrogels, nanofibers, and 3D printing [77].

Natural polymers from plants, microorganisms, or animals are commonly used to make wound dressings because, unlike most synthetic polymers, they are compatible with the body and break down in a way that the body can handle, which helps reduce immune system reactions. Among the natural polymers, chitosan, whose properties and remarkable characteristics have been mentioned in this chapter, stands out.

Koosehgol et al. [41] developed chitosan/polyethylene glycol fumarate/thymol (CS/PEGF/TH) films using thymol (0.6–1.8% v/v) and PEGF (20% w/w) as a crosslinker via solvent casting. The formulation containing 1.8% thymol exhibited optimal physicochemical and antimicrobial properties. The films showed suitable mechanical strength, high permeability to water and air, and a porous morphology. In vitro assays demonstrated strong antimicrobial activity against both *S. aureus* and *E. coli*, confirming the potential of CS/PEGF/TH films for wound dressing applications.

Rieger and Schiffman [68] developed chitosan/poly(ethylene oxide) (CS/PEO) mats containing cinnamaldehyde (CAL) via electrospinning, demonstrating strong antibacterial activity against *E. coli* and *P. aeruginosa*, suggesting their potential as CAL delivery systems. In a subsequent study, Rieger et al. [69] investigated the influence of chitosan's molecular weight and degree of acetylation (DA) on mat morphology and solution behavior. They found that lower DA improved CS/PEO/CAL solution properties, while higher DA favored greater incorporation of hydrocinnamic alcohol (H-CIN).

Vitamins such as E, C, A, and B_3 exhibit antioxidant and anti-inflammatory properties that make them promising agents for wound-healing applications [14]. Motealleh [56] incorporated vitamin C and phenytoin sodium (1%) into a N, O-carboxymethyl chitosan (CMCS) and PEO blend via electrospinning. In vivo studies demonstrated enhanced wound closure and healing, with histological analyses confirming effective epidermal regeneration. Similarly, Madni et al. [48] formulated vitamin C within chitosan/polylactic acid (PLA) membranes, using glycerol and polyethylene glycol (PEG) as compatibilizers to address membrane fragility, further supporting vitamin C's utility in advanced wound dressing systems.

Vitamin E (α-tocopherol), commonly found in olive and sunflower oils, has shown significant wound-healing potential. Ehterami et al. [26] encapsulated α-tocopherol in porous cross-linked chitosan/alginate hydrogels and demonstrated enhanced wound closure in a rat model, suggesting its efficacy for treating skin injuries. Similarly, Bonferoni et al. [12] incorporated α-tocopherol into chitosan oleate (CS-OA) using nanoemulsions. Their results indicated that both CS-OA and α-tocopherol promoted keratinocyte and fibroblast proliferation. Additionally, ex vivo skin biopsies confirmed the suitability of CS-OA-α-tocopherol nanoemulsions for wound healing applications, highlighting the antioxidant's therapeutic potential in regenerative medicine.

Aloe vera (AV), a perennial plant from the *Aloe* genus, is widely cultivated in tropical regions for its agricultural and medicinal applications. It is commonly used in treating skin disorders and promoting wound healing by enhancing fibroblast growth and proliferation. Several studies have examined AV's incorporation into marine-based wound dressing materials [75]. Pathalamuthu et al. [64] explored a novel Spirograph-Based Mechanical System (SBMS) for collecting chitosan/polyethylene oxide/Aloe vera fiber mats, demonstrating superior mechanical properties and improved wound healing in vitro and in vivo compared to conventional methods. Additionally, Naseri-Nosar et al. [60] developed polyvinyl alcohol/chitosan (PVA/CS) nanofibers incorporating AV via wet electrospinning, subsequently coated with recombinant human erythropoietin-loaded chitosan nanoparticles. In vivo studies revealed significantly enhanced wound closure with these non-toxic, dual-loaded fibers. These findings underscore the promising role of AV-based composite systems (PVA/CS) in advanced wound care applications.

Chen et al. [18] developed and characterized an injectable thermosensitive chitosan/quaternary chitosan/$NaHCO_3$/β-glycerophosphate hydrogel, loaded with polyhexamethylene biguanide (PHMB), to promote wound healing in a marine environment. The hydrogel was able to release up to 79.9% PHMB in 6 h and sustained release for more than 48 h, with a high swelling rate (794%) and an excellent porous structure promoting wound healing. The hydrogel also showed excellent biocompatibility, antibacterial and adhesive properties, and mechanical properties, including Young's modulus, which support its viability as a wound dressing material, especially for the treatment of microbial infections associated with seawater-contaminated wounds.

Chen et al. [19] developed an injectable self-healing hydrogel based on quaternized chitosan (HACC) via electrostatic interaction between HACC and carboxymethyl chitosan (CMCS) to enhance wound treatment for older adults. The study evaluated the influence of different HC hydrogel compositions on the physicochemical characteristics of the hydrogel and showed exceptional water absorption capacity, adequate degradation time, excellent self-healing properties, and outstanding injectability, making it particularly suitable for application in infected wounds in aged rats.

Recently, a group of researchers specifically created the therapeutic nanoformulation OXY-Exo Aloe to address diabetic wounds. This formulation uses nanovesicles derived from aloe vera (*Aloe barbadensis*), ginger (*Zingiber officinale*) and the fruit of the neem tree (*Azadirachta indica*), which are encapsulated in a nanomembrane based on chitosan and polyvinyl alcohol (PVA) [55]. This biological macromolecule is known for its biocompatibility, biodegradability, and structural integrity, so its combination with the nanomembrane optimizes its performance, improving its protective and healing properties in diabetic patients. The chitosan/PVA matrix plays a key role in ensuring structural integrity, moisture retention, and stabilization of the medicinal nanovesicles, creating a bioactive and sterile environment that promotes healing. By reducing oxidative stress, improving oxygenation, and promoting a sterile, bioactive environment, this nanomembrane offers a promising therapeutic approach for the management of chronic diabetic wounds.

Murtaza et al. [57] employed nanotechnology using nanoparticles that contribute to improving the healing process by facilitating the complete transition of the healing phases. They synthesized nickel nanoparticles using an environmentally friendly process with turmeric plant extract (*Curcuma longa*), doping with chitosan and PEG. The results for these CS-Ni showed a larger particle size (944 nm) and a more negative zeta potential (−2.55 mV) compared to PEG-Ni (460 nm, −4.19 mV). PEG-Ni (smaller particle size) showed slight efficacy against *E. coli* in antibacterial tests compared to CS-Ni. In the same sense, Lei et al. [58] reported a chitosan-based oriented hydrogel of *Periostracum cicadae*; it was made by an electro assembly method, and silver nanoparticles (AgNP) formed in situ on the O-CH hydrogel without the need for additional reducing agents or stabilizers, obtaining O-CH@Ag_2O, which showed excellent biocompatibility and enhanced antibacterial properties. In a rat model with deep skin wounds infected with *S. aureus*, using O-CH@Ag20 had a strong positive effect on the early stages of healing.

Bhardwaj et al. [8] made a biocompatible film based on chitosan/alginate (CAT) incorporating the drug ursodeoxycholic acid (UDC) and an iron oxide/selenium nanocomposite (SeIO) for applications in wound healing, and they found that the films showed good antioxidant (90%) and bactericidal activity against *E. coli* (86%) and *S. aureus* (89%). In addition to its antibacterial properties, the film showed approximately 91%, thus demonstrating its efficacy as a promising material to facilitate infectious wound healing.

On the other hand, Azari et al. [6] investigated the co-electrospinning of polyvinyl alcohol-chitosan (PVA-CS) with cerium oxide nanoparticles (CeNPs) and silk fibroin-collagen (SF-Col) with vitamin D3 for diabetic wound healing applications. This combination promotes the development of versatile scaffolds that not only support tissue engineering but also facilitate targeted drug delivery, enhance cellular interactions, and promote overall tissue regeneration. The importance of this study lies in its innovative exploration of hybrid SF-Col and PVA-CS nanofibers, fabricated by dual electrospinning techniques to improve their morphology and cytocompatibility in tissue engineering and wound healing applications. The results obtained showed that higher concentrations of SF-Col resulted in larger-diameter fibers. In SF-Col nanofibers, increasing vitamin D3 concentration also increased fiber diameter, which is consistent with previous studies on polymer-dopant interactions.

Moreover, Nazemoroaia et al. [59] developed a novel asymmetrical wound dressing to mimic the complex structure and texture of natural skin. The top layer of the dressing consisted of electrospun PCL and SS, while the bottom layer contained CS-Alg hydrogel loaded with 10-HAD. This study introduces an asymmetrical electrothreaded membrane that enhances mechanical stability and ensures optimal hydration of the wound dressing. It serves as a physical barrier against microbial infiltration while also demonstrating biodegradability and effective moisture control. It is non-toxic to fibroblasts while effectively accelerating wound healing in rat models.

A three-dimensional quaternized chitosan/pullulan/citric acid quaternized nanofiber sponge (3D-NS) was prepared by electrostatic spinning and gas foaming techniques after quaternizing the MCS. In vitro tests indicated that 3D-NS could inhibit the release of inflammatory factors, promote angiogenesis, accelerate collagen deposition, and promote wound healing [42]. These effects significantly facilitated wound healing in rats with full-thickness skin damage; the high bioactivity and physicochemical properties of 3D-NS make it an ideal candidate for the development of new wound dressings.

Finally, Borse et al. [13] developed an advanced biocomposite film by combining chitosan (CS) with gum kondagogu (GKG) to form a polyelectrolyte complex (PEC) and incorporating zinc oxide nanoparticles (ZnO NP) and barbaloin (BB). In the study, the optimized CS/GKG/BB-3 biocomposite film showed suitable mechanical strength (8.9 ± 0.30 MPa), high water absorption ($451.1 \pm 6.02\%$), and outstanding water retention potential. The film exhibited remarkable antioxidant properties and effective antimicrobial activity against *Staphylococcus aureus* and *Escherichia coli.* In vitro drug release studies showed a favorable release profile of BB, with an initial explosive release followed by sustained release, ideal for wound healing applications. We demonstrate the various applications of chitosan biopolymer in biomedical fields, particularly in wound healing.

7 Conclusions

Chitosan's multifunctional characteristics make it a highly promising biopolymer in the field of wound healing. As previously described in this chapter, chitosan has unique physicochemical and biological properties, such as biodegradability, biocompatibility, non-toxicity, and broad-spectrum antimicrobial activity, which make it ideal for incorporation into advanced wound dressing systems. Chitosan's structural closeness to glycosaminoglycans allows it to interact favorably with mammalian cells, promoting tissue regeneration and influencing immunological responses. Furthermore, its antimicrobial mechanism, which includes microbial membrane breakdown, nucleic acid interference, and interaction with fungal and microbial components, increases its therapeutic potential in both acute and chronic wounds.

This chapter has highlighted the significance of important parameters such as molecular weight, degree of deacetylation, ambient pH, and the microbial species involved, all of which influence chitosan's biological activity. Furthermore, the combination of chitosan with natural and synthetic polymers, as well as bioactive agents, has resulted in the development of novel wound dressings with tailored properties such as increased mechanical strength, moisture retention, controlled drug release, and synergistic antimicrobial activity.

The novel application of chitosan in multifunctional composite dressings represents a substantial development in wound care technology. These systems not only operate as a physical barrier against pathogens but also actively participate in all stages of wound healing, including hemostasis, cellular proliferation, and tissue remodeling. Chitosan-based wound dressings represent a change from passive to bioactive healing techniques.

Future research should concentrate on optimizing chitosan formulations, scalability of production processes, and clinical validation of these systems to ensure their successful translation from laboratory to bedside. In summary, chitosan is a versatile and novel material that has great potential for the development of next-generation wound treatment products.

Acknowledgments The authors would like to thank Patricia del C. Madera-López for her support in preparing the figures and diagrams in this chapter.

References

1. Abedian, Z., Jenabian, N., Moghadamnia, A. A., Zabihi, E., Tashakorian, H., Rajabnia, M., Sadighian, F., & Bijani, A. (2019). Antibacterial activity of high-molecular-weight and low-molecular-weight chitosan upon oral pathogens. *Journal of Conservative Dentistry and Endodontics, 22*, 169–174. https://doi.org/10.4103/JCD.JCD_300_18
2. Abruzzo, A., Zuccheri, G., Belluti, F., Provenzano, S., Verardi, L., Bigucci, F., Cerchiara, T., Luppi, B., & Calonghi, N. (2016). Chitosan nanoparticles for lipophilic anticancer drug delivery: Development, characterization and in vitro studies on HT29 cancer cells. *Colloids Surfaces B, 145*, 362–372. https://doi.org/10.1016/j.colsurfb.2016.05.023

3. Aider, M. (2010). Chitosan application for active bio-based films production and potential in the food industry: Review. *LWT - Food Science and Technology, 43*, 837–842. https://doi.org/10.1016/j.lwt.2010.01.021
4. Al-Musawi, S., Albukhaty, S., Al-Karagoly, H., Sulaiman, G. M., Alwahibi, M. S., Dewir, Y. H., Soliman, D. A., & Rizwana, H. (2020). Antibacterial activity of honey/chitosan nanofibers loaded with capsaicin and gold nanoparticles for wound dressing. *Molecules, 25*(20), 4770. https://doi.org/10.3390/molecules25204770
5. Arenas, J. (2003). Las heridas y su cicatrización. *Offarm., 22*(5), 126.
6. Azari, Z., Sadeghi-Avalshahr, A., Alipour, F., Kondori, B. J., Askari, V. R., Mollazadeh, S., Nazarnezhad, S., Nasiri, S. N., Kermani, F., & Ranjbar-Mohammadi, M. (2025). *International Journal of Biological Macromolecules, 310*(Part 2), 143099. https://doi.org/10.1016/j.ijbiomac.2025.143099
7. Bautista-Baños, S., Ventura-Aguilar, R. I., Correa-Pacheco, Z., & Corona-Rangel, M. L. (2016). Chitosan: A versatile antimicrobial polysaccharide for fruit and vegetables in postharvest—A review. *Revista Chapingo Serie Horticultura, 23*(2), 103–122. https://doi.org/10.5154/r.rchsh.2016.11.030
8. Bhardwaj, D., Chawla, V., Gupta, A., Dhiman, A., Singh, Y., & Agrawal, G. (2025). Fe/Se nanocomposite-loaded chitosan/alginate films for wound healing. *International Journal of Biological Macromolecules, 306*(Part 3), 141626. https://doi.org/10.1016/j.ijbiomac.2025.141626
9. Baxter, R. M., Dai, T., Kimball, J., Wang, E., Hamblin, M. R., Wiesmann, W. P., McCarthy, S. J., & Baker, S. M. (2013). Chitosan dressing promotes healing in third degree burns in mice: Gene expression analysis shows biphasic effects for rapid tissue regeneration and decreased fibrotic signaling. *Journal of Biomedical Materials. Research A, 101*(2), 340–348. https://doi.org/10.1002/jbm.a.34328
10. Beck, B., Yildirim Aksoy, M., Shoemaker, C., Fuller, A., & Peatman, E. (2019). Antimicrobial activity of the biopolymer chitosan against *Streptococcus iniae*. *Journal of Fish Disease, 42*(3), 371–377. https://doi.org/10.1111/jfd.12938
11. Binder, L., de Sousa, S. F., & Ferreira, T. (2024). The influence of molecular weight on the anticorrosion properties of chitosan coatings. *International Journal of Biological Macromolecules, 278*, 134912. https://doi.org/10.1016/j.ijbiomac.2024.134912
12. Bonferoni, M. C., Riva, F., Invernizzi, A., Dellera, E., Sandri, G., Rossi, S., Marrubini, G., Bruni, G., Vigani, B., Caramella, C., & Ferrari, F. (2018). Alpha tocopherol loaded chitosan oleate nanoemulsions for wound healing.Evaluation on cell lines and ex vivo human biopsies, and stabilization in spray driedTrojan microparticles. *European Journal of Pharmaceutics and Biopharmaceutics, 123*, 31–41. https://doi.org/10.1016/j.ejpb.2017.11.008
13. Borse, S., Nangare, S., Bafna, P., Jain, P., & Zawar, L. (2025). Barbaloin loaded chitosan-gum kondagogu polyelectrolyte complex based biocomposites film for enhanced antibacterial, antioxidant and wound healing activity. *Surfaces and Interfaces, 56*, 105506. https://doi.org/10.1016/j.surfin.2024.105506
14. Burgess, C. (2008). Topical vitamins. *Journal of Drugs in Dermatology, 7*(7 Suppl), s2–s6.
15. Castrillón-Rivera, L. E., Palma-Ramos, A., & Padilla-Desgarennes, M. C. (2011). Interferencia de las biopelículas en el proceso de curación de heridas. *DermatologíaRevista Mexicana, 55*(3), 127–139.
16. Chang, S.-H., Lin, H.-T. V., Wu, G.-J., & Tsai, G. J. (2015). pH effects on solubility, zeta potential, and correlation between antibacterial activity and molecular weight of chitosan. *Carbohydrate Polymers, 134*, 74–81. https://doi.org/10.1016/j.carbpol.2015.07.072
17. Cheung, R. C. F., Ng, T. B., Wong, J. H., & Chan, W. Y. (2015). Chitosan: An update on potential biomedical and pharmaceutical applications. *Marine Drugs, 13*, 5156–5186. https://doi.org/10.3390/md13085156
18. Chen, Z., Zhou, D., Fu, Q., & Wang, X. (2025). Enhanced wound healing and antimicrobial efficacy of PHMB-loaded injectable chitosan/quaternary chitosan/ NaHCO3/β-glycerophosphate hydrogels for seawater-infected wounds. *International Journal*

of Biological Macromolecules, 310(Pt 3), 143418. https://doi.org/10.1016/j.ijbiomac.2025.143418
19. Chen, G., Xun, X., Ao, H., Chen, Z., Wang, D., Wang, M., Zhang, D., Liu, M., & Guo, G. (2025). Quaternized chitosan-based injectable self-healing hydrogel for improving wound management in aging populations. *Colloids and Surfaces B: Biointerfaces, 253*, 114721. https://doi.org/10.1016/j.colsurfb.2025.114721
20. Croisier, F., & Jérôme, C. (2013). Chitosan-based biomaterials for tissue engineering. *European Polymer Journal, 49*(4), 780–792. https://doi.org/10.1016/j.eurpolymj.2012.12.009
21. Das, S., & Baker, A. B. (2016). Biomaterials and nanotherapeutics for enhancing skin wound healing. *Frontiers Bioengineering and Biotechnology, 4*, 82. https://doi.org/10.3389/fbioe.2016.00082
22. De la Mora-López, D., Madera-Santana, T. J., Olivera-Castillo, L., Castillo-Ortega, M. M., López-Cervantes, J., Sánchez-Machado, D. I., Ayala-Zavala, J. F., & Soto-Valdez, H. (2024). Production and performance evaluation of chitosan/collagen/honey nanofibrous membranes for wound dressing applications. *International Journal of Biological Macromolecules, 275*(2), 133809. https://doi.org/10.1016/j.ijbiomac.2024.133809
23. Dutta, P. K., Tripathi, S., & Mehrotra, G. K. (2009). Perspectives for chitosan based antimicrobial films in food applications. *Food Chemistry, 114*, 1179–1182. https://doi.org/10.1016/j.foodchem.2008.11.047
24. ECDC. European Centre for Disease Prevention and Control. Available online: https://www.ecdc.europa.eu/en/news-events/33000-people-die-every-year-due-infections-antibiotic-resistant-bacteria. Accessed on 15 April 2025.
25. El-Saadony, M. T., Saad, A. M., Alkafaas, S. S., Dladla, M., Ghosh, S., Elkafas, S. S., Hafez, W., Ezzat, S. M., Khedr, S. A., Hussien, A. M., Fahmy, M. A., Elesawi, I. E., Salem, H. M., Mohammed, D. M., El-Mageed, T. A. A., Ahmed, A. E., Mosa, W. F. A., El-Tarabily, M. K., AbuQamar, S. F., & El-Tarabily, K. A. (2025). Chitosan, derivatives, and its nanoparticles: Preparation, physicochemical properties, biological activities, and biomedical applications—A comprehensive review. *International Journal of Biological Macromolecules, 313*, 142832. https://doi.org/10.1016/j.ijbiomac.2025.142832
26. Ehterami, A., Salehi, M., Farzamfar, S., Samadian, H., Vaez, A., Ghorbani, S., Ai, J., & Sahrapeyma, H. (2019). Chitosan/alginate hydrogels containing Alpha-tocopherol forwound healing in rat model. *Journal of Drug Delivery Science and Technology, 51*, 204–213. https://doi.org/10.1016/j.jddst.2019.02.032
27. Enoch, S., & Leaper, D. J. (2008). Basis science of wound healing. *Surgery (Oxford), 26*(2), 31. https://doi.org/10.1016/j.mpsur.2007.11.005
28. FAO. *The State of World Fisheries and Aquaculture 2024— Blue Transformation in action.* Development Goals, Rome. Available online: https://openknowledge.fao.org/items/06690fd0-d133-424c-9673-1849e414543d. Accessed on 30 April 2025.
29. Fu, X., Shen, Y., Jiang, X., Huang, D., & Yan, Y. (2011). Chitosan derivatives with dual-antibacterial functional groups for antimicrobial finishing of cotton fabrics. *Carbohydrate Polymers, 85*, 221–227. https://doi.org/10.1016/j.carbpol.2011.02.019
30. Galván Marquez, I., Akuaku, J., Cruz, I., Cheetham, J., Golshani, A., & Smith, M. L. (2013). Disruption of proteinsynthesis as antifungal mode of action by chitosan. *International Journal of Food Microbiology, 164*, 108–112. https://doi.org/10.1016/j.ijfoodmicro.2013.03.025
31. Goy, R. C., de Britto, D., & Assis, O. B. (2009). A review of the antimicrobial activity of chitosan. *Polímeros: Ciência e Tecnologia, 19*, 241–247. https://doi.org/10.1590/S0104-14282009000300013
32. Guarín-Corredor, C., Quiroga-Santamaría, P., & Landínez-Parra, N. S. (2014). Procesos de cicatrización de heridas de piel, campos endógenos y su relación con las heridas crónicas. *Revista Facultad de Medicina, 61*(4), 441.
33. Hamed, I., Özogul, F., & Regenstein, J. M. (2016). Industrial applications of crustacean by-products (chitin, chitosan, and chitooligosaccharides): A review. *Trends in Food Science andTechnology, 48*, 40–50. https://doi.org/10.1016/j.tifs.2015.11.007

34. Huang, K.-X., Zhou, L.-Y., Chen, J.-Q., Peng, N., Chen, H.-X., Gu, H.-Z., & Zou, T. (2023). Applications and perspectives of quaternized cellulose, chitin and chitosan: A review. *International Journal of Biological Macromolecules, 242*(3), 124990. https://doi.org/10.1016/j.ijbiomac.2023.124990
35. Jiang, T., James, R., Kumbar, S. G., & Laurencin, C. T. (2014). *Chitosan as a biomaterial: Structure, properties, and applications* (pp. 91–113). Elsevier.
36. Kadam, S., Shai, S., Shahane, A., & Kaushik, K. S. (2019). Recent advances in non-conventional antimicrobial approaches for chronic wound biofilms: Have we found the 'chink in the armor'? *Biomedicine, 7*(2), 35. https://doi.org/10.3390/biomedicines7020035
37. Karakeçili, A. G., Satriano, C., Gümüşderelioğlu, M., & Marletta, G. (2008). Enhancement of fibroblastic proliferation on chitosan surfaces by immobilized epidermal growth factor. *Acta Biomaterialia, 4*, 989–996. https://doi.org/10.1016/j.actbio.2008.02.003
38. Kirsner, R. S., & Eaglstein, W. H. (1993). The wound healing process. *Dermatology Clinics, 11*(4), 629.
39. Kondo, T. (2007). Timing of skin wounds. *Legal Medicine, 9*(2), 109.
40. Kong, M., Chen, X. G., Xing, K., & Park, H. J. (2010). Antimicrobial properties of chitosan and mode of action: A state of the art review. *International Journal of Food Microbiology, 144*(1), 51–63. https://doi.org/10.1016/j.ijfoodmicro.2010.09.012
41. Koosehgol, S., Ebrahimian-Hosseinabadi, M., Alizadeh, M., & Zamanian, A. (2017). Preparation and characterization of in situ chitosan/polyethylene glycol fumarate/thymol hydrogel as an effective wound dressing. *Materials Science and Engineering C, 79*, 66–75. https://doi.org/10.1016/j.msec.2017.05.001
42. Li, Q., Lai, X., Duan, Y., Jiang, F., Li, Y., Huang, Z., Liu, S., Wang, Y., Jiang, C., Zhang, C., & Pan, X. (2025). 3D nanofiber sponge based on natural insect quaternizedchitosan/pullulan/citric acid for accelerating wound healing. *Carbohydrate Polymers, 348*(Part A), 122827. https://doi.org/10.1016/j.carbpol.2024.122827
43. Li, X. F., Feng, X. G., Yang, S., Fu, G. G., Wang, T. P., & Su, Z. X. (2010). Chitosan kills Escherichia coli through damage to be of cell membrane mechanism. *Carbohydrate Polymers, 79*, 493–499. https://doi.org/10.1016/j.carbpol.2009.07.011
44. Li, Y., Chen, X. G., Liu, N., Liu, C. S., Liu, C. G., Meng, X. H., Yu, L. J., & Kennedy, J. F. (2007). Physicochemical characterization and antibacterial property of chitosan acetates. *Carbohydrate Polymer, 67*, 227–232.
45. Liu, X., Xia, W., Jiang, Q., Yu, P., & Yue, L. (2018). Chitosan oligosaccharide-N-chlorokojic acid mannich base polymer as a potential antibacterial material. *Carbohydrate Polymer, 182*, 225–234. https://doi.org/10.1016/j.carbpol.2017.11.019
46. Lopes, C., Antelo, L. T., Franco-Uría, A., Alonso, A. A., & Pérez-Martín, R. (2018). Chitin production from crustacean biomass: Sustainability assessment of chemical and enzymatic processes. *Journal of Cleaner Production, 172*, 4140–4151. https://doi.org/10.1016/j.jclepro.2017.01.082
47. Madera-Santana, T. J., Herrera-Méndez, C. H., & Rodríguez-Núñez, J. R. (2018). An overview of the chemical modifications of chitosan and their advantages. *Green Materials, 6*, 131–142. https://doi.org/10.1680/jgrma.18.00053
48. Madni, A., Khan, R., Ikram, M., Sohaila Naz, S., Khan, T., & Wahid, F. (2019). Fabricationand characterization of chitosan–Vitamin C–Lactic acid composite membrane forpotential skin tissue engineering. *International Journal of Polymer Science, 2019*, 1–8. https://doi.org/10.1155/2019/4362395
49. Malone, M., Bjarnsholt, T., McBain, A. J., James, G. A., Stoodley, P., Leaper, D., Tachi, M., Schultz, G., Swanson, T., & Wolcott, R. D. (2017). The prevalence of biofilms in chronic wounds: A systematic review and meta-analysis of published data. *Journal of Wound Care, 26*, 20–25. https://doi.org/10.12968/jowc.2017.26.1.20
50. Másson, M. (2024). The quantitative molecular weight-antimicrobial activity relationship for chitosan polymers, oligomers, and derivatives. *Carbohydrate Polymers, 337*, 122159. https://doi.org/10.1016/j.carbpol.2024.122159

51. Matica, M. A., Aachmann, F. L., Tøndervik, A., Sletta, H., & Ostafe, V. (2019). Chitosan as a wound dressing starting material: Antimicrobial properties and mode of action. *International Journal of Molecular Sciences, 20*(23), 5889. https://doi.org/10.3390/ijms20235889
52. Mawazi, S. M., Kumar, M., Ahmad, N., Ge, Y., & Mahmood, S. (2024). Recent applications of Chitosan and its derivatives in antibacterial, anticancer, wound healing, and tissue engineering fields. *Polymers (Basel), 16*(10), 1351. https://doi.org/10.3390/polym16101351
53. Mehata, A. K., Singh, C. S., Malik, A. K., Vaishali, V., Viswanadh, M. K., Setia, A., & Muthu, M. S. (2024). *Chitosan nanoparticles in wound healing and dressing application*. Elsevier eBooks. https://doi.org/10.1016/b978-0-443-14088-4.00008-3
54. Mir, M., Ali, M. N., Barakullah, A., Gulzar, A., Arshad, M., Fatima, S., & Asad, M. (2018). Synthetic polymeric biomaterials for wound healing: A review. *Progress in Biomaterials, 7*(1), 1–21.
55. Miya, M. B., Ashutosh, M., Dey, D., Pathak, V., Khare, E., Kalani, K., Chaturvedi, P., Singh, V., Chaturvedi, P., & Kalani, A. (2025). Accelerated diabetic wound healing using a chitosan-based nanomembrane incorporating nanovesicles from *Aloe barbadensis*, *Azadirachta indica*, and *Zingiber officinale*. *International Journal of Biological Macromolecules, 310*(Part 2), 143169. https://doi.org/10.1016/j.ijbiomac.2025.143169
56. Motealleh, B. (2015). Drug release, cell adhesion and wound healing evaluations of electrospun carboxymethyl chitosan/polyethylene oxide nanofibres containing phenytoinsodium and vitamin C. *IET Nanobiotechnology, 9*(4), 191–200(9). Retrieved from https://digital-library.theiet.org/content/journals/10.1049/iet-nbt.2014.0030.
57. Murtaza, H., Ali, S., Mubeen, A., Khalid, N., Ahmad, N., & Farrukh, M. A. (2025). Green synthesized (*Curcuma longa*) Ni nanoparticles doped chitosan and PEG for wound healing and anti-bacterial activity. *International Journal of Biological Macromolecules, 308*(Part 4), 142631. https://doi.org/10.1016/j.ijbiomac.2025.142631
58. Lei, Y., Wang, S., Liu, Q., Zeng, D., Wang, K., Chang, C., Zhang, B., Zhang, L., Shi, Z., & Meng, Y. (2025). Construction of AgNPs-loaded oriented hydrogel based on *Periostracum Cicadae* chitosan by electro-assembly for rapid hemostasis and wound healing. *Carbohydrate Polymers, 358*, 23500. https://doi.org/10.1016/j.carbpol.2025.123500
59. Nazemoroaia, M., Bagheri, F., Mirahmadi-Zare, S. Z., Eslami-Kaliji, F., & Derakhshan, A. (2025). Asymmetric natural wound dressing based on porous chitosan-alginate hydrogel/electrospun PCL-silk sericin loaded by 10-HDA for skin wound healing: *In vitro* and *in vivo* studies. *International Journal of Pharmaceutics, 668*, 124976. https://doi.org/10.1016/j.ijpharm.2024.124976
60. Naseri-Nosar, M., Farzamfar, S., Salehi, M., Vaez, A., Tajerian, R., & Azami, M. (2017). Erythropoietin/aloe vera-releasing wet-electrospun polyvinyl alcohol/chitosansponge-like wound dressing: *In vitro* and *in vivo* studies. *Journal of Bioactive and Compatible Polymers, 33*(3), 269–281. https://doi.org/10.1177/0883911517731793
61. No, H. K., Lee, K. S., & Meyers, S. P. (2000). Correlation between physicochemical characteristics and binding capacities of chitosan products. *Journal of Food Science, 65*(7), 1134–1137. https://doi.org/10.1111/j.1365-2621.2000.tb10252.x
62. Oliveira Junior, E., Melo, I., & Franco, T. (2012). Changes in hyphal morphology due to chitosan treatment in somefungal species. *Brazilian Archives of Biology and Technology, 55*(5), 637–646. https://doi.org/10.1590/S1516-89132012000500001
63. Park, S. C., Nam, J. P., Kim, J. H., Kim, Y. M., Nah, J. W., & Jang, M. K. (2015). Antimicrobial action of water-soluble b-chitosan against clinical multi-drug resistant bacteria. *International Journal of Molecular. Science, 16*, 7995–8007. https://doi.org/10.3390/ijms16047995
64. Pathalamuthu, P., Siddharthan, A., Giridev, V. R., Victoria, V., Ramar, T., Sivasubramanian, S., Savariar, V., & Hemamalini, T. (2019). Enhanced performance of *Aloe vera*incorporated chitosan-polyethylene oxide electrospun wound scaffold producedusing novel Spirograph based collector assembly. *International Journal of Biological Macromolecules, 140*, 808–824. https://doi.org/10.1016/j.ijbiomac.2019.08.158

65. Pohling, J., Hawboldt, K., & Dave, D. (2025). Influence of chitin purity on its dissolution behaviour in alkaline solvent. *Carbohydrate Polymers, 356*, 123414. https://doi.org/10.1016/j.carbpol.2025.123414
66. Radoor, S., Karayil, J., Jayakumar, A., Siengchin, S., & Parameswaranpillai, J. (2021). A low cost and eco-friendly membrane from polyvinyl alcohol, chitosan and honey: Synthesis, characterization and antibacterial property. *Journal of Polymer Research, 28*, 82. https://doi.org/10.1007/s10965-021-02415-2
67. Rajinikanth, B. S., Rajkumar, D. S. R., & K K, Vijayaragavan, V. (2024). Chitosan-based biomaterial in wound healing: A review. *Cureus, 16*(2), e55193. https://doi.org/10.7759/cureus.55193
68. Rieger, K. A., & Schiffman, J. D. (2014). Electrospinning an essential oil: Cinnamaldehydeenhances the antimicrobial efficacy of chitosan/poly(ethylene oxide) nanofibers. *Carbohydrate Polymers, 113*, 561–568. https://doi.org/10.1016/j.carbpol.2014.06.075
69. Rieger, K. A., Birch, N. P., & Schiffman, J. D. (2016). Electrospinning chitosan/poly(ethylene oxide) solutions with essential oils: Correlating solution rheology to nanofiberformation. *Carbohydrate Polymers, 139*, 131–138. https://doi.org/10.1016/j.carbpol.2015.11.073
70. Rodríguez-Núñez, J. R., Madera-Santana, T. J., Sánchez-Machado, D. I., López-Cervantes, J., & Soto-Valdez, H. (2014). Chitosan/hydrophilic plasticizer-based films: Preparation, physicochemical and antimicrobial properties. *Journal of Polymers and the Environment, 22*, 41–51. https://doi.org/10.1007/s10924-013-0621-z
71. Rodríguez-Vázquez, M., Vega-Ruíz, B., Ramos-Zúñiga, R., Saldaña-Koppel, D. A., & Quiñones-Olvera, L. F. (2015). Chitosan and its potential use as a scaffold for tissue engineering in regenerative medicine. *BioMed Research International, 2015*, 1. https://doi.org/10.1155/2015/821279
72. Sánchez-Machado, D. I., López-Cervantes, J., Martínez-Ibarra, D. M., Escárcega-Galaz, A. A., & Vega-Cázarez, C. A. (2022). The use of chitosan as a skin-regeneration agent in burns injuries: A review. *E-Polymers, 22*(1), 75–86. https://doi.org/10.1515/epoly-2022-0011
73. Sánchez-Machado, D. I., López-Cervantes, J., & Rodríguez-Núñez, J. R. (2015). Antimicrobial effect of different chitosan preparations against selected food borne pathogens. *International Journal of Pharma and Bio Sciences, 6*, 204–212.
74. Shi, C., Zhu, Y., Ran, X., Wang, M., Su, Y., & Cheng, T. (2006). Therapeutic potential of chitosan and its derivatives in regenerative medicine. *Journal of Surgical Research, 133*, 185–192. https://doi.org/10.1016/j.jss.2005.12.013
75. Silva, S. S., Popa, E. G., Gomes, M. E., Cerqueira, M., Marques, A. P., Caridade, S. G., Texeira, P., Sousa, C., Mano, J. F., & Reis, R. L. (2013). An investigation of the potential application of chitosan/aloe-basedmembranes for regenerative medicine. *Acta Biomaterialia, 9*(6), 6790–6797. https://doi.org/10.1016/j.actbio.2013.02.027
76. Singh, S., Singh, S. K., Chowdhury, I., & Singh, R. (2017). Understanding the mechanism of bacterial biofilms resistance to antimicrobial agents. *Open Microbiology Journal, 11*, 53–62. https://doi.org/10.2174/1874285801711010053
77. Su, J., Liu, C., Sun, A., Yan, J., Sang, F., Xin, Y., Zhao, Y., Wang, S., & Dang, Q. (2025). Hemostatic and antimicrobial properties of chitosan-based wound healing dressings: A review. *International Journal of Biological Macromolecules, 306*(Pt 2), 141570. https://doi.org/10.1016/j.ijbiomac.2025.141570
78. Suhaimi, N. A., Amirul, N. B., Hasman, A. A., Shahri, N. N., Roslan, N. N., Hock, H. L. L., Ying, K. C. P., Kusrini, E., & Usman, A. (2025). Insights into depolymerization of chitosan using acid hydrolysis, direct photolysis, and photocatalysis: A review. *Results in Chemistry, 13*, 102044. https://doi.org/10.1016/j.rechem.2025.102044
79. Sun, B. K., Siprashvili, Z., & Khavari, P. A. (2014). Advances in skin grafting and treatment of cutaneous wounds. *Science, 346*(6212), 941–945. https://doi.org/10.1126/science.1253836

80. Valencia-Gómez, L. E., Martel-Estrada, S. A., Vargas-Requena, C. L., Rodríguez-González, C. A., & Olivas-Armendariz, I. (2016). Apósitos de polímeros naturales para regeneración de piel. *Revista Mexicana Ingeniería Biomédica, 37*(3), 235–249.
81. Wang, Q. Z., Chen, X. G., Liu, N., Wang, S. X., Liu, C. S., Meng, X. H., & Liu, C. G. (2006). Protonation constants of chitosan with different molecular weight and degree of deacetylation. *Carbohydrate Polymers, 65*, 194–201. https://doi.org/10.1016/j.carbpol.2006.01.001
82. Yadav, P., Bandyopadhyay, A., Chakraborty, A., & Sarkar, K. (2018). Enhancement of anticancer activity and drug delivery of chitosan-curcumin nanoparticle via molecular docking and simulation analysis. *Carbohydrate Polymers, 182*, 188–198. https://doi.org/10.1016/j.carbpol.2017.10.102
83. Yang, Y.-C., Cheng, S.-H., Zhang, Y.-X., Mao, J., Zheng, B.-D., Zhang, N., Huang, Y.-Y., Zhang, X.-Q., Ye, J., & Xiao, M.-T. (2025). Highly effective preparation of low molecular weight chitosan by enhanced chemical degradation applying high gravity technology. *Innovative Food Science and Emerging Technologies, 100*, 103938. https://doi.org/10.1016/j.ifset.2025.103938
84. Yien Ing, L., Zin, N., Sarwar, A., & Katas, H. (2012). Antifungal activity of chitosan nanoparticles and correlation with their physical properties. *International Journal of Biomaterials, 2012*, 632698. https://doi.org/10.1155/2012/632698
85. Younes, I., Sellimi, S., Rinaudo, M., Jellouli, K., & Nasri, M. (2014). Influence of acetylation degree and molecular weight of homogeneous chitosans on antibacterial and antifungal activities. *International Journal of Food Microbiology, 185*, 57–63. https://doi.org/10.1016/j.ijfoodmicro.2014.04.029
86. Zhao, J., Qiu, P., Wang, Y., Wang, Y., Zhou, J., Zhang, B., Zhang, L., & Gou, D. (2023). Chitosan-based hydrogel wound dressing: From mechanism to applications, a review. *International Journal of Biological Macromolecules, 244*, 125250. https://doi.org/10.1016/j.ijbiomac.2023.125250
87. Zheng, L.-Y., & Zhu, J.-F. (2003). Study on antimicrobial activity of chitosan with different molecular weights. *Carbohydrate Polymers, 54*, 527–530. https://doi.org/10.1016/j.carbpol.2003.07.009

Index

S. Jana (ed.), *Chitosan-Based Drug Delivery Systems*, AAPS Introductions in the Pharmaceutical Sciences 16, https://doi.org/10.1007/978-3-032-08347-0

MIX
Papier aus verantwortungsvollen Quellen
Paper from responsible sources
FSC® C105338

If you have any concerns about our products,
you can contact us on
ProductSafety@springernature.com

In case Publisher is established outside the EU,
the EU authorized representative is:
Springer Nature Customer Service Center GmbH
Europaplatz 3, 69115 Heidelberg, Germany

Printed by Libri Plureos GmbH
in Hamburg, Germany